ALLE ZEIT WACH
1842

Natural Product Chemistry

Proceedings
of the First International Symposium
and Pakistan-U.S. Binational Workshop,
Karachi, Pakistan

Edited by
Atta-ur-Rahman

Springer-Verlag
Berlin Heidelberg New York
London Paris Tokyo

Professor Atta-ur-Rahman
H.E.J. Research Institute of Chemistry
University of Karachi
Karachi-32, Pakistan

With 142 Figures and Numerous Schemes

ISBN-13: 978-3-642-71427-6 e-ISBN-13: 978-3-642-71425-2
DOI: 10.1007/978-3-642-71425-2

Library of Congress Cataloging-in-Publication Data. International Symposium and Pakistan-U.S. Binational Workshop on Natural Product Chemistry (1st : 1984 : Karachi, Pakistan). Natural product chemistry. "The first International Symposium and Pakistan-U.S. Binational Workshop on Natural Product Chemistry ... February 1984"–Pref. Includes index. 1. Natural products–Congresses. I. Rahman, Atta-ur-, 1942- . II. Title. QD415.A1I47 1984 547.7 86-20420.

Softcover reprint of the hardcover 1st edition 1986.

Offsetprinting: Druckhaus Beltz, Hemsbach/Bergstr.

2131/3130-543210

Preface

The First International Symposium and Pakistan-U.S. Binational Workshop on Natural Product Chemistry represents the first international conference ever to be organised in Pakistan on chemistry. For many years I had aspired to organise such a conference, and finally when it materialised in February 1984, it brought together the very best in the field. The result was as expected - a most stimulating series of lectures and discussions, all of which were recorded on videotape for circulation to universities and research organisations in Pakistan. The fields covered during the Symposium include such diverse areas as plant chemistry, marine natural products, insect hormones, glycoproteins and peptides, synthetic methodology and spectroscopy. I am confident that the papers published here would prove invaluable to the natural product chemists and biochemists working in the fascinating area of natural products.

I wish to record here my indebtedness to Professor Philip W. Le Quesne and Professor Wolfgang Voelter for their untiring efforts in organising the visits of the U.S. and German scientists. I am grateful to the National Science Foundation (USA), Deutsche Forschungsgemeinschaft, U.N.E.S.C.O., Hamdard National Foundation, Pakistan Science Foundation, University Grants Commission, Ministry of Science and Technology, British Council, B.A.S.F. and E. Merck for providing financial support for the conference. But above all, the conference owes its success to the students and staff of the Institute who worked so very hard.

I dedicate these Proceedings to Professor Salimuzzaman Siddiqui, F.R.S., who at his "young" age of 88, provides a very rare example of selfless devotion to his field.

Karachi, June 1986 ATTA-UR-RAHMAN

Contents

List of Contributors

Chemical Constituents of Some Medicinal Plants of Pakistan

Viqar Uddin Ahmad
H. E. J. Research Institute of Chemistry, University of Karachi, Karachi-32, Pakistan

The folkloric medicine has played a key role in the span of human civilization and human wilderness. Based on judicious use of roots, leaves, fruits and flowers of herbs and plants, spread on all widths and breadth of the world, man has been mitigating his sufferings and ailments. Modern medicine can enrich itself immensely from the experiences of antiquity. Nature's abundant renewable supply of plants and herbs could be a great source of cheap drugs, which can be complimentary to modern medicines for alleviating the suffering of the third world's population.

In the following pages a summary of some of the work carried out by our group on the isolation and structure elucidation of chemical constituents of medicinal plants of Pakistan is presented.

1. *Primula denticulata*

Primula (primrose) is used as an expectorant in bronchial cattarh, pneumonia, as diuretic and in folklore in rheumatism, arthritis, migraine and similar diseases [1]. Primula denticulata (Primulaceae) occurs as a common weed in mountains of North West Frontier province of Pakistan at a height of 7000 to 13000 ft [2]. In view of the medicinal importance attributed to many Primula species [3,4], Primula denticulata was chosen for the study of sapogenins and saponins. The saponins isolated from the alcoholic extract of the plant was hydrolysed and the sapogenin

Natural Product Chemistry
Ed. by Atta-ur-Rahman

mixture so obtained contained several triterpenoids. Five sapogenins were isolated [5] through column chromatography. They were named as pridentigenin A, pridentigenin B, pridentigenin C, pridentigenin D and pridentigenin E in increasing order of polarity on the TLC plate.

a) Pridentagenin A

This was the least polar sapogenin and crystallized from ether in the form of colourless needles m.p. 308-10°. The nmr spectrum ($CDCl_3$) of pridentigenin A showed the presence of six tertiary methyl groups through singlets at δ 0.70 (3H, $1xCH_3$), δ 0.84 (6H, 2x CH_3) δ 0.97 (3H, $1xCH_3$) δ1.03 (3H x 1 x CH_3), δ 1.24 (3H, $1xCH_3$). There was a singlet equivalent to six protons at δ 3.51 indicating the presence of two equivalent OCH_3 groups in compound. Another singlet at δ 4.30 was ascribed to H-30.

The molecular ion peak in the mass spectrum appeared at m/z 516 corresponding to the molecular formula $C_{32}H_{52}O_5$. Other important peaks were at m/z 485(base peak) M-OCH_3),484 (M-CH_3OH), 466 (M^+ - $2CH_3OH$), 454 (M^+ - (OCH_3 + CH_2OH), 452 (M^+ - $2CH_3OH$), 308 (retro Diels Alder fragment a) [6], 277 (a-OCH_3) 276 (a-CH_3OH) 207 (retro-Diels-Alder fragment b), 189 (b-H_2O).

The compound pridentigenin A (1) closely resembled cyclamigenin D reported by Dorchai et al. [7] from authentic sample was available for but no direct comparison.

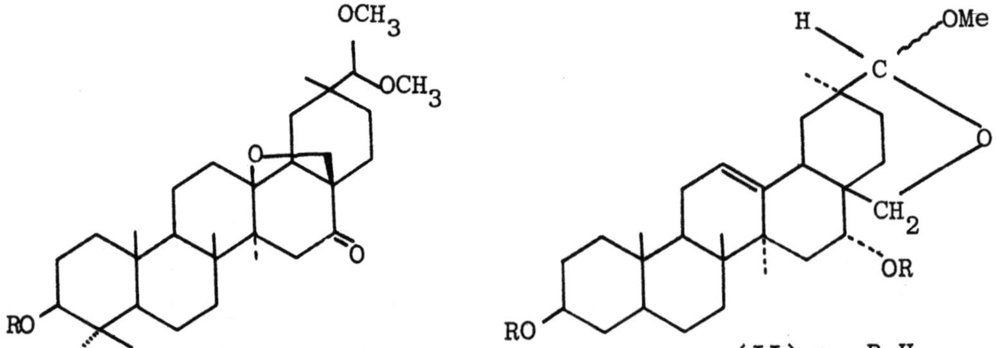

(I) a, R=H
b, R=Ac

(II) a, R=H
b, R=Ac

b) Predintigenin B

Pridentigenin B analysed for $C_{31}H_{50}O_4$. The mass spectrum of this compound showed a very low intensity molecular ion peak at m/z 486 which became somewhat stronger when the field desorption technique was used. Its UV spectrum has λ_{max} at 212-213 nm in methanol which showed the absence of conjugated double bonds in the compound. In the i.r. spectrum (KBr) a strong peak at 3367 cm^{-1} (OH) and a peak of medium intensity at 1630 cm^{-1} (C=C stretching) were visible. No carbonyl bands were present. It forms a diacetate m.p.222°C.

On the basis of the spectral data and differnt reactions we proposed that preditigenin B has structure (II). It was therefore a homologue of cyclamigenin A (or C) isolated by Dorchai et al. [7]. The stereochemistry C-30 could not be determined with certainly but it was most likely the thermodynamically more stable equatorial 30βepimer. The structure proposed for pridentigenin B found a measure support from ^{13}C n.m.r. spectra in $CDCl_3$ + CD_3OH. Thus is analogy with the spectra of olean-12-enes [8] peaks due to C(12) and C(13) at 121.5 and 142.1 ppm were present. The C(3) and C(16) bearing the hydroxyl groups showed peaks 78.93 and 77.43 ppm respectively. Whereas the peaks of C(28) attached to an ether function occurred at 70.1 ppm. The peak due to C(30) which was attached to two oxygen atoms was shifted to 109.7 ppm. The assignment of other ^{13}C nmr peaks were reported in our earlier communcation [9].

c) Pridentigenin C

Pridentigenin C was crystallized from methanol as colourless crystals m.p.152-154°. The mass spectrum of pridentigenin C showed a very weak molecular ion peak at m/e 486. The nmr spectrum of pridentigenin C ($CDCl_3$) showed the presence of six methyl groups at δ 0.69 (3H, 1 x CH_3), δ 0.81 (6H, 2 x CH_3),δ 0.87 (6H, 1 x CH_3),δ 1.2 (3H, 1xCH_3). A methoxy peak appeared at δ 3.51 (3H, OCH_3), a proton in the ring at 4.13 (1H) was attached at C-30, another broad signal at δ 5.35 was due to proton at C-12.

The nmr spectrum and mass spectral fragmentation pattern of pridentigenin C was very similar to that of pridentigenin B, so it appeared that pridentrigen C was an stereoisomer of pridentigenin B at C-30.

d) Pridentigenin D

This was found to be a new sapogenin and was crystallized from ether into small colourless crystals m.p. 170°. The i.r. spectrum showed a broad band at 3360 cm^{-1} due to OH group and a sharp peak at 1700 cm^{-1} due to carbonyl function. The mass spectrum shows a medium intensity molecular ion peak at m/e 472.35097 corresponding to molecular formula $C_{30}H_{48}O_4$. Calc. 472.35520.

The nmr spectrum (400 MHz) of pridentigenin D ($CDCl_3$) showed five singlets at δ 0.78 (3H, s, 24-Me), δ 0.88 (3H, s, 26-Me), δ 0.94 (3H, s, 25-Me), δ 0.97 (3H, s, 29-Me), δ 1.05 (3H, s, 23-Me), and δ 1.23 (3H, s, 27-Me) representing six methyl groups. The double doublet at δ 3.19 (J = 11.5 Hz, 5.5Hz) was due to H-3α and therefore the compound had a 3 β-OH. There were two doublets at δ 2.14 and δ 2.71 (J = 16Hz) assigned to the two H-15 protons in the vicinity of carbonyl group at C-16, the higher field doublet showed a further long range coupling (J = 2Hz) with a methyl groups and each wing was splitted into a quartet. A double doublet at δ 3.47 and δ 3.87 (J = 8Hz each) was assigned to the two H-30 protons whereas the two H-30 protons were magnetically non-equivalent indicating a hindered rotation of the C(20)-C(30) bond. Dreiding model showed that the hinderance in rotation was possible only if the OH group would be attached to C-30 and not to C-29. There was no signal due to olefinic protons.

Pridentigenin D was found to form diacetate with molecular peak at m/z 556.

On the basis of the spectral data we proposed the structure (III) for the compound. This structure found a measure support from biogenic point of view because other compounds containing closely related structures had been isolated by us from this plant.

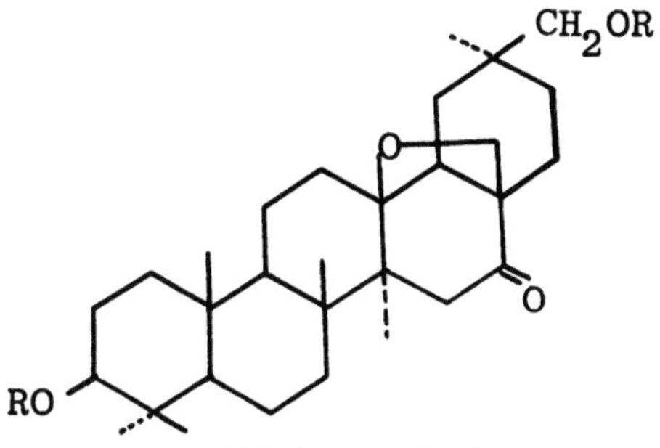

III) a, R=H
b, R=Ac

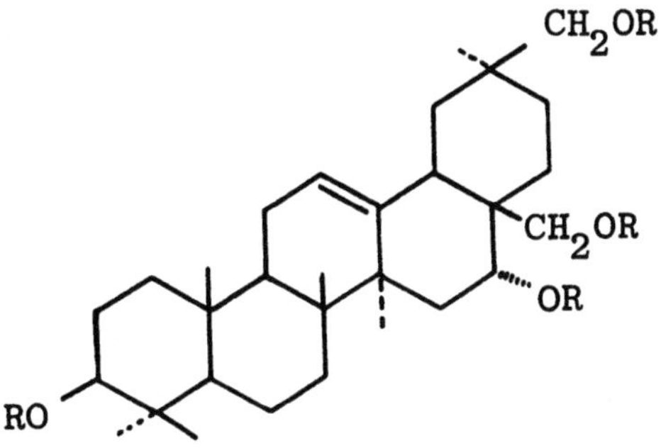

IV) a, R=H
b, R=Ac

e) Pridentigenin E

This compound was crystallized from methanol, has m.p. 268-270 and analyzed for $C_{30}H_{50}O_4$. The mass spectrum showed molecular ion peak at m/z 474.

The nuclear magnetic resonance spectrum in C_5H_5N showed four signals in upper field which corresponded to six methyl groups at δ1.08 (6H, 2 x CH_3), δ1.2 (3H, 1 x CH_3), δ 1.31 (3H, 1 x CH_3) and δ 1.84 96H, 2 x CH_3) two signals at δ3.71 and δ 3.97 corresponding to four protons due two CH_2OH, the later signal had been assinged to functional group at C-28, a broad signals at δ 4.64 was due to vinyl proton at C-12.

From nmr and other spectral data, it was concluded that pridentigenin E was a pentacyclic triterpenoid. The nmr spectral showed six methyl groups without coupling and mass spectral fragmentation pattern in which fragment ion at m/e 203 is stronger than the peak at m/z 191 showed that pridentigenin E was β-amyrin type of triterpene [6]. It was suggested that pridentigenin E had the structure (IV) identical to dihydrocyclamiratin D, prepared by Barton [10] et al. and Tschesche [11]. The compound had been isolated by Ito et al. [12] from Cyclamen europea.

2. *Nepeta hindostana*

The geneus Nepeta belongs to the family Labiatae and about 250 species of this genus are known. Nepeta hindostana (Roth) Haines is an important medicinal plant of the Indo-Pakistan subcontinent. It is known in indigenous system as badrangboya and is used to cure fever, as cardiac tonic in sore throat [13]. Its extract is reported to lower the blood cholesterol level in animals [14]. The whole plant of another species Nepeta cilliaris in the powdered form is administered as syrup to curve fever and cough [15]. Leaves are also chewed to relieve toothache.

Isolation of triterpenes

Two new triterpenes were isolated by us from Nepeta hindostana. They were named nepeticin [16] and nepetidin.

f) Nepeticin (lup-20(29)-ene-3β ,11 α-diol)

It was crystallized from methanol m.p. 215°. The molecular formula, according to high resolution of the molecular ion peak was $C_{30}H_{50}O_2$. On the basis of spectroscopic and chemical studies the structure of nepeticin was proposed as lup-20(29)-ene 3β,11α diol (V). The nmr spectrum (250 MHz) was typical of lup-20 (29)-ene [17] indicating that nepeticin was a member of this class of triterpenes. The methyl singlets were present at δ 0.78, δ0.79, δ 0.96, δ 0.98, δ 1.03 (2 x CH_3) and δ1.68 (broad s, C (20)-Me). The triplet like signal of H-3 was centred at δ 3.2 (J = 7.5 Hz) . Thus from the position and shape of this signal as well as from biogenetic point of view, it was conluded that one hydroxyl group was present in 3β position. The C-11 proton signal was a hextet centred at δ3.93 showed two diaxial (J = 10.5 Hz) and one axial equatorial (J = 5 Hz) spin spin couplings. This clearly indicated that the second hydroxyl group was in 11α position. The olefinic protons of the isopropenyl group appeared as two doublet at δ 4.59 and δ 4.72 (J = 2.2 Hz). The mass spectrum of nepeticin showed important peaks at m/z 442 (M^+), 427 (M^+-CH_3), 424 (M^+-H_2O), 406 (M^+-HO_2), 391 (M^+-CH_3-2 x H_2O), 225, 237, 231, 216, 189, 175. The last two peaks showed that rings D and E were not substituted [18].

The proposed structure was also supported by ^{13}C nmr. The assignments were made on the basis of the known ^{13}C chemical shifts of lupeol and related compounds [19,20], as well as the observed multiplicities in the off-resonance spectrum of nepeticin. It may be noted that the compounds containing lup-20 (30)-ene skeleton generally shown a peak at about 20.9 ppm for C-11 which was shifted to 70.46 ppm in nepeticin due to the presence of a hydroxyl group at this carbon atom.

g) Nepetidin (lup-20 (29)-ene-1β 3β 11 α-triol)

The field ionization mass spectrum (M^+ at 458) of this new triterpene suggested the molecular formula $C_{30}H_{50}O_3$ which was further confirmed through elemental analysis. The i.r. spectrum ($CHCl_3$) showed the bands at 1650 cm^{-1} and 882 cm^{-1} which are typical for isopropenyl group of lup-20(29) ene series of triterpenes. The pmr spectrum of nepetidin in $CDCl_3$ at 400 MHz showed two olefinic protons of isopropenyl group at δ4.58 and δ4.71 (each d, J = 1Hz). Two carbinylic methine proton signals of nepetidin similar in shape were visible as four line pattern each centred at δ 3.20 and δ 3.50. Both of these carbinylic methine protons showed on axial-axial and one axial-equatorial coupling (Jaa = 11.3, Jae = 5.5 Hz). This suggested that both hydroxylic groups were equatorial. The signal centre at δ 3.20 possessed an identical chemical shift and coupling constants of proton geminal to βOH group attached to C-3 in triterpenes. The third carbonyl proton appeared as sextett centred at δ 3.89 (J = 9.5, 9.5, 6.3 Hz) showed that hydroxyl group was equitorial position with two axial and one equitorial protons at neighbouring carbon atoms. The methyl singlets were present at δ0.70 (1 x CH_3), δ0.72 (1 x CH_3), δ 0.90 (3 x CH_3) and δ 0.95 (1 x CH_3). There was a broad singlet at δ 1.68 due to the vinylic methyl group.

The spectroscopic data suggested that nepetidin belongs to the lup-20 (29) ene series of triterpenes, and we proposed the structure (VI) [2] for it. This structure is supported by the mass and ^{13}C-nmr spectra of this compound.

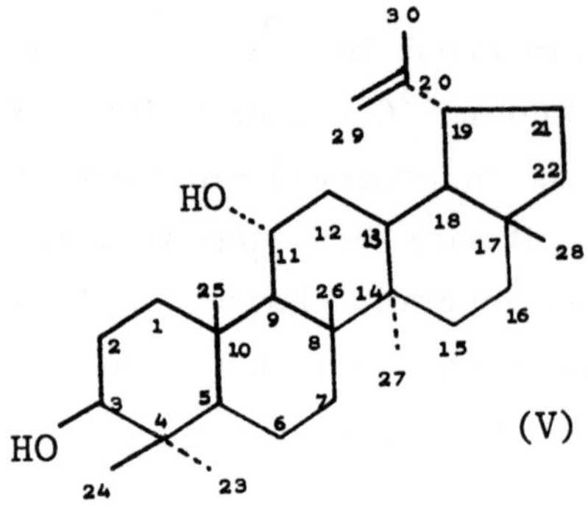

(V)

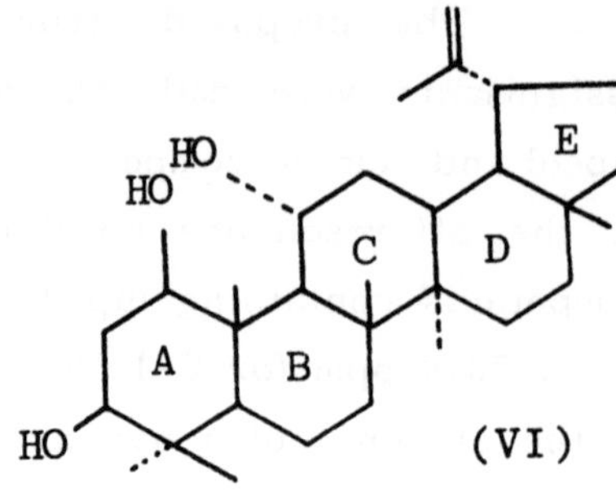

(VI)

3.Prosopis juliflora

Prosopis juliflora Swartz DC (varn.Jand) belongs to the family Mimosaceae. It grows abundantly as a weed in Sind and Punjab provinces of Pakistan [22].

The fruits of Prosopis juliflora contains patulitrin which shows significant activity against lung carcinoma in vivo [23] Alkaloids of Prosopis africana have been found to act on central and autonomic nervous system and to have antibioitc action [24]. The alkaloids prosopinine and prosopine from P.africana are used for the treatment of angina, laryngitis, rhmitis, hemorrhoids and local anaesthetics, for dentistry and minor surgery [25]. An infusion prepared by boiling the leaves with water in taken internally for the treatment of rheumatism and employed externally to treat open sores on skin [24]. The pods are astringent and the bark is used as remedy in rheumatism and scorpion sting [27]. The pod and bark are used in Punjab and central provinces as astringent [29]. The flowers pounded and mixed with sugar are eaten by women during pregnancy as safeguard against miscarriage. Prosopis spicegera is reported to be employed against snake bite [29]. In view of the medicinal properties of Prosopis species we worked on the mesquite plant (Prosopis juliflora) which grows abundantly as a weed in Karachi. As a results of this work we reported [31] the isolation of three new alkaloids from this plant. They were named as juliflorine, julifloridine and julifloricine.

h) Juliflorine

The mass spectrum of juliflorine [31] showed molecular peak at m/z 629.5857 ($C_{40}H_{75}N_3O_2$ requires 629.5858). We published the partial structure of this alkaloid along with i.r. ^1H nmr, ^{13}C nmr and mass spectroscopic data. Prof. Hesse and coworkers subsequently reported [32] the isolation of the alkaloid juliprosopin from this plant. They proposed structure (VII) for this alkaloid. A careful comparison of the spectroscopic data of juliprosopin of Hesse et al. with already published data of juliflorine revealed that these two alkaloids were identical.

i) julifloridine [30]

Julifloridine analysed for $C_{18}H_{37}NO_3$ with 3 active hydrogen atoms. The nmr spectrum ($CDCl_3$) of the base showed a doublet (3H) at δ 1.12 due to the methyl group attached to piperidine ring. No other methyl peaks were visible. The signal due to methylene protons occurred as a singlet at δ 1.30. There was another broad singlet at δ 2.33 (2H) which disappeared on shaking with CD_3OD and therefore appeared to be due to the two OH groups. A distorded triplet (2H) at δ 3.65 was due to $-CH_2$ -OH protons in the side chain. In addition there were other signals due to protons on the piperidine ring. The protons at C-4 and C-5 absorbed as multiplet in the region δ 1.5-1.8. On the other hand the protons at C-2 and C-6 at α-position to piperidine nitrogen atom as well as at C-3 appeared as a broad signal at δ 2.56-3.15.

Recently we published [33] the configuration of this alkaloid. The 3-hydroxy-2-methyl piperidine alkaloids isolated so far mostly exist in all cis configuration ie the absolute configuration at C-2, C-3 and C-6 are R,R and S. for example in spectaline and juliflorine (values in parenthesis) at 57.0 (57.2), 67.6 (67.8) and 55.7 (55.7) ppm respectively. The two piperidine alkaloids iso-6-cassine and iso-6-carnavaline isolated from Cassia spectalis have different stereochemistry at C-2, C-3 and C-6 i.e. R,R and R. This has been confirmed by x-ray diffraction studies. The change of stereochemistry at these centres leads to change of chemical shifts of C-2, C-3 and C-6 in the ^{13}C-nmr spectrum which shows signals in iso-6-cassineat 50.4, 68.9 and 49.5 ppm respectively.

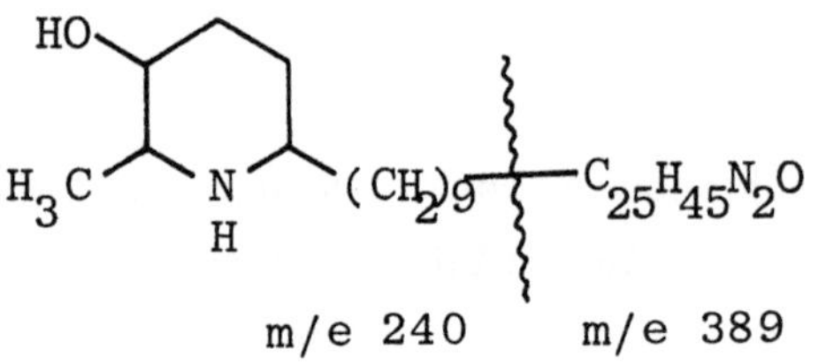

(VII)

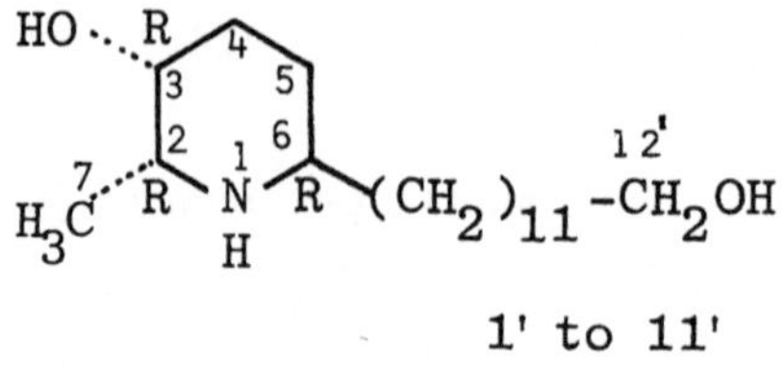

(VIII)

The comparison of the values for C-2, C-3and C-6 with those of juliflorine (juliprosopin) and spectaline [34] on one hand and of iso-6-cassine on the other shows that julifloridine belongs to the latter series. Therefore structure (VIII) was proposed, having the absolute configuration at C-2 C-3 and C-6 as R,R and R. This configuration finds support from pmr spectrum (100 MHz) of julifloridine.

In spectaline the H-2 signal is a broad multiplet at δ 2.90 and H-3 signal is broad singlet at δ 3.55 with a width at half height ($W \frac{1}{2}$) of 6Hz. In iso-6-cassine on the other hand H-2 signal is an Octet at δ 3.06 whereas the H-3 signal is a quintet at δ 3.66 with $J_{3,4}$ = 6.5Hz, $J_{3,4}$ = 3.2 Hz. This indicated the clear difference between H-2 and H-3 signals in the two series. In julifloridine the H-2 signal was an octet centred at δ 3.10 with $J_{2,7}$ = 6.5 Hz and $J_{2,3}$ = 3Hz which resembles to that of iso-6-cassine rather than that of spectaline. The H-3 signal was masked by H-12' triplet centred at δ 3.63.

j) julifloricine

The mass spectrum of julifloricine with M^+ at m/z 629 resembles greatly to that of juliflorine, indicating that these two alkaloids are isomeric. It is suppposed that these two alkaloids differ only with respect to the absolute configuration at C-2, C-3 and C-6 of the piperidine ring (s). Further work on this alkaloid is in progress.

Antibacterial and antifungal activity of juliflorine:*

The antibacterial and antifungal activity of juliflorine dihydrochloride was tested. It was found that this alkaloid shows antibacterial activity against some pathogenic bacteria in concentrations of 0.5 gm/ml onwards. Table (1). At 20 mg/ml there was complete inhibition of growth. Juliflorine inhibits the growth of T. rubrum, T.mentagrophytes, M.avis, E.flocculum at a concentrations of 2 µg/m onward (table II).

*Effect of juliflorine on Aedes aegypte**:*

The compound was tested against 4th instar larvae of A.aegypte and it was found to be effective as a number of larvae could not survive due to toxic effect, while the others have morphological abnormalities as white pupae with shrunken body as compared to normal. Among adults emergence was also low and many of them were found dead and attached with their exuvae.

4. Blepharis sindica

Blepharis sindica T.anders (syn.B.Linariafolia pers. N.O.Acanthaceae) is a herb. or under shurb growing widely in Sind and Baluchistan provinces of Pakistan. In the folkloric medical system the seeds have been used as a cure of earache [35]. Since very little work has been done on this species, the chemical constitiuents of the seeds of Blepharis sindica were studied.

The flavone glycosides terniflorin (cosmosin 6"-o-p-coumarate) and (prunine-6-o-coumarate) were isolated, in addition to other compounds [36].

---

*The work was performed in collaboration with Dr.Khurshid Ali Khan, Department of Microbiology, University of Karachi, Karachi-32,Pakistan.

**This work was performed in collaboration with Prof. Naim Naqvi, Department of Zoology, University of Karachi, Karachi-32, Pakistan.

Table 1

ZONE OF INHIBITION OF VARIOUS PATHOGENIC BACTERIA TO DECREASING CONCENTRATIONS OF JULIFLORINE DIHYDROCHLORIDE

S.No.	Name of Bacteria	Zone of inhibition in (mm)			
		20 mg/ml	1 mg/ml	0.5 mg/ml	0.25 mg/ml
1.	Salmonella typhi	20 mm	18 mm	16 mm	No effect
2.	Solmonella paratyphi A	18 mm	16 mm	14 mm	""
3.	Salmonella paratyphi B	18 mm	1.65 mm	14.2 mm	""
4.	Escherichia coli	10 mm	8 mm	6mm	""
5.	Corneybacterium diph theriae	19 mm	12 mm	10 mm	""
6.	Corneybacterium citreus	30 mm	25 mm	20 mm	""
7.	Staphylococcus aureus	35 mm	22 mm	20 mm	""
8.	Staphyloccus albus	30 mm	23 mm	20 mm	""
9.	Streptococcuslactics	20 mm	16 mm	12 mm	""
10.	Candida albicans	16 mm	13 mm	12 mm	""

Table 2

EFFECT OF JULIFLORINE DIHYDROCHLORIDE ON THE GROWTH OF PATHOGENIC FUNGI - NO GROWTH + TO ++++ GROWTH OF INCREASING INTENSITIES

S.No.	Name of Fungus	Sensitivity of Dermatophylis to juliflorine dihydrochloride							
		3 g/ml	2.5 g/ml	2 g/ml	1.5 g/ml	1 g/ml	0.5 g/ml	0.1 g/ml	Control
1.	Trichophyton rubrum	-	-	-	-	-	+++	+++	+++
2.	Trichophyton mentagrophytes	-	-	-	++	++	+++	+++	+++
3.	Microsporum canis	-	-	-	+	++	+++	+++	+++
4.	Epidermophyton floccosum	-	-	-	+	++	++	++++	++++

k) Terniflorin

The compound was crystallized in methanol m.p. 240. The field desorption spectrum of the compound showed peaks at m/z 601 (M^++Na), 579 (M^++1) and 578 (M^+) 432 (M^+-p-coumaroyl) and 270 (aglucose). The F.I. spectrum showed highest peak at m/z 270. On hydrolysis with dilute hydrochloric acid in methanol-water the compound yielded apigenin, identified by direct comparison with an authentic sample and glucose, p-coumaric acid as well as its methyl ester. The compound was therefore glucoside of apigenin with one of the OH group esterified with p-coumaric acid. The lack of shift of U.V. maxima in presence of sodium indicate that glucose moiety is linked to 7-OH group.

The ^{13}C nmr spectrum of the flavone glycoside in DMSO-d_6 was compared with the ^{13}C nmr data of apigenin 7-o-glucoside published by Redaelli, Formentini and Santaniello. The comparison showed that in the flavone glycoside isolated by us the C-6 signal was shifted downfield by δ2.9 ppm whereas C-5 signal was shifted upfield by δ 3.2 ppm indicating that coumarate moiety was attached to C-6 oxygen of apigenin glucoside. It was concluded with help of the spectral data that flavone glycoside isolated from Blepharis sindica has the structure (IX) apigenin -7-O-β -D-(6" coumaroyl) glucoside (terniflorine). The compound had been isolated earlier form Clematis terniflora [37] Salix alba [38] (as an artefact) and Pogostemon cablin [39].

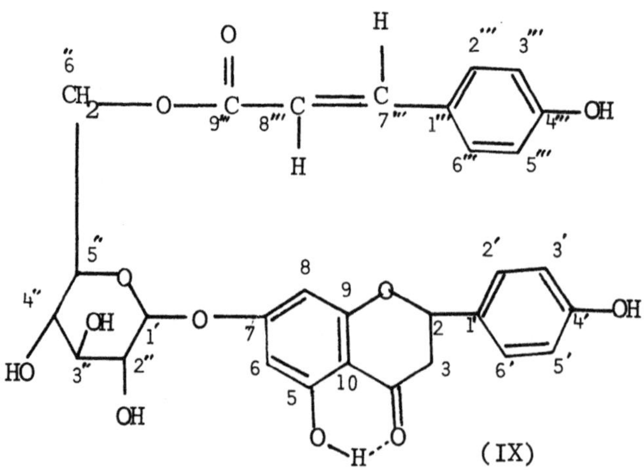

(IX)

l) Prunine

The compound was crystallized in methanol m.p. 170°. The field desorption mass spectrum showed peaks at m/z 581 (M^{+}-1) and 580 (M^{+}). The field ionization spectrum showed highest peak at m/z 272. There was a peak at m/z 164 which was attributed to the p-coumarate moiety and other at m/z 120.

On refluxing the compound with methanolic HCl, p-coumaric acid methyl ester (m.p. 130°) narringenin m.p. 250° and D-glucose were obtained.

The assignments of ^{13}C nmr spectrum of the flavone glycoside was based on the published data of narringenin [40] p-coumarate moiety [41] and of glucose moiety [42].

All the spectral data showed that the compound was narringenin-7-O-(6"-O-p-coumaroyl) -β-D-glucoside (6"-O coumaroyl prunine (X). This compound had been isolated only once earlier by W. Rehman et al.[43] from Anacardium occidentale L.

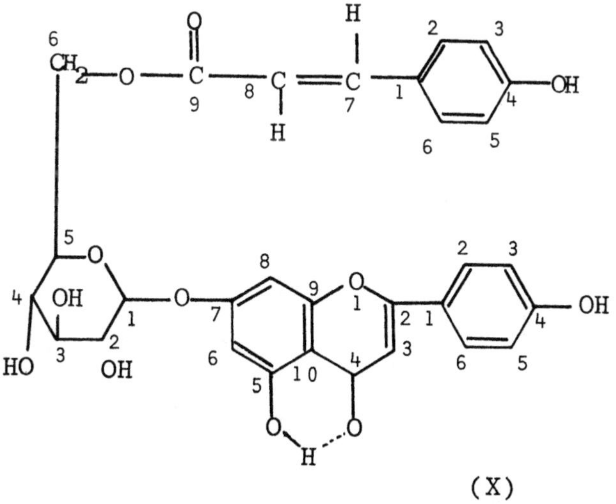

(X)

5. *Guaiacum officinale L.*

Guaiacum officinale L. (lignum vitae, Zygophyllaceae) is an evergreen tree which is cultivated for ornamental purpose. Its wood is used as diuretic, diaphoretic, sudorific, sialagogue and for mouth wash [44] Acid hydrolysis of the saponins from the bark of Guaicum officinale L. yielded new saponin 3β , 20 ξ - dihydroxy-30-norolean-12-en-28-oic acid (XI) and the artifacts 3 β-hydroxy-30-norolean-12, 19-dien-28-oic acid (XII)and its methyl ester.(XIII). Besides these new compound, larreagenin, which was first isolated from Larrea divaricata [45], β-sitosterol and oleanolic acid were also isolated.

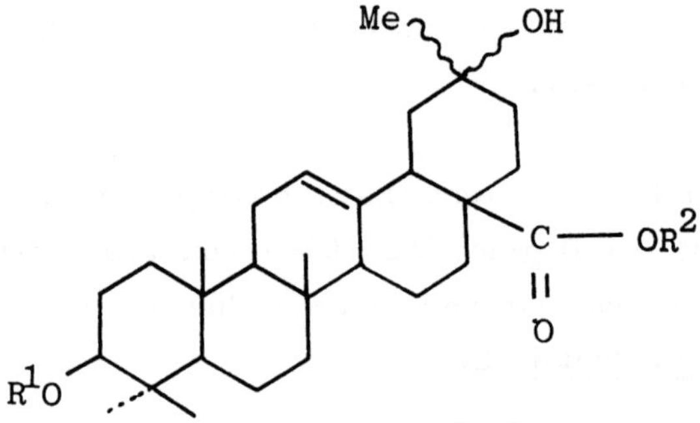

XI $R^1=R^2=H$

XIa $R^1=H, R^2=Me$

XIb $R^1=Ac, R^2=Me$

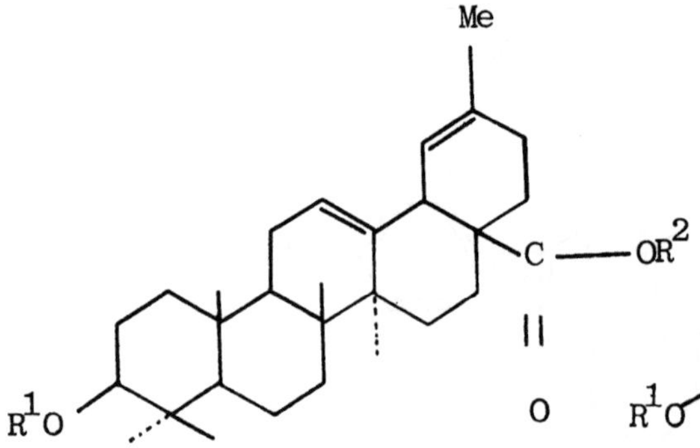

XIII $R^1=H, R^2=Me$

XIIIa $R^1=Ac, R^2=Me$

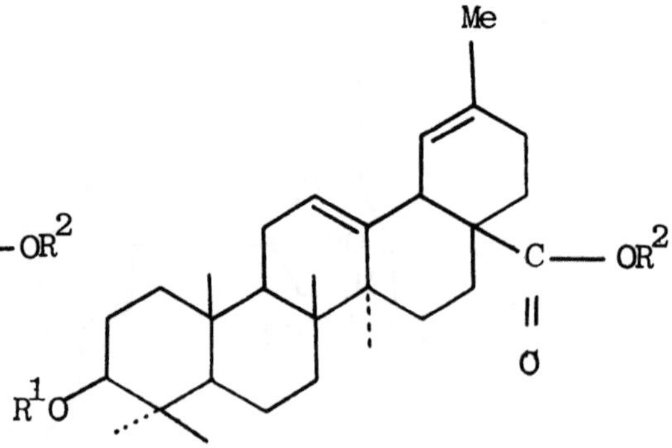

XII $R^1=H, R^2=H$

m) Methyl-3 β-Hydroxy-30-norolean-12,19-dien-28-oate (XIII)

This compound m.p. 220-222°, exhibited a molecular ion peak at m/z 454 corresponding to the formula $C_{30}H_{46}O_3$. Its u.v. showed only end absorption and ir spectrum revealed the presence of OH (3420 cm^{-1}), ester (1720 cm^{-1}). The pmr spectrum displayed signals at δ 0.77, 0.79, 0.91, 0.96, 1.01, 1.60 (each 3H, s, $6xCH_3$), 3.25 (1H, m, H-3 and H-18), 3.63 (3H, s, OCH_3), 5.12 (1H, brs, H-19), 5.4 (1H, t, J= 3.75 Hz, H-12). From i.r. pmr spectra, this compound appeared to be a methyl ester of a nor-triterpenic acid. The downfield shift of H-18 signal in this compound as compared to methyl oleanolate (δ 2.86) is appearently due to its allylic position to the addition double bond at C-19. The mass spectrum reveals the strong peak m/z 246 & 247 due to retro-Diels-Alder fragmentation. The mass spectrum also shows that the double bond in ring E is not present between C-20 and C-21 because in that case retro-Diels-Alder fragmentation of this ring would give rise to M^+ -68 peak [46]. From these spectroscopic data, the structure proposed for this triterpene is 3β -hydroxy-30-norolean-12, 19-dien-28-oate. This structure is also supported by ^{13}C nmr data.

n) 3β-Hydroxy-30-norolean-12,19-dien-28-oic acid (XII)

This compound m.p. 215-218°,showed a molecular ion peak at m/z 440, corresponding the formula $C_{29}H_{44}O_3$. This was identified as an acid corresponding to XII because on treatment with diazomethane it a converted into the latter compound.

o) 3 β ,20 ξ Dihydroxy-30-norolean-12-en-28-oic acid (XI)

This new triterpene, m.p.234-236° exhibited a molecular ion peak at m/z 458 corresponding to the formula $C_{29}H_{46}O_4$. Thus (XI) differs from (XII) only by the elements of a molecule of water. Its i.r. spectrum showed bands at 3440 (OH), 1695 (CO, acid). The pmr spectrum of its methyl ester (XI a) showed signals at δ 0.72, 0.78, 0.90, 0.98, 1.14, 1.27,(each 3H, S, Me), 3.20 (1H, m, H-3), 2.80 (1H, dd, J=5, 13.75

Hz, H-18), 5.32 (1H, t, J=3.75 Hz, H-12), 3.63 (3H, S, OCH_3). The i.r. spectrum of its monomethyl ester acetate (XIb) also showed OH absorption due to the tetriary OH group, the mass spectrum of (XI) shows peak due to the RDA fragmentation at m/z 207 & 250, the latter losing water easily to yield the base peak at m/z 232. This indicate that the second tertiary OH group is located in ring D/E, obviously at C-20. On the basis of the above spectra, structure (XI) was suggested to the 3 β,20 ξ-dihydroxy-30-norolean-12-en-28-oic acid. It was supported by the ^{13}C-nmr spectrum. There were two carbon atoms bearing oxygen function at 78.94 (C-3) and 70.93 ppm (C-20). The latter peak was inverted in the GASPE spectrum indicating that this carbon atom was quarternary & hydroxyl group attached to this carbon atom was tertiary. The absolute configuration at C-20 is not known so far.

On refluxing (XI) in methanolic HCl, it was completely converted into a mixture of (XII) and (XIII), which suggest that XI is genuine sapogenin of G.officinale, whereas (XII) & (XII) are artefact formed during acid catalysed hydrolysis of the saponins.

Acknowledgement

I sincerely thank Prof. Voelter and Prof. Fuchs for collaboration and I am also thankful to my students Qazi Najmus Saqib, K.Usman Ghani, Wasimul Haq, Anwar Basha, Shaheen Bano, Zubaida G.Mohammad Sabiha Qazi, Iffat Mahmood and Nasreen Bano, without their assistance this work could not be possible.

The financial support from Pakistan Science Foundation and University Grants Commission is gratefully acknowledged.

References

1. Heinz A.Hoppe, **Drogenkunde** vol.I, Angiospermen, Walter de Gruyter Berlin, 8. Auflage, New York 1975 page 872.
2. R.R.Stewart. An Annotated catalogue of vascular plants of West Pakistan & Kashmir (Ed. E.Nasir and S.I.Al), 1972
3. W.Dymock,C.J.Warden D.Hand Hooper, **Pharmacographia indica** 345 (1890)
4. W.H.Lewis and M.P.F. Elvin, Lewis, **Medical Botany**, John Wiley & Sons p. 84 (1977)
5. V.U.Ahmad, Q.N.Saqib, K.U.Ghani and W.Voelter and W.Fuch, **J.Chem.Soc.Pak.**, vol.5 No. 1 (1983)
6. H.Budzikiewicz, J.M.Wilson and C.Djerassi, **J.Amer.Chem.Soc.**, 85, 3688 (1963)
7. R.O.Dorchai, H.H.Rubalcava, J.B.Thomson & B.Zeeh, **Tetrahedron** 24, 5649 (1968)
8. K.Tori, S.Seo, A.Shimaoka and Y.Tomita, **Tetrahedron Lett.**, 4227 (1974)
9. V.U.Ahmad, Q.Najmus Saqib, K.Usmanghani, W.Fuchs and W.Voelter, **Z.Naturforsch**, 35b, 511 (1980)
10. D.R.H.Barton, A.Hameed, J.F.McGhee, **J.Chem.Soc.**, 5176 (1962)
11. R.Tschesche, H.Striegler, H.W.Fahlhaber, **Liebigs.Ann.Chem.** 691, 165 (1966)
12. S.Ito, M.Kodaman, M.Sunagawa, T.Oba, **Tetrahedron Lett.**, 2905 (1969)

13. R.N.Chopra, S.L.Nayar, I.C.CHopra, **Glossary of Indian Medicinal Plant**, Council of Scientific and Industrial Research, New Delhi (1956) 0.175.

14. Hakim Mohammad Said (Editor), **Hamdard Pharmacopoeia of Eastern Medicine**, Hamdard Foundation, Karachi (1969) p.446.

15. K.M.Nadkarni, **Indian Materia Medica**, Popular Prakashan Private Limited, 1, 845-846 (1976).

16. Viqar Uddin Ahmad ans Shaheen Bano, W.Voelter and W.Fuchs, **Tetrahedron Lett.**, 22 (18) 1715 (1981)

17. J.M.Lehn and G.Ourisson, **Bull.Soc.Chim.Fr.**, 1137 (1962)

18. E.L.Ghisalberti, P.R.Jefferies and M.A.Sefton, **Phytochemistry**, 12, 1125 (1973)

19. E.Wenkert, G.V.Baddeley, I.R.Burfilt and L.N.Moreno, **Org.Mag. Res.**, 11(7), 337 (1978)

20. M.Sholichin, K.Yamasaki, R.Kasai and O.Tanaka, **Chem.Pharm.Bull.** (Tokyo) 28(3), 1006 (1980)

21. Viqar Uddin Ahmad and Shaheen Bano, W.Voelter and W.Fuchs, **Z.Naturforsch**, 37b, 1675 (1982)

22. E.Naser and S.I.Ali (Editors), **Flora of West Pakistan** (Fakhri Printing Press, Karachi)

23. P.Shankar Murthy and S.Siddiqui, **J.Sci and Ind.Res**, 7B(11) 188 (1948)

24. M.M.El-Merzabani, A.A.El-Aaser, M.A.Attia, A.K.El Durweini and A.M.Ghazala, **Planta Medica** 36, 158 (1979)

25. K.Jewers, N.J.Nagler, K.A.Zirvi, F.Amer. **Phalvi Medical Journals**, 5 (1) (1974)

26. G.Ratle, X.Monseur, B.C.Das, J.Yassi, Q.Khuong-Hu and R.Goutaral, **Bull.Soc.Chim.Belg.**, 2945 (1966)

27. K.M.Nadkarni, **Indian Matria Medica**, Popular Prakashan Private Limited, Bombay (1954) p. 1011

28. R.N.Chopra, I.C.Chopra, K.L.Handa, L.D.Kapur, **Indigenous Drugs of India** (U.N.Dhur & Sons, Calcutta) p. 521 608, 610 (1958)

29. K.R.Kirtikar and B.D.Basu, **Indian Medicinal Plants**, Published by Sunhindra Nath Basu, Panini Office Bhuneswari Asram, Bahadurgang, Allahabad (1981) p. 494

30. R.N.Chopra, S.L.Nayar, I.C.Chopra, **Glossary of Indian Medicinal Plants**, C.S.I.R., New Delhi, p. 204 (1956)

31. V.U.Ahmad, Anwar Basha and Wasimul Haque, **Z.Natuforsch**, 33b, 247-348 (1978)

32. Viqar Uddin Ahmad Z.G.Mohd, **J.Chem.Soc.Pak.**, 1(2), 137 (1979)

33. Hesse, **Helv.Chim. Acta**, 63, 2119 (1980)

34. Viqar Uddin Ahmad and Sabiha Qazi, **Z.Naturforsch**, 38b, 660 (1983)

35. I.Chrostofidis, A.Welter, J.Jadot, **Tetrahedron**, 33, 977 (1977)

36. S.R.Baquar and M.Tasnif, **Medicinal Plants of South West Pakistan**, PCSIR bulletin/monograph No.3 (1967) page 57.

37. Viqar Uddin Ahmad, Akhtar Mohd Burki, Iffat Mahmood and David Smith, **J.Chem.Soc.Pakistan**, 6(4), 217 (1984)

38. C.Radaelli, L.Formentin and E.Santaniello, **Phytochemistry**, 19, 985 (1980)

39. M.Aritoni, **Chem.Pharm.Bull.**(Tokyo), 11 (10) 1225 (1963)

40. C.Karl, G.Muller, P.A.Padersen, **Phytochemistry**, 15, 1084 (1976).

41. H.Itokawa, K.Suto, J.Takaya, **Chem.Pharm.Bull.**,29 (1), 254 (1981)

42. H.Duddeck, G.Snatzke and S.S.Yemul, **Phytochemistry**, 17, 1369 (1978)

43. V.M.Chari, M.Jordan and H.Wagner, **Planta Medica**, 34, 93 (1978)

44. K.R.Markham, B.Tenai, R.Stanley, H.Geili and T.J.Marby, **Tetrahedron**, 34, 1389 (1978)

45. W.Rehman, K.Ishratullah, H.Wagner, O.Seligmann, V.M.Chari and B.G.Osterdahl, **Phytochemistry**, 17,1064 (1978)

46. Heinz, A.Hoppe (1975), **Drogenkunde** Vol. I, (Angiosperms) 8th Edition pp-551-552, Walter de Gruyter, Berlin, New York.

47. G.Habermehl and H.Moeller, (1974) **Justus Liebigs Ann.Chem** (2), 169 (1974)

48. H.Budzikiewicz, J.M.Willson and C.Djerassi, **J.Am.Chem.Soc.**, 85, 3688 (1963)

Hypcholesterolemic Effect of *Allium sativum* Linn. and its Potential Protective Action Against Coronary Heart Disease

A Model for Research on Drugs of Traditional Medicine

Yusuf Ahmad

Pharmaceutical and Fine Chemicals Research Division, Pakistan Council of Scientific and Industrial Research, Karachi-39, Pakistan

Abstract: In the literature of traditional medicine garlic (Allium sativum, Linn) is reported to be very effective against heart disease.

There have been conflicting reports on blood cholesterol lowering effects of garlic and garlic oil. It necessitated a more detailed study on lipid metabolism by garlic and/or its fractionated components.

White Leghorn pullets were fed for 4 weeks either a control diet based on corn and soybean meal or an experimental diet containing either 3.8% garlic paste, a solvent extract (petroleum ether, methanol and water in sequence) of garlic paste, the residue or commercial garlic oil. Significant decreases in hepatic 3-hydroxy-3-methylglutaryl-CoA reductase (79-83%), cholesterol-7-hydroxylase (43-51%), fatty acid synthetase (17-29%) and in representative pentose-phosphate pathway (23-39%) activities accompanied the feeding of the petroleum ether-, methanol- and water-soluble fractions of garlic. Garlic paste and oil also suppressed these activities. Significant decreases in serum lipids occurred in response to the feeding of these garlic fractions: serum total cholesterol by 20-25%, low density lipoprotein cholesterol by 28-41% and triglycerides by 10-26%. The residue remaining after solvent fraction had little influence on these parameters. These findings were substantiated by a second study in which pullets of a commercial broiler line were fed the garlic fractions. The results confirm that garlic oil and odorous components of garlic lower cholesterol levels. However, an odourless water-soluble component of

Natural Product Chemistry
Ed. by Atta-ur-Rahman

garlic also has an almost equally potent effect. The mechanism of the hypocholesterolemic action is at the level of the suppression of cholesterol biosynthesis.

The need for research on Traditional Medicine on modern scientific lines has been examined, and thereby to discover from it potent medicinal agents comparable to modern drugs acceptable to current medical profession.

For centuries people have revered the extraordinary medicinal properties of garlic (Allium sativum, Linn). The Babylonians used it to treat disease as early as 3000 B.C. In Egypt, garlic was believed to give slaves the strength to endure hard labour and to give soldiers courage in the battle field. After intense fatigue a clove of garlic slowly chewed, and swallowed, has been reported to act as a very powerful restorative [1].

In traditional Medicine, too, garlic has been extensively used, and prescribed. In 1893 Dymock [1] summarised the old literature on garlic. Under its history and use, he records, among other things, "Garlic is the (thum) of the Arabians and (sir) of the Persian.......A decoction of garlic in milk is given in small doses in hysteria, flatulence, sciatica, and heart disease....."These claims in the past have been made on the basis of personal experiences of old physicians. However, during the last few decades hypocholesterolemic properties of garlic have been widely reported [2-15]. The magnitude of this action ranges from a modest 14% lowering of serum cholesterol in humans [12] to a lowering of 80% in cholesterol-fed hypercholesterolemic rabbits [3]. Garlic was fed in these various studies as the raw cloves, steamed or fried cloves, as garlic (oil) pearls, powder, as juice or extract or as the essential oil.

In most of the studies made so far it has been assumed that hyocholesterolemic activity in garlic resides only in the very small quantity (about 0.25%) of the (essential) oil, which is responsible for the flavour and taste of the herb. In 1944 C.J. Cavallito and J.H. Bailey [16] isolated this oil, which in its pure form is pale yellow to colourless is of a very repulsive taste and has the true concentrated odour of garlic.

They succeeded in isolating from it 'allicin' a sulphur compound, and determined its structure as diallyl disulphide-oxide (I). An amino acid alliin is also present in garlic. Its structure has been determined to be S-allylcysteine sulphoxide (II). In the intact clove 'alliin' and the enzyme allinase are kept apart by the cell walls. However, when the clove is cut the enzyme allinase reacts with alliin and converts it into allicin, which has the characteristic garlic odour.

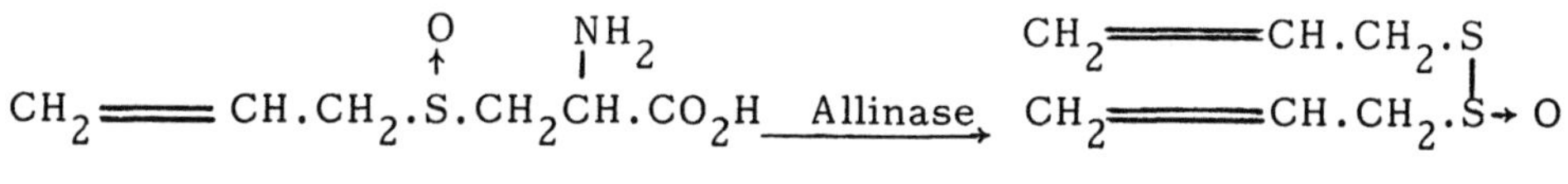

I *Alliin* (S-allylcysteine sulfoxide)

II *Allicin* (diallyldisulfide-oxide)

The effects of maturation and of the postharvest changes in garlic on this hypochlolesterolemic property have not been studied. Such effects might explain some of the inconsistencies in the results of various studies of different garlic products. Alliin [17] and allicin [18] two pure products isolable from garlic have been shown to lower cholesterol levels of cholesterol-fed [17] and normal [18] rats. But this activity does not account for the total activity of the whole garlic paste. Besides, according to recent Japanese work*, allicin taken in large quantities can destroy the membranes of the red blood cells, and inflame the gastro-intestinal tract. These reports prompted us to look for hypochlolestrolemic activity of garlic in its various extractives in non-polar, to progressively more polar solvents, and to confirm this activity through a study of their action on various enzymes which have been well studied and well established for the synthesis of cholesterol in the body.

The previous studies have been mostly made using rabbit or rat as the model animal. The overall effects of garlic and its constituents on lipid metabolism in chicken have not been described. Work done on rats

---

*Personal Communication (Prof. Kitabara, S.Kunamoto University, Japan)

and rabbits is of value but chicken (or quail) is a better model [19] to study atherosclerotic changes, because the lipoproteins of chicken (or quail) are more closely related to those of the human [19].

In this study we describe the effects of feeding garlic or garlic extracts on hepatic 3-hydroxy-3-methyl glutrayl-coenzyme A (HMG-CoA) reductase, cholesterol-7α-hydroxylase, pentose-phosphate pathway enzymes and fatty acid synthetase (FAS) activities. These enzymatic activities reflecting the control of overall lipid metabolism were measured, along with serum lipid components: total cholesterol, LDL-chol (low and very low density lipoprotein bound cholesterol), HDL-chol (high density lipoprotein bound cholesterol), and triglycerides, in layer and broiler pullets after a fasting-refeeding regimen that induced lipogenic activity.

MATERIALS AND METHODS

Materials:

Experimental materials and enzymes reagents were obtained from renowned suppliers. Cholesterol was recrystallised fresh twice from glacial acetic acid, and dried in vacuum. Steam-distilled commercial garlic oil was of J.Manheimer, Inc., Long Island, NY. Garlic bulbs and the diet components were purchased locally. All other chemicals were of analytical grade.

Pullets, diets, experimental design:

White Leghorn pullets (layers) 33-37 days of age and the crossbred broiler pullets (broilers) 21 days of age were purchased from a local hatchery. The pullets were randomly distributed into groups (8 layers; 6 broilers) which were placed in a three tier battery. Commercial grower mashes formulated for layers and broilers were fed prior to the start of the experiments. At 8 weeks of age, pullets in each group were weighed and returned to the battery. The experimental diets were based on commercial formulations. The layer diet contained 78% corn, 16% soybean meal (44% protein), 2% alfalfa meal (17% protein) and 1.5% meat and bone meal.

The broiler diet consisted of 70% corn, 23.5% soybean meal (44%) protein), 2% alfalfa meal (17% protein) and 2% meat and bone meal. Mineral and vitamin supplements are given as footnotes to table 1&2 (all tables are given at the end of the paper). The experimental diets were modified with the substitution of 3.8% garlic paste or a solvent fraction or garlic oil equivalent to 3.8% garlic paste for an equal quantity of corn. The groups of pullets, which had been established 2 or 4 weeks prior to the experiment, were randomly assigned the experimental diets. The layers were fed the diets ad libitum for 24 days. The control group of broilers was fed ad libitum for 24 days; the experimental broiler groups were fed isogravically with the control. Feed consumption of the broilers was recorded. Water was given ad libitum at the end of the 24-days feeding trial, the pullets were fasted for 24 hours and then refed for 72 hours. Blood samples were taken from wing veins of the broilers at the end of the fasting period. After refeeding, the pullets were weighed, killed by severing the carotid arteries, and blood samples were collected. The livers were removed, washed, chilled on ice, weighed and then prepared for analyses.

Preparation of the garlic paste and solvent fractionations:

Garlic bulbs (1 kg) with the outer and inner husks removed yielded 760 g cloves. The cleaned cloves were homogenized to a paste with a commercial Waring blender. One-half of the garlic paste was mixed with sufficient corn to provide 7800 g of mixture for the formulation of the "garlic paste" experimental diet for the layers or 7000 g for the broilers. The remainder of the paste was stirred with light petroleum ether (500 ml) for 2 hours. After stannding for 1 hour, the solvent was decanted. The procedure was repeated twice. The last extraction was completed by filtering the solvent through a sintered glass funnel under vacuum. The three extracts were combined and then concentrated to dryness under vacuum at 60°. The semisolid garlic residue was lyophilized, and the resulting powder was extracted successively as described above with methyl alcohol and water. The final residue was dried overnight in an oven at 60°. The fractionation yielded 1.4 g (1.1%) petroleum ether-soluble fraction (PESF), 34.2 g (26.4%) methyl alcohol-soluble fraction (MESF), 79.3 g

(62.4%) water-soluble fraction (WASF) and 12.1 g (9.5%) of residue. The values in parentheses correspond to the percent dry weight in each fraction. Each fraction was taken up in a minimal volume of the appropriate solvent and added to sufficient corn to give 7800 g or 7000 g. The experimental diets thus contained the specific solubles in quantities equivalent to those in a 5% garlic bulb (3.8% garlic paste) diet. The solvent was removed from each diet by air-drying the mixture spread thinly in a pan overnight in a fume hood.

Preparation of liver homogenates and assay of enzymes:

The liver homogenates were prepared in 0.1 M Potassium phosphate buffer, pH 7.4, containing 4 mM $MgCl_2$. 1 mM EDTA, and 2 mM dithiothreitol. The tissue was chopped and suspended in the buffer (1:2, wt:vol), and homogenization was done at 0-4° with a Polytron homogenizer. The 100,000 x g supernates (cytosols) and precipitates (microsomes) were stored at -20° prior to assay for enzymatic activities.

Assays for HMG-CoA reductase and cholesterol 7 α-hydroxylase have been described (20,21). The activities of glucose-6-phosphate dehydrogenase, 6-phospho-gluconate dehydrogenase, malic enzyme, citrate-cleavage enzyme and fatty acid synthetase in the cytosol were assayed spectrophotometrically at 25° [20,21]. Protein concentrations were estimated by a modification of the biuret method with bovine serum albumin as a standard [22].

Estimation of serum cholesterol and triglyceride concentrations

Cholesterol and triglyceride concentrations in serum samples were estimated by using Worthington "Cholesterol Reagent" and "Triglyceride Reagent", kits.

Low density (LDL) and very low density lipoproteins (VLDL) were isolated from the serum (100 μl) by precipitation with a mixture of phosphotungstic acid, 9.7 mM(10 μl) + $MgCl_2$ 0.4 M (10 μl). After standing for 5 minutes at room temperature, the mixture were centrifuged at 2000 x g for 10 minutes, the supernatent was removed and used to determine the level of cholesterol in high density lipoprotein (HDL). The precipitate was dissolved in 0.1 M sodium citrate (100 µl), and the level of cholesterol (LDL + VLDL) was estimated by using the above method. This measure was used to verify the accuracy of the estimates of LDL-chol which were calculated from (total-chol) - (HDL-chol + triglycerides/5) according to previous reports (23,24).

Expression of data and statistical methods

Enzyme data are presented as specific activities units/(miligram cytosolic or microsomal protein. minute). Statistical comparison of results was performed by a one-way analysis of variance. When the F test indicated a significant effect, the differences between the means were analysed by a protected least squares difference test [25].

RESULTS

The average weight gain of the groups of layers fed the experimental diets (384.0 $\pm$ 13.3 g) was equal to that of the control diet group (381 g) table 1). Although feed consumption was not monitored in this experiment, the above data suggest that the incorporation of the garlic products into the diets did not decrease feed consumption. Significant treatment effects were shown for all aspects of cholesterol and lipid metabolism save that of the HDL-chol concentrations (tables 3,4). HMG-CoA reductase activity in all experimental groups except the group fed the garlic residue exhibited less than 30% of the control activity ($P < 0.01$). The solvent extracts PESF, MESF and WASF - were the more potent sources of the HMG-CoA reductase inhibitor ($P < 0.01$) compared to garlic paste and garlic oil (table 3). A similar but less pronounced pattern of inhibition was found for cholesterol-7 α-hydroxylase (MESF, 49%), fatty acid synthetase (PESF, 71%), malic enzyme (MESF, 73%), glucose-6-phosphate dehydrogenase (MESF, 61%) and 6-phosphogluconate dehydrogenase (WASF, 68%). In the parentheses are shown the most effective treatment and enzyme activity as percentage of control (table 3).

TABLE 1

Composition of the layer diets and layer weight gain

Groups	Diet[1,2]		Body wt		
	Corn	Garlic fractions	Initial[3]	Final[4]	Gain in wt
	%	%	g		%
Corn (control)	78.000	—	459 ± 64[5]	840 ± 95[5]	83
Corn + garlic paste	74.200	3.8	508 ± 84	900 ± 101	77
Corn + garlic PESF	77.980	0.014	487 ± 69	880 ± 97	81
Corn + garlic MESF	77.658	0.341	486 ± 72	853 ± 85	76
Corn + garlic WASF	77.207	0.793	482 ± 59	868 ± 98	80
Corn + garlic residue	77.879	0.121	494 ± 32	892 ± 102	81
Corn + garlic oil (commercial)	77.986	0.014	456 ± 60	824 ± 86	81

TABLE 2

Percent composition of broiler diets, initial and final body weights and feed consumption

Groups	Diet[1,2]		Body weight			
	Corn	Garlic fractions	Initial[3]	Final[4]	Gain in wt	Feed consumption
	%		g		%	kg
Corn (control)	70.000	—	1414 ± 128	2169 ± 145[5]	53	18.85
Corn + garlic paste	67.200	3.800	1587 ± 135	2219 ± 115	39[a]	18.60
Corn + garlic PESF	69.986	0.014	1552 ± 74	2296 ± 82	48	18.94
Corn + garlic MESF	69.658	0.342	1640 ± 48	2418 ± 92	47	18.96
Corn + garlic WASF	69.207	0.793	1583 ± 51	2411 ± 99	52	18.95
Corn + garlic residue	69.879	0.121	1516 ± 108	2183 ± 121	44[a]	18.64
Corn + garlic oil (commercial)	69.986	0.014	1598 ± 150	2281 ± 182	43[a]	18.93

[1]PESF, MESF and WASF stand for petroleum ether–, methanol– and water–soluble fractions of garlic, respectively. Five grams of garlic bulb yielded 3.89 garlic paste. Fractionation of 3.8 g garlic paste yielded 14 mg PESF, 342 mg MESF, 793 mg WASF and 121 g residue. Commercial garlic oil was added at the level of the PESF. [2]Each diet also contains soybean meal (44% protein), 23.5%; alfalfa meal (17% protein), 2.0%; meat and bone meal, 2.0%; dicalcium phosphate, 1.0%; calcium carbonate, 0.5%; iodized salt, 0.5%; vitamin and mineral mixture, 0.5%. Vitamin and mineral mixture provides per kilogram diet: vitamin A, 3000 IU; cholecalciferol, 500 IU; riboflavin, 2.5 mg; calcium pantothenate, 3.0 mg; vitamin B-12 0.005 mg; zinc sulfate ($ZnSO_4$), 70 mg; and manganese dioxide (MnO_2), 25.0 mg. Grit (5.0%) was also incorporated at the expense of each diet. [3]Eight weeks of age. [4]Twelve weeks of age.
[5]Mean ± SD; n = 8 chickens per layers and 6 broiler group. [a]Significantly different from control at $P < 0.01$.

TABLE 3.

Effects of garlic fractions on hepatic enzyme activities in 12-week-old layers[1,2]

Diet	HMG-CoA reductase[3]	Cholesterol 7α-hydroxylase[4]	Fatty acid synthetase[5]	Malic enzyme[6]	Glucose-6-phosphate dehydrogenase[6]	6-Phosphogluconate dehydrogenase[6]
Corn (Control)	909 ± 70^{a} (100)[7]	1.14 ± 0.08^{a} (100)	175 ± 14.0^{a} (100)	450.8 ± 42.0^{a} (100)	11.8 ± 1.8^{a} (100)	45.0 ± 7.0^{a} (100)
Corn + garlic paste	253 ± 40^{b} (28)	0.72 ± 0.04^{b} (63)	158 ± 12.0^{ac} (90)	396.6 ± 30.0^{ab} (88)	8.7 ± 0.8^{b} (74)	31.5 ± 3.0^{b} (70)
Corn + garlic PESF	193 ± 30^{c} (21)	0.61 ± 0.03^{c} (54)	124 ± 11.0^{b} (71)	407.6 ± 46.0^{a} (90)	9.1 ± 1.9^{a} (77)	32.4 ± 2.0^{b} (72)
Corn + garlic MESF	159 ± 30^{c} (17)	0.56 ± 0.04^{c} (49)	141 ± 11.0^{c} (81)	331.2 ± 37.0^{b} (73)	7.2 ± 0.7^{b} (61)	33.3 ± 4.0^{b} (74)
Corn + garlic WASF	192 ± 28^{c} (21)	0.65 ± 0.05^{bc} (57)	145 ± 8.0^{c} (83)	368.4 ± 32.0^{b} (82)	7.9 ± 1.8^{b} (67)	30.5 ± 6.0^{bc} (68)
Corn + garlic residue	781 ± 62^{a} (86)	0.93 ± 0.06^{d} (82)	172 ± 13.0^{a} (98)	450.2 ± 45.0^{a} (100)	10.8 ± 2.3^{ab} (92)	38.6 ± 2.0^{ac} (86)
Corn + garlic oil (commercial)	230 ± 30^{b} (26)	0.67 ± 0.05^{c} (59)	160 ± 4.0^{a} (91)	412.4 ± 40.0^{ab} (92)	7.7 ± 1.2^{b} (65)	36.0 ± 4.0^{abc} (80)

[1]Experimental period was 4 weeks; time of killing was 0800. Data expressed as means ± SD; $n = 8$, HMG-CoA reductase, 3-hydroxy-3-methylglutaryl-CoA reductase. PESF, MESF and WASF stand for petroleum ether–, methanol– and water–soluble fractions of garlic, respectively. [2]Values not sharing a common superscript letter are different at $P < 0.01$. [3]Picomoles of mevalonic acid synthesized per minute per milligram of microsomal fraction. [4]Picomoles of [^{14}C]cholesterol into 7α-[^{14}C]hydroxycholesterol per minute per milligram of microsomal fraction. [5]Nanomoles of NADPH oxidized per minute per milligram of cytosolic fraction. [6]Nanomoles of NADP^{+} reduced per minute per milligram of cytosolic fraction. [7]Percentages of respective control activity data are in parentheses.

TABLE 4

Effects of garlic fractions on serum lipids in 12-week-old layers[1,2]

Diet	Concentration in serum					
	Cholesterol	Triglycerides	HDL-chol	LDL-chol	Total cholesterol / HDL-chol	LDL-Chol / HDL-chol
			mg/100 ml			
Corn (control)	168.0 ± 2.0^{a} (100)[3]	125.1 ± 12.0^{a} (100)	57.0 ± 4.0^{a} (100)	86.0 ± 7.0^{a} (100)	2.95 (100)	1.50 (100)
Corn + garlic paste	132.8 ± 10.0^{b} (79)	98.1 ± 10.0^{b} (78)	55.5 ± 8.0^{a} (97)	57.7 ± 5.0^{b} (67)	2.39 (81)	1.03 (69)
Corn + garlic PESF	128.6 ± 8.0^{b} (76)	113.1 ± 10.0^{a} (90)	56.2 ± 8.0^{a} (99)	59.4 ± 4.0^{b} (69)	2.29 (78)	1.05 (70)
Corn + garlic MESF	126.5 ± 4.0^{b} (75)	92.6 ± 11.0^{b} (74)	57.0 ± 5.0^{a} (100)	51.0 ± 4.0^{b} (59)	2.22 (75)	0.89 (59)
Corn + garlic WASF	134.7 ± 11.0^{b} (80)	96.3 ± 10.0^{b} (77)	54.7 ± 6.0^{a} (96)	60.7 ± 5.0^{b} (71)	2.46 (83)	1.10 (73)
Corn + garlic residue	152.2 ± 14.0^{a} (91)	121.7 ± 10.0^{a} (97)	53.5 ± 4.0^{a} (94)	74.4 ± 7.0^{a} (87)	2.84 (96)	1.39 (93)
Corn + garlic oil (commercial)	128.9 ± 7.0^{b} (77)	115.1 ± 7.0^{a} (92)	54.5 ± 3.0^{a} (96)	61.2 ± 6.0^{b} (71)	2.36 (80)	1.12 (75)

[1]Experimental period was 4 weeks; time of killing was 0800 hours. Data are means ± SD; $n = 8$. PESF, MESF and WASF stand for petroleum ether–, methanol– and water–soluble fractions of garlic, respectively. [2]Values not sharing a common superscript letter are different at $P < 0.01$. [3]Percentages of respective control activity data are in parentheses.

The above pattern of enzyme inhibition by the various garlic factors was expressed in the serum levels of cholesterol and triglycerides (table 4). All garlic products except the residue lowered the serum cholesterol level, specifically the LDL-chol fraction. The MESF was most potent, lowering cholesterol and LDL-chol by 25 and 41%, respectively. HDL-chol was not affected by the treatments. These lowerings of the total cholesterol and LDL-chol are reflected in the ratios of total cholesterol: HDL-chol and LDL-chol: HDL-chol shown in table 4.

A similar experiment was carried out by using pullets of a commercial broiler strain bred for high rates of gain and feed efficiency. The basal diet fed the broilers contained 23.5% soybean meal (44%) and 70% corn (table 2). Garlic (3.8%) or the 3.8% equivalent garlic extract was added to the diet at the expense of corn. Feed consumption (18.84 $\pm$ 0.17 kg) did not vary between groups; the control group gained an average of 755 g, a 53% increase in body weight. The group fed the WASF gained 828 g, a 52% increase in body weight. Gains by the groups fed PESF and MESF were equal to that of the control group, whereas groups fed the garlic residue and garlic oil gained 667 and 683 g, respectively, increase of 43% in body weight. The major deviation in body weight gain was recorded for the group fed garlic paste. The average weight gain of this group was 632 g, only a 40% increase in body weight. Feed consumption by this group was 18.60 kg; whether or not the 3.8% dilution in the energy density of this diet influenced this weight performanc could not be determined.

The results shown in table 5 confirm that factors extracted from garlic suppress hepatic HMG-CoA reductase. Feeding the MESF and WASF of garlic reduced this activity by 66-68%. HMG-CoA reductase was lowered by 50% or more by each of the remaining garlic treatments except that of the extracted residue. These suppressions of HMG-CoA reductase are reflected in cholesterol levels of serum taken from fasted and from refed pullets and in the LDL-chol levels of refed birds (table 5). The MESF, the most potent source of this suppressor, lowered total cholesterol by 25% and LDL-chol by 43%. HDL-chol was not affected by the cholesterol-

TABLE 5

Effects of garlic fractions on hepatic HMG-CoA reductase activity and serum cholesterol in 12-week-old broilers[1,2]

Diet	HMG-CoA reductase[3]	Serum cholesterol: Total (fasted)	Serum cholesterol: Total (refed)	Serum cholesterol: HDL-chol	Serum cholesterol: LDL-chol
		mg/100 ml			
Corn (control)	890 ± 44^a (100)[4]	158.3 ± 5^a (100)	163.1 ± 6^a (100)	59.3 ± 4^a (100)	74.6 ± 6^a (100)
Corn + garlic paste	448 ± 29^b (50)	147.2 ± 4^b (93)	149.3 ± 4^b (92)	56.2 ± 3^a (95)	50.9 ± 7^b (68)
Corn + garlic PESF	380 ± 27^c (43)	139.7 ± 3^b (88)	142.6 ± 3^b (87)	56.7 ± 2^a (96)	46.8 ± 6^b (63)
Corn + garlic MESF	280 ± 27^d (32)	122.8 ± 7^c (78)	122.2 ± 4^c (75)	58.6 ± 3^a (99)	42.2 ± 4^b (57)
Corn + garlic WASF	300 ± 25^d (34)	129.6 ± 2^c (82)	132.5 ± 8^c (81)	57.3 ± 2^a (97)	44.7 ± 5^b (60)
Corn + garlic residue	720 ± 40^e (81)	152.6 ± 4^a (96)	153.7 ± 6^a (94)	58.1 ± 5^a (98)	66.8 ± 6^a (90)
Corn + garlic oil (commercial)	410 ± 25^b (46)	137.4 ± 6^b (87)	133.8 ± 3^c (82)	55.8 ± 4^a (94)	57.9 ± 5^b (78)

[1]Experimental period was 4 weeks; time of killing was 0800 hours. Data are means ± SD; $n = 6$. HMG-CoA reductase, 3-hydroxy-3-methylglutaryl-CoA reductase; PESF, MESF and WASF stand for petroleum ether–, methanol– and water–soluble fractions of garlic, respectively. [2]Values not sharing a common superscript letter are different at $P < 0.01$. [3]Picomoles of mevalonic acid synthesized per minute per milligram of microsomal fractions. [4]Percentages of respective control activity data are in parentheses.

TABLE 6

Effects of garlic fractions on hepatic lipogenic enzymes in 12-week-old broilers[1,2]

Diet	Fatty acid synthetase[3]	Glucose-6-phosphate dehydrogenase[4]	6-Phosphogluconate dehydrogenase[4]	Malic enzyme[4]	Citrate-cleavage enzyme[5]
Corn (control)	246 ± 19.0^a (100)[6]	15.8 ± 0.5^a (100)	48 ± 0.4^a (100)	307.9 ± 26^a (100)	8.6 ± 0.2^a (100)
Corn + garlic paste	182 ± 11.5^b (74)	10.7 ± 0.6^b (68)	35 ± 0.3^b (68)	268.7 ± 22^a (87)	6.7 ± 0.3^b (78)
Corn + garlic PESF	171 ± 8.0^b (70)	9.2 ± 0.4^b (58)	31 ± 0.7^c (65)	262.3 ± 24^{ab} (85)	6.2 ± 0.4^b (72)
Corn + garlic MESF	118 ± 7.5^c (48)	6.7 ± 0.2^c (42)	24 ± 0.6^d (50)	224.6 ± 17^b (73)	5.0 ± 0.2^c (58)
Corn + garlic WASF	122 ± 9.0^c (50)	7.0 ± 0.3^c (44)	28 ± 1.0^e (58)	263.5 ± 16^{ab} (86)	5.2 ± 0.1^c (60)
Corn + garlic residue	226 ± 18.2^a (92)	13.7 ± 0.6^d (87)	40 ± 1.4^f (83)	267.4 ± 22^a (87)	6.9 ± 0.2^b (80)
Corn + garlic oil (commercial)	167 ± 7.0^b (68)	8.8 ± 0.2^b (56)	30 ± 0.6^c (63)	240.0 ± 24^{ab} (63)	5.9 ± 0.2^b (69)

[1]Experimental period was 4 weeks; time of killing was 0800 hours. Data are means ± SD; $n = 6$. PESF, MESF and WASF stand for petroleum ether–, methanol– and water–soluble fractions of garlic, respectively. [2]Values not sharing a common superscript letter are different at $P < 0.01$. [3]Nanomoles of NADPH oxidized per minute per milligram of cytosolic fraction. [4]Nanomoles of $NADPH^+$ reduced per minute per milligram of cytosolic fraction. [5]Nanomoles of product formed per minute per milligram of cytosolic fraction. [6]Percentages of respective control activity data are in parentheses.

suppressive agents of garlic. Compared to the control, fatty acid synthetase activity was significantly lower in each of the experimental groups except that fed the extracted residue. The MESF and WASF of garlic were the more potent suppressors of this activity (table 6). This impact of garlic on fatty acid biosynthesis was also manifested in the inhibition of two pentose-phosphate path enzymes, malic enzyme and citrate cleavage enzyme (table 6). Glucose-6-phosphate dehydrogenase and 6-phospho-gluconate dehydrogenase activities were low compared to fatty acid synthetase and malic enzyme activities. The low pentosephosphate pathway activities in the avain liver (table 2,3) have also been reported by others [26,27].

In general, the impact of garlic and of soluble fractions of garlic on lipid metabolism in layers was confirmed by using a broiler strain of chicks. All garlic treatment except that involving the extracted residue were associated generally with reduced lipogenic cholesterolopoietic activities and with lower serum total cholesterol and LDL-chol levels.

Discussion

Serum cholesterol levels fall when chickens as well as other species (2-15,17,18) are fed garlic supplemented diets. Neither the mechanism of the hypocholesterolemic action nor its agent has been clearly defined. When fed in conjunction with cholesterol, garlic decreased the incorporation of acetate into cholesterol [10,11] and increased fecal bile acid and neutral sterol excretion [14]. Both actions are consistent with the lowering of serum cholesterol. Anomalously, the activity of HMG-CoA reductase was increased and that of cholesterol 7 α-hydroxylase decreased [10]; both responses would be consistent, in cholestrol fed animals, with the elevation of serum cholesterol.

In chickens fed diets essentially free of cholesterol, the suppressive action of garlic is clearly at the level of cholesterol biosynthesis. The suppression of HMG-CoA reductase is manifested in the decreased serum concentrations of LDL-chol and total cholesterol. The garlic-mediated change in the LDL-chol:HDL-chol ratio of the pullets (-30%) is very similar to the change reported in the cholesterol-fed rat [14].

Addition of steam-distilled garlic oil, a product widely reported to be a cholesterol-suppressive agent (2-4,6,10), at a level (0.014%) equivalent to the addition of 3.8% garlic paste elicited similar responses.

The suppressor, readily extracted from garlic paste with the sequential application of solvents of increasing polarity, was not sharply defined in terms of polarity.The broad distribution suggests either the presence in garlic of polar and nonpolar mediators of cholesterol biosynthesis or alternatively the presence of an active substance conjugated in various forms to the extent that it is distributed among the solvents employed. An inhibitor of fatty acid oxygenases, present in onion and garlic and in steam-distilled garlic oil, exhibits a similar lack of definition in terms of polarity [28]. Alliin [17] and allicin[18], the highly polar, odorous compounds in garlic, reported to lower plasma cholesterol, are readily extracted in the methanol step of the fractionation sequence. These compounds conjugated in less polar form could be the active agents in the PESF. However, the demonstration that the odourless WASF is as effective as the MESF in suppressing HGM-CoA reductase implies that compounds other than those previously identified [17,18] are involved in the garlic-mediated suppression of cholesterol biosynthesis.

Key lipogenic enzymes were suppressed to varying degrees by the garlic additives. Differences in the potency of the lipogenic suppressor in each of the solvent fractions were recorded. This discrimination was not expressed towards HMG-CoA reductase, perhaps due to the presence of sufficient suppressor in each solvent extract to dampen this enzyme to some residual level of activity, a level 35% of control in broilers and 25% of control in layers. In layers, HMG-CoA reductase, cholesterol 7-α-hydroxylase, total serum cholesterol and serum LDL-chol values were each clustered about two points, a high point for control and garlic residues treatments and a low point for the five garlic treatments.

The above study establishes the cholesterol lowering properties, on modern scientific lines, of a traditional herbal medicine for which in the past claims had been made to be effective in heart disease purely on the basis of experience and tradition only.

Cholesterol is a normal constituent of the body. Its daily turn over in an average human body is about 200 mg, a good part of which is produced endogenously (mainly by the liver). Thus in a life span of 70 years, the body produces about 50 kg of cholesterol. Its major concentration in the body is in the brain, spinal chord, liver, and blood. It forms an integral of cell membrances. It is the precursor of bile acids, adrenal and sexhormones. It is therefore difficult to imagine that this endogenous cholesterol should heavily contribute to the destruction of our cardiovascular system. It has been reported [29] that in a variety of pathological conditions, in combination with calcium, fibrin, collagen etc., it forms large part of the lesion (scars, tubercles, gumata, old fibroids, thrombi, cholesteatomata), and performs a repair function. Thus it is possible cholesterol starts accumulating as a consequence of the lesion and not as its cause. This idea finds strength from recent findings [30] that cholesterol even if briefly exposed to atmosphere picks oxygen, resulting in oxygenated products, the presence of which in micro quantities can only be demonstrated by extremely sensitive techniques developed very recently. Further it has been shown [30] that of these 25-hydroxycholesterol when consumed even in pico-gram quantities causes lesions in aorta of rabbit.

We have recently started a project to look:

i How garlic (or its fractions) can regulate the delicate balance between LDL and HDL (the serum lipoproteins) in the body. As LDL is responsible to provide cholesterol through its circulation wherever needed in the body, whereas HDL directs it to liver where it is converted into bile acids and is excreted.

ii Whether garlic (and its fractions) can prevent formation of lesions, or reverse the process where lesions have already been formed on account of the intake of oxygenated products of cholesterol (or oxygenated lipids in the process of prolonged frying of foods).

This study, we hope, will place in our hands an effective natural dietary product to counter arteriosclerosis in humans. The serum cholesterol-lowering synthetic drugs (triparanol, estrogens, chlofibrate etc.)

at present in use in modern medicine, have serious side-effects, and their prolonged use turns out to be harmful [29].

The Reasons for Decline in Research on Traditional Medicinal Plant Drugs

Until the beginning of the current century most of drugs in use were of vegetable, mineral, or animal origin. Most of them had to be procured from the tropical areas of the world. With the advent of Organic Chemistry in Europe a good deal became known about the chemical structure of the pure compounds which could be relatively easily isolated and luckily also turned out to be the active principles (active in much smaller dosage) responsible for the major total effect of the bulky herbal drug. This gave fillip to the synthesis of drugs - using structures of the active principles as templates - from simple and locally available raw materials from the coal and petroleum distillation industries. This also obviated the dependance on tropics for costly herbal drug materials, and long wait involved in their import.

The synthesis made possible the production of drugs on a much larger scale in a factory, and at a much cheaper cost. Simple screening and quality control procedures were soon evolved, and even some very simple structure molecules proved to be equally effective. The number of drugs started multiplying and it gave rise to a very profitable, and profit-oriented pharmaceutical industry. This way very powerful, and fast-acting tools became available to the medical profession. But in this process both medical profession and pharmaceutical indsutry relegated from a noble human service to a purely materialistic business.

However, within a few decades of the use synthetic drugs, many started showing very serious side-effects -- a few were even irreversible resulting in fatalities. This necessitated strict government controls in Europe, and more so in America. Today to put a new research drug on the market by a pharmaceutical company in the U.S.A., it has been estimated, it requires an expenditure of nearly 60 million dollars,and a time lag of 3-7 years to get the final FDA approval. After this huge expenditure (research + FDA + manufacturing/packaging/ marketing /promotional and patent coverage costs) the firm is in hurry to recover costs and to make profits, as inspite of all the precautions, the drug is likely to manifest side-effects, or face a competition from a better or faster acting drug within a decade or even earlier. The pharmaceutical firms in the industrialized countries are therefore reluctant to take up research on herbal drugs, as they may be safe but are usually slower acting and require huge expenditure enumerated above, and are likely to take longer to get FDA approval.

There is far too great a stress on animal trials even in the case of inoccuous dietary herbals for their investigation to regulate some body processes, first for the fear that humans may not be used as guinea-pigs, although human beings have been using these herbals in their diet for the last many centuries, without suffering any ill effects. On the other hand when very potent synthetic drugs are withdrawn on showing side-effects have the human beings been not used as guinea-pigs in the intervening periods?

A few of the drugs, which have been withdrawn, or put on very restricted careful use in developed countries, are still being marketed by multinationals in developing countries, just to reap fat profits.

In this whole process, the worst sufferers are the developing countries, which are dependent on industrialized countries for drugs, or drug raw material supplies at a cost which they cannot afford, and at the same time they have ceased to investigate their own herbal drugs, which at least could be developed within the countries, and after establishment of their activity on modern scientific lines, could possibly be used in the vast rural areas where they could be grown.

What has been Wong in Research on Herbal Drugs in the Developing Countries ?

In doing research on medicinal plants in developing countries, we have been copying the example of Europe and America. Instead of establishing the clinical activity/efficacy for which the drug had the reputation of curing a disease, we in most cases have been resorting to extraction procedures, whereby we succeed in isolating some crystalline product(s), and call it "Active Principle"; determine its structure with the powerful tools now available, and are content with publications on novel and complicated structures only. In a still fewer cases these crystalline (or pure) products are subjected to general pharmacological screening and finding no or very poor activity we declare the herbal drug of no use. It is quite possible that in the extraction process we may have missed the actual "Active Principle".

Probably for the same reason, the huge programmes, at a cost of millions of dollars at N.I.H. (U.S.A.) and similar insitutes in Europe or elsewhere on medicinal plants to discover curative agents against some of the killer diseases (like cancer and heart disease etc.) have met with very little success, and have been dropped.

What Can or Should Now be Done ?

These days traditional doctors lack the modern knowledge, and assistance of available sophisticated techniques for diagnosis of disease, and facilities for research on drugs. They rely on their imperical visual observations, and on descriptions of the use of herbal drugs for the treatment of disease developed a few centuries back. These descriptions, no doubt, are the sum total of personal experiences spread over many centuries. Some of the descriptions may have been disorted through translations. However, if properly investigated, they may prove to be treasure houses worth rediscovery.

In a W.H.O. report it has been observed [31] that chances of clinical verification of claims of herbal drugs are almost 50:50 if investigations are made in collaboration with traditional healers. In Asia and Africa where traditional medicine is still practiced and people go to these healers of their own free will, suitable observation teams comprising traditional healer, clinicians, and scientists should work together to provide modern aids in diagnosis of disease, and the course of disease and cure should be monitored together. The observations should be made with a sympathetic rather than the common preconceived antagonistic attitude [32]. Then the obsevations showing definite promise should be repeated by other teams and on larger number of patients, taking care that botanically the herbs used are properly and correctly identified and recorded [32]. For the curative herbs so discovered quality control procedures should be developed and these standardized products could be used in health-care programmes, at least in the developing countries. For further study extractions may be made carefully monitoring where the activity is going. Most active fractions then should be investigated in detail for isolation of the active principles, and determination of their structures.

The start should be made with herbal drugs which are dietary, and hence through their long use have proved to be nontoxic, and hence there should be no hesitation in their trials directly on humans, as is being proposed to be done in establishing the hypocholestrolemic properties of garlic in the treatment of arteriosclerosis in an extension of the above reported study.

With the growing cost of health-care --almost becoming unbearable for the common man--and manifestation of serious side-effects on prolonged use of some of the synthetics, there is already a growing tendency even in America, Europe, and other developed countries to go back to the natural drugs, which is particularly manifest from a very rapid growth in health-food stores there. Many of the herbs are being sold and used as health-foods to avoid long, rigorous, cumbersome, and costly FDA approvals, which certainly are unnecessary, at least, in the case of dietary herbs.

Logically, to correct the (natural) disorders of the human body, one should look to the natural diets (including herbs) with which man has interacted for centuries, and has accummulated the empirical knowledge. The scientists (including medical scientists) should get together at the sites where such herbals grow and are used, first to confirm on modern scientific lines, the empirical knowledge which exists there, then record it and standarize it for common use. It may then be extended for extractions, isolations, structure determination, and study of the mechanism of the observed actions. Through such an approach, I am sure, we shall soon discover many effective and safe medicinal agents even for the so far unconquered killer diseases. Besidesaffordable health-care programmes, particularly for the poorer nations can be developed. In the past we have done the process in the reverse order, and therefore have achieved little success.

LITERATURE

1. Dymock W., Warden C.J.H. & Hooper D. (1893) **Pharmacographia Indica**, Vol.III, p.488. Publisher Kegan Paul, Trench, Trubner & Co. Ltd. London Reprinted by Hamdard Foundation, Karachi in 1972.

2. Krtichevsky, D. (1975) Effect of garlic oil on experimental atherosclerosis, **Artery**,1, 319-323.

3. Bordia. A., Arora, S.K. Kothari, L.K., Jain, R.C.Rathore, B.S., Rathore, A.S., Dobe, M.K. & Bhu.N. (1975). The protective action of essential oils of onion and garlic in cholesterol fed rabbits. **Atherosclerosis**, 22, 103-109.

4. Jain, R.C. (1976) Onion and garlic in experimental cholesterol induced atherosclerosis . **Indian J.Med.Res.,** 64, 1509-1515.

5. Jain,R.C.(1977) Effect of garlic on serum lipids, coagulability and fibrinolytic activity of blood (letter).**Am.J.Clin.Nutr.,** 30, 1380-1381.

6. Jain R.C. & Konar, D.B. (1978) Effect of garlic oil in experimental cholesterol atherosclerosis. **Atherosclerosis**, 29, 125-129.

7. Jain, R.C.(1978) Effect of alcoholic extract of garlic in atherosclerosis. **Am.J.Clin.Nutr.**, 31, 1982-1983.

8. Brodia, A.K. & Verma, S.K. (1978) Garlic on the reversibility of experimental atherosclerosis. **Indian Heart J.**, 30, 47-50.

9. Bordia, A.K., & Verma, S.K. (1980) Effect of garlic feeding on regression of experimental atherosclerosis in rabbits, **Artery**, 7, 428-437.

10. Kritchevsky, D., Tepper, S.A. Morrisey, R. & Klurfeld, D. (1980) Influence of garlic oil on cholesterol metabolism in rats. **Nutr.Rept. Int.**,22, 641-645.

11. Chang, M.W. & Johnson, M.A. (1980) Effect of garlic on carbohydrate metabolism and lipid synthesis in rats. **J.Nutr.**, 110, 931-936.

12. Bordia, A. (1981) Effect of garlic on blood lipids in patients with coronary heart disease. **Am.J.Clin.Nutr.**, 34, 2100-2103.

13. Arora, R.R. & Arora, S. (1981) Comparative effects of clofibrate, garlic and onion on alimentary hyperlipemia, **Atherosclerosis**, 39, 447-452.

14. Chi,M.S.Koh, E.T. & Stewart, T.J. (1982) Effects of garlic on lipid metabolism in rats fed cholesterol or lard. **J.Nutr.**, 112, 241-248.

15. Kamanna, V.S. & Chandrasekhara, N. (1982) Effect of garlic (Allium sativum Linn) on serum lipoproteins and lipoprotein cholesterol levels in albino rats rendered hypercholesteremic by feeding cholesterol. **Lipids**, 17, 483-486.

16. (a) Cavalito, C.J. & Bailey, J.H. (1944) Allicin the Antibacterial Principle of Allium sativum. 1. Isolation, Physical Properties and Antibacterial Action. **J.Amer.Chem.Soc.**, 66, 1950-51.

(b) Cavalito, C.J., Buck, J.S. & Suter, S.M. (1944) Allicin, the Antibacterial Principle of Allium sativum II. Determination of the Chemical Structure. **J.Am.Chem.Soc.**, 66, 1952-54.

(c)Stoll, A., & Seebeck, E. (1950) The Synthesis of Natural Alliin. **Experientia**, 6, 330; and **Helv.Chim.acta.**,32, 197 & 866 (1949).

17. Itokawa, Y., Inoue, K., Sasagawa, S. & Fujiwara, M. (1973) Effect of S-methylcysteine sulfoxide, S-allyl-cysteine sulfoxide and related sulphur-containing amino acids on lipid metabolism of experimental hypercholesterolemic rats. **J.Nutr.**, 103, 88-92.

18. Augusti, K.T. & Mathew, P.T. (1974) Lipidlowering effect of allicin (diallyl disulfide-oxide) on long term feeding to normal rats. **Experientia**, 30, 468-470.

19. (a) Shih J.C.H. (1982) Avian models for research in atheroscherosis and cholesterol metabolism. **Fed.Amer.Soc.Exptl.Biology.**, (April 15,1982) p.2475.

(b)Shih J.C.H. (1983) Atherosclerosis in Japanese quails and the effect of lipvic acid. **Federation Proc.** 42, 2494-2497.

20. Qureshi, A.A., Burger, W.C., Elson, C.E. & Benevenga, J.N. (1982) Effects of cereals and culture filtrate of Trichoderma viride on lipid metabolism of swine. **Lipids**, 17, 924-934.

21. Qureshi, A.A., Abuirmeileh, N., Burger, W.C., Din Z.Z. & Elson, C.E. (1982) Effect of AMO 1618 on cholesterol and fatty acid metabolism in chickens and rats. **Atherosclerosis**, 46, 203-216.

22. Gornall, A.G., Bardawill, C.J. & David, M.M. (1949) Determination of serum proteins by means of the biuret reaction. **J.Biol.Chem.**, 177, 751-766.

23. Kostner G.M. (1976) Enzymatic determination of cholesterol in high density lipoprotein fractions prepared by polyanion precipitation **Clin.Chem.**, 22, 695-698.

24. Lobos-Virella, M.F., Stone, P., Ellis, S. & Colwell, J.A.(1977) Cholesterol determination in high density lipoproteins separated by three different methods. **Clin.Chem.**, 23, 882-884.

25 Snedecor, G.W.& Cochran, W.G. (1971) Statistical Methods, 6th ed., pp. 258-298. The Iowa State University Press, Ames, IA. U.S.A.

26. Leveille, G.A., Romsos, D.R., Yeh, Y.Y. & O'Hea, E.K. (1975) Lipid biosynthesis in the chick. A consideration of site of synthesis, influence of diet and possible regulartory mechanisms. **Poult.Sci.**, 54, 1075-1093.

27. O'Hea, E.K. & Leveille, G.A. (1968) Lipogenesis in isolated adipose tissue of the domestic chick (Gallus domesticus). **Comp.Biochem. Physiol.**, 26, 111-120.

28. Vanderhoek, J.Y., Makheja, A.N. & Bailey, J.M. (1980) Inhibition of fatty acid oxygenases by onion and garlic oils. Evidence for the mechanism by which these oils inhibit platelet aggregation **Biochem.Pharmacol.**, 29, 3169-3173.

29. Kaunitz, H. (1977) Repair Function of Cholesterol Versus the Lipid Theory of Arteriosclerosis. **Chemistry and Industry**, 761-763.

30. Imai, H., Werthessen, N.T., Subromanyam, V., Le Quesne, P.W. & Soloway, A.H. (1980) Angiotoxicity of Oxygenated Sterols and Possible Precursors. **Science** (U.S.A.) 207, 651-653.

31. Ampofo, O. (1977) Plants that heal. **World Health** (W.H.O., Geneva, Switzerland) pp 26-30 (November issue).

32. Farnsworth, N.R. & Bingel A.S. (1977). Problems and Prospects of Discovering New Drugs from High Plants by Pharmacological Screening **New Natural Products and Plant Drugs with Pharmacological, Biological or Therapeutical Activity**. Editors Wagner, H. & Wolf, P., pp, 1-22. Springer-Verlag, Berlin, Heidelberg, New York.

Synthesis, Structure Activity Relationships and Mode of Action of Insect Pheromones

A Contribution for the Transmission of Chemical Information in Nature

Hans Jürgen Bestmann

Organic Chemistry Institute, University Erlangen-Nürnberg, Henkestr. 42, D-8520 Erlangen, FRG

Insect pheromones are signal substances which effect the transmission of chemical information between the individua of an insect species. Female butterflies and moths in many cases use mono- and bisolefinic aldehydes, alcohols and acetates[1)] as their species specific sex pheromones. The corresponding male moths possess specialized sensory cells in their antennae, the dendrites of which end in the sensory hairs (sensilla). The odor molecules impinge on the antennal cuticle and reach the dendritic membrane by diffusion via pores and tubuli, where they interact with specific acception sites.

This interaction produces in a still unknown way the opening of an ion channel and hence a change of the membrane potential. The sum of the slow receptor potentials of all sensory cells of the antenna stem, stimulated by pheromone molecules, can be recorded electrophysiologically in the form of an electroantennogram (EAG)[2)]. This technique permits the comparison of efficacies of different compounds with each other and thus the study of structure activity relationships.

The need of numerous pheromones and pheromone analogues made necessary the development of a unitized construction principle for the synthesis of these chemicals. (Z)-Monoolefinic pheromones of type 3 are easily available by (Z)-stereospecific Wittig reaction from phosphonium ylides 1 and aldehydes 2 using sodium silazide as the base for the ylide generation from the corres-

Natural Product Chemistry
Ed. by Atta-ur-Rahman

ponding phosphonium salts[3]. (For the synthesis of (Z)-unsaturated monoolefinic pheromones see refs.[1) 3) 4) 5) 6)].

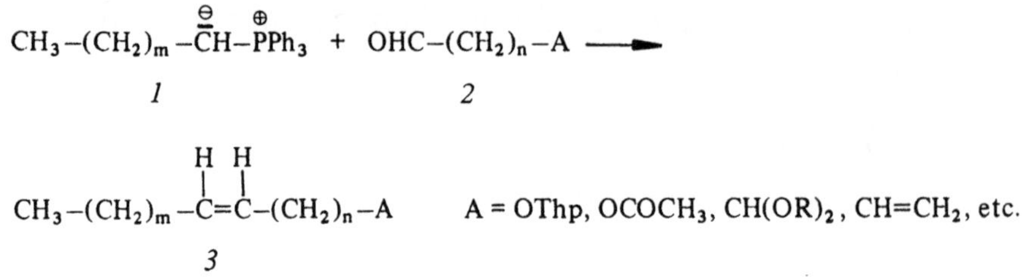

(E)-olefins 7 can be obtained in certain cases reacting trimethyl silylated phosphoranes 4 with 5. In the course of this reaction the (Z)-vinyl silanes 6, the trimethylsilyl group of which can be substituted facilely by hydrogen on acid treatment affording 7[7], are formed stereoselectively.

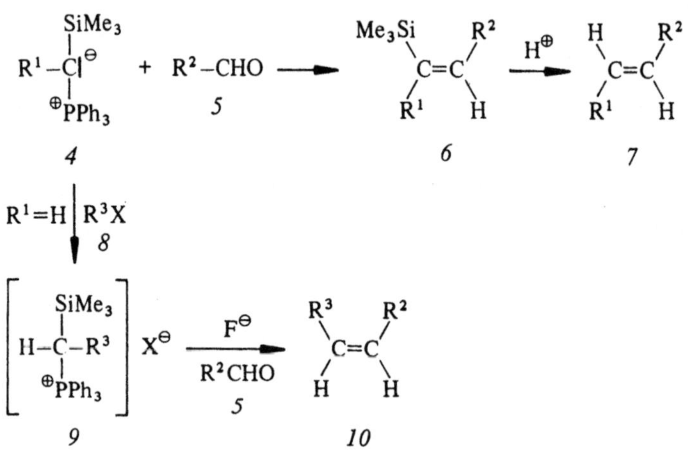

When the compound 4 (R^1=H) is reacted with alkyl halides 8, the phosphonium salts 9 are formed. By treating 9 with CsF in the presence of an aldehyde 5, (Z)-olefins 10 are generated[8].

Two acetylene syntheses proved to be very successful, allowing the formation of (Z)- as well as (E)-olefinic pheromone systems. Ylides 11 may be acylated by use of thioesters 12[9]. The resulting acylated ylides 13 can be oxidized to 1,2-diketones 14 which are transformed into bis-hydrazones 15. When oxygen is introduced into a solution of 15 in pyridine in the presence of CuCl acetylenes 16 are formed[10].

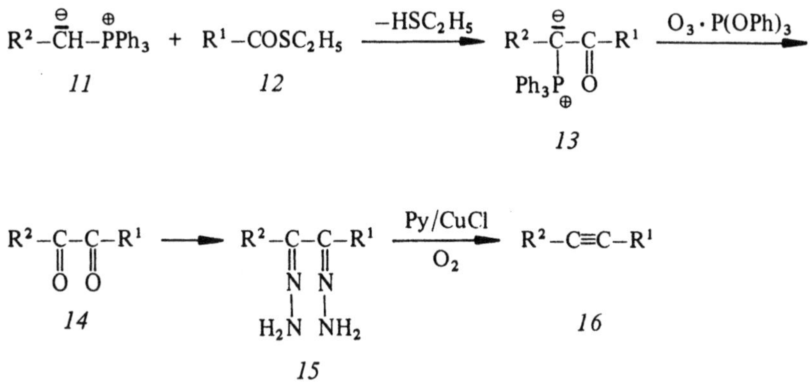

Acyl ylides 13 and the anhydride of trifluoromethane sulfonic acid yield the salts 17, which are cleaved into acetylenes 16, triphenyl phosphane 18 and sodium salt 19, when reacting with sodium amalgam[11].

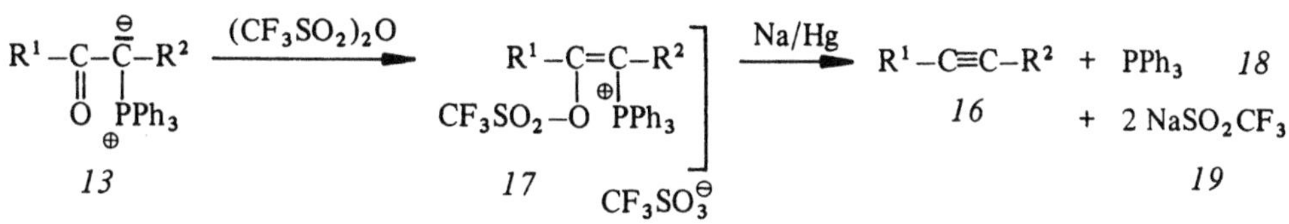

Furthermore, we have developed a synthetic approach for acetylenes starting with two different aldehydes 20 and 5[12].

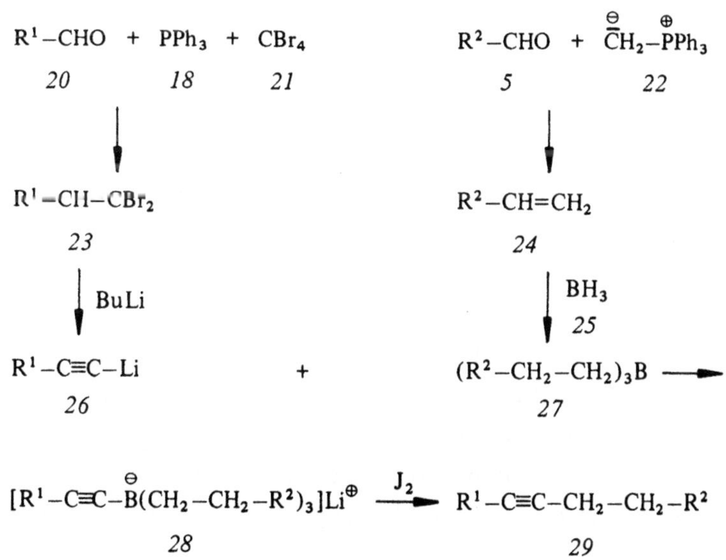

Reacting aldehyde 20 with 18 and CBr_4 21[13)14)15)] yields the dibromo olefins 23, which can be converted into the lithium acetylides 26 with butyl lithium. The second aldehyde 5 in the meantime is reacted with ylide 22 to the olefin 24. Its hydroboration with 25 gives the trialkyl borane 27 which is converted into boronate 28 with 26[16)]. By treating 28 with iodine, the acetylenes 29 are obtained in high yield and purity.

We reported the possibility of converting ylides 30 using 25 into the monoalkyl borane triphenyl phosphane adduct 32[17)]. From these compounds 34 can be obtained with 2,4,6-trimethylphenol 33[18)] which reacts with bromoacetylenes 35 to give vinyl borane 36 in the presence of methyl iodide. Treating 36 with methanolate and consequent protonolysis gives (E)-olefins 37[18)].

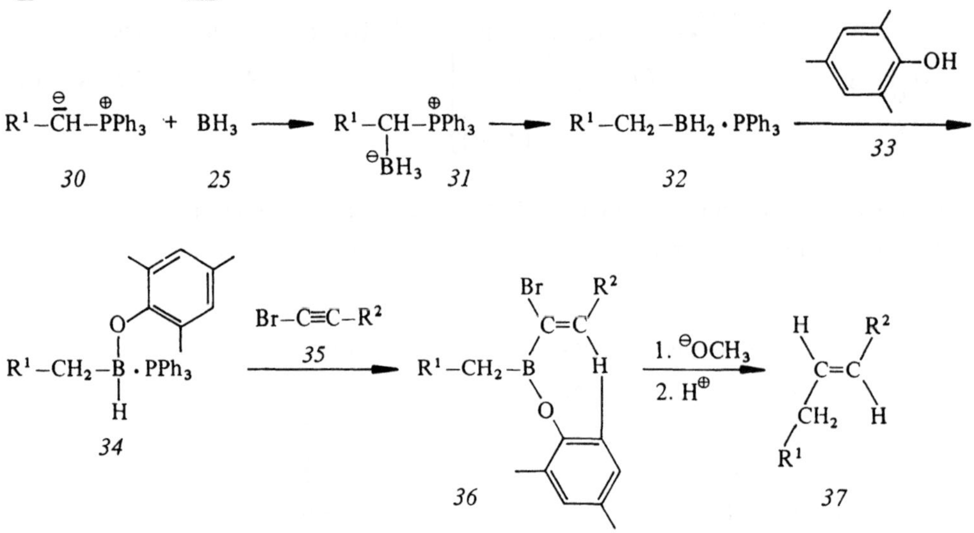

Conjugated dienes of defined geometry often are found in a series of lepidopterous pheromones. Their synthesis succeeds when obeying the rules about the stereoselectivity of the Wittig-reaction[19)20)]. 5 reacts with Ylide 38 because of the electron withdrawing carbonyl group (CHO in 38) (E)-stereoselectively to (E)-α,ß-unsaturated aldehydes 36, which now (Z)-selectively give the (E,Z)-conjugated dienes 40 with ylide 30[21)].

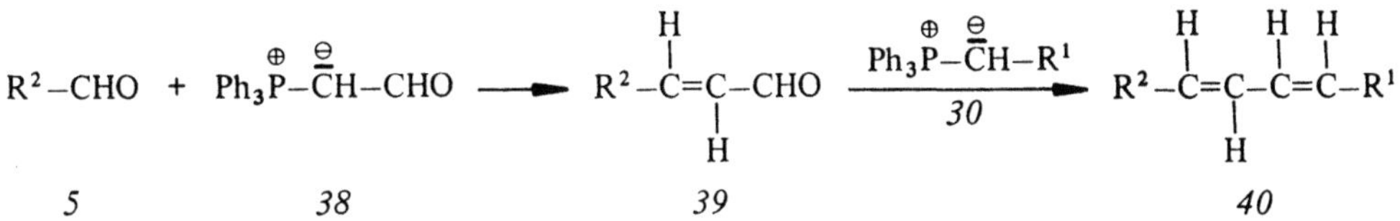

For the synthesis of (Z)-α,β-unsaturated aldehydes 45 the ylide 38 is alkylated with ethyl bromid affording 41. 41 is deprotoned by a base, yielding the vinylidene phosphorane 42. 42 adds ethanol forming the ylide 43 which reacts (Z)-stereoselectively with aldehydes 20 giving rise to the formation of the acetal 44[22]. 44 is also available from the reaction of ylides 30 (R^1 = alkyl) with the monoacetal of glyoxal 46[23]. The acetal 44 can be cleaved to aldehydes 45, which react with ylide 11 yielding (Z,Z)-dienes 47[21,24].

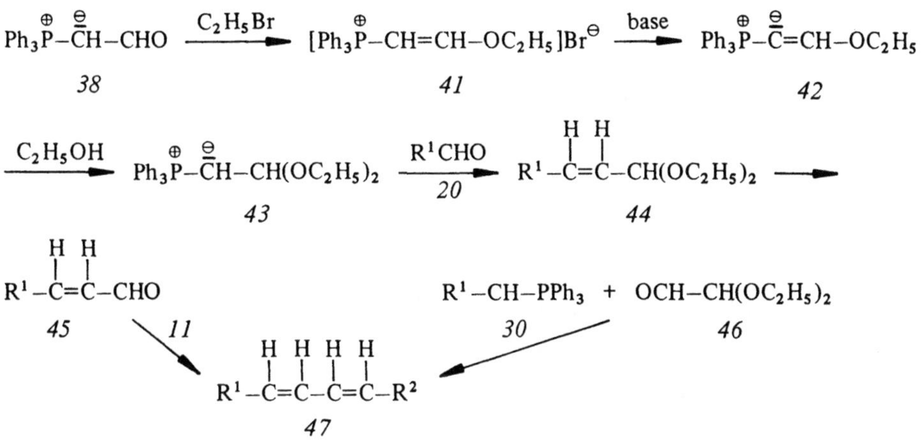

Methylene-interrupted polyolefinic pheromones have been found in *Lepidoptera* recently quite often[1]. The following method for the preparation of (Z,E)-compounds of that type has been developed[25]:

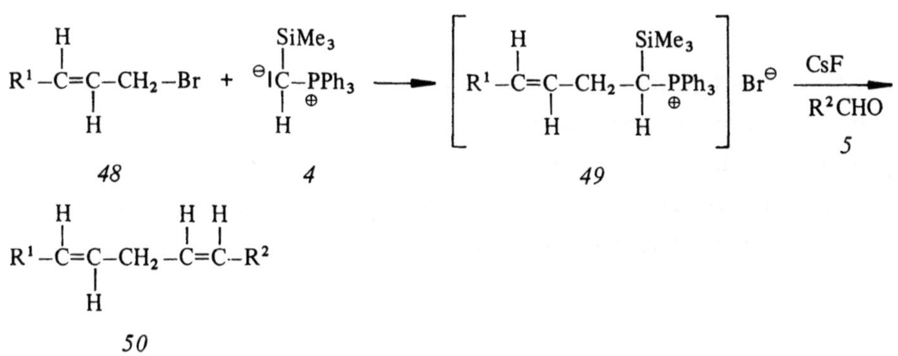

Alkyl halides (e.g. (E)-configurated 48), were reacted with silylated ylide 4 (R^1=H) to give the salt 49; desilylation with CsF in the presence of aldehyde 5 yields the (E,Z)-1,4-dienes 50.

The (E,E)-1,4-dienes 56 are available by the reaction of the bromides 51 with triethyl phosphite 52, alkylation of the phosphonate 53 with butyl lithium and alkyl halides 54 to give 55 and reaction of 55 with $LiAlH_4$[26].

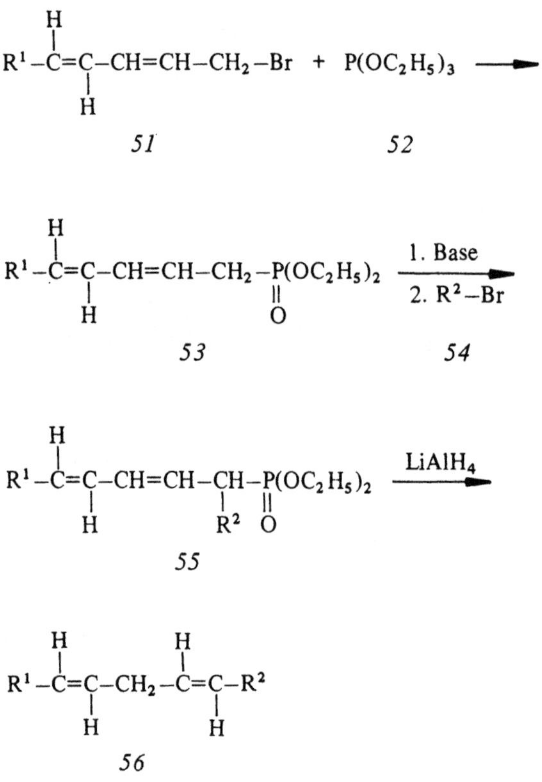

(6E,11Z)-6,11-Hexadecadienylacetate 67 and hexadecadienal 68 are the sex attractant components of the female *Antheraea polyphemus* moth. Both compounds can be prepared by a combination of Wittig reaction and acetylene synthesis, according to the following formula scheme:[27]

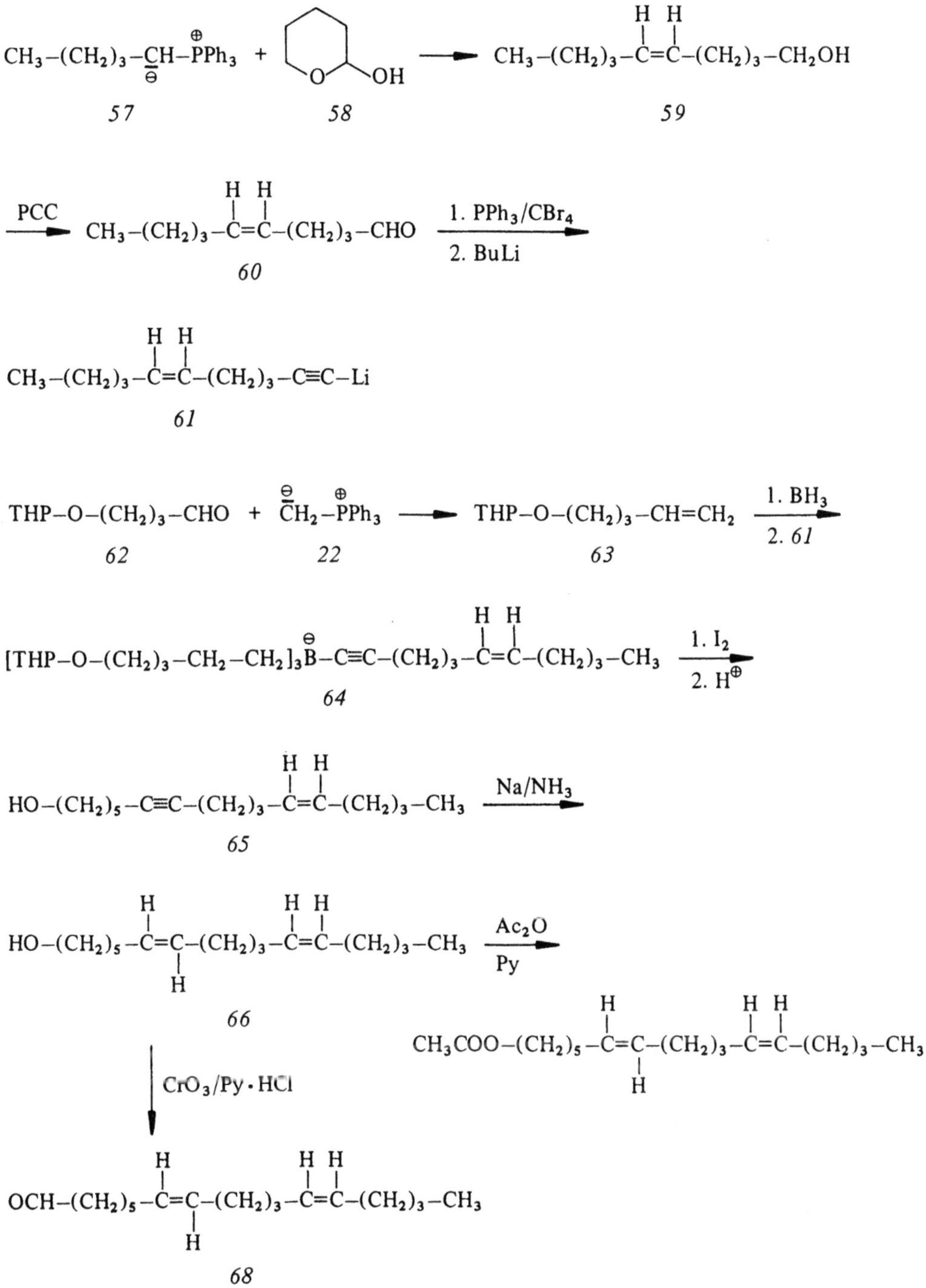

Ylide <u>57</u> is converted with <u>58</u> into (<u>Z</u>)-enol <u>59</u> which is oxidized to aldehyde <u>60</u>. From <u>60</u> the acetylide <u>61</u> is formed by treatment with PPh_3/CBr_4 and subsequently butyl lithium.

The aldehyde 62 is olefinated with 22 to 63, the hydroboration of which and subsequent reaction with 61 giving 64. Treating 64 with iodine affords the (Z)-enyne 65 with sodium in liquid ammonia. Acetylation and oxidation gives rise to the pheromone components of A. polyphemus[27].

For more detailed studies of pheromone receptor interactions we have also synthesized chiral pheromone analogs[28]. The following formula scheme presents some of the prepared test compounds.

$$CH_3-\overset{\overset{C_2H_5}{|}}{\underset{*}{C}H}-CH=CH-(CH_2)_{10}-OCOCH_3 \qquad 69$$

$$CH_3-\overset{\overset{C_2H_5}{|}}{\underset{*}{C}H}-CH=CH-CH=CH-(CH_2)_8-CH_2OH \qquad 70$$

$$CH_3-\overset{\overset{C_2H_5}{|}}{\underset{*}{C}H}-CH=CH-CH=CH-(CH_2)_8-CHO \qquad 71$$

All these systematically varied pheromone structures and pheromone analogs were tested using the electroantennogram (EAG) technique[2]. Thus, the sensory efficacy of a compound is taken to be the amplitude of the EAG response related to the respective stimulus source loading (10^{-3} to 1000 µg).

Recent investigations resulted in the following structure activity rules: i) Starting from the naturally occuring pheromone of the general formula 72, elongating or shortening the carbon chain by one methylene unit $-CH_2-$, results in a decrease of

$$CH_3-(CH_2)_n-CH=CH-(CH_2)_m-O-COCH_3 \qquad 72$$

efficacy by a factor of 3 to 30, two methylene groups give a 10 to 100 fold loss of activity. In succeeding alterations of the molecule length often a more drastic diminution of efficacy can be observed for elongations than for shortening[29][30].

ii) Varying the alkyl part of the molecule $-(CH_2)_n-$ (here called the n-part) produces a more drastic loss of efficacy in the EAG test than when equivalent changes in the $-(CH_2)_m-$ element (m-part) are made. E.g., when elongating the carbon chain of the pheromone by two methylene groups $-(CH_2)_n- \longrightarrow -(CH_2)_{n+2}-$, the double bond must be shifted also for two CH_2-groups to elicit maximal response so that $[-(CH_2)_{n+2}-]$ is converted into $[-(CH_2)_n]$, i.e., the end alkyl part $CH_3(CH_2)_n-$ of the pheromone analog becomes identical to that of the natural signal substance[29)30)].

From this principle it could be concluded, that the end alkyl section (n-part) of the molecule is of more importance for the strength of the stimulus than the distance between double bond and ester function (m-part). As a consequence we have further varied the structure of the n-part, introducing methyl, ethyl or propyl groups into the α, ß- or γ-position relatively to the double bond[28)]. The EAG measurements gave the following results:

The antennal receptors of different moth species, maximally responding to molecules of typ 72, can be classified according to the number n of the methylene groups of the end alkyl section of the species own pheromone. With equivalent n and different m the various pheromone receptors respond relatively adequately over a range of variations in the n-part. With differences in the m-part qualitative gradations can arise. This is valid for receptors maximally reacting to pheromones with n=1,3 and 5[28)31)].

Elongating the n-part of a pheromone with n=1 (compound 73 in table 1) by two methylene groups, (74), causes the EAG activity to be lowered by a factor of 30 to 300 for the corresponding receptors. Introduction of a methyl group in the α-position relative to the double bond in the elongated molecule, (75), results in an efficacy increase compared to the unbranched compound 74 by a factor of 3 (cf. table 1, section A).

Table 1. Relative electroantennogram efficacies of alkyl branched pheromone analogs (activity of pheromone = 100)

compound		rel. efficacies		compound		rel. efficacies		
Section A: receptors, maximally responding to type 72, n=1				Section B: receptors, maximally responding to type 72, n=3				
		m = 8	m = 10			m = 6	m = 8	m = 10
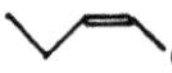 $(CH_2)_m$-O-$COCH_3$	73	100	100	 $(CH_2)_m$-O-$COCH_3$	74	100	100	100
 $(CH_2)_m$-O-$COCH_3$	74	3	0.3	 $(CH_2)_m$-O-$COCH_3$	75	3	3	3
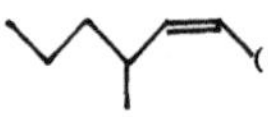 $(CH_2)_m$-O-$COCH_3$	75	10	1	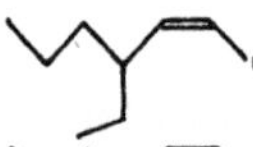 $(CH_2)_m$-O-$COCH_3$	78	3	3	3
 $(CH_2)_m$-O-$COCH_3$	76	0.1	-	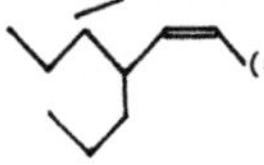 $(CH_2)_m$-O-$COCH_3$	79	3	30	3
 $(CH_2)_m$-O-$COCH_3$	77	30	-					

The activating effect of an α-positioned methyl group in the n-part of elongated pheromone analogs is especially remarkable in the comparison of activities of the compound 76 and 77 (section A). Presentation of molecule 76, elongated by four CH_2-groups, to receptors maximally responding to the signal substance (Z)-9-dodecenyl acetate (73, m=8), gives an activity decrease by a factor of 300 to 1000. The α-methyl branched analog 77 however shows a loss of efficacy by a factor of only 3 compared with 69 (m=8)[28)31)].

Similar results characterize our investigations with moth maximally responding to pheromone molecules 74 (72, n=3). Introducing a methyl group in the α-position, (75), lowers the efficacy of the molecule by a factor of 10 to 30. Substituting the methyl group by an ethyl group (78, section B) effects the activity only a little. For m=6 and m=10 no activity change could be observed when the side chain was further elongated for one CH_2-group, (79), whereas in the case of m=8 an increase of efficacy by a factor of 10 could be observed for this propyl group when compared with structures 75 and 78 (table 1, B)[28)31)]. As long as the odor molecule retains the propyl group necessary for a correct fitting into the receptor site, the latter responds to the introduction of an α-positioned side chain with a loss of activity almost independent on the size of this alkyl group. With increasing n in 72 the number of possible conformations is also increased. During the fitting of a defined conformation of the pheromone to the receptor, one should therefore expect increasing differences between the activity produced by different structures as the number of methylene groups n, is increased. This prediction is supported by our measurements.

All these structure activity investigations indicate that the stimulation by interactions between molecule and receptor site is based neither on a rigid "key and lock principle", nor a three-point interaction. Instead we assume that the pheromone molecule is inserted flexibly into the whole receptor region, as it is shown schematically in figure 1.

For further confirmation of this hypothesis we have synthetically varied the pheromone of the saturnid Nudaurelia cytheraea[32) which has been structurally identified as (Z)-5-decenyl isovalerate (80). In the case of this rather uncommon attractant structure we synthetically modified the molecular section between the ester function and the double bond, (m-section), as well as the isovalerate group. The EAG results, given in figure 2, clearly demonstrate that when the chain length of the m-part is altered, it is the isovalerate which is best recognized by the receptor, thus reflecting the flexible winding of the pheromone into the receptor sphere for its whole length.

It has to be assumed that the receptor site contains some elements of chirality and that the sections of the multiple prochiral pheromone molecule (e.g. A or B in fig. 3) can only properly be adjusted from one of the two enantiotopic sides of each centre of prochirality. This would again mean that the signal molecule is embedded only in a defined conformation. In the course of this process conformational changes

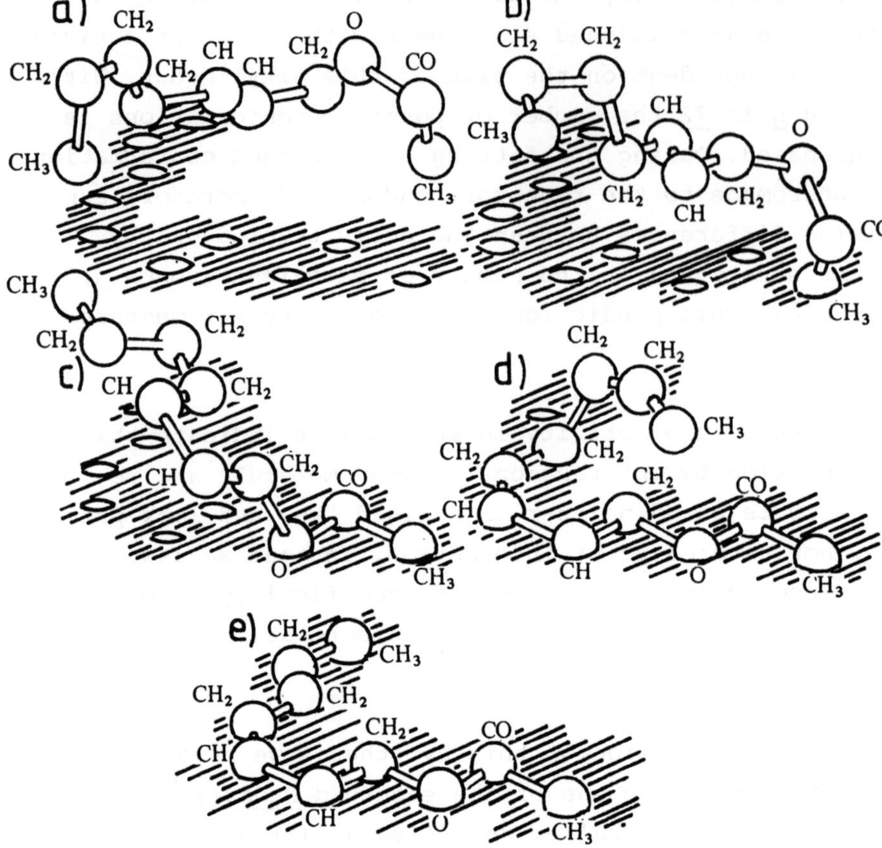

Figure 1, a-e. schematic presentation of the flexible insertion of a pheromone molecule into the receptor sphere. a) Approach of the molecule; b) first contact between pheromone and receptor; c) and d) dynamic insertion into the receptor sphere; e) final phase; synchronously occuring changes of the receptor region are not considered

are also to be expected in the receptor sphere. If these model ideas were right, receptors responding to achiral molecules of type 72 should differentiate between the enantiomers of chiral analogs 81 because of the appearance of additional diastereomeric interactions and because of the fact that only one of the two enantiomers can properly adopt the conformation of section A of the signal molecule required by the receptor site.

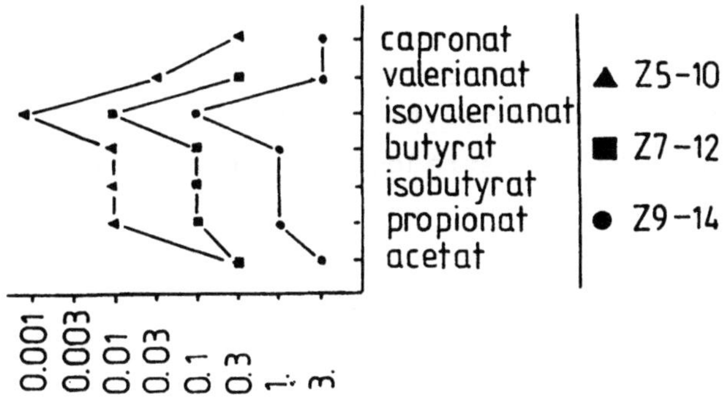

Figure 2. Comparing the EAG efficacy of various alkenol esters with N. cytheraea[32)]

$$CH_3-(CH_2)_n-CH{=}CH-(CH_2)_m-O-COCH_3 \qquad 72$$

←A→ ←B→

$$CH_3-(CH_2)_{n-1}-\overset{R}{\underset{*}{C}}H-CH{=}CH-(CH_2)_m-O-COCH_3 \qquad 81$$

$$CH_3-\overset{C_2H_5}{\underset{*}{C}}H-CH{=}CH-(CH_2)_{10}-O-COCH_3 \qquad 82$$

$$CH_3-\overset{C_2H_5}{\underset{*}{C}}H-CH{=}CH-CH{=}CH-(CH_2)_8-CH_2OH \qquad 83$$

$$CH_3-\overset{C_2H_5}{\underset{*}{C}}H-CH{=}CH-CH{=}CH-(CH_2)_8-CHO \qquad 84$$

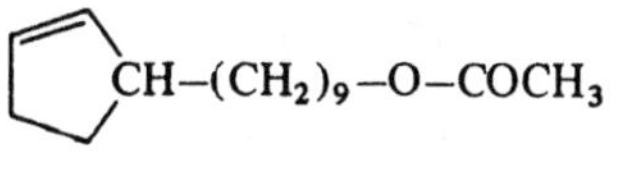

85

Figure 3. Optically active lepidoptera pheromone analogs

The results of the EAG test of these compounds with the European corn borer Ostrinia nubilalis and the red banded leaf roller Argyrotaenia velutiana are shown in table 2. Both species use (Z)-11-tetradecenyl acetate (Z-11-TDA) as the main pheromone component. 82 is an 13-ethyl derivative of the species own attractant. 83 and 84 were electrophysiologically tested with the silk worm moth Bombyx mori, the female of which produces bombykol and bombykal as sex pheromones.

The enantiomers and the racemate of the cyclic pheromone analog 85 synthesized by Chapman et al.[33] were also included into our EAG measurements. Unlike 81, in the analog 85 the A-part of the molecule cannot assume every conformation because of the ring-structure. The centre of chirality is positioned in the B-part of 85. The relative EAG efficacy values in table 2 demonstrate that the receptors could differentiate the enantiomers in all cases investigated.

Table 2. Relative EAG activities of chiral pheromone analogs. Activity of the naturally occuring pheromone ≙ 100

Compound no.	Moth species		Compound no.	Moth species
	O. nubilalis	A. velutinana		Bombyx mori
Z-11-TDA	100	100	bombykol	100
(R)(-)-82	10	10	(S)(+)-83	10
(R,S)- 82	1	1	(R,S)- 83	3-10
(S)(+)-82	0.1	0.1	(R)(-)-83	3
(S)(-)-85	0.3	0.1	bombykal	100
(R,S)- 85	0.03	0.03	(S)(+)-84	10
(R)(+)-85	0.01	0.01	(R,S)- 84	3-10
			(R)(-)-84	3

With Ostrinia and Argyrotaenia, both using the same pheromone molecule but belonging to a different lepidoptera family, (R)-82 was a hundred fold more effective than the (S)-isomer. Also the enantiomers of 85 were distinguished by both insect species (by factors of 10 to 30). The activity value of the corresponding racemate was always between the enantiomer values, when the statistical method of activity determination was applied[34].

Furthermore table 2 shows that for both species the efficacies of the mobile (R)-82 and the rigid (S)-85 differ by 30 to 100 fold. Compared with the species own pheromone, (R)-82 gives only a 10 fold loss of activity, whereas (S)-85 is less effective by a factor of 300 to 1000. Bombyx mori also distinguished the (R)- and (S)-form of 83 as well as of 84. In both cases, however, the (S)-enantiomer was 3 fold more effective than the (R)-isomer, the value of the racemate laying in between[34].

Using single cell recording techniques it could be demonstrated that with variation of the structure of bombykol or bombykal in the length of the carbon chain but with retention of the functional group (i.e. the alcohol or aldehyde funtion) only the specific alcohol or aldehyde cell, respectively, was stimulated. For equivalent structural changes in bombykol and bombykal the corresponding cell responded with the same efficacy difference compared with the natural pheromone. This result is also reflected in the electroantennogram (which is the sum of all cells) as shown in table 2. From these findings one must suppose that the functional group is of particular importance for the primary recognition by the receptor site. Therefore we attribute haptophoric characters to the functional end group, whereas more intrinsic properties should be attributes to the other molecule sections. These sections are thus responsible for the excitation function of the molecule; they might affect, for example, the width of opening of the ionic channel.
Furthermore, we must conclude that proteins as well as lipid regions belong to the receptor sphere interacting with the signal molecule. At the same time, we believe that the boundary

layer between protein and lipid is of special importance. In this region the lipid layer is of particular rigidity. This "solid domain" floats in a liquid double layer[35). We suppose that interactions, especially those of the intrinsic parts of the pheromone molecule with the crystalline lipid layer adjacent to the protein, lead to a change in the degree of order of the lipid phase.

We have now studied the effect of pheromone molecules on the degree of order of synthetic dipalmitoyl lecithine micelles. these lecithine model membranes, ultrasonically suspended, were investigated by fluorescence spectroscopic methods[36) for their phase transition points. A pre-phase transition point at a transition temperature PT_U of 30°C could be observed. It was followed by a main phase transition point at T_U= 41°C (a) in figure 4). Adding long chain, pheromone-like alcohols (C_{12}, C_{14}, C_{16}, etc.), at first did not effect the phase transition and the course of the graph remained the same, but after a critical concentration was reached the pre-phase transition point PT_U disappeared and a graph of the form b) in figure 4[37) resulted. This critical concentration depended on the chain length of the alcohol added, for example

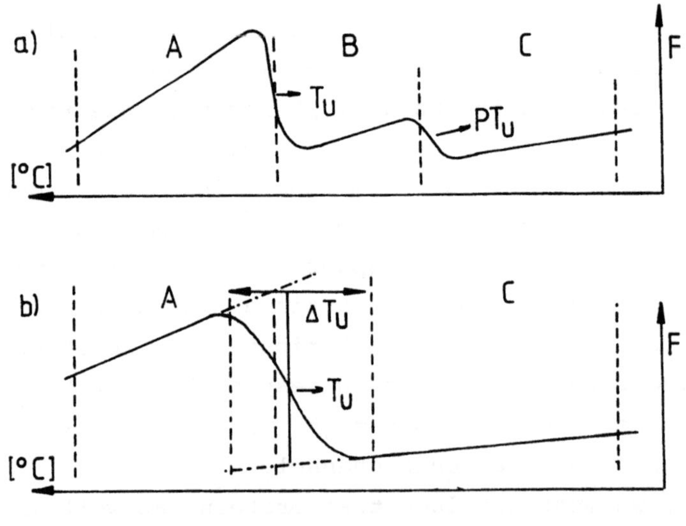

Figure 4. Schematical drawings of fluorescence graphs of type I(a) and type II(b)

for tetradecanole it was 27-29%, and for hexadecanole 10-12%[37]. These findings show that long chain compounds can influence the degree of order of the lipid layer.

If lipid regions are involved in the molecular mechanism resulting in the opening of an ionic channel, one should expect a temperature dependence of the EAG response resulting in an optimum temperature for maximum amplitude. Furthermore if in this mechanism cooperative effects such as the participation of domain structures or short lived metastable phase states are involved, something like a hysteresis phenomenon should be observed on measuring the temperature dependence[38]. Therefore we have recently investigated the temperature dependence of the electroantennogram response of various lepidopterous species. Figure 5 represents a typical example. The appearance of a maximum at a distinct temperature can be clearly seen. In addition, we noted that the downward curve (from higher to lower temperature) was of higher amplitude than the upward curve (measuring from lower to higher temperatures). This suggests a "hysteresis" phenomenon which is currently under investigation[39].

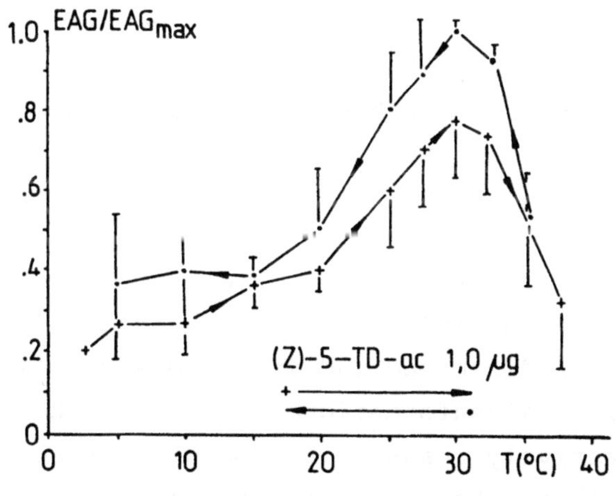

Figure 5. EAG amplitude of Scotia exclamationis (Noct.): dependence on the temperature. Stimulus pheromone (Z)-5-tetradecenyl acetate

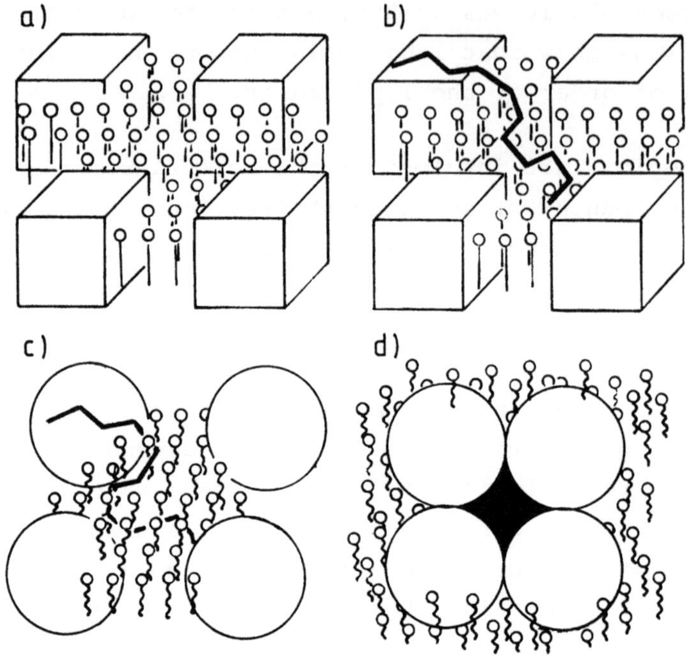

Figure 6. Hypothetic model of ionic channel formation

Considering our results thus far we would like to propose the following hypothetic model for the mechanism of the generation of an ionic channel by participation of proteins and lipids (fig. 6). The receptor regions consist of four or more protein subunits, which are surrounded by a crystalline lipid layer (fig. 6a). The unhindered movable molecule "reacts" with a protein subunit using its haptophoric group and begins to penetrate into the lipid sphere (fig. 6b). This produces a conformational change in the protein as well as an interaction of the intrinsic parts of the pheromone molecule with the lipid region. The degree of order of the lipid region is disturbed and it becomes more fluid (6c). This makes possible a conformational change in the other subunits and cooperatively leads to an even greater softening of the crystalline lipid phase, so that the protein subunits can drift together and form an ionic channel (6d).[4o].

References

1. H.J. Bestmann and O. Vostrowsky, in Chemie der Pflanzenschutz-Schädlingsbekämpfungsmittel (R. Wegler, ed.), Vol. 6 p. 29, Springer Verlag Berlin-Heidelberg (1980).
2. D. Schneider, Z. vergl. Physiol. 40, 9 (1957).
3. H.J. Bestmann, O. Vostrowsky and W. Stransky, Chem. Ber. 109, 1964.
4. H.J. Bestmann, O. Vostrowsky, W. Stransky and P. Range, Chem. Ber. 108, 3582 (1975).
5. H.J. Bestmann, I. Kantardjiew, P. Rösel, W. Stransky and O. Vostrowsky, Chem. Ber. 111, 248 (1978).
6. H.J. Bestmann, K.H. Koschatzky and O. Vostrowsky, Chem. Ber. 112, 1923 (1979).
7. H.J. Bestmann and A. Bomhard, unpublished.
8. H.J. Bestmann and A. Bomhard, Angew. Chem. 94, 562 (1982); Intern. Ed. Engl. 21, 545 (1982).
9. H.J. Bestmann and B. Arnason, Chem. Ber. 95, 1513 (1962).
10. H.J. Bestmann, K. Kumar and L. Kisielowski, Chem. Ber. 116, 2378 (1983).
11. H.J. Bestmann and K. Kumar, Angew. Chem. 95, 154 (1983).
12. H.J. Bestmann and K. Li, Chem. Ber. 115, 828 (1982).
13. F. Ramirez, N.B. Desai and N. McKelvie, J. Amer. Chem. Soc. 84, 1745 (1962).
14. E.J. Corey and P.L. Fuchs, Tetrahedron Lett. 1972, 3769
15. H.J. Bestmann and H. Frey, Liebigs Ann. Chem. 1980, 2061.
16. H.C. Brown, J.A. Sinclair and M.M. Midland, J. Amer. Chem. Soc. 95, 3080 (1973).
17. H.J. Bestmann, K. Sühs and T. Röder, Angew. Chem. 93, 1098 (1981); Int. Ed. Engl. 20, 1038 (1981).
18. H.J. Bestmann and T. Röder, Angew. Chem. 95, 812 (1983); Intern. Ed. Engl. 22, 782 (1983).
19. H.J. Bestmann, Pure Appl. Chem. 51, 515 (1979); 52, 77 (1980).
20. H.J. Bestmann and O. Vostrowsky, in Topics in Current Chemistry 109, 85 (1983).
21. H.J. Bestmann, J. Süss and O. Vostrowsky, Liebigs Ann. Chem. 1981, 2117.
22. H.J. Bestmann, K. Roth and M. Ettlinger, Chem. Ber. 115, 161 (1982)

23. H.J. Bestmann and P. Erman, Chem. Ber. 116, 3269 (1983).
24. H.J. Bestmann and J. Süss, Liebigs Ann. Chem. 1982, 363.
25. H.J. Bestmann, A. Bomhard, R. Dötzer and O. Vostrowsky, unpublished.
26. H.J. Bestmann, K.H. Koschatzky, A. Plenchette, J. Süss and O. Vostrowsky, Liebigs Ann. Chem. 1982, 536.
27. H.J. Bestmann and K. Li, Tetrahedron Lett. 1981, 4941.
28. H.J. Bestmann, P. Rösel and O. Vostrowsky, Liebigs Ann. Chem. 1979, 1189.
29. O. Vostrowsky, H.J. Bestmann and E. Priesner, Nachr. Chem. Techn. 21, 501 (1973).
30. E. Priesner, M. Jacobson and H.J. Bestmann, Z. Naturforsch. 30C, 283 (1975).
31. E. Priesner, H.J. Bestmann, P. Rösel and O. Vostrowsky, Z. Naturforsch. 32c, 979 (1977).
32. H.J. Bestmann, K.H. Koschatzky, O. Vostrowsky and E. Priesner, in preparation.
33. O.L. Chapman, K.C. Mattes, R.S. Sheridan, and J.A. Klun, J. Amer. Chem. Soc. 100, 4878 (1978).
34. H.J. Bestmann, H.L. Hirsch, H. Platz, M. Rheinwald and O. Vostrowsky, Angew. Chem. 92, 492; Int. Ed. Engl. 19, 475 (1980).
35. E. Sackmann, in Biophysik (W. Hoppe, W. Lohmann, H. Markl and H. Ziegler, eds.), Springer Verlag, Berlin Heidelberg, New York, p. 316 (1978).
36. H. Träuble, Naturwissenschaften 58, 277 (1971).
37. R. Wax, Diplomarbeit University Erlangen-Nürnberg (1979).
38. E. Neumann, Angew. Chem. 85, 430 (1973).
39. H.J. Bestmann and K. Dippold, Naturwissenschaften 70, 47 (1983).

New Constituents from Compositae

Ferdinand Bohlmann

Institut für Organische Chemie der Technischen Universität Berlin,
Straße des 17. Juni 135, 1000 Berlin 12, FRG

Several years ago we have decided to concentrate our efforts on the isolation and structure elucidation of the constituents of the large family Compositae with more than 20.000 species. This family is rich in acetylenic compounds, but also in all kinds of other natural products, especially sesquiterpenes and other terpenoids. All together we have isolated more than 4.000 new compounds, many of them showing interesting biological activity. Furthermore interesting chemotaxonomic aspects are clearly visible. Modern techniques allow the structure elucidation with very small amounts of material and allow the isolation of minor compounds, which often indicate possible biogenetic pathways.

The most powerful method of structure elucidation still is the ^{1}H NMR spectroscopy with all its sophisticated techniques as X-ray analysis still is limited as suitable crystals often can not be obtained easily, especially if only minute amounts of material are available.

In the last year we again have isolated a considerable number of new compounds and I would like to show you some examples.

From a Villanova species we have isolated a mixture of six different

Natural Product Chemistry
Ed. by Atta-ur-Rahman

VILLANOVA TITICAENSIS

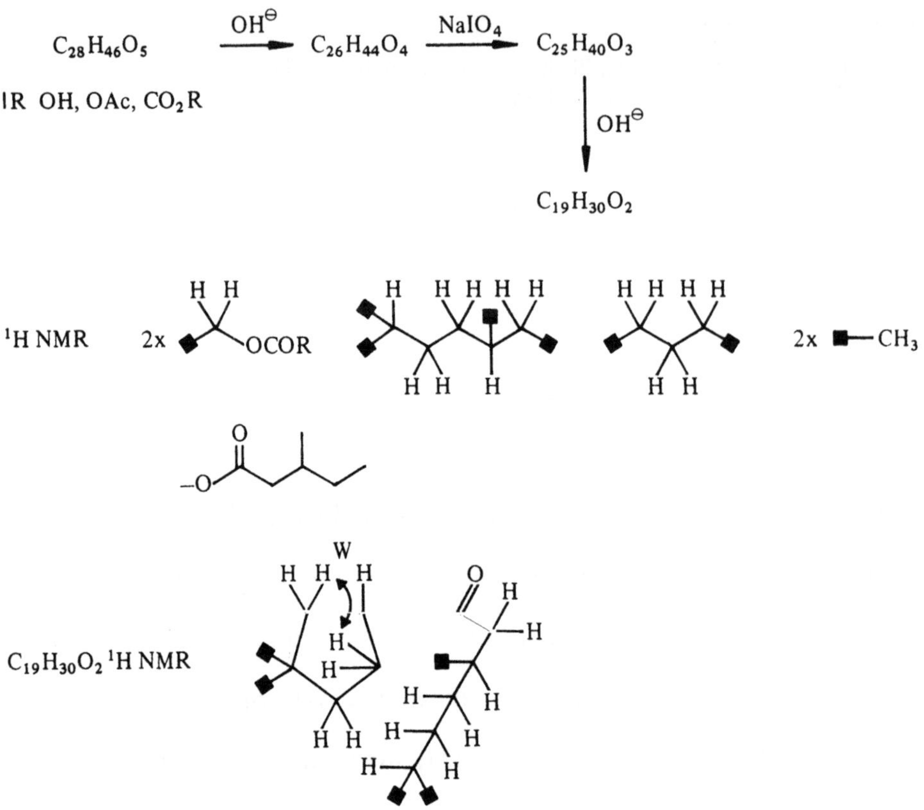

Fig.1

esters. One of these had molecular formula $C_{28}H_{46}O_5$ (Fig.1). The IR and NMR spectra indicated the presence of an acetate group and a further ester group as well as a hydroxy group. Thus the nature of all oxygen functions was clear. Partial saponification gave a diol which could be transformed to a ketone by periodate splitting. The molecular formula show that after hydrolysis of the acetate group one carbon was lost. More drastic saponification then gave $C_{19}H_{30}O_2$, indicating that the second ester group had six carbons. The structure of this group clearly followed from the NMR data. Spin decoupling led to some partial sequences. The molecular formula of the

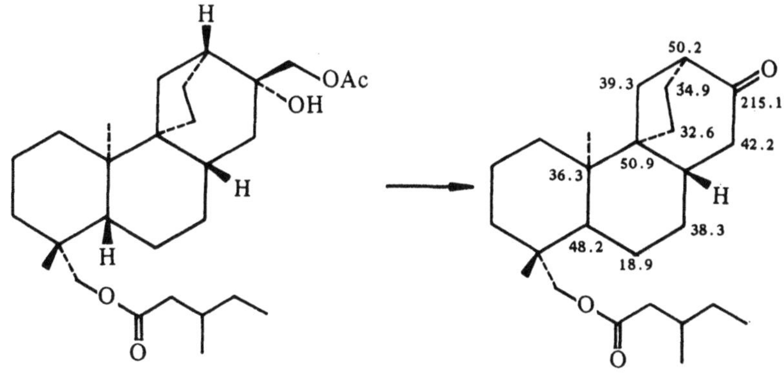

Fig.2

degradation product show that we are dealing with a tetracyclic carbon skeleton as no double bonds seem to be present. This could be established by the ^{13}C NMR. The presence of kaurene derivatives could be excluded as the IR spectrum and also the ^{13}C data showed that no cyclopentanone derivative was present. All data therefore only agree with the structure shown in Fig.2.

The stereochemistry follows from the NMR data and from the NOE's. The new carbon skeleton we have named villanovane. A further derivative of this carbon skeleton we have isolated from a Viguiera species (Fig.3). The position of the oxygen functions indicates that these compounds may be derived from the helifulvanes, a further new diterpene type.

The Fig.4 shows some data of two isomeric diterpenes from an Aspilia species, which both gave the same ketone by MnO_2 oxidation. Again careful NMR investigations gave several partial structures. The single sequences could be combined as long range couplings were present, thus leading to structures shown in Fig.5.

These diterpenes obviously are degradation products of kaurene derivatives. Fig.6 shows some labdane derivatives from a Gutierrezia species.

VIGUIERA PAZENSIS

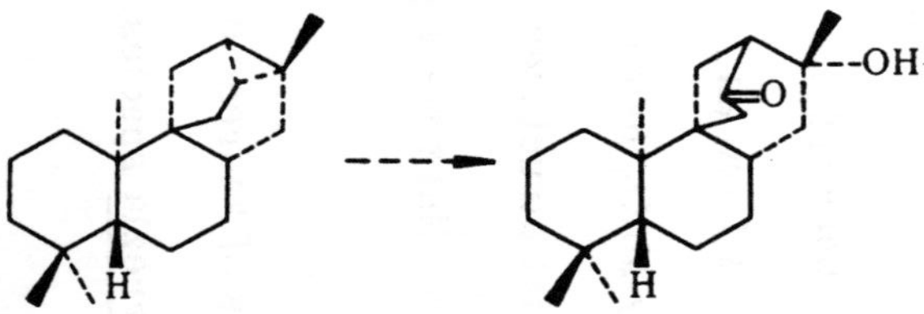

Fig.3

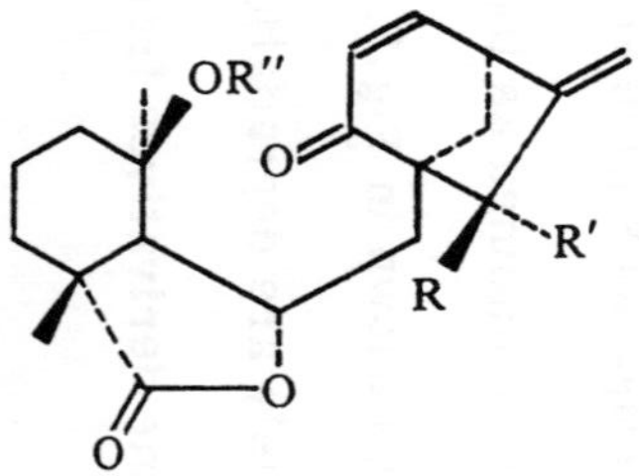

R	H	OH	OAc	OAc
R′	OH	H	H	H
R″	H	H	H	$COCH_2CO_2Me$

Fig.5

ASPILIA

$C_{20}H_{26}O_6$ two isomers

MnO_2 → $C_{20}H_{24}O_6$ (from both)

Ac_2O → $C_{22}H_{28}O_7$

CD: $\Delta\epsilon_{333}$ −3.2

IR: 1770(γ-lactone), 3600(OH), 1660cm^{-1} (C=C−C=O)

MS: M^+ 346 ⟶ 328 ⟶ 310

174($C_9H_{10}O_2$, Mc Lafferty)

^{1}H-NMR:

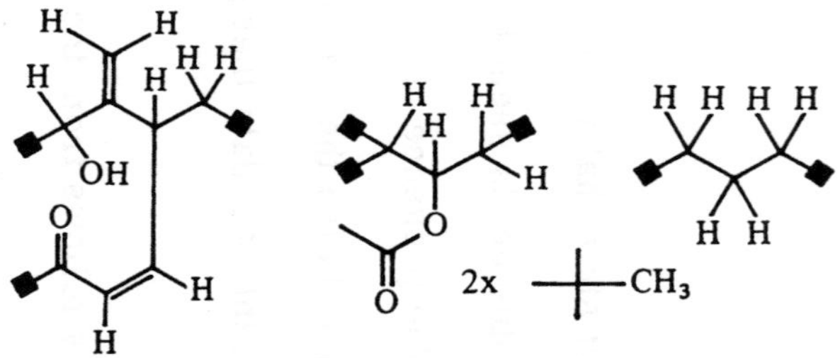

Fig.4

GUTIERREZIA SAROTHRAE

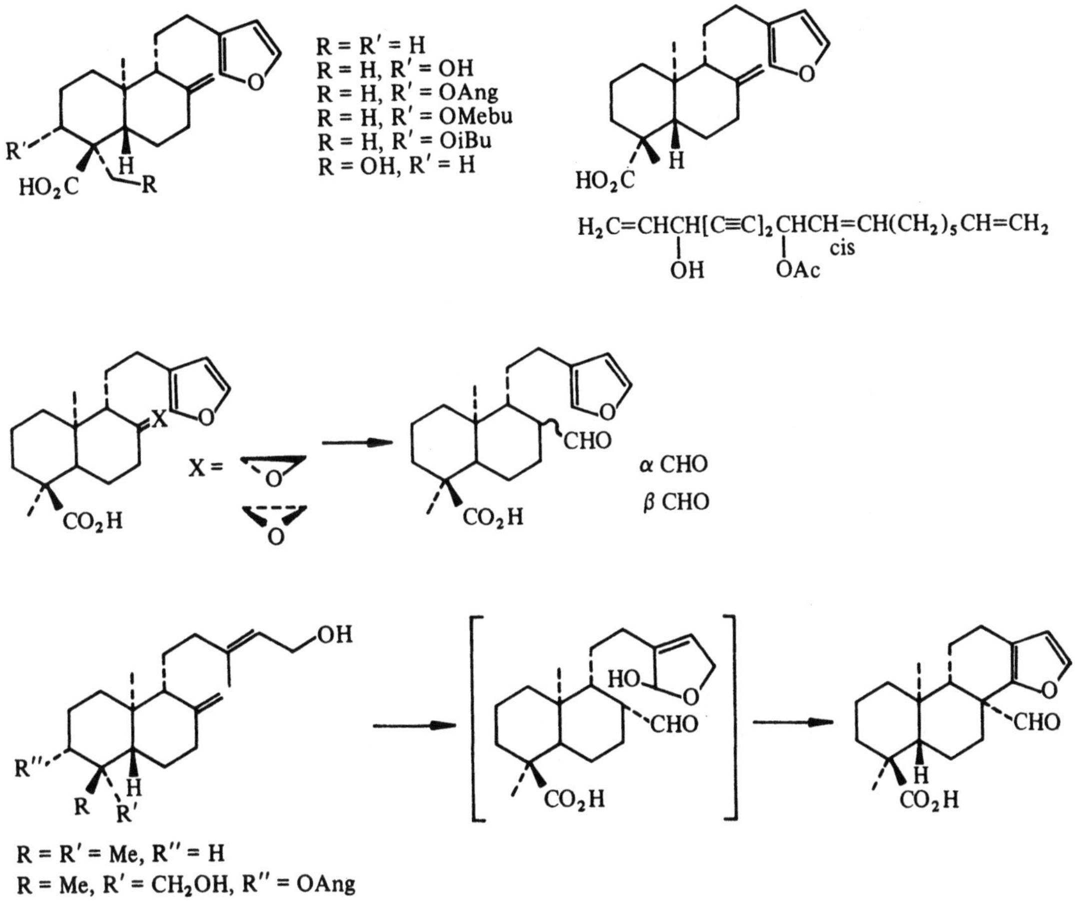

Fig.6

The carbon skeleton of the tetracyclic furane derivative is a new one. Most likely this aldehyde is formed via the hemi acetal, an intermediate in the formation of the furane moiety. In an already published paper we have described several diterpenes from Conyza scabrida. Most likely all compounds are formed by electrocyclic reactions shown in Fig. 7. Many diterpenes derived from geranylgeraniol also have been isolated from Compositae. An example is the lactone present in an Acanthospermum species (Fig.8).

DITERPENES DERIVED FROM CLERODANES

CONYZA

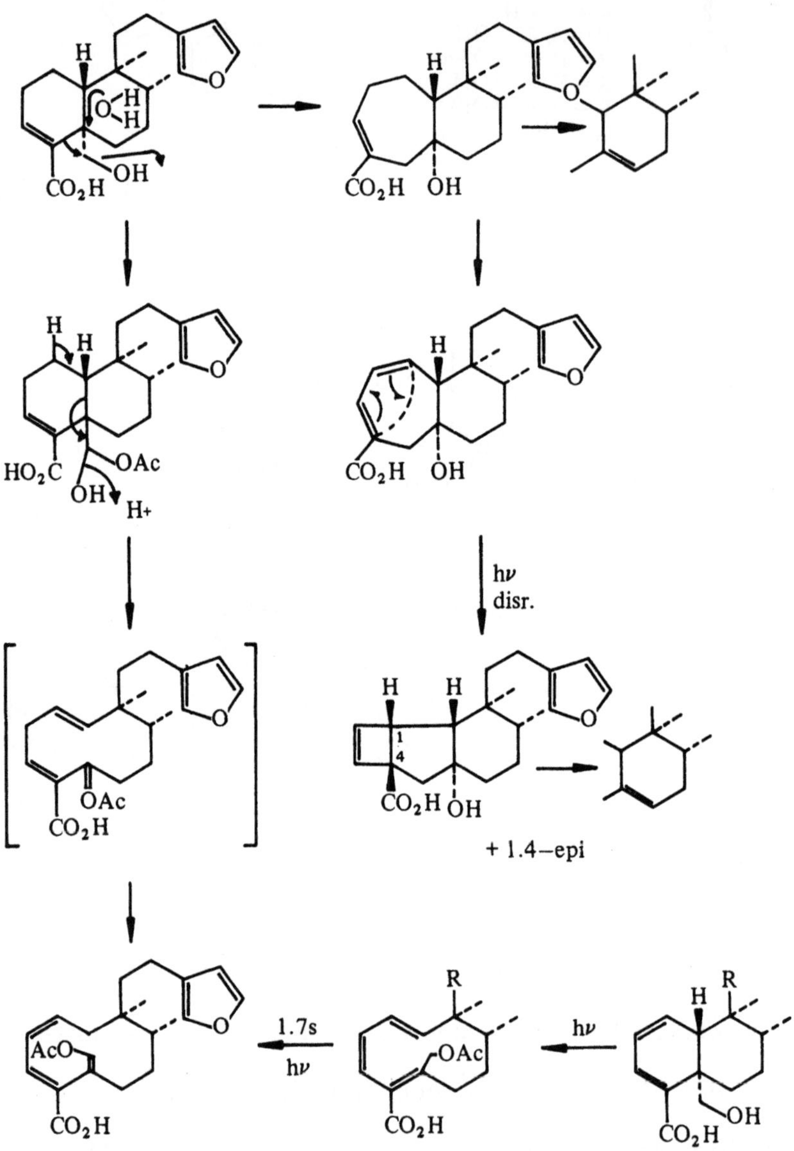

Fig. 7

ACANTHOSPERMUM AUSTRALE

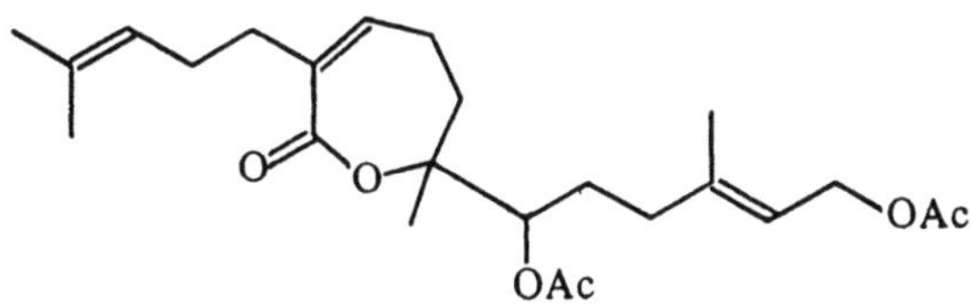

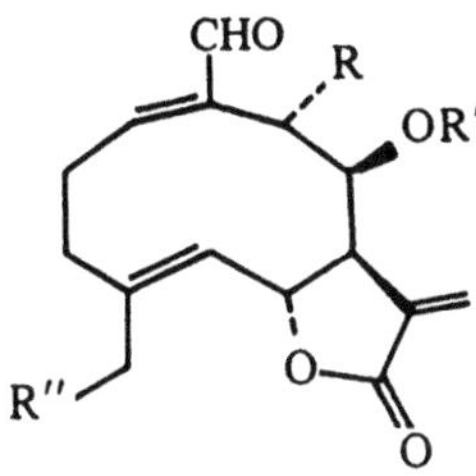

R = palmitate, linolenate, linolate, stearate

R' = iBu, Mebu, 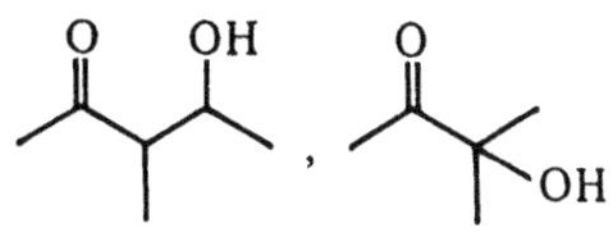

R'' = H, OH

Fig.8

In addition to high field NMR spectroscopy NOE difference spectra are very helpful to get clear stereochemical informations. Fig.9 shows as an example a lactone obtained from a Helenium species
A clear assignment of the relative configuration, especially of epoxides, in five membered rings often is difficult. However, by using this technique it is easy to show that H-2 and H-3 are β-orientated. Irradiation of H-15 further show that a 10α-methyl group has to be assumed. This species further contains a number of arabinosides, a glycoside type which has not been observed previously (Fig.10).
Similary in a case of a lactone from Stevia isomeca the NOE difference spectra clearly allow the assignment of the complete stereochemistry. (Fig.11).

Using these techniques the structure of an unusual lactone from Dugaldia hoopesii could be elucidated too. At 400 MHz all signals could be assigned

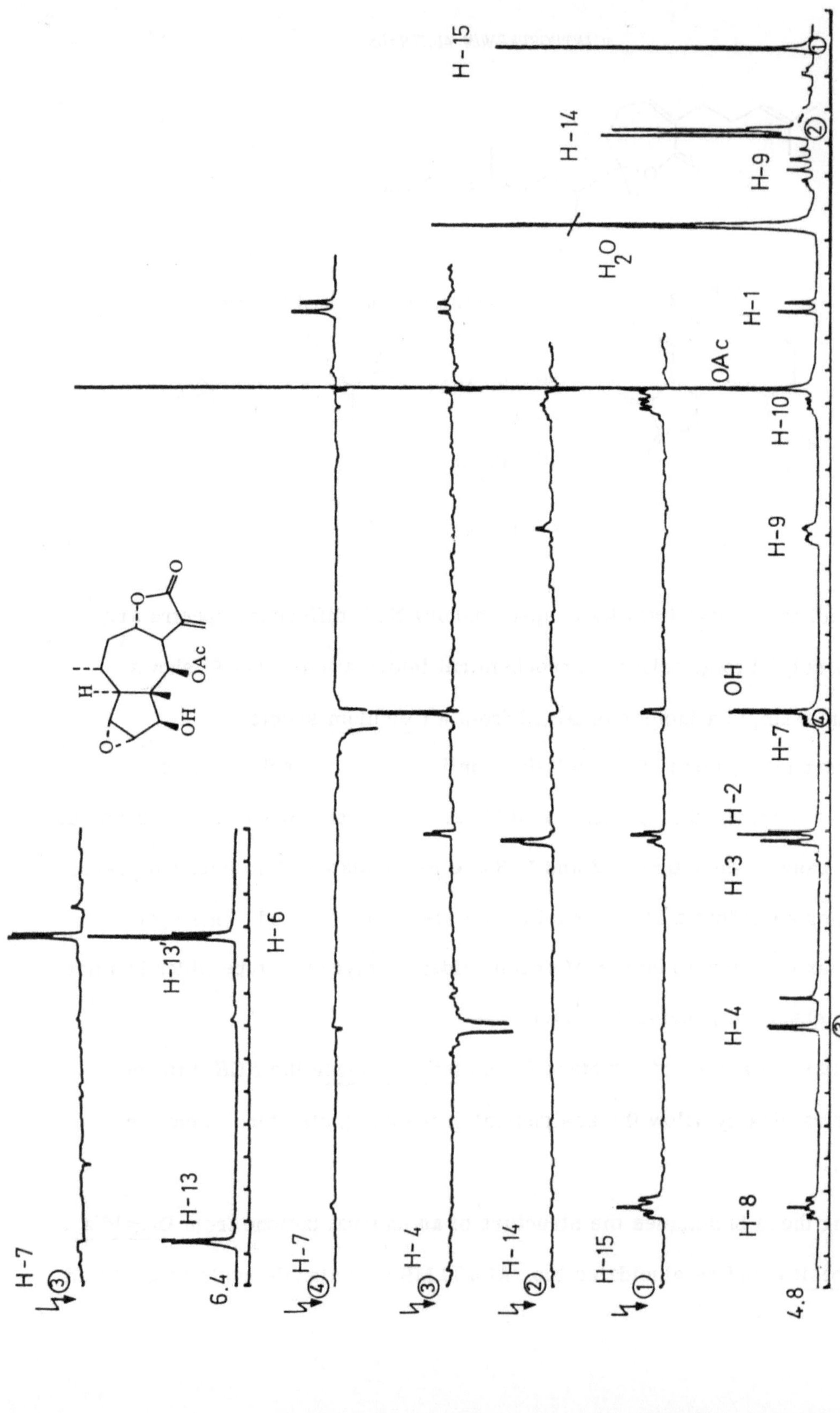

H-15
H-14
H-9
H_2O
H-1
OAc
H-10
H-9
OH
H-7
H-2
H-3
H-4
H-8
4.8
H-7
H-13'
H-6
H-13
6.4
H-7
H-4
H-14
H-15

Fig. 9

HELENIUM TENUIFOLIUM

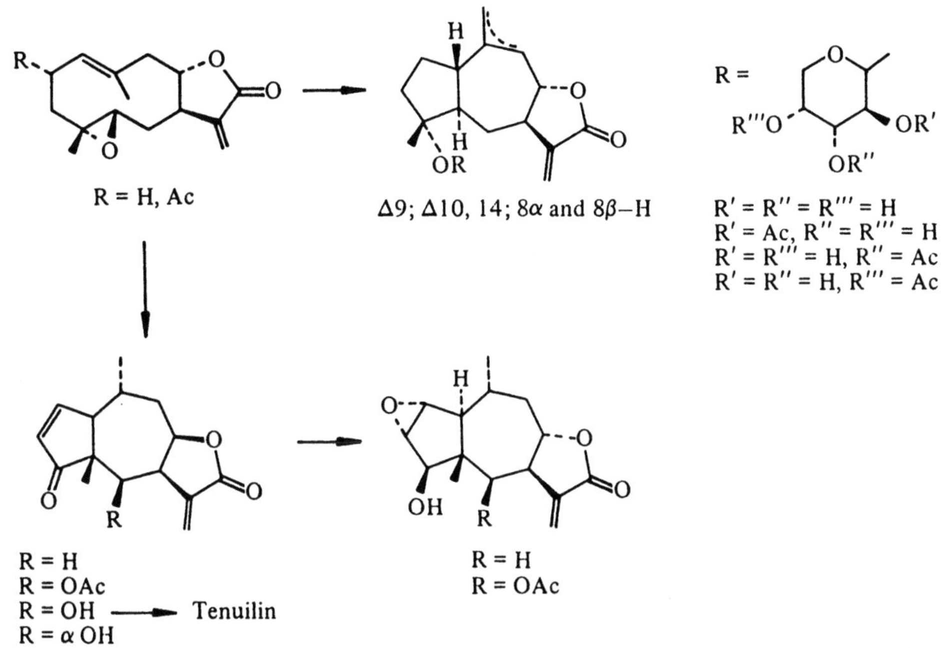

Fig.10

by spin decoupling while the stereochemistry followed from the NOE's (Fig.12).

The same lactone together with several others we have isolated from <u>Gaillardia pulchella.</u> The possible biogenetic pathways are present in (Fig.13).

In some cases the NMR spectra, especially of medium sized rings, can display heavily broadened signals which do not allow clear spin decouplings. In these cases sometimes simple reactions are helpful.

As shown in Fig.14 methylene lactones easily can be transformed to pyrazolines by addition of diazomethane. Germacranolides can be transformed by Cope-rearrangement to elemane derivatives or methanol can

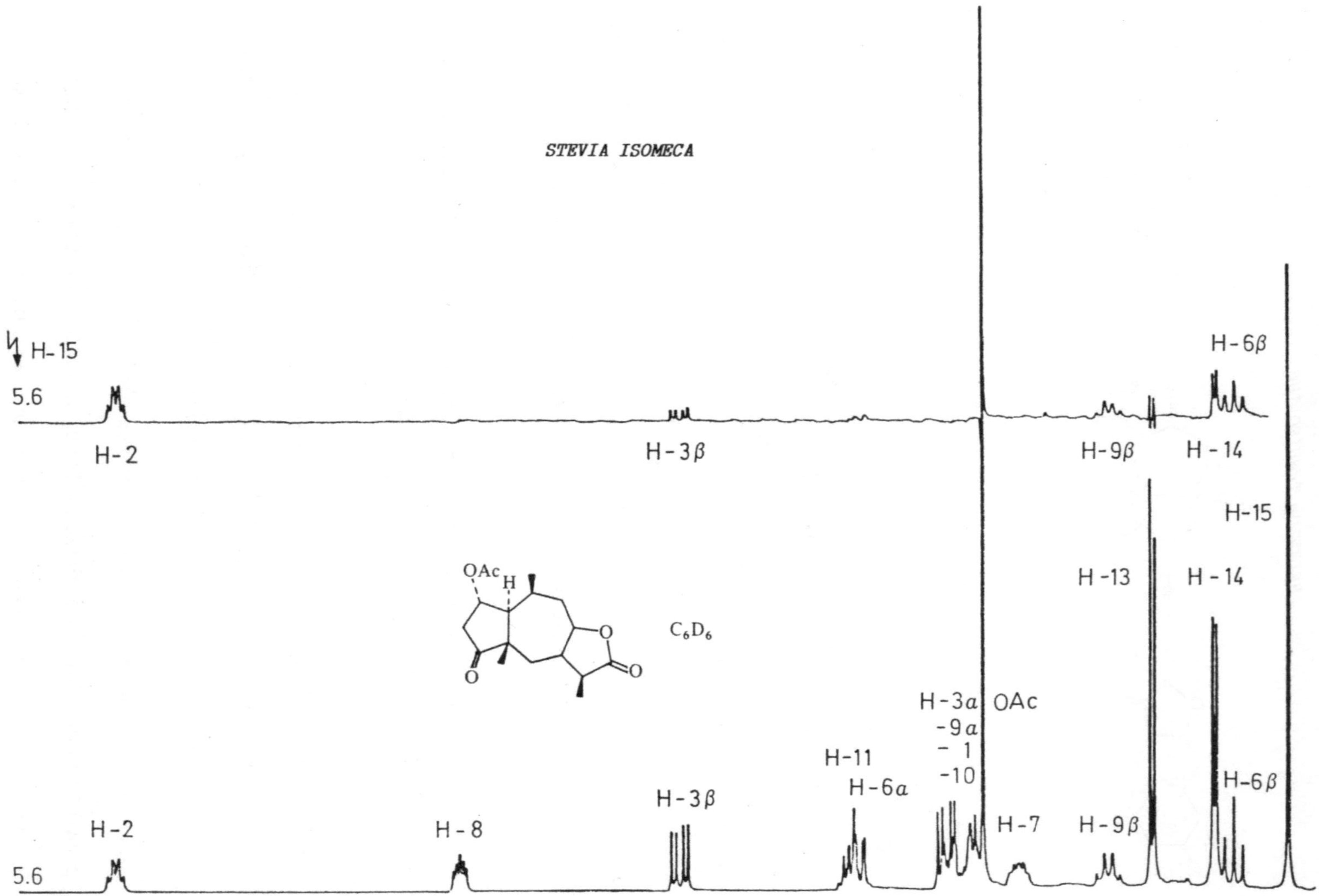
STEVIA ISOMECA
H-15
5.6
H-2
H-3β
H-9β
H-14
H-6β
H-15
H-13
H-14
OAc
H
O
O
O
C_6D_6
H-3α
-9α
- 1
-10
OAc
H-11
H-6α
H-6β
5.6
H-2
H-8
H-3β
H-7
H-9β

Fig.11

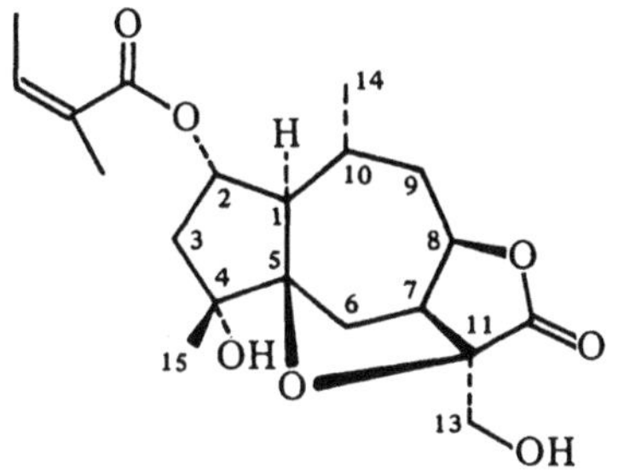

NOE: ↓ H–14 ⟶ H-1, H–2, H–9β
↓ H–15 ⟶ H–3β, H–6β, H–13′
↓ ester Me ⟶ H–2

Fig.12

be added to the exocyclic double bond of the lactones. Normally the derivatives obtained show clear NMR spectra which allow the assignment of all signals. In addition to this, the observed preferred addition often are an indication of the conformation. As can be seen in Fig. 14 the addition of methoxide proceeds from different sides with the Δ^4isomeric lactones.

From a Gochnatia vernonioides we have isolated several new lactones. Again the NMR spectrum of the 10-carboxylic acid shows very broad signals. In order to get a more clear spectrum we have tried to achieve a Cope-rearrangement. However, in this case we observed a new type fo cyclization. As follows from the NMR spectrum a bis lactone was obtained. Most probably it is formed by autocatalysis as shown Fig.15 & 16.

Especially rich in a large variety of sesquiterpene lactones is the genus Vernonia. From two Peruian species we again got several new ones. Vernonia patens contains several glaucolides and hirsutinolides, its structure and stereochemistry now has been established by X-ray analysis. The configuration of the bis epoxide was established by NOE difference spectroscopy. Furthermore a guaianolide was isolated (Fig.17).

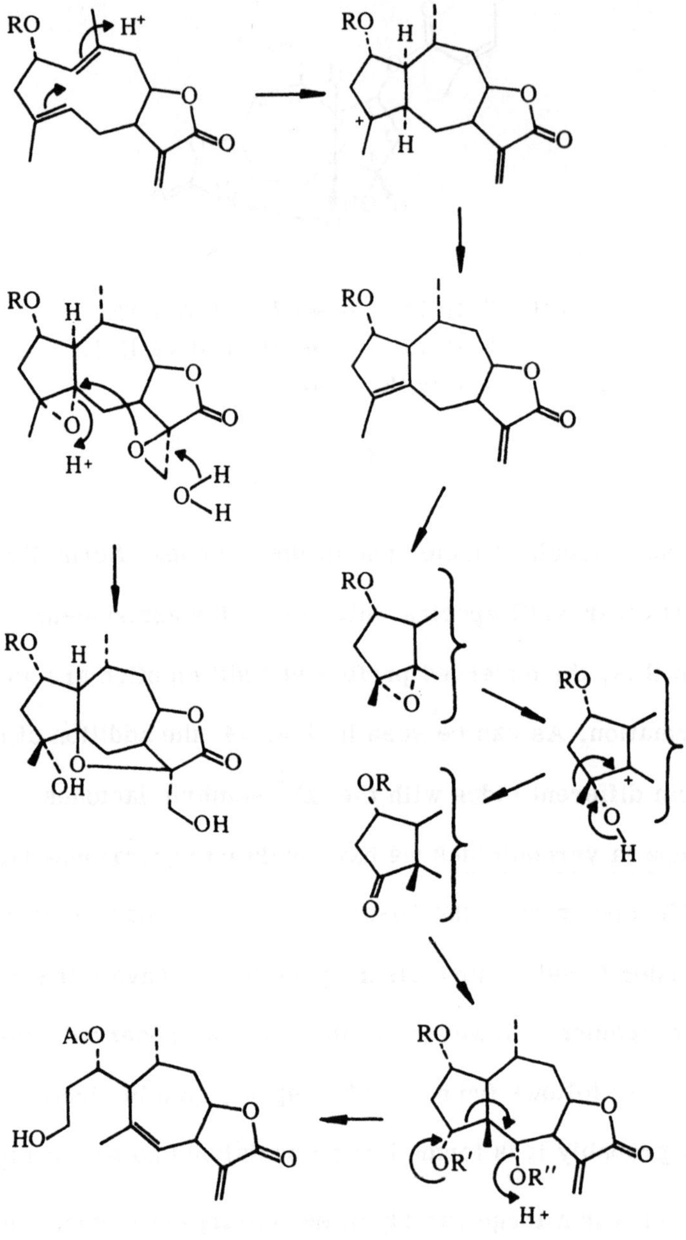

Fig.13

CRONQUISTIANTHUS CHACHAPOYENSIS

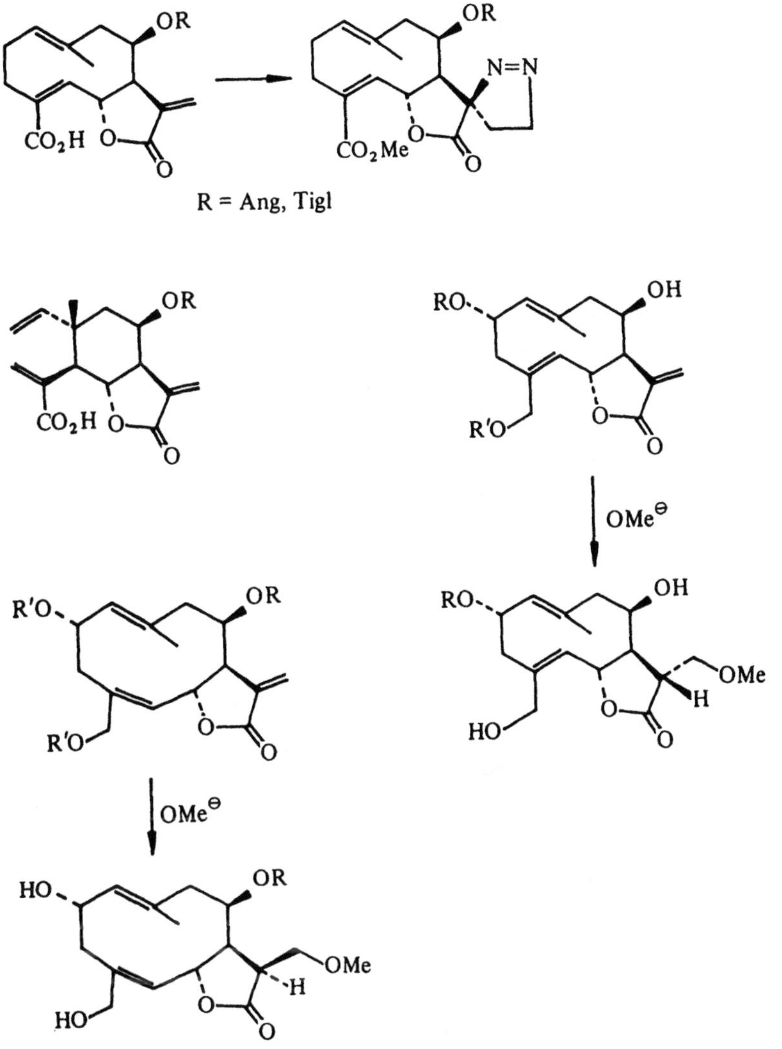

Fig.14

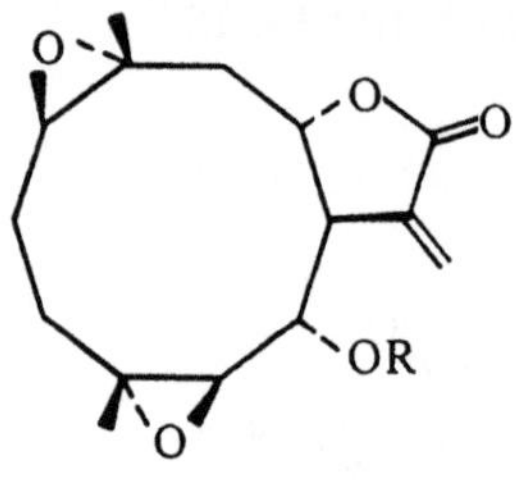

R = Ang
R = COCH$_2$Ph

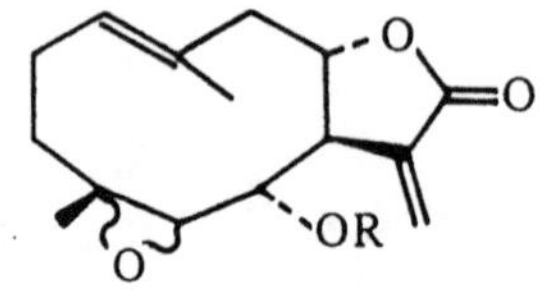

α + β

α + β

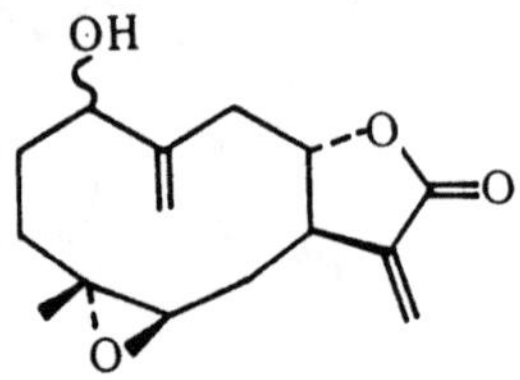

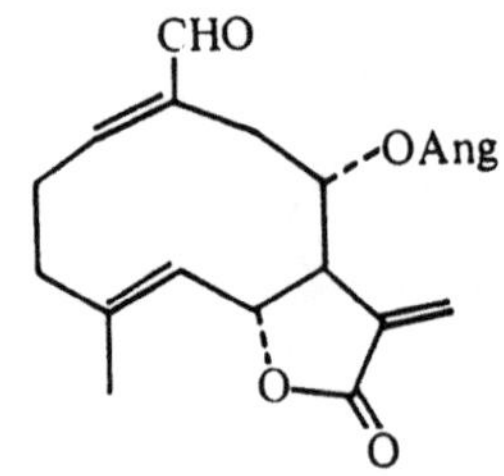

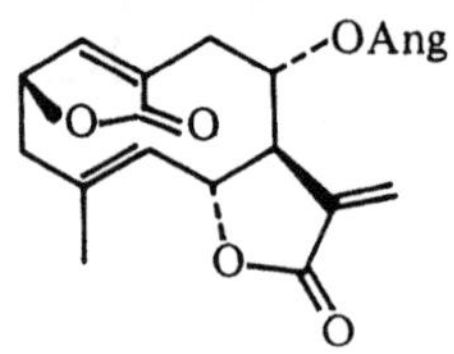

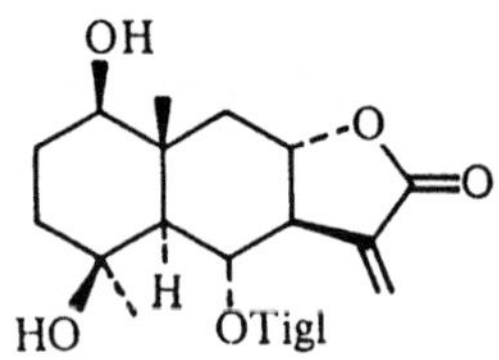

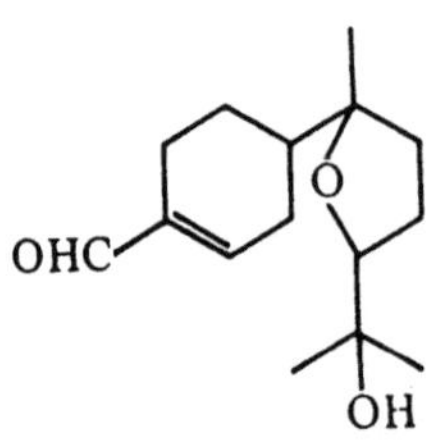

Fig.15

GOCHNATIA VERNONIOIDES

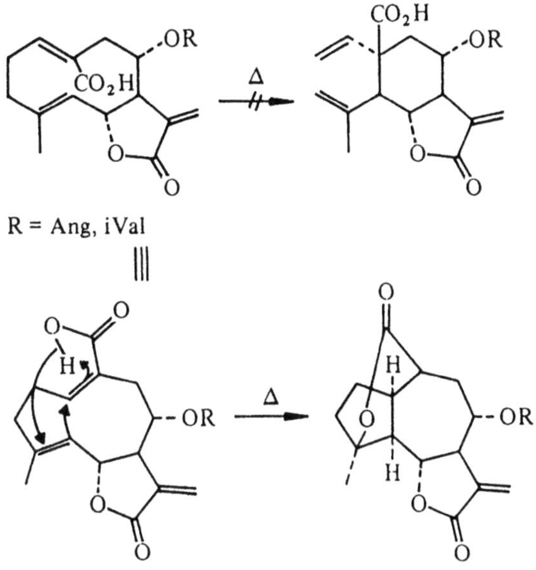

Fig.16

The same type also is present in Vernonia jalcana together with a further one and three lactones derived from cadinane (Fig.18).

This type has not been observed previously. Perhaps these compounds are formed as shown in Fig.19.

In order to get informations about the relative position of the acetate and the angelate group we have studied the partial saponification of several of these lactones. In all cases the ester group at C-13 was hydrolized first. But in addition to this reaction we observed several other transformations. The keto epoxide without an oxygen function at C-10 simply is converted by treatment with K_2CO_3 in dioxane/water to the enol lactone as shown in Fig.20. The same reaction with the corresponding 10-acetoxy compound proceeds much more complicated.

VERNONIA PATENS

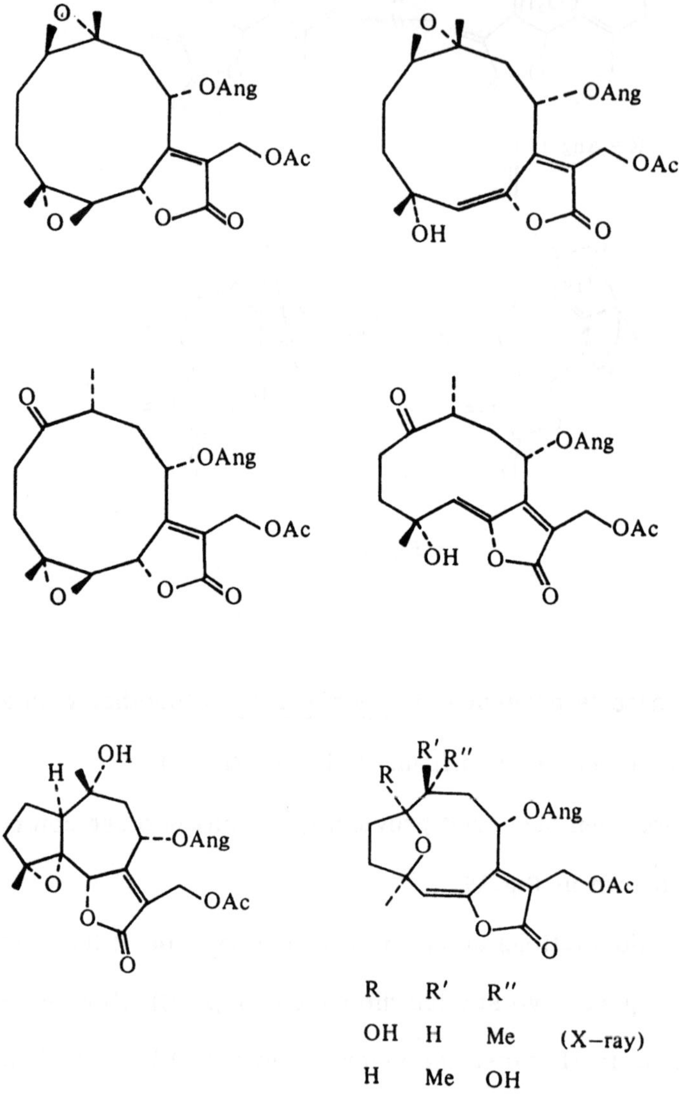

Fig.17

VERNONIA JALCANA

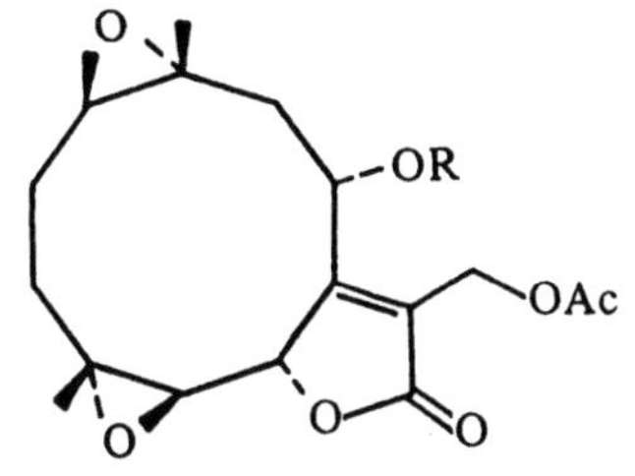

R = Ac, Mebu, iBu

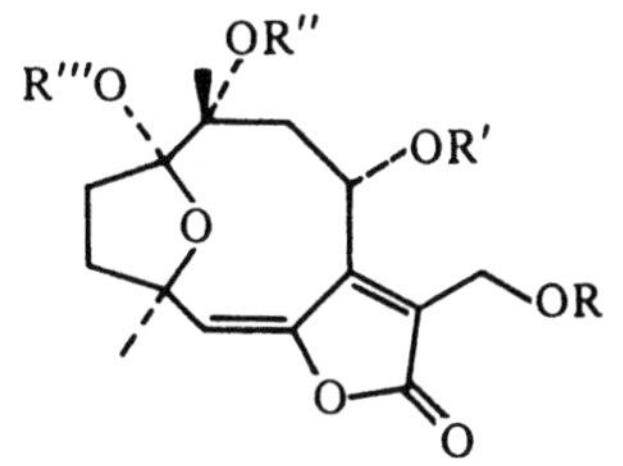

6 different esters

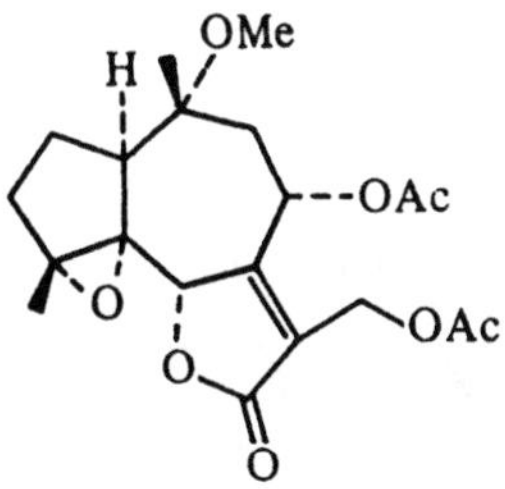

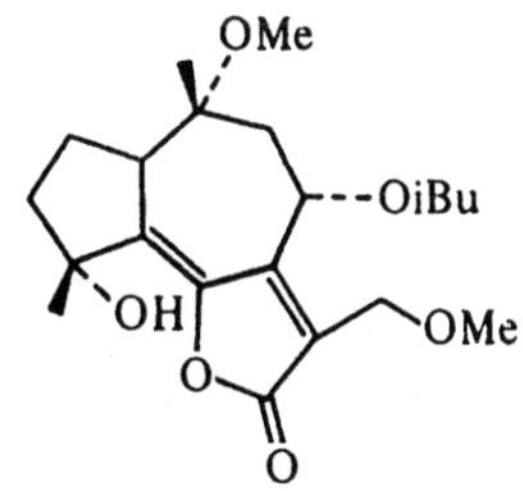

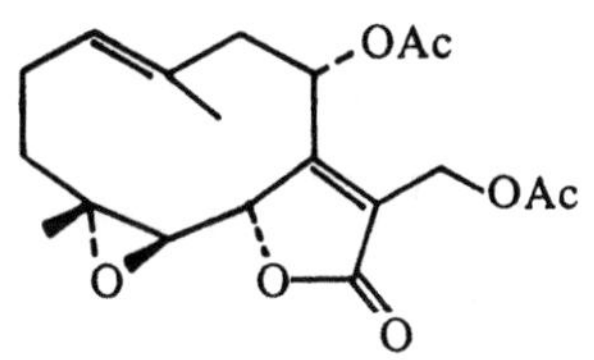

Glaucolide A + B

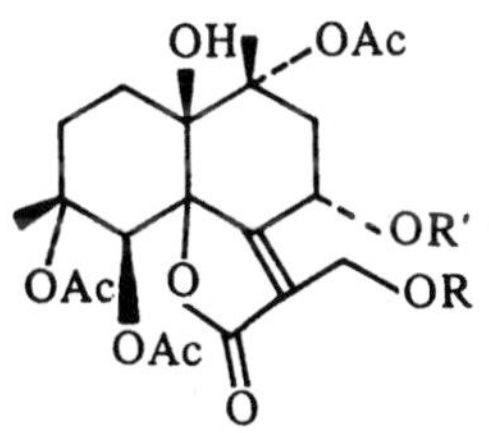

R	H	Me	Me
R'	Ac	Ac	Meacr

Fig.18

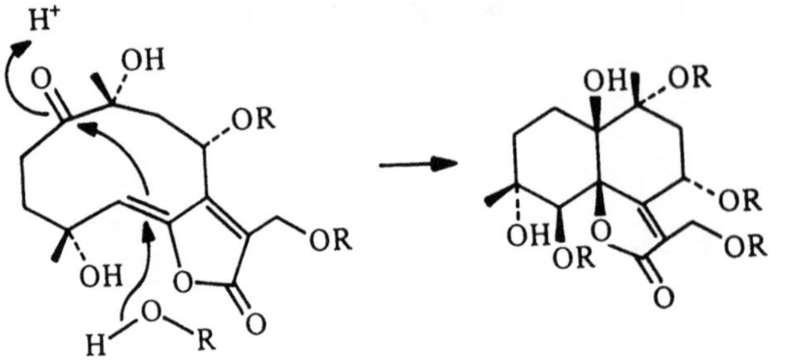

Fig.19

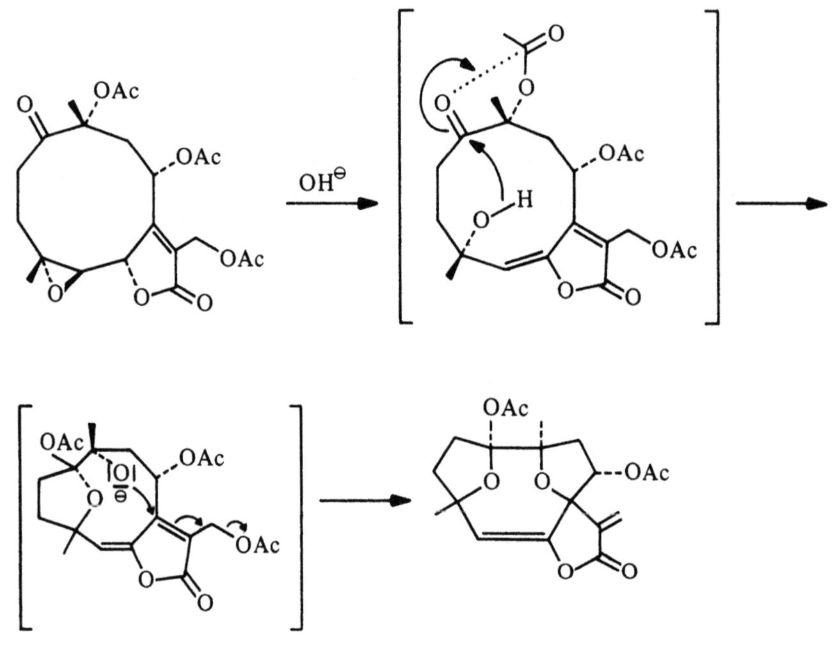

Fig.20

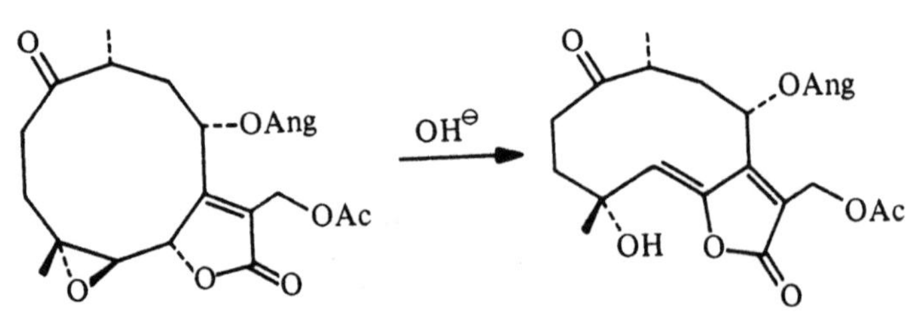

Fig.21

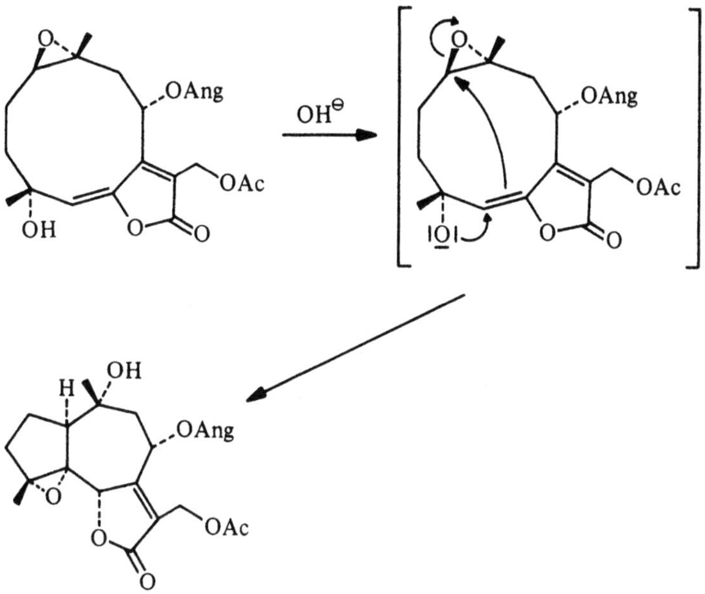

Fig.22

Careful NMR studies led to the structure presented in Fig.21. Most likely this compound is formed as shown in the Scheme. Inspection of models indicated that the hemi acetal formation requires an E-configurated Δ^5double bond. The last step then is an addition-elimination reaction.

Reaction of the enol lactone with an epoxide at C-1/C-10 gave a guaianolide. A possible pathway is presented in Fig.22.

The corresponding diepoxide on reaction with K_2CO_3 gave two products, a guaianolide and a eudesmanolide, which may be formed as shown in Fig.23. In addition to glaucolides and hirsutinolides we have isolated from some Vernonia species also bourbonolides. It therefore is interesting that the corresponding monoepoxide with a $\Delta^{1(10)}$double bond gave a mixture of two prodcuts, a bourbonolide, which was identical with the natural product, while the second lactone was an isomeric cyclobutane derivative (Fig.24).

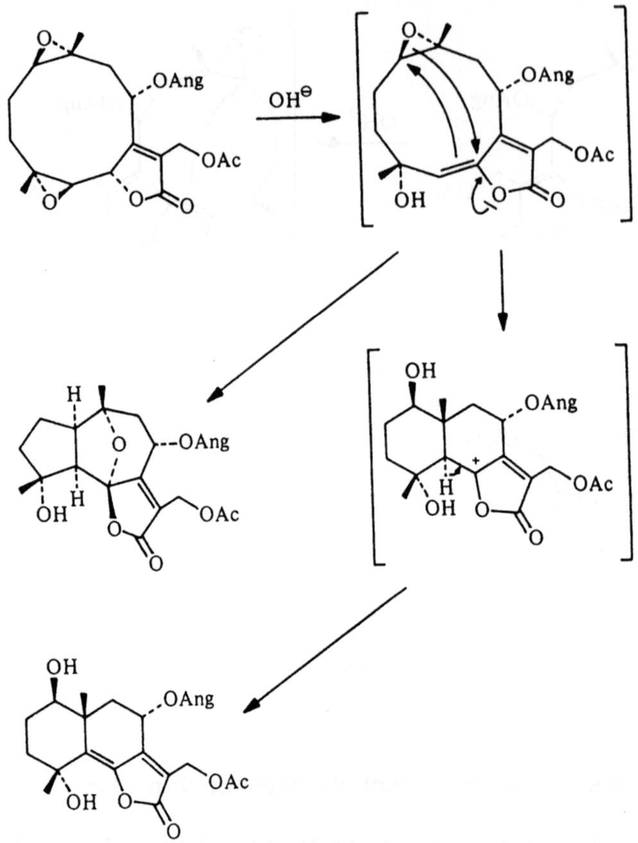
OAng
OH⊖
OAc
OH
H
O
OAng
OAc
O

Fig.23

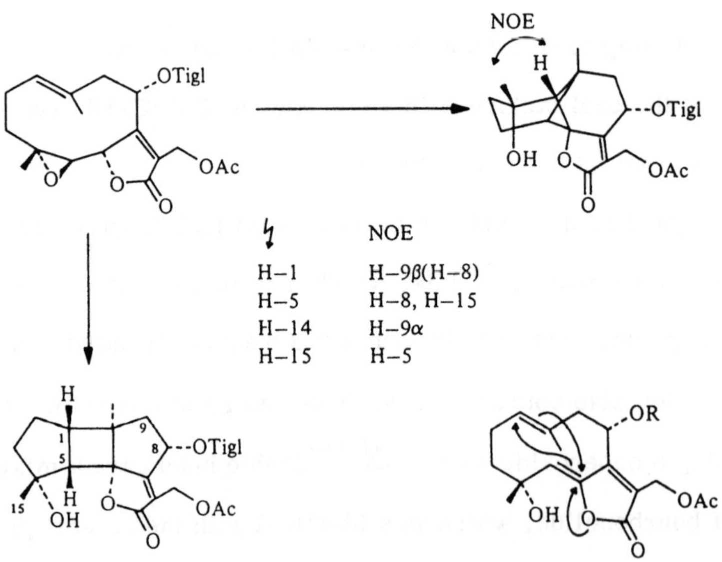
NOE
H
OTigl
OAc
OH
O
NOE
H-1 H-9β(H-8)
H-5 H-8, H-15
H-14 H-9α
H-15 H-5
1
5
9
8
15
OR

Fig.24

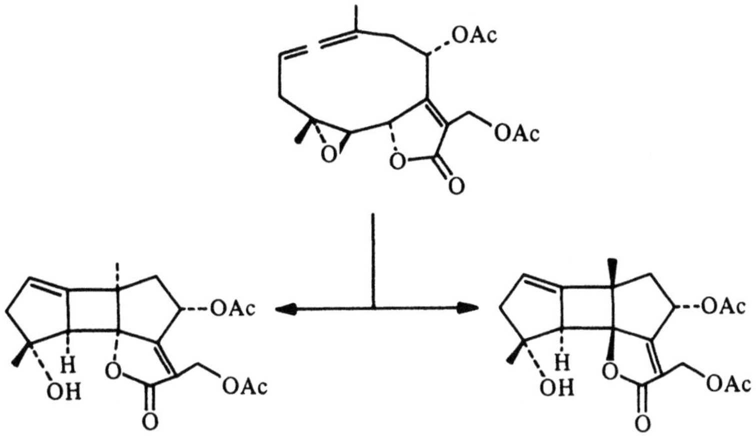

ϟ	NOE	ϟ	NOE
H–14	H–2, H–9α	H–14	H–2, H–9β
H–15	H–8, H–2β	H–15	H–3β
OH	H–5	OH	H–5

Fig.25

WUNDERLICHIA MIRABILIS

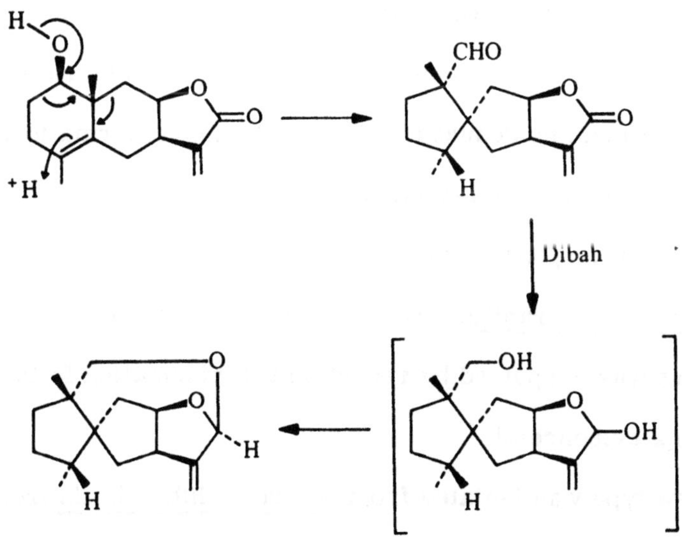

Fig.26

MONTANOA HIBISCIFOLIA

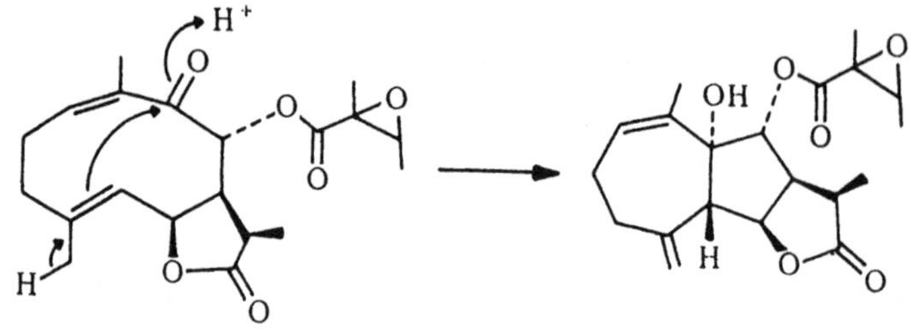

Fig.27

The configurations of these compounds could be established again by NOE difference spectroscopy. The corresponding allenic lactone, also present in some Vernonia species, afforded under the same conditions the two isomeric bourbonolides with an additional double bond(Fig.25). Again the stereochemistry was established by NOE difference spectroscopy. The formation of the cyclobutane derivatives is probably due to a thermal 2+2 addition. As can be deduced from models the conformations would allow these reactions (suprafacial-antarafacial).

An unusual sesquiterpene lactone together with more common ones we got from Wunderlichia mirabilis. Its formation is indicated in Fig.26.

The structure was supported by reduction transformation followed by acid treatment to a cylcoacetal.

A further new type was isolated from Montanoa hibiscifolia from Costa Rica. Most likely it is formed as shown in Fig.27.

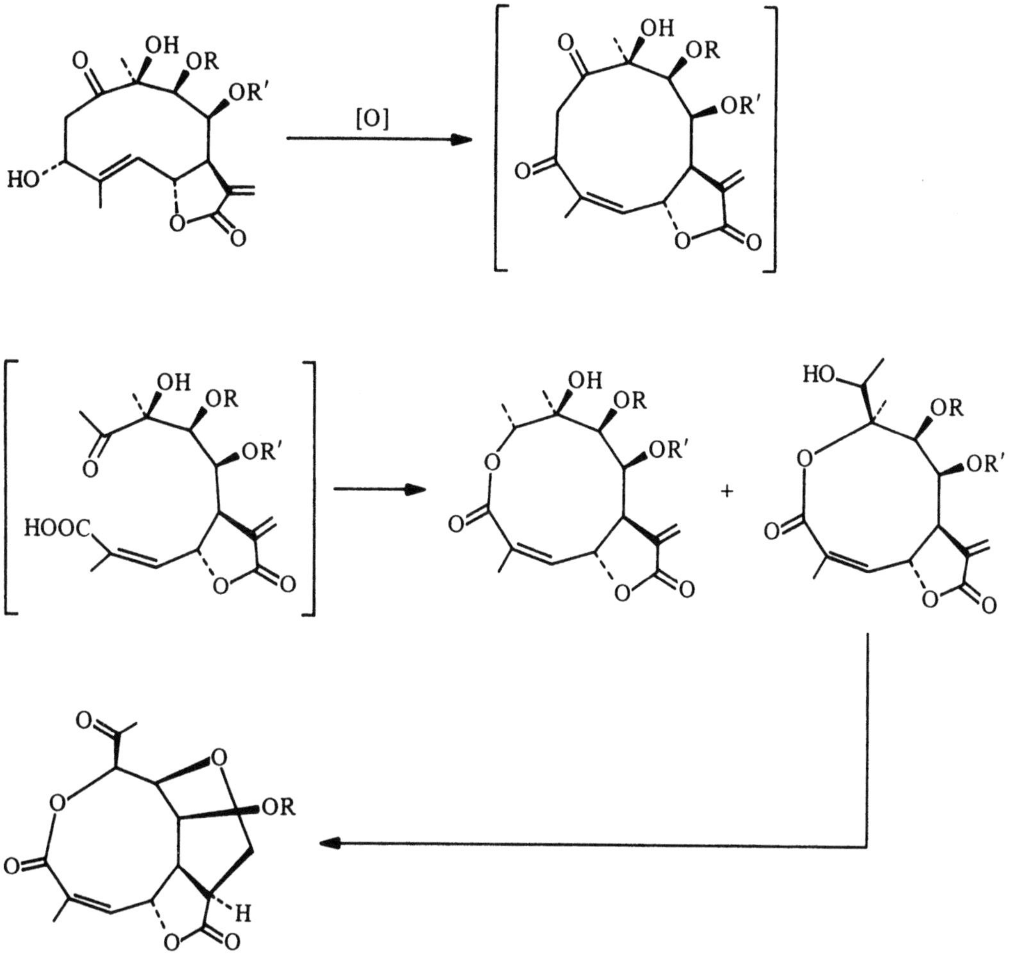

Fig.28

DECACHAETA THIELEANA

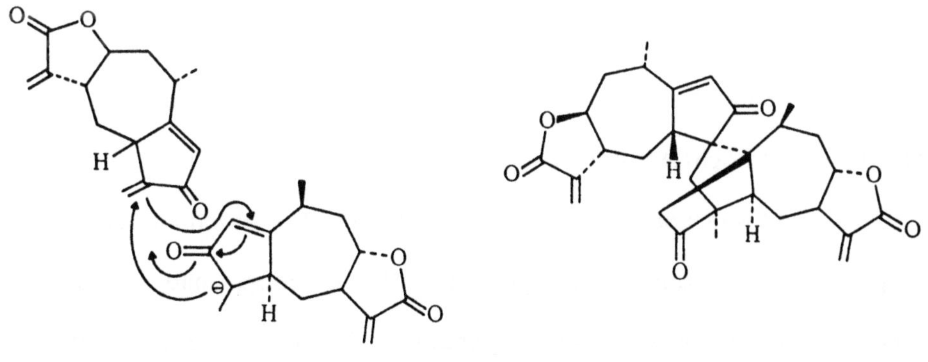

Fig.29

ARTEMISIA PECTINATA

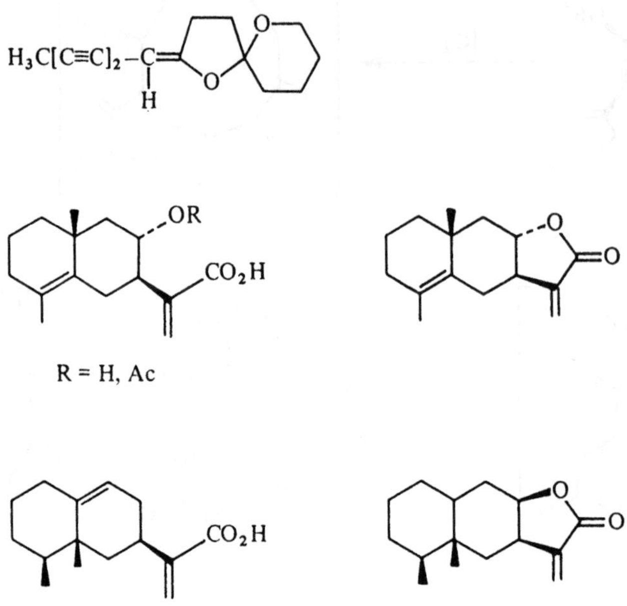

Fig.30

Complex lactones are also present in Trichogonia species, which may be formed as outlined in Fig.28.

An unusual dimeric lactone we have isolated from a Decachaeta thieleana which may be formed by a double Michael-addition (Fig.29).

From Artemisia pectinata from the Mongolei we got in addition to the typical acetylenic spiroketal enolether three precursors of sesquiterpene lactones as well as an eremophilanolide (Fig.30).

The hydroxy acid was transformed to a eudesmanolide which so far has not been reported from nature. It is interesting that eremophilene derivatives have been isolated from an Artemisia species. The cooccurrence with eudasmanes, which are the precursors of the eremophilanes also is noteable.

Many more compounds including very different types of natural products have been isolated too. However, time does not allow to discuss all these compounds. The investigation of the biological activities of the new compounds is in progress.

Finally I wish to thank my co-workers. The investigations were performed by Mrs. Zdero, Mrs. Schuster, Dr. Jakupovic, Dr. Castro, Dr. Umemoto, the Ph.D. students G. Schmeda-Hirschmann, G. Ludwig, T. Gerke, Dr. Misra and the graduate student Mrs. C. Trinks.

X-ray Diffraction and Marine Natural Products

Jon Clardy
Department of Chemistry, Baker Laboratory, Cornell University, Ithaca, NY 14853, USA

Introduction

While natural products research has expanded into many new areas of inquiry, the traditional goal of describing the molecular structure of a naturally occurring substance has not diminished in importance. A secure molecular structure is still the common starting point for synthetic, biosynthetic, and even some biological endeavors. In this paper I will explain why x-ray diffraction is increasingly the method of choice for structure elucidation and illustrate its utility with some recent applications to marine natural products.

X-ray Diffraction

There is an analogy between the x-ray analysis of a molecular structure and the observation of a much larger object with visible light. Molecules have characteristic distances between atoms of roughly one Angstrom. The x-rays used in a typical experiment are light waves with wavelengths of roughly one Angstrom. The electrons in a molecule interact with the x-rays and scatter them, much as visible light is scattered from this page. In this way, x-rays allow us to observe the electrons, and only the electrons, in a sample. Unfortunately the analogy is far from perfect. When an object and the illuminating light are roughly the same size, the scattered light pattern does not resemble the object. This new pattern is called the diffraction pattern. It is a rather different looking pattern, and informa-

Natural Product Chemistry
Ed. by Atta-ur-Rahman

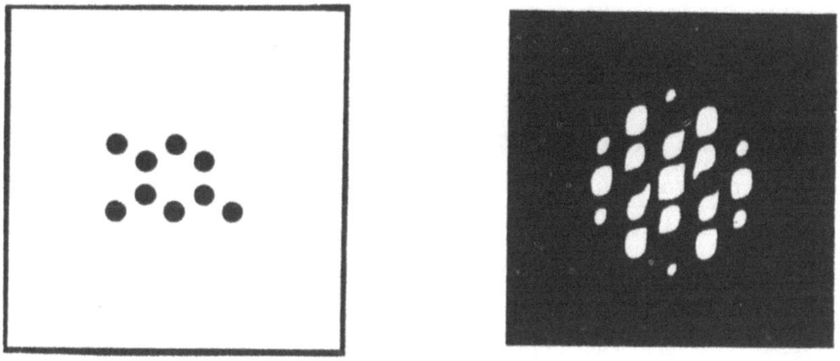

Figure 1. The pattern on the right results if the object on the left is illuminated with light roughly the same size as the object. From Atlas of Optical Transforms by G. Harbury, C.A. Taylor, and T.R. Welberry. Copyright © 1975 by G. Bell & Sons Ltd. Used by permission of the publisher, Cornell University Press

tion from all parts of the original object are dispersed throughout the diffraction pattern. A simple illustration of this is given in Figure 1. The laws governing the diffraction of x-rays from an object are well understood, and it is a simple, if sometimes tedious, matter to calculate the diffraction pattern of an atomic assembly. The analogy is also imperfect in the sense that it is not technically feasible to illuminate a single molecule with x-rays and obtain a measurable pattern. If a random assemblage of molecules is illuminated with x-rays, a rather messy and generally uninterpretable pattern results. Crystals are used because they contain large numbers of molecules which are held in a regular pattern. This regular pattern is characterized by a repeating block, called the unit cell, and a description of the internal symmetry of the block, called the space group. The minimum chemical entity that must be specified to completely describe the crystal is called the asymmetric unit. The necessity of using a crystal introduces a relatively minor problem in that the pattern, which was continuous for one molecule, can now only be observed at certain locations.

The most serious drawback to the analogy is that, while visible light rays can be focused to form an image, there are no lenses for x-rays. The crystallographer is forced to record the scattered intensities and then try to calculate what the focused image would be. A digital computer is the crystallographer's lens. Unfortunately, currently available techniques of recording the scattered intensities do not measure the phases of the scattered waves. Without the phases, it is impossible to calculate an image since the waves would be combined in an effectively infinite number of ways, and only one of these ways corresponds to an image of the original object. Crystallographers have assiduously developed relatively few, but usually reliable, techniques to recover or estimate these lost phases. The methods are not foolproof and, in difficult cases, require some experience and ingenuity.

The final stage of an analysis is a careful least squares fitting of the experimental diffraction data to an atomic model. A typical organic structure determination gives bond lengths with an accuracy of approximately 0.01 Å and bond angles within 0.5°. This is a level of detail greater than most natural products chemists care to have.

The advantages of an x-ray analysis are: a very small sample size, a nondestructive means of analysis, an enormous amount of useful data, and a prejudice-free fashion of analysis. A typical crystal weighs only a few micrograms and can be used for other experiments following the x-ray experiment. A typical x-ray experiment will have on the order of 50 reflections for every non-hydrogen atom in the molecule. There are other techniques which also give a lot of data, but the x-ray experiment is unique in that the use of each piece of data in the analysis is clearly specified. The most important advantage, in my opinion, is that the analysis makes essentially no use of proposed molecular structures or substructures. It is no more difficult to determine a completely unknown structure of molecular weight 400 than it is carry out an analysis to ascertain the stereochemistry at one center of an otherwise known steroid. There are several examples from my laboratory of an analysis which has been successfully completed starting only with an incorrect molecular formula. The constraints in fitting a molecular model to the large number of experimental observations are quite severe. The number of observations exceeds the number of parameters by roughly a factor of ten.

There are of course disadvantages to the x-ray experiment. One of the most vexatious is that it requires crystals. Preparing a suitable single crystal can often be the rate limiting step in an analysis. Another disadvantage is that if an x-ray analysis is unsuccessful, nothing of chemical value is learned. There is a very nearly absolute quality about an x-ray analysis, it is either absolutely right or it is absolutely useless.

The most challenging crystallographic problems in organic chemistry are those involving chiral molecules with only light atoms – the usual characteristics of a natural product. The most enjoyable problems are natural products with brand new types of structures. Marine organisms are less well explored than terrestrial ones, and we have found their secondary metabolites to be a fruitful field for the application of x-ray diffraction.

The Aplysiatoxins

The marine blue-green alga Lyngbya majuscula contains two highly inflammatory agents, aplysiatoxin (1) and debromoaplysiatoxin (2), which cause a severe contact dermatitis that sometimes afflicts swimmers and bathers during the summer months in Hawaii.[1] These two toxins were first isolated by Scheuer and Kato, at the University of Hawaii, from the digestive gland of the sea hare Stylocheilus longicauda, a gastroped mollusk that feeds preferentially on L. majuscula.[2] Professor Moore and his group, also

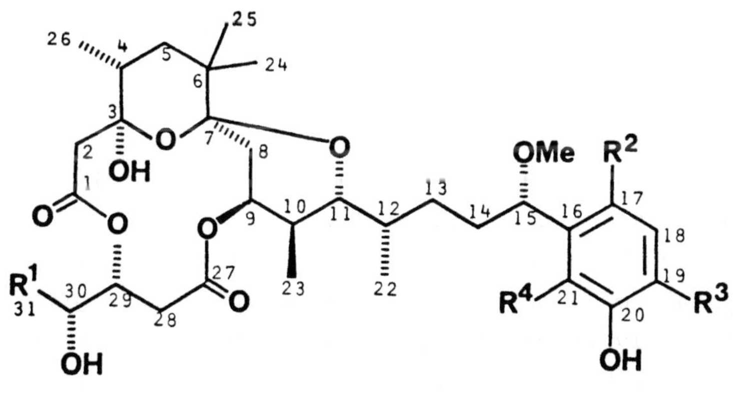

	R^1	R^2	R^3	R^3
1	Me	Br	H	H
2	Me	H	H	H
3	Me	Br	Br	Br

at the University of Hawaii, succeeded in locating some of the algal sources of these toxins.[3] Debromoaplysiatoxin (2) is present in several blue-green algae belonging to the Oscillatoriaceae. For example it has been isolated from an inseparable mixture of two blue-green algae found on the seaward side of Enewetak Island, viz. *Schizothrix calcicola* and *Oscillatoria nigrovirdis*.

Debromoaplysiatoxin displays some antineoplastic activity with a T/C value of 140 at the 0.6 mg/kg dose level in the P-388 assay.[1] Recently the aplysiatoxins have also been found to be powerful tumor promoters.[4] Scheuer and Kato were able to propose a two dimensional structure for aplysiatoxin as a result of chemical degradation and spectral studies.[2] The aplysiatoxin structure was remarkable when it was first proposed, and it is no less remarkable today. Moore and his collaborators extended this work and, through extensive ^{1}H NMR decoupling experiments, assigned most of the stereochemistry. In spite of these chemical and spectral studies, some workers felt that correctness of the structure had not been illustrated in a completely convincing manner. We obtained crystalline samples of debromoaplysiatoxin from Professor Moore and tried on several occasions to solve the structure with x-ray diffraction. In spite of a great deal of effort, we never succeeded. The difficulty appeared to be that debromoaplysiatoxin crystallizes in such a way that all of the atoms in two independent molecules had to be located. During the last year, we tried a different approach. When debromoaplysiatoxin is treated with bromine in aqueous methanol buffered at pH 6; 19,21-dibromoaplysiatoxin (3) is produced along with other products. So far 3 has not been identified as a natural product. Efforts to crystallize this material were successful, and we began a new x-ray structural analysis.

The crystals of 3 belong to the triclinic space group P1, which is the only triclinic space group compatible with a chiral, optically active molecule. The cell constants were a=9.859, b=9.929(5), c=11.351(7) Å, α=101.132(44), β=112.072(41), and γ=74.941(39)°, and this indicated that only one molecule of $C_{32}H_{45}Br_3O_{10}$ formed the asymmetric unit. This simplest possible arrangement, only a single molecule in each unit cell, is rather rare. The presence of three bromine atoms in 3 has both helped and hindered the analysis. Since only electrons scatter, a good first approximation of the structure was simply three bromine atoms. Thus arriving at a crude picture of 19,21-dibromoaplysiatoxin (3) was not unduly difficult. The bromines have hindered the refinement of the model because their scattering tended to overwhelm the scattering from all of the other, more interesting atoms in the structure. The fit of the model to experimental data was also complicated by an unknown amount of disordered solvent of crystallization which occupied voids in the crystal lattice of 3. Nevertheless the analysis has progressed far enough that we are completely confident of the structure of 3. This x-ray derived model is shown in Figure 2. It fully confirms the original two dimensional structure proposed by Scheuer and Kato, and also agrees with all of the stereochemical assignments made by Moore. The presence of the bromine atoms allowed us to do an experiment that is not usually possible for natural products. The normal x-ray experiment introduces a center of symmetry into the data. Thus the derived structure represents the relative and not the absolute stereostructure. In this case the interaction of the bromine atoms with the x-rays broke the center of symmetry and allowed us to determine the absolute configuration. The technique used is called anomalous scattering and usually requires the presence of a heavy atom such as bromine. This absolute configuration has been independently confirmed by Woodward and Craig at the University of California, San Francisco through circular dichroism studies.[5]

Pectenotoxin

Sporadic appearances of toxic shellfish are a widespread but imperfectly understood phenomena.[6] The usual source of the toxin is a dinoflagellate. For example, paralytic shellfish poisoning (PSP) involves the highly potent toxins of the saxitoxin family produced by *Gonyaulax* spp. or *Pyrodinium bahamense* var. *compressa*. Recently another form of shellfish poisoning, which is distinctly different from PSP in both symptomatology and etiology, was described.[7] The poisoning was named diarrhetic shellfish poisoning (DSP) after its predominant human symptom. Unlike PSP,

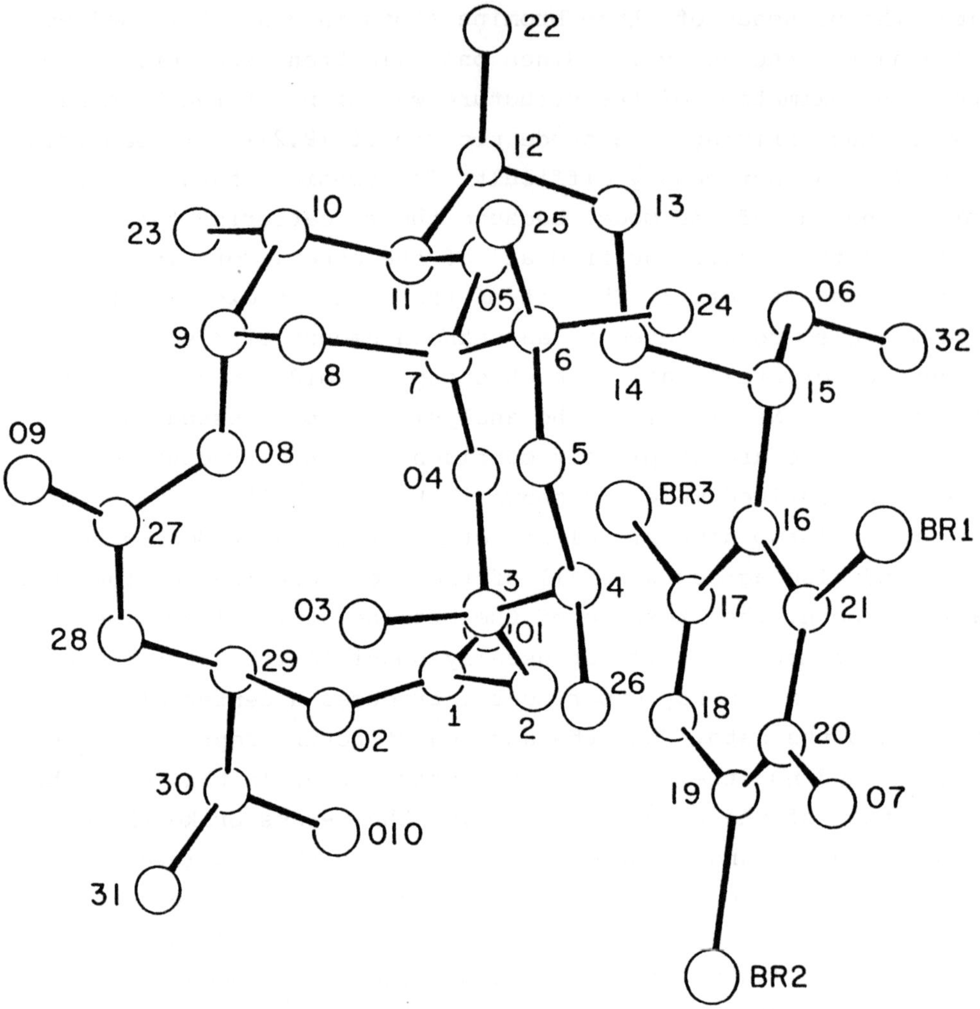

Figure 2. A computer generated perspective drawing of the current x-ray model of 19,21-dibromoaplysiatoxin (3). Hydrogens are omitted for clarity

DSP does not appear to be fatal, but its high morbidity rate and worldwide distribution, as revealed by surveys in Japan, Spain, The Netherlands, and Chile, make DSP a serious threat to public health and the shellfish industry. The toxin producing dinoflagellate was identified as Dinophysis fortii and the toxins were named dinophysistoxins (DTX's).[7] Earlier work has established the structures of DTX_1 and DTX_3 to be 35(S)-methyl okadaic acid and its 7-O-acyl derivatives respectively.[8] Okadaic acid is a C_{38} polyether fatty acid first isolated from sponges and later from the dinoflagellate Prorocentrum lima.[9] The structural assignments of DTX_1 and DTX_3 were done by spectral analysis, but the analysis ultimately relied on the established structure of okadaic acid – an x-ray derived result.[9] It is often the case that x-ray diffraction is used to establish the first structure in a new family, and other members are identified by spectral methods. Further work using contaminated scallops rather than dinoflagellate cultures has led to the isolation and characterization of another causative agent, pectenotoxin 1 (PTX_1), as the novel polyether shown as 4.

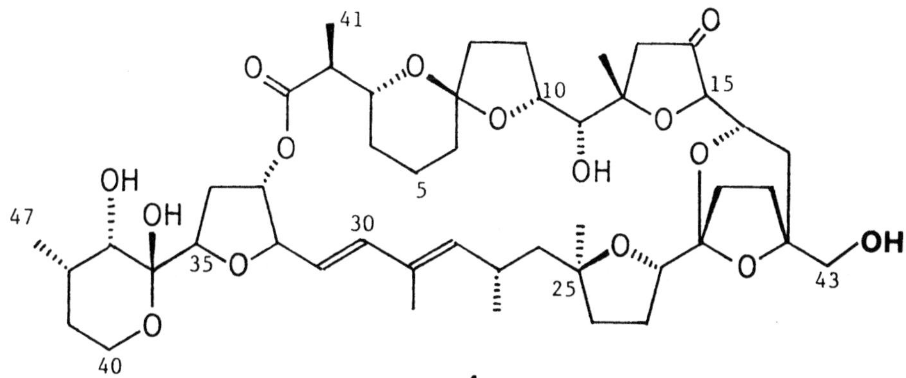

4

The isolation was carried out by Professor Yasumoto's group at Tohoku University and the scallop Patinopecten vessoensis was used as the starting source. Crystallization of PTX_1 from MeCN-H_2O furnished a white crystalline solid, m.p. 208-209°, $[\alpha]_D^{20}$ +22.4 (c 0.001, MeOH). PTX_1 was lethal to mice at 250 µg/kg (ip). IR bands at 3400, 1760, 1740 and 970 cm^{-1}, combined with the UV maximum at 235 nm, indicated the presence of hydroxyl, ketone, ester and conjugated diene groups. The ^1H NMR (360 MHz, acetone-d_6) spin decoupling and NOE measurements confirmed the trans-trans conjugated diene. The ^{13}C NMR (CD_3OD) indicated 47 carbons: 2 carbonyl singlets, 4 olefinic carbons, 3 ketal or hemiketal singlets, 15 oxycarbons. The remaining 23 highfield signals ranged from 51.2 to 12.8 and included 3 methines and 6 methyls. A secondary ionization mass spectrum gave a m/z 875 (M^++H), indicating a formula $C_{47}H_{70}O_{15}$.

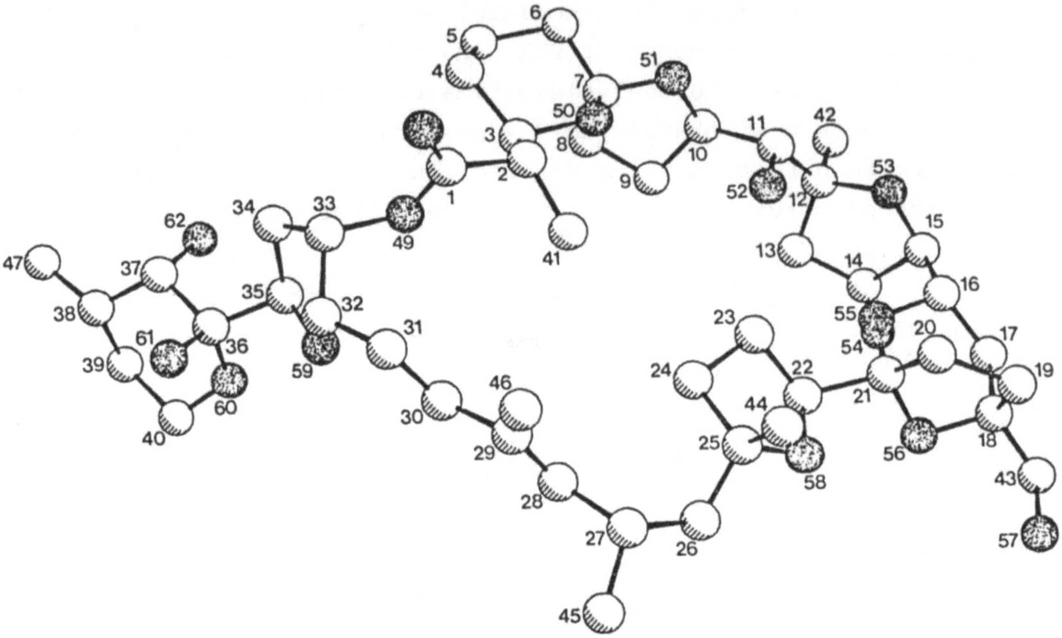

Figure 3. A computer generated perspective drawing of pectenotoxin-1 (PTX_1, 4). Hydrogens are omitted for clarity and no absolute configuration is implied

The relative stereostructure of PTX_1 (4) was secured by single crystal x-ray diffraction techniques. The crystals belonged to the monoclinic system with a=31.551(5), b=6.219(7), c=25.350(3) Å and β=97.78(1)°. Systematic extinctions, the presence of chirality and crystal density indicated space group C2 with one molecule of PTX_1 (4) forming the asymmetric unit. In this case there were no heavy atoms and a phasing model was achieved using direct methods. Direct methods make use of equations which relate the phases of the experimental data. Unfortunately these equations are only probably, not absolutely, correct. A computer generated perspective drawing of the final x-ray model is given in Figure 3.[10] In this case, we have not been able to assign the absolute stereochemistry, and the enantiomer shown is an arbitrary choice. This structure is completely consistent with the spectral data.[10]

PTX_1 is substantially different from previously described dinoflagellate toxins including DTX_1 and DTX_3. Specifically, it differs from the latter in having a longer carbon backbone (C_{40}), a C_{33} lactone ring rather than an open structure, and a novel dioxabicyclo ring. It also has some familiar features such as a spirocyclic ketal ring. The coexistence of compounds with such highly varied structural features raises interesting questions about the biosynthesis of polyether compounds in dinoflagellates. The large, oxygen-rich internal cavity is grossly similar to the cavities found in the polyether ionophores from terrestrial microorganisms. Efforts are now under way to elucidate PTX_1's mode of action on various biological systems.

Zoanthamine

Coral reefs are often covered with thick mats of colonial anemones or zoanthids, which can be recognized by the regular appearance of oral disks, each fringed with a tiny ring of tentacles. Not much chemical work has been reported for the zoanthids, although the genus Palythoa has been the source of very interesting toxin palytoxin. Zoanthus of sociatus occurs on the east coast of India and has recently been examined by Professor Rao of Andhra University and Professor Faulkner of the Scripps Institution of Oceanography. When provoked these zoanthids squirt a fluid which causes severe inflammation of the eye. Since the isolation was not guided by a bioassay, we are not able to say what the inflammatory compounds are. One of the compounds isolated from this source was crystalline and was trivially named zoanthamine. A molecular formula of $C_{30}H_{41}NO_5$ was indicated by mass spectrometry and by the off-resonance decoupled ^{13}C NMR. An initial analysis of the ^{13}C NMR data revealed the presence of a ketone, and α,β-unsaturated ketone, and an ester. The molecular formula required that zoanthamine be heptacyclic and contain both a tertiary amine and a cyclic ether. A ^{13}C NMR signal at δ 101.6 (s) suggested that there was at least one sp^3 carbon attached to two heteroatoms. From these data we had absolutely no idea what zoanthamine might be, but we were certain that it was something quite unusual. Zoanthamine crystallized in the orthorhombic space group $P2_12_12_1$ with a=11.775(5), b=13.88(1), and c=15.877(6) Å. In this case we also used direct methods to obtain our first view of the molecule. One of the

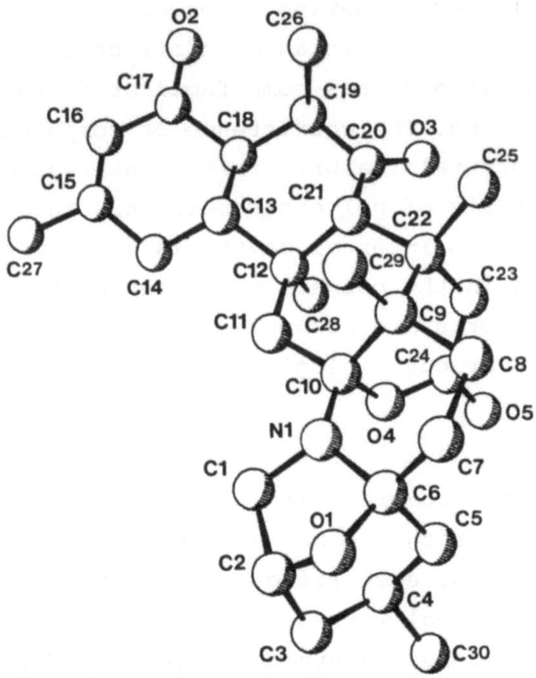

Figure 4. A computer generated drawing of zoanthamine. Hydrogens are omitted for clarity and no absolute steroechemistry is implied

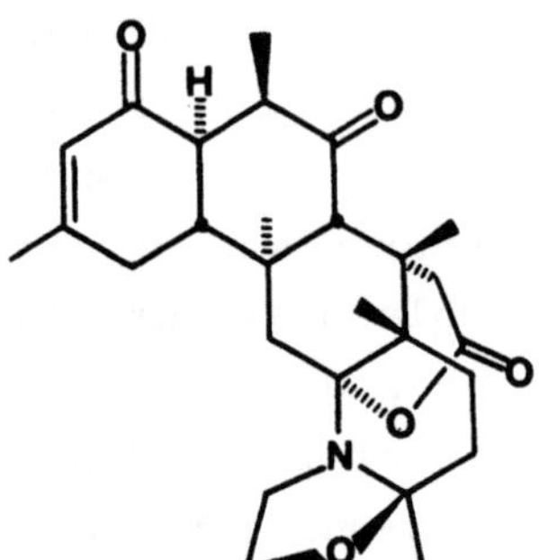

5

problems with zoanthamine was that it was such a novel structure that we were not certain how to assign the atoms. Sometimes an x-ray experiment does not clearly indicate what kind of atom is present and the attribution of atom types can be difficult. The relatively few incorrect x-ray structure that have appeared in the literature have usually been confusions of atom types such as calling an oxygen a nitrogen. In principle, the location of hydrogen atoms would clarify the problem, but, since hydrogens contribute relatively little to the scattering, their location can be troublesome. The zoanthamine situation was further complicated by the fact that we initially had an incorrect molecular formula to work with. By carefully considering the molecular geometry and locating as many hydrogens as we could, we eventually arrived at the structure shown as 5 and in Figure 4 for zoanthamine. This structure fit all of the spectral data including a new mass spectrum which gave the molecular formula above. Zoanthamine appears to be the first member of a family of unusual marine alkaloids, and two additional structures are being worked on by spectroscopic methods at Scripps. The biogenesis of this family is obscure.

An Antifungal Agent from A Bryozoan

The structures that have been discussed so far have been rather complicated, and I would like to show that x-ray diffraction can also be very useful in determining the structures of relatively simple molecules. Such an example is provided by the work of Professor Andersen at the University of British Columbia. On a recent collecting trip he noted that the bryozoan, *Phidolopora pacifica* was uncontaminated by other organisms. Lack of contamination is often a clue that a chemical defense is being employed. Such casual field observations have played an important role in marine natural products chemistry.[11] In this particular case, chemical defense is not yet proven, but an extremely active antifungal and antibiotic substance has been isolated. It is very active against the fungi *Pythium ultimum*, *Helminthosporium sativum*, and *Rhizoctonia solani*. It also has very good antibiotic activity against *Staphylococcus aureus* and *Bacillus subtilis*. It became clear rather early in the investigation that the active substance was made up of a substituted purine and an aromatic piece, but the nature of the substituents, their placement on the rings, and the nature of the linkage between the two fragments were obscure. This is probably a case where the ^{1}H NMR was too simple and could not be used to connect large

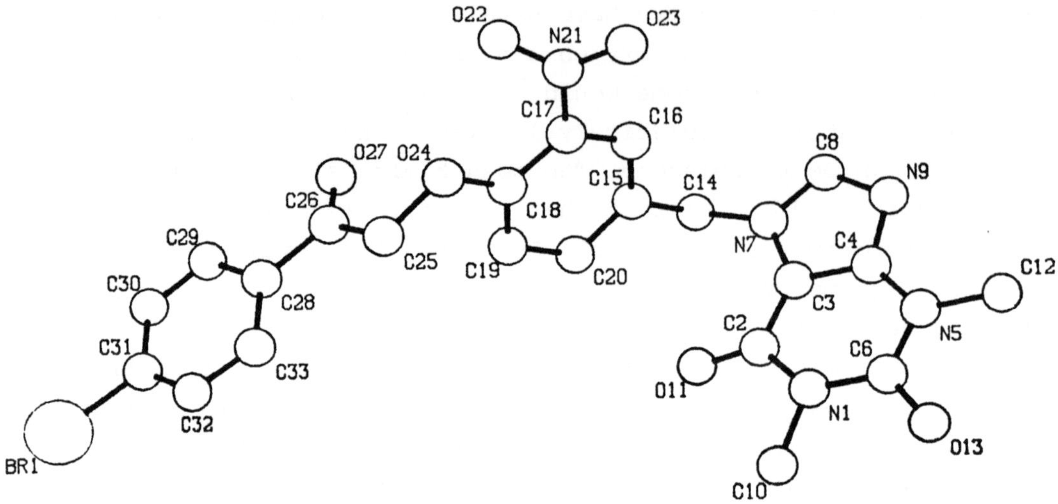

Figure 5. A computer generated perspective drawing of the anti-fungal agent isolated from Phidolopora pacifica. Hydrogens are omitted for clarity

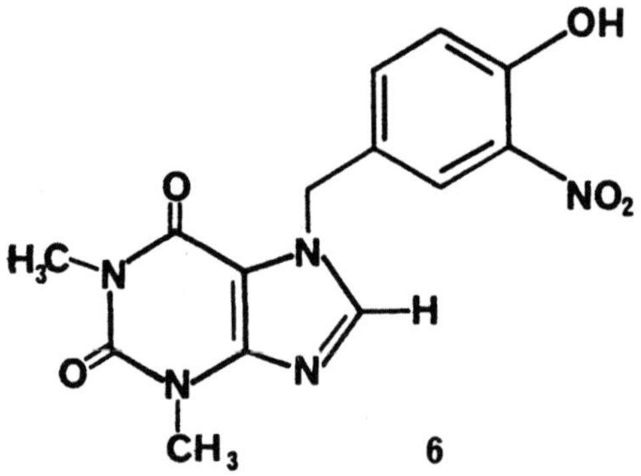

6

fragments of the molecule. The native material did not crystallize well – a common problem in natural products crystallography. A p-bromophenacyl derivative was prepared, and fortunately it crystallized nicely. The crystals belonged to the triclinic system with a=9.549(3), b=9.514(1), c=15.212(3) Å, α=72.69(1), β=82.95(2), and γ=81.87(2)°. The naturally occurring material had the formula $C_{14}H_{13}N_5O_5$, and the derivative, $C_{22}H_{18}N_5O_6Br$. From this information it is possible to calculate the density of the crystal, and the density of organic molecules is remarkably constant. Using a standard density led us to conclude that two molecules were in each unit cell. There are two triclinic space groups, and they differ by having either no point group symmetry or an inversion center. The usual case for a natural product would be the space group with no symmetry, since the inversion center would require a chiral molecule and its enantiomer to be present in equal proportion. Of course if the molecule were achiral, this argument is invalid, and the more probable choice would be the centrosymmetric space group. Difficulties in assigning space groups are relatively rare, but when they occur they can be very frustrating. If the initial determination of the space group is not correct, all subsequent calculations can be wasted. We guessed that the molecule would be achiral, and this turned out to be correct. Our guess was based on the molecular formula, which suggested that the molecule was highly unsaturated, and on the fact that it was the easier of the two choices. A computer generated perspective drawing is shown in Figure 5 and a chemical drawing as 6. As expected the molecule contains both a modified purine and an aromatic piece. The nitro group and the C-nucleoside bond are noteworthy. A simple model of the aromatic portion, *m*-nitrophenol, is mildly toxic with an oral LD_{50} of 1.3 g/kg in mice. Compound 5 may be the bryozoan's version of the old-fashioned antimicrobial carbolic acid.

Novel $(C_5)_n$ Compounds from *Laurencia chilensis*

Many of the unusual compounds that first indicated the exciting chemistry to be discovered in marine natural products came from red algae of the genus *Laurencia*. Many workers believed that this source had been well explored, and that profitable future investigations would deal with other organisms. We recently began a collaborative program that indicates that there is still interesting chemistry in *Laurencia*. Two independent groups in Chile, directed by Professor A. San Martin and M. Silva, began an investigation of *L. chilensis*. All of the new compounds had formulas with multiples of C_5, but the spectral data could not be reconciled with known terpenes. In fact the spectra of these relatively small compounds did not clearly suggest any well known class of natural products. I will briefly discuss two of them because they have led us into some interesting synthetic chemistry.

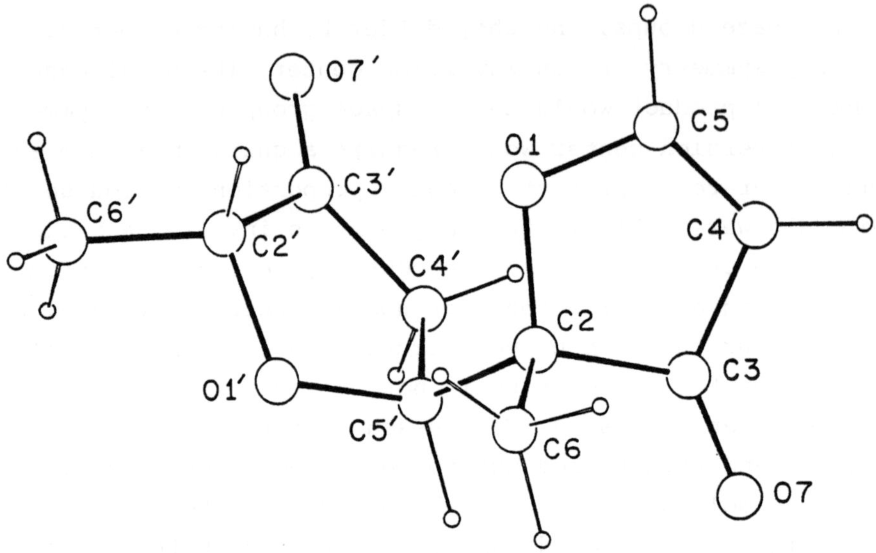

Figure 6. A computer generated drawing of the final x-ray model of chilenone A (7). The material is a naturally occurring racemate

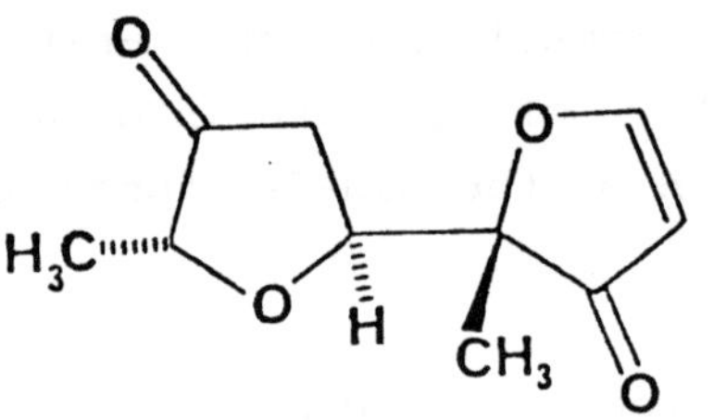

7

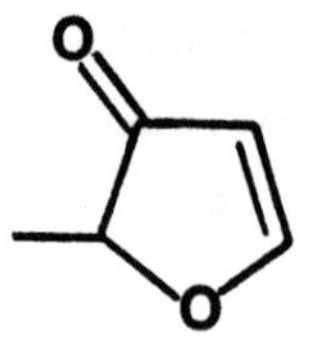

8

The first is chilenone A from A. San Martin's laboratory.[12] Chilenone A had the formula $C_{10}H_{12}O_4$ from mass spectrometry, but, despite this rather simple formula, it was difficult to unambiguously formulate a molecular structure based on the spectral data. Fortunately the molecule was nicely crystalline. The crystals were orthorhombic with a=7.916(2), b=24.676(4), and c=10.054(2) Å. The most interesting point at this stage of the analysis was that the space group was clearly indicated to be Pbca, which required that chilenone A either be achiral or a racemic mixture. The x-ray deduced structure is shown in Figure 6 and chemically as 7.

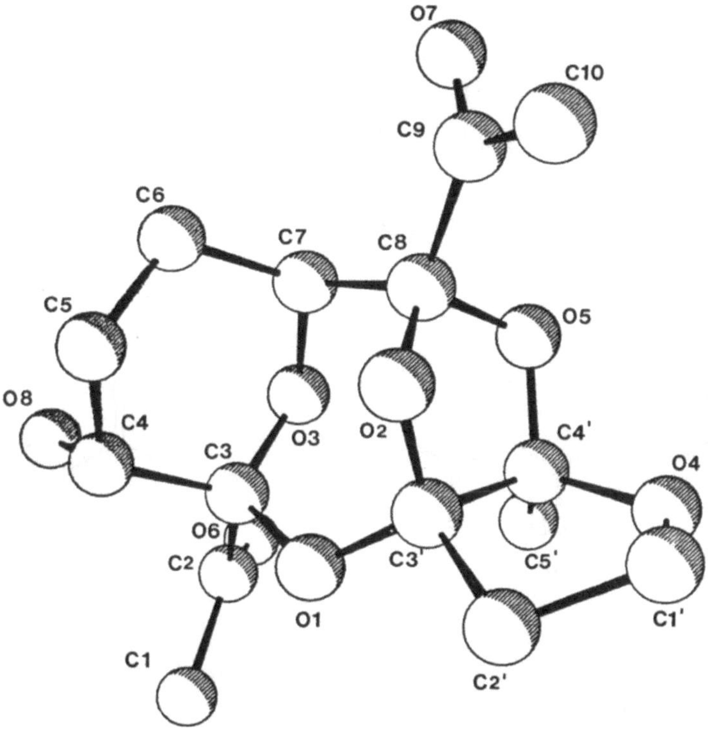

Figure 7. A computer generated perspective drawing of 9. Hydrogens are omitted for clarity and the material is a naturally occurring racemate

The structure of chilenone A is unusual and we are not aware of any published precedents. It could be very plausibly arise from the dimerization of 2-methyl-3(2H)-furanone (8) by a conjugate addition of the derived enolate. This proposed biogenesis is especially attractive because of the racemic nature of chilenone A and the trans relationship of the substituents on the saturated ring. A chemical synthesis based on this idea is proceeding nicely. The potential precursor (8) has been known since 1929 when it was prepared in poor yield from $CH_3CO(CHOH)_3CO_2H$.[13] We can find no reports of its isolation from a natural source. There are also several closely related metabolites whose structures we hope to elucidate through a combination of synthesis and spectroscopy.

The second Laurencia compound comes from M. Silva's laboratory and its formula, $C_{15}H_{20}O_8$, suggested that it might be

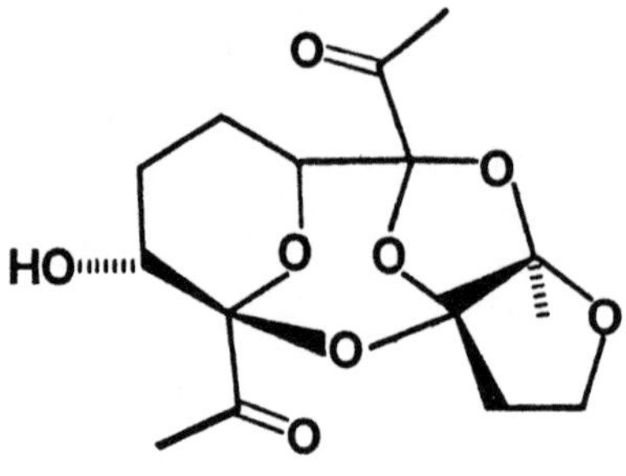

9

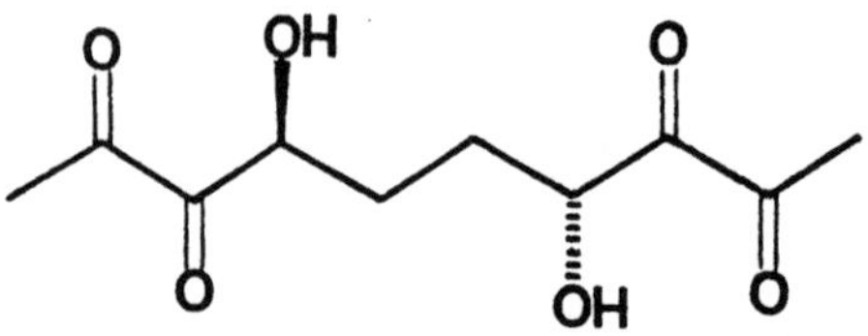

10

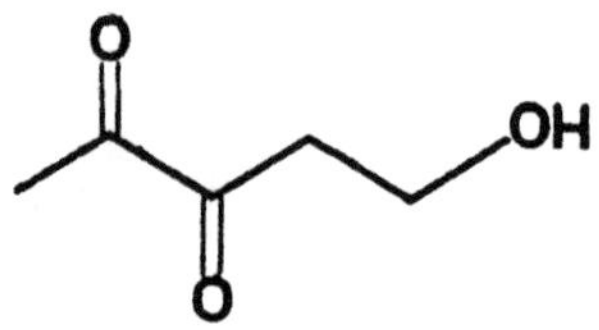

11

closely related. This material also crystallized in a space group that is appropriate for a racemate or an achiral molecule. The space group was $P2_1/c$ with a=16.224(3), b=6.157(1), c=15.757(3) Å and β=108.93(8)°. The x-ray derived structure is shown in Figure 7 and chemically as 9.

The remarkable tetraketal 9 is racemic in spite of the presence of six chiral centers. It can be formally dissected into the simpler components shown as the meso diol 10 and the achiral diketone 11. Since 10 and 11 are achiral, it is not surprising that 9 is also achiral. We are currently working on a synthesis of 9 patterned on this dissection. We do not know if there is any biological activity associated with these Laurencia chilensis metabolites, and this is also being investigated.

Conclusion

The possibilities of finding dramatically new and challenging metabolites in marine organisms are far from exhausted. X-ray diffraction is a very efficient way to determine their structures. It is also a great deal of fun.

Acknowledgment: This work would not have been possible without enthusiastic collaborators in other laboratories, skilled and hardworking associates at Cornell, and research funds. I have indicated my outside collaborators in the text. The work at Cornell was done by Gayle K. Matsumoto, Marie H.M. Chen, and Greg Van Duyne. It has been my great pleasure to work with each of them. The research was supported by NIH CA24487-01 to 06 and the New York State Sea Grant.

References

1. J.S. Mynderse, R.E. Moore, M. Kashiwagi, and T.R. Morton, Science 196, 538 (1977).
2. Y. Kato and P. Scheuer, J. Am. Chem. Soc. 96, 2245 (1974).
3. J.S. Mynderse and R.E. Moore, J. Org. Chem. 43, 2301 (1978).
4. L. Eliasson, B. Kallin, M. Patarroyo, G. Klein, H. Fujiki, and T. Sugimura, Int. J. Cancer 31, 7 (1983).
5. R.E. Moore, A.J. Blackman, C.E. Cheuk, J.S. Mynderse, G.K. Matsumoto, R.W. Woodard, and J.C. Craig, J. Am. Chem. Soc., manuscript submitted.
6. Y. Shimizu, 'Marine Natural Products' vol. 1, P.J. Scheur Ed., Academic Press, New York, 1978, pp. 1-42.
7. T. Yasumoto, Y. Oshima, W. Sugawara, Y. Fukuyo, H. Oguri, T. Igarashi, and N. Fujita, Bull. Jpn. Soc. Sci. Fish. 46, 1405 (1980).
8. M. Murata, M. Shimatani, H. Sugitani, Y. Oshima, and T. Yasumoto, Bull. Jpn. Soc. Sci. Fish. 48, 549 (1982).
9. K. Tachibana, P.J. Scheuer, Y. Tsukitani, H. Kikuchi, D. Van Engen, J. Clardy, Y. Gopichand, and F.J. Schmitz, J. Am. Chem. Soc. 103, 2469 (1981).
10. T. Yasumoto, M. Murata, Y. Oshima, M. Sano, G.K. Matsumoto, and J. Clardy, Tetrahedron, 1984, manuscript submitted.
11. R.R. Izac, S.E. Poet, W. Fenical, D. Van Engen, and J. Clardy, Tetrahedron Lett. 23, 3743 (1982).
12. A. San Martin, J. Rovirosa, O. Muñoz, M.H.M. Chen, R.D. Guneratne, and J. Clardy, Tetrahedron Lett. 24, 4063 (1983).

Biosynthesis of Riboflavin

Heinz G. Floss[1], Paul J. Keller[1], Quang Le Van[2], Peter Nielsen[2], Gerhard Neuberger[2], and Adelbert Bacher[2]

[1]Department of Chemistry, The Ohio State University, Columbus, Ohio 43210, USA
[2]Lehrstuhl für Organische Chemie und Biochemie der Technischen Universität München, D-8046 Garching, FRG

Riboflavin (III, Fig. 1) was discovered 50 years ago by Kuhn and coworkers (1) as an essential vitamin; it is today produced in large quantities industrially, both by chemical synthesis and by fermentation, and is widely used as a food additive and as part of most vitamin supplements. Considering these facts it is surprising that the biosynthetic origin of almost half of the carbon atoms in the molecule is very poorly understood.

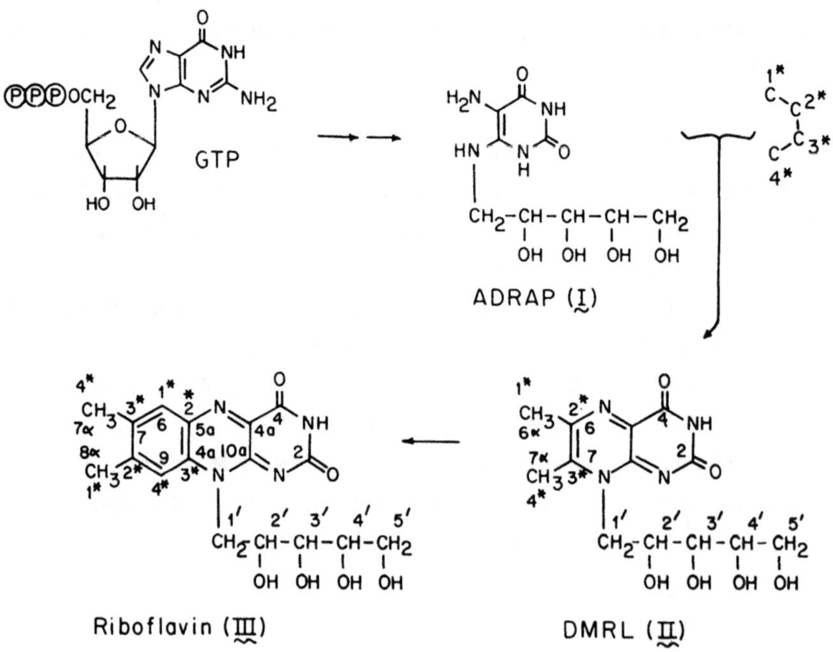

Figure 1. Conversion of guanosine triphosphate into riboflavin

Natural Product Chemistry
Ed. by Atta-ur-Rahman

As summarized in Figure 1, riboflavin arises from guanosine triphosphate by a reaction sequence which is initiated by the enzyme cyclohydrolase. Opening of the imidazole ring with expulsion of carbon 8 and removal of two of the phosphate groups is followed by replacement of the nitrogen function at C-2 by an oxygen function and by reduction of the ribosylamino to the ribitylamino linkage to give the key intermediate I as its phosphate ester. This compound or its dephosphorylation product then condenses with an unknown substrate which provides four carbon atoms to give dimethylribityllumazine (II). The enzyme riboflavin synthase finally reacts 2 molecules of II to produce one molecule of riboflavin with regeneration of a molecule of I. In this process the 4-carbon unit of one molecule of II is transferred to a second molecule, with the regiochemistry shown, to give the o-xylene ring of III (for reviews see ref. 3-5). In order to facilitate discussion of biosynthetic data an arbitrary numbering system (1* through 4*) for the carbons of the 4-carbon unit was devised.

The 8 carbon atoms of the o-xylene moiety of III are thus derived by duplication of the 4-carbon moiety of II. However, despite extensive studies in a number of laboratories the origin of this 4-carbon unit has remained elusive. Various authors have proposed diacetyl (6), aceton (7), a tetrose (8,9), a pentose (10) or a hexose (11) as the ultimate precursor of this 4-carbon moiety. Recently, it has been suggested that the pyrazine ring of II arises by donation of the ribityl moiety of I to a second molecule of I to give, after removal of one carbon atom, II (12,13). This hypothesis is formally analogous to the known formation of III from II; it is, however, at variance with earlier results showing that [ribose-^{14}C]GTP only labels the ribityl moiety but not the ring system of III (14).

In view of this multitude of conflicting hypothesis on the origin of the 4-carbon unit, we decided to take a new look at the problem using an approach which so far had not been brought to bear on this question. We chose to feed to cultures of a riboflavin-producing organism, the flavinogenic fungus Ashbya gossypii, a variety of general precursors labeled in single or multiple positions with ^{13}C and to analyze the resulting III by ^{13}C-NMR spectroscopy for the ^{13}C distribution and the ^{13}C-^{13}C coupling patterns. The ribityl moiety of III would serve as an internal reference, reflecting the labeling pattern of the pentose pool and, more generally, the carbohydrate pool; comparison of the labeling pattern in the o-xylene moiety with that in the ribityl portion, we hoped, would provide clues to the origin of the 4-carbon unit. In addition, the fact that the 4-carbon unit ap-

C-Atom	Chemical Shift(ppm)		$^{1}J_{C-C}{}^{a}$(Hz)
	Riboflavin[c]	Riboflavin tetraacetate[b]	
2	155.5	154.6	
4	159.9	159.3	77 (4a)
4a	136.8	136.0	77 (4); 53 (10a)
5a(2*)	134.0	134.5	67 (6); 59 (9a)
6(1*)	130.7	132.8	67 (5a); 61 (7)
7(3*)	135.7	136.9	44 (7a); 61 (6); 54 (8)
7α(4*)	18.8	19.4	44 (7)
8(2*)	146.0	148.0	43 (8α); 62 (9); 54 (7)
8α(1*)	20.8	21.4	43 (8)
9(4*)	117.4	115.5	66 (9a); 62 (8)
9a(3*)	132.1	131.1	66 (9); 59 (5a)
10a	150.8	150.6	53 (4a)
1'	47.3	44.9	42 (2')
2'	68.8	69.3	41 (1'); 43 (3')
3'	73.6	70.4	46 (4'); 43 (2')
4'	72.8	68.9	42 (5'); 46 (3')
5'	63.4	61.8	42 (4')
$\underline{CH_3}CO$		20.3 - 21.0	
$CH_3\underline{CO}$		169.7 - 170.6	

(a) Coupling partners are indicated in brackets. These data are compiled from several biosynthetically labeled riboflavin samples which we have examined.
(b) In $CDCl_3$.
(c) In DMSO-$\underline{d}_6$.

Figure 2. ^{13}C-NMR spectral data for riboflavin and riboflavin tetraacetate

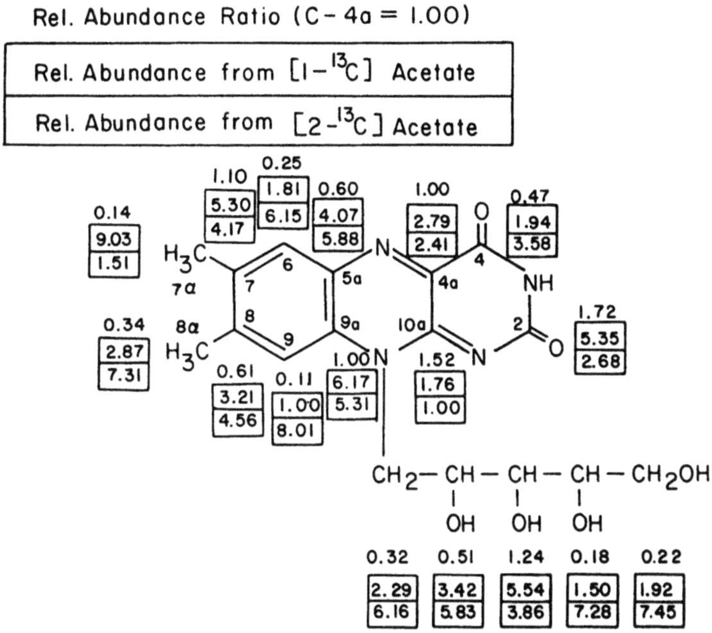

Figure 3. Labeling pattern of riboflavin from [1-^{13}C]- and [2-^{13}C] acetate

pears twice in the o-xylene moiety would allow a check on the accuracy of our quantitative analyses. For reasons of solubility and in order to provide a natural abundance reference signal we decided to convert III into its tetraacetate for ^{13}C-NMR analysis. As the first task we had to arrive at unambiguous assignments for all the signals in the ^{13}C-NMR spectrum of III-tetraacetate. This was accomplished (Figure 2) by comparison with literature data, by various two-dimensional analyses of multiple labeled samples and by a correlation between riboflavin tetraacetate and III itself, for which spectra of synthetic species labeled in specific positions were available (15,16).

The first general precursors we decided to feed were [1-^{13}C]- and [2-^{13}C]-acetate. The labeling patterns observed, summarized in Figure 3, indicate that as expected acetate is not a direct precursor. The agreement in the enrichments of pairs of equivalent carbons of the 4-carbon unit is quite good. In addition, simple pattern recognition indicates a correspondence of C-1* of the 4-carbon unit with C-1' of the ribityl moiety, C-2* correspond to C-2' and C-3* to C-3'. C-4* could correspond to either C-4' or C-5' of the ribityl moiety, which show virtually identical labeling (17). The results suggest a close biogenetic relationship of the 4-carbon unit to carbohy-

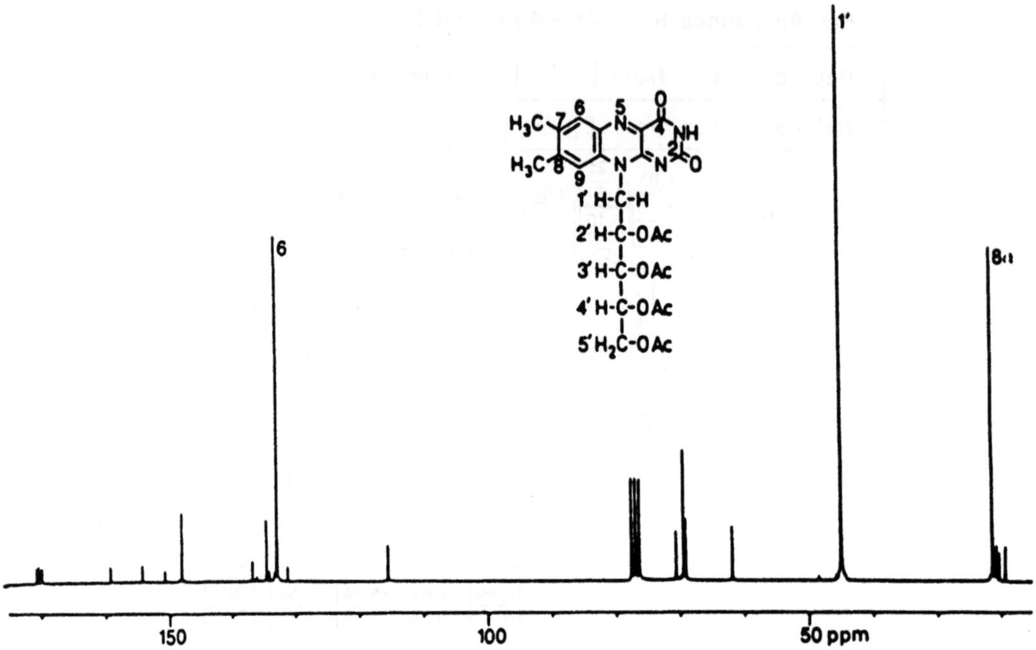

Figure 4. ^{13}C-NMR spectrum of riboflavin tetraacetate biosynthesized from [1-^{13}C-]ribose

drate, particularly pentose, metabolism. This relationship was immediately confirmed by feeding [1-^{13}C]ribose. The spectrum of the resulting III-tetraacetate (Figure 4) displayed heavy labeling at C-1' (20.3% enrichment), C-6 (10.2%) and C-8α (8.9%), indicating that C-1 of the pentose labels specifically C-1' of the ribityl moiety as well as the C-1* positions in the o-xylene moiety. Secondary labeling (via metabolic breakdown of the pentose) appears at C-2' and also in the C-2* positions. The results thus point to a precursor role of a pentose, or its biochemical equivalent, for both the ribityl and the 4-carbon moiety. At the same time the fact that C-1' and the C-1* positions show very different enrichments, well beyond the error range of the measurements, rules out the hypothesis (12,13) that two molecules of I react to give one molecule of II, one molecule of I donating its ribityl moiety to generate the 4-carbon unit. If this hypothesis were correct, C-1', C-6 and C-8α should have all shown the same enrichment (15).

To check if the 4-carbon unit comes from an intact carbohydrate chain, we next fed [U-^{13}C]glucose and we analyzed the ^{13}C-^{13}C coupling pattern of the resulting III. The rationale of such an experiment is based on the fact that the labeled glucose will be diluted with large amounts of ^{12}C-glucose from the culture medium upon addition to the fermentation. If in the en-

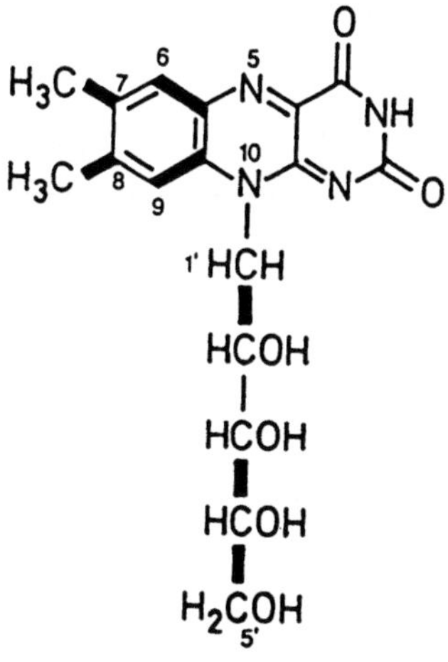

Figure 5. Homonuclear coupling pattern in riboflavin biosynthesized from [U-^{13}C]glucose

suing metabolism of the glucose a carbon-carbon bond is broken and a bond is subsequently resynthesized from the fragments, the probability of reconnecting two ^{13}C nuclei becomes very small because of the large dilution with ^{12}C. The absence of strong ^{13}C-^{13}C one-bond coupling between two connected carbon atoms derived from the same precursor indicates that the bond between these two carbon atoms has been broken during the conversion of precursor into product. Conversely, the presence of a strong one-bond ^{13}C homonuclear coupling indicates that the two carbons showing the coupling are derived from the same precursor molecule, usually by a process which preserves the bond between them. The coupling pattern which was deduced by one-dimensional, two-dimensional and double-quantum coherence NMR techniques is shown in Figure 5 (16,18). Clearly, the 4-carbon unit is derived from glucose in the form of two intact 2-carbon pieces corresponding to C-1*-C-2* and C-3*-C-4*. The absence of significant coupling between C-2* and C-3* at first glance would suggest that the 4-carbon unit is not derived from an intact pentose. However, such a conclusion is negated by the finding that there is also no coupling between C-2' and C-3' of the ribityl moiety, i.e., formation of pentose from the glucose proceeds with cleavage of the bond between the carbons giving rise to C-2 and C-3 of the pentose. Most likely, the hexose is shuttled through the glycolytic pathway several times before being drained off to produce pentose. A feeding experiment with [2,3-$^{13}C_2$]ribose is now underway to establish unequivocally whether the intact carbon chain of a pentose gives rise to the 4-carbon unit.

Relative Enrichments

Carbon atom	[1-^{13}C]Glucose	[6-^{13}C]Glucose	[2-^{13}C]Glycerol
2	1.1	1.6	5.4
4	1.2	1.2	1.4
4a	1.3	1.1	1.9
5a(2*)	2.4	2.2	15.9
6 (1*)	5.5	3.2	12.9
7 (3*)	1.4	1.6	3.0
7a(4*)	3.3	9.0	2.4
8 (2*)	1.8	1.9	12.1
8a(1*)	5.0	2.7	12.3
9 (4*)	3.4	9.0	2.1
9a(3*)	1.2	1.6	3.0
10a	0.7	0.9	1.1
1'	4.5	2.3	8.5
2'	1.8	1.6	7.4
3'	1.5	1.7	2.1
4'	1.9	1.9	12.9
5'	3.2	6.8	1.6
<u>C</u>H$_3$CO	1.0[a,b]	1.0[a,b]	1.0[a,b]
CH$_3$<u>C</u>O	0.9[a]	0.8[a]	0.9[a]

[a]Average of 4 signals.
[b]Natural abundance standard.

Figure 6. Labeling pattern of riboflavin from [1-^{13}C]- and [6^{13}C]glucose and from [2-^{13}C]glycerol

Conversion of a 5-carbon chain into a 4-carbon chain requires removal of one carbon atom; most logically this would be a carbon at the end of the chain. Since incorporation of C-1 of the pentose into C-1* has already been demonstrated, loss of C-5 would be the most plausible option. This notion was tested by feeding [1-^{13}C]- and [6-^{13}C]glucose (17). As expected based on the standard pathways of carbohydrate metabolism, [6-^{13}C]glucose labels predominantly C-5' of the ribityl moiety and to a lesser extent C-1' (Figure 6). In the o-xylene moiety these labels are reflected in highest en-

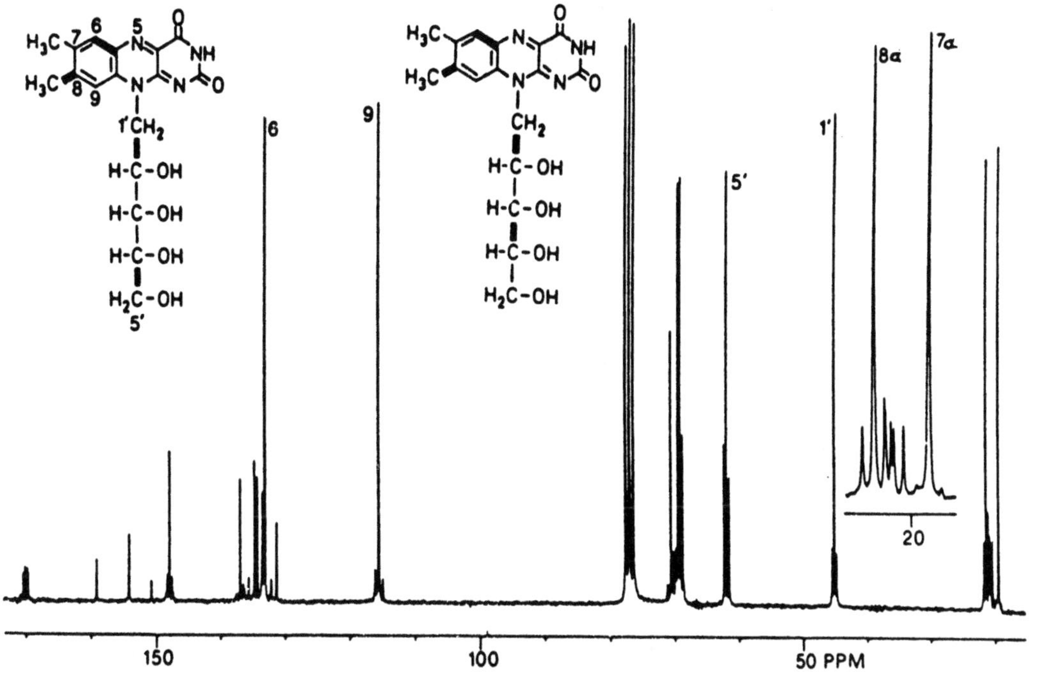

Figure 7. ^{13}C-NMR spectrum of riboflavin tetraacetate from [2,3-$^{13}C_2$]succinate

richments in the C-4* positions and additional labeling in the C-1* positions. Conversely, [1-^{13}C]glucose gives highest enrichments at the C-1* positions and secondary labeling at C-5' and C-4*. Hence it is clear that C-5' gives rise to C-4* and that a carbon atom corresponding to a position within the pentose chain must be eliminated.

Incorporation of C-2 of the pentose had already been demonstrated by the acetate and [1-^{13}C]ribose experiments and hence, either C-3 or C-4 must be the carbon atom that is removed. The acetate experiments also argued for retention of C-3. However, since the [U-$^{13}C_6$]glucose experiment had shown that C-3* and C-4* come from the same glucose molecule, presumably as an intact 2-carbon unit, and since an attractive mechanism could be written to account for removal of the carbon corresponding to C-3, we decided to test more rigorously the possibility that the 4-carbon unit is derived from the "top" (C-1 + C-2) and "bottom" (C-4 + C-5) carbons of a pentose or its biochemical equivalent. To this end we fed [2,3-^{13}C]succinate and determined the coupling patterns in the resulting III. Such a succinate sample is expected to give, via the Krebs cycle and phosphoenolpyruvate, a triose showing coupling

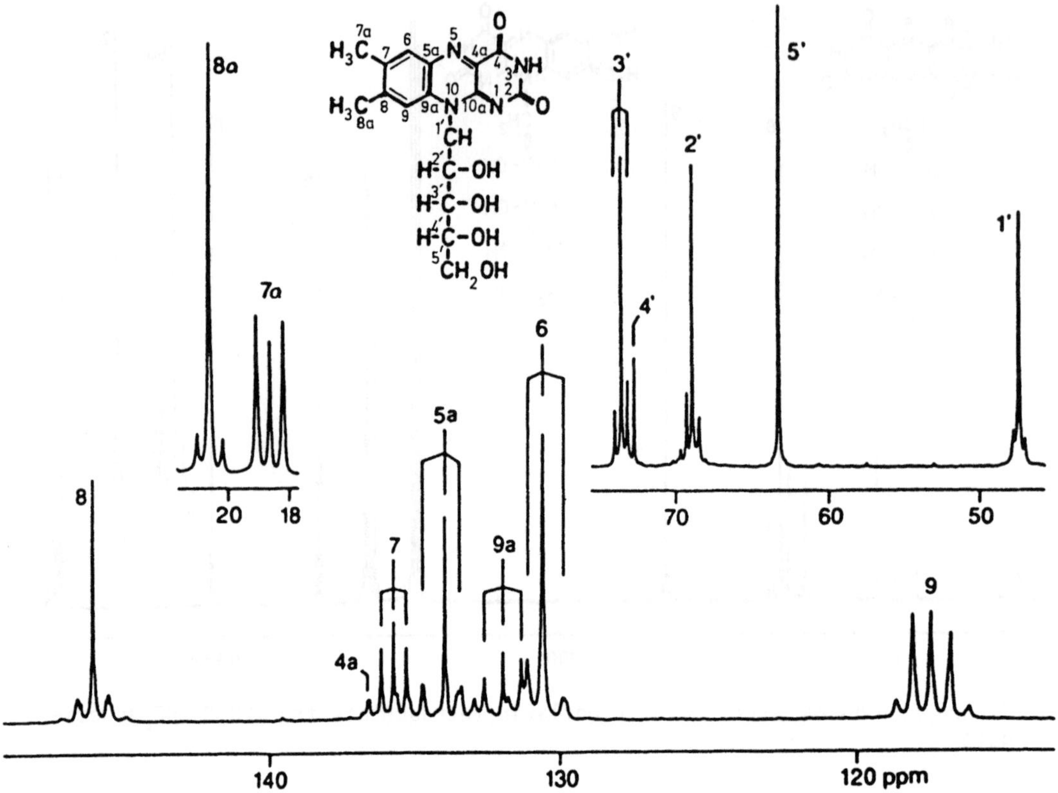

Figure 8. Selected regions of the ^{13}C-NMR spectrum of riboflavin tetraacetate from [1,3-^{13}C]glycerol

between C-2 and C-3 and between C-1 and C-2. Further metabolism of this triose will lead to a pentose showing coupling between C-4 and C-5, as well as additional coupling between C-3 and C-4 and C-1 and C-2. The spectrum of the riboflavin tetraacetate from this experiment (Figure 7) showed the predicted coupling pattern in the ribityl moiety, particularly the pronounced coupling between C-4' and C-5'. However, the o-xylene moiety showed only coupling between connected C-1*/C-2* positions, but not between C-3* and C-4* (16). This result unequivocally demonstrates that C-3* + C-4* are not derived from C-4 + C-5 of a pentose or its biochemical equivalent, and hence, that C-3 cannot be the carbon atom that is removed.

By way of exclusion the results point to the alternative that C-4 of the pentose is eliminated. However, since C-3* and C-4* in the [U-$^{13}C_6$]glucose experiment showed strong coupling, the process must occur in such a way that C-3 and C-5 are intramolecularly reconnected. Proof for this pathway comes from two experiments with labeled glycerol. In the first of these [2-^{13}C]-

Substrate (0.95 mM)	Riboflavin (uM)
none	1.6
Arabinose 5-phosphate	6.1
Fructose	1.7
Fructose 6-phosphate	1.9
Glucose	1.7
Glucose 6-phosphate	1.9
Ribose	1.6
Ribose 5-phosphate	10.6
Rubulose	1.5
Ribulose 5-phosphate	7.0
Xylulose 5-phosphate	9.8

Figure 9. Enzymatic synthesis of riboflavin from I and a precursor of the 4-carbon unit by cell-free extracts of C. guillermondii

glycerol was fed to give, as expected, heavy labeling at C-4' of III (Figure 6) (17), with additional label appearing at C-1' and C-2'. However, in the o-xylene moiety the heavy label of C-4 of the pentose did not result in heavy labeling of C-3* or C-4*; these reflect only the low levels of enrichment of C-3 and C-5 of the pentose, confirming that C-4 is indeed not incorporated into the 4-carbon unit. Finally, intramolecular reconnection of C-3 and C-5 of the pentose was demonstrated unequivocally by feeding [1,3-^{13}C]-glycerol (19). As expected the spectrum of the product (Figure 8) shows high enrichment at C-3' and C-5', but these signals exhibit no one-bond ^{13}C-^{13}C coupling since they are separated by the unlabeled carbon atom 4'. However, in the o-xylene moiety the two pairs of C-3*/C-4* carbons show pronounced one bond coupling.

The results with Ashbya gossypii thus point to a biosynthetic pathway in which the 4-carbon unit used to convert I into II is derived from a pentose, or its biochemical equivalent, by a process involving excision of C-4 and intramolecular reconnection of C-3 and C-5. Such a process is rather unprecedented; the perhaps closest analogies, the formation of apiose in higher plants and L-streptose in Streptomyces, involve a different reconnection of carbon atoms (20). Since there is evidence for variations in the early steps of the riboflavin biosynthetic pathway in different organisms (21), it is necessary to establish that the pathway demonstrated in the fungal system also operates in other organisms. We are therefore carrying out feeding experiments with a prokaryotic system, a mutant of Bacillus subtilis producing II. The results of a first experiment with [1-^{13}C]ribose parallel completely those in A. gossypii, i.e., II is heavily labeled at C-1' and at C-6α.

Shavlovskii and coworkers (22) recently reported a cell-free system from

Substrate	Specific activity (dpm/nmol)		
	Ribulose 5-phosphate	Riboflavin	Lumichrome
[1-^{14}C] Ribulose 5-phosphate	403	516	504
[3,5-^{14}C] Ribulose 5-phosphate	457	535	499
[4-^{14}C] Ribulose 5-phosphate	290	9.6	9.1
[5-^{14}C] Ribulose	194	238	220

Figure 10. Incorporation of ^{14}C from different positions of labeled ribulose-5-phosphate into III by cell-free extracts of *C. guillermondii*

Candida guillermondii capable of synthesizing riboflavin from I and a suitable sugar phosphate. We were able to reproduce and further extend their observations. As shown in Figure 9, only pentose phosphates serve as second substrate for the synthesis of III, hexose phosphates or free sugars are inactive. Incubations with ribulose-5-phosphate samples labeled in different positions with ^{14}C indicate incorporation of carbon atoms 1, 3 and 5, but not of C-4, into the riboflavin formed (Figure 10). Photolytic degradation to lumichrome showed that all the ^{14}C incorporated into III resides in the ring system. In a further experiment the cell-free system was incubated with I and [1-^{13}C]ribose-5-phosphate. The resulting III was shown by NMR spectroscopy to be labeled exclusively at the C-1* positions, C-6 and C8α. All these results are in agreement with a pathway to the 4-carbon unit paralleling that established in *A. gossypii*.

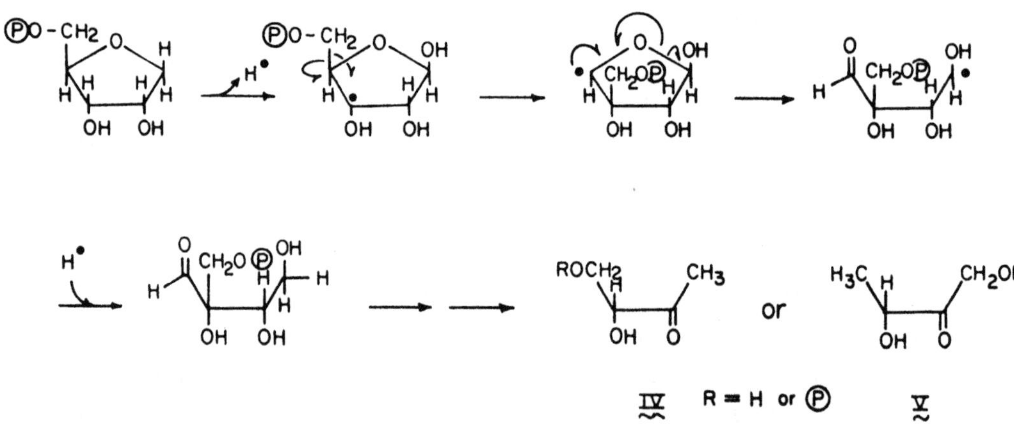

Figure 11. Hypothetical mechanism for the formation of the 4-carbon unit of II and III from a pentose phosphate

A considerable amount of work needs to be done to unravel the mechanism of the unusual reactions leading from a pentose phosphate to the 4-carbon unit of II and III. At this point it is not even clear whether the rearrangement process takes place before or after condensation with I. From a purely mechanistic point of view, compound IV or its isomer V would appear to be attractive candidates for the condensation with I to give II. Their formation from ribose-5-phosphate could be rationalized by the sequence shown in Figure 11. The rearrangement could be initiated by coenzyme B_{12} and proceed as a radical process, as shown, or it could be initiated by a pyridine nucleotide and proceed as a cationic process. Obviously, a great many possibilities are open at this point, and the only merit of such a hypothetical mechanism is that it provides a rationale for further experiments.

Acknowledgements. This work was supported by the USPHS National Institutes of Health, the Deutsche Forschungsgemeinschaft, and the Alexander von Humboldt Foundation. The services of the Los Alamos Stable Isotope Resource are greatfully acknowledged.

Note added in proof. Preliminary data from incorporation of [2,3-$^{13}C_2$]ribose into riboflavin by A. gossypii indicate that the pentose C2-C3 linkage is preserved in C-2*-C-3*.

References

1. R. Kuhn, P. György, T. Wagner-Jauregg, Ber. Dt. Chem. Ges. 66, 1034 (1933).
2. F. Foor, G.M. Brown, J. Biol. Chem. 250, 3545 (1975).
3. G.W.E. Plaut, Comprehensive Biochemistry 21, 11 (1971).
4. G.W.E. Plaut, C.M. Smith, W.L. Alworth, Ann. Rev. Biochem. 43, 899 (1974).
5. G.M. Brown, J.M. Williamson, Adv. Enzymol. 53, 345 (1982).
6. K. Bryn, F.C. Stormer, Biochim. Biophys. Acta 428, 257 (1976).
7. T.W. Goodwin, D.H. Treble, Biochem. J. 70, 14P (1958).
8. S.N. Ali, U.A.S. Al-Khalidi, Biochem. J. 98, 182 (1966).
9. W.L. Alworth, H.N. Baker, M.F. Winkler, A.N. Keenan, G.W. Gobel, F.L. Wood, Biochem. Biophys. Res. Commun. 40, 1026 (1970).
10. W.L. Alworth, M.F. Dove, H.N. Baker, Biochemistry 16, 526 (1977).
11. M.T. Jabasani, U.A.S. Al-Khalidi, Int. J. Biochem. 6, 735 (1975).

12. S.E. Bressler, D.A. Perumov, T.P. Chernik, E.A. Glazunov, Genetika 12, 83 (1976).
13. I.J. Hollander, J.C. Braman, G.M. Brown, Biochem. Biophys. Res. Commun. 94, 515 (1980).
14. B. Mailänder, A. Bachner, J. Biol. Chem. 251, 3623 (1976).
15. A. Bacher, Q. Le Van, M. Büher, P.J. Keller, V. Eimicke, H.G. Floss, J. Amer. Chem. Soc. 104, 3754 (1982).
16. P.J. Keller, Q. Le Van, A. Bacher, H.G. Floss, Tetrahedron 39, 3471 (1983).
17. A. Bacher, Q. Le Van, P.J. Keller, H.G. Floss, J. Biol. Chem. 258, 13431 (1983).
18. P.J. Keller, Q. Le Van, A. Bacher, J.F. Kozlowski, H.G. Floss, J. Amer. Chem. Soc. 105, 2505 (1983).
19. H.G. Floss, Q. Le Van, P.J. Keller, A. Bacher, J. Amer. Chem. Soc. 105, 2495 (1983).
20. H. Grisebach, Adv. Carbohydrate Chem. 35, 81 (1978).
21. G. Klein, A. Bacher, Z. Naturforschg. 35b, 482 (1980), and references therein.
22. E.M. Logvinenko, G.M. Shavlovskii, A.E. Zakal'skii, I.V. Zakhodylo, Biokhim. (Moscow) 47, 931 (1982).
23. P. Nielsen, G. Neuberger, H.G. Floss, A. Bacher, Biochem. Biophys. Res. Commun., in press.

Conversion of Anhydrovinblastine into Vinblastine in *Catharanthus roseus* Tissue Culture

Mashooda Hasan
Department of Chemistry, Quaid-i-Azam University, Islamabad, Pakistan

This paper deals with some of the efforts made [1,2] towards the partial synthesis of vinblastine 1, a dimeric indole alkaloid using Catharanthus roseus tissue cultures*. In a relatively short period of time several bis-indole alkaloids, especially those of C. roseus G.Don (periwinkle) have become highly valuable agents in chemotherapy. Actually 1 and vincristine 3 have been in clinical use for more than 20 years in the treatment of various leukaemias, Hodgkins disease and solid tumors (Fig.1). The natural abundance of these compounds is rather low, some g./ton of dried plant material. These compounds are consequently among the most expensive drugs on pharmaceutical market. It is therefore, no way surprising that a considerable research effort has been directed to the partial or total synthesis of these precious chemicals. The subject matter is rather vast, where new grounds are being broken by various groups working on different approaches.

As it is a well known fact that despite the development of impressive total syntheses, natural products synthesis is still a dilemma. Syntheses suitable for the technical production of natural products are only available in relatively few cases. Most total synthesis of complicated natural products

---

*This work was done in Professor A.I.Scott's laboratories, Department of Chemistry, University of Edinburgh, Scotland, U.K. in the year 1981-82 in collaboration with Dr. R.L. Baxter, Dr. Neil E. Mackenzie and Dr. W.R.McLauchlan. Author is grateful to the Royal Society for the grant of Bursary during this period.

Natural Product Chemistry
Ed. by Atta-ur-Rahman

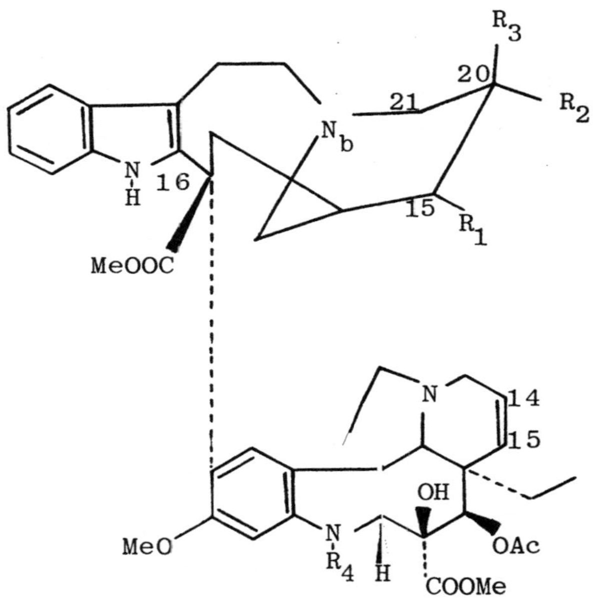

$\underline{1}$: R_1=H R_2=Et R_3=OH R_4=Me :Vinblastine

$\underline{2}$: R_1-R_2=O R_3=Et R_4=Me :Leurosine

$\underline{3}$: R_1=H R_2=Et R_3=OH R_4=CHO :Vincristine

$\underline{4}$: R_1=H R_2=OH R_3=Et R_4=Me :Leurosidine

$\underline{5}$: R_1=Et, $\Delta^{15'(20')}$ R_3=H, R_4 = Ac, R_5=CH_3 :Anhydrovinblastine

Fig.1

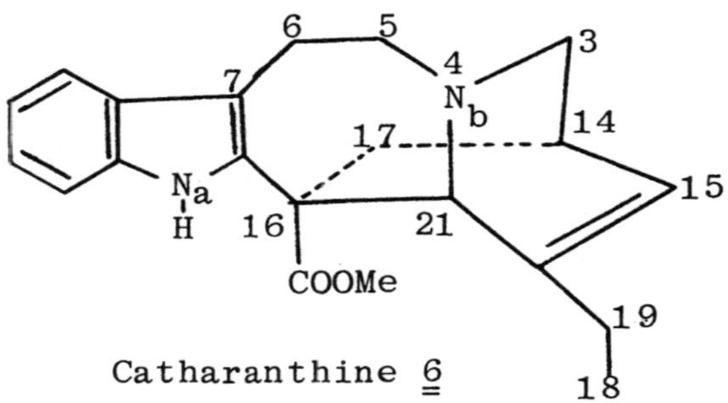

Catharanthine $\underline{\underline{6}}$

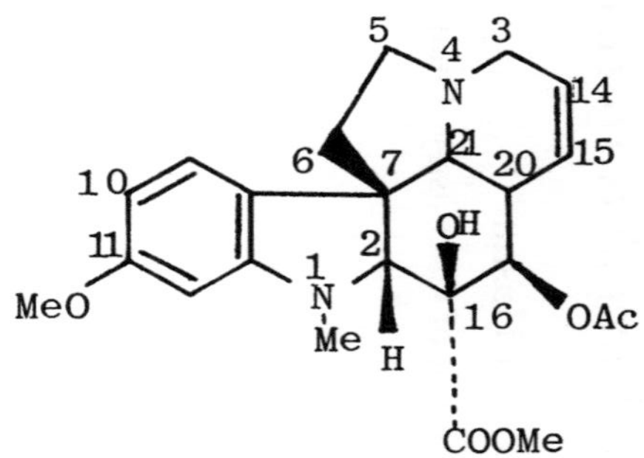

Vindoline $\underline{\underline{7}}$

Fig.2

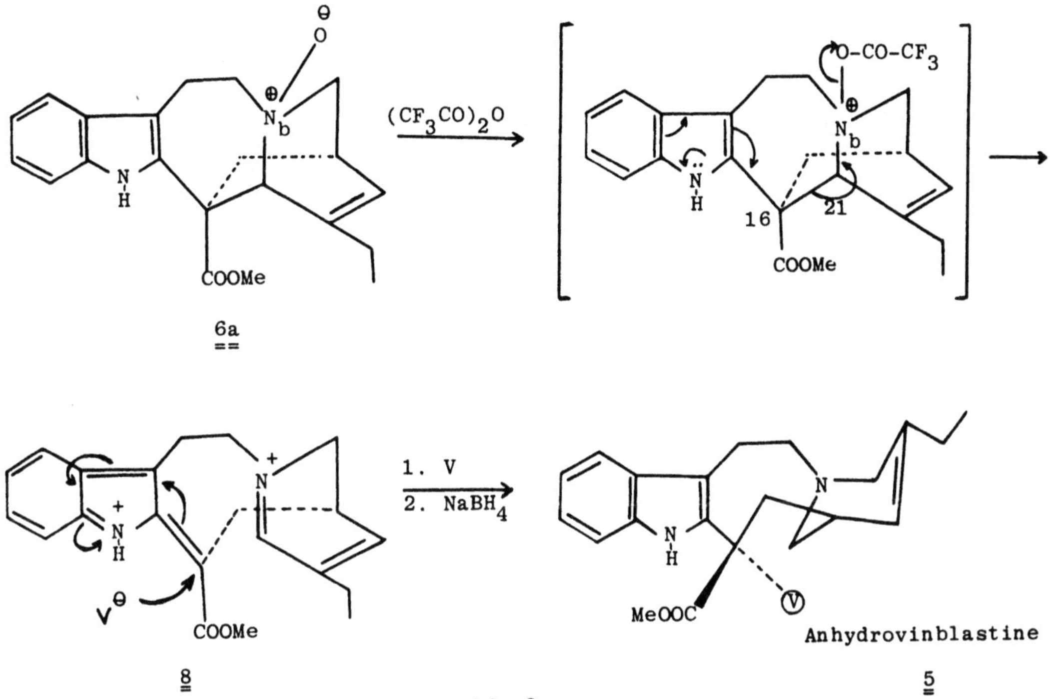

Fig. 3

require so many reaction steps that the total yield becomes vanishingly low. The requirement of high yield can only be met if natural product syntheses are based on biosynthetic models. Such biomimetic syntheses are generally characterized by a small number of steps and a high total yield.

Working on this principle, it was initially assumed that vinblastine type compounds could well be derived from their two obvious precursors - catharanthine 6 and vindoline 7 (Fig. 2), which are the major monomeric tertiary bases present in the aerial parts of various C.roseus spp. It has been experimentally demonstrated that biosynthesis of 1 proceeds in nature by the enzymic coupling of 7 and 6 leading [3] to anhydrovinblastine 5 which is later on enzymically converted [4] to 1. Since 7 and 6, which are readily available as the major alkaloids of C.roseus, can be chemically coupled [5] by the modified Polonovski reaction and reduced to 5 in good yield (Fig.3) which could later be chemically converted to 1 (Fig. 4), it was considered that enzymic hydration of 5 may prove important as a final step in the future production of the dimeric antitumour alkaloids. Positive results have been achieved [6] in this direction using cell free

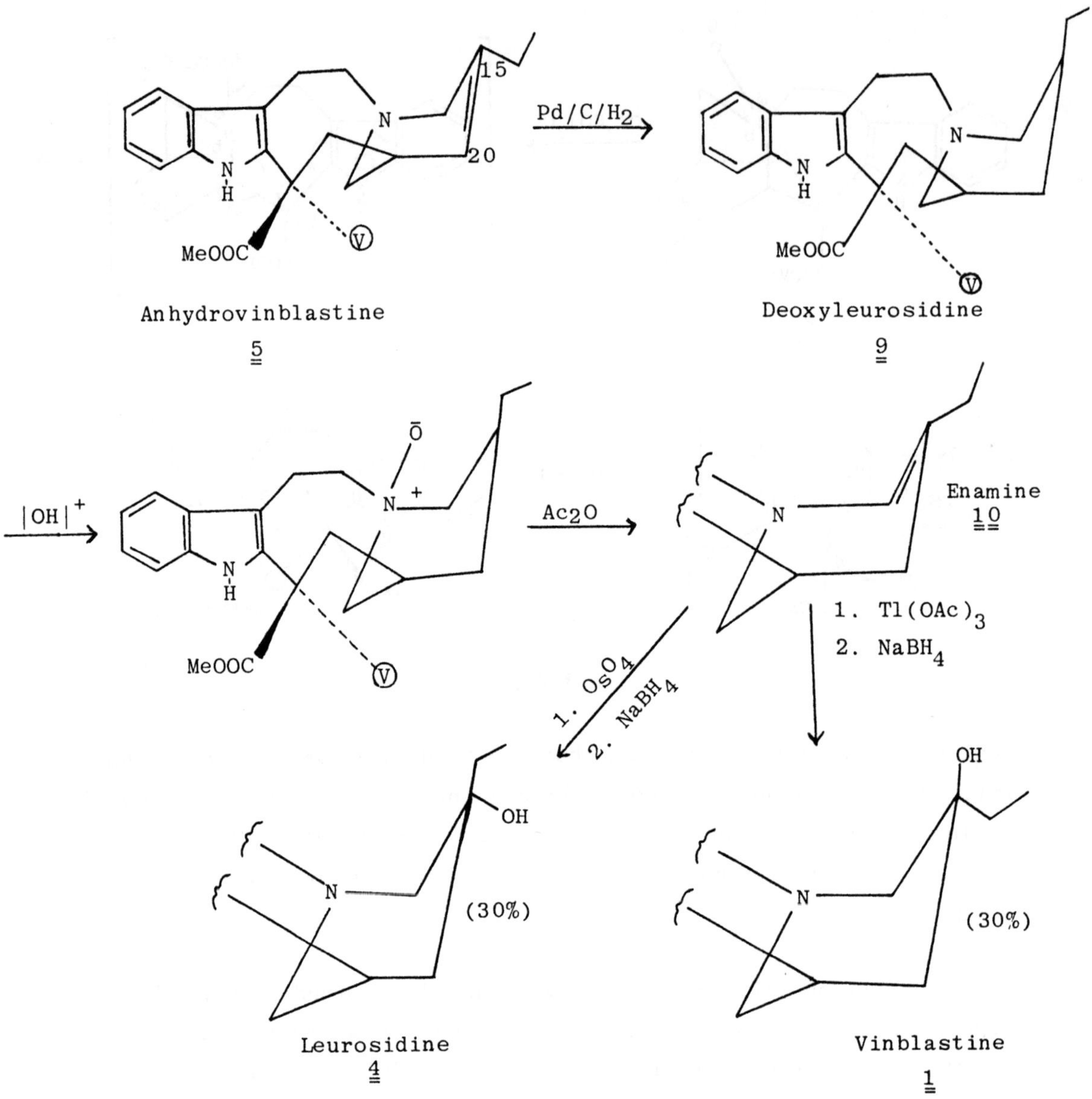

15
20
N
N-H
MeOOC
V
Pd/C/H2
Anhydrovinblastine
5
Deoxyleurosidine
9
|OH|+
Ō
N+
Ac2O
Enamine
10
1. Tl(OAc)3
2. NaBH4
1. OsO4
2. NaBH4
OH
(30%)
Leurosidine
4
Vinblastine
1

Fig.4

extract of the whole plant. The second difficulty lay in the rapid inactivation of the biosynthetic enzymes by the phenolic compounds when the cells are ruptured. The tissue culture of C.roseus produces approximately ten alkaloids, mainly of Corynanthe type, greatly simplifying the analytical problem. At the same time, a low abundance of phenolic compounds contributes to a higher retention of enzyme activity. These advantageous factors contributed to the disclosure of a method [8] for the cell free study in C.roseus in 1975, which not only led to substantial understanding of the early stages of alkaloid biosynthesis, but as promising cell lines were selected, provided a burgeoning source of later alkaloids, for example 6 [9]. It has also been observed that 7, which is the major alkaloid present in the whole plant has not been isolated from any of the culture strains, although a feeding of tryptophan to the suspension cultures indicated the production [9] of this alkaloid.

Encouraged by these results, it was considered worthwhile to test the capability of these cell lines to produce the dimeric alkaloids, particularly 1. It may be mentioned here that dimeric alkaloids have not been detected in any cell - line developed so far. For this purpose, tritiated 5 [6], which is the precursor of 1 was fed to the cell - free homogenate derived from C.roseus (strain EU 4A) suspension culture.

Results and Discussion

$21^1\alpha$ - ^3H, 5 was prepared [5] by the modified Polonovski coupling of 6 Nb oxide 6 a and 5 and reduction of the resultant imine 8 with sodium borotritide.

When radiolabelled 5 was administered to suspension cultures of C.roseus, virtually no uptake of the alkaloid into the cells could be detected. TLC radioscans of the dichloromethane extracts of the medium 6 hours after administration showed that almost all of the added alkaloid had been degraded to materials the principals of which coeluted with leurosine 2 and catharine. Oxidation of 5 to 2 in aerated solvents has been described previously [10]. Faced with the fact that the rate of degradation of the precursor greatly exceeded uptake by the cells, it was necessary to examine the possible conversion in cell - free extracts where transport of 5 would not be a rate determining factor.

Accordingly $21^1\alpha$- ^3H 5 was fed to a cell - free homogenate of suspension culture cells of C.roseus, (EU4A) prepared in a similar manner to that used for leaf tissue [11]. At the end of a 3 hour incubation 1 - sulphate was added as carrier and the 1 reisolated and rigorously purified by successive TLC, reversed phase HPLC and crystallisation to constant specific activity. The results of two independent experiments carried out with different batches of cells and independent preparations of $21^1\alpha$ - ^3H-5 are shown in Table II. No incorporation of activity was detected when the homogenates were boiled for 15 minutes prior to incubation showing that no non-enzymic hydration of the 15^1-20^1 double bond of 5 occurs.

Methanolysis of the isolated 1 afforded the desacetyl derivative which was purified by TLC and subsequently reacetylated with Ac_2O/pyridine to give 1 which showed no loss of activity, thus precluding a possible exchange of label via chemical degradation of AVLB and reincorporation of tritium into the acetyl moiety. Exchange of the acetyl group of AVLB in homogenates from leaf - tissue had been shown to occur in earlier studies when $21^1\alpha$ - ^3H-acetyl - ^{14}C - 5 was employed as a substrate [13]. Enzymic deacetylation of 6 by suspension cultures of C.roseus has recently been reported [12].

The results clearly show that the enzymatic activity responsible for the formal hydration of 5 to 1 is present in this strain of C.roseus cell culture despite the fact that no dimeric alkaloids are produced in detectable amounts by cell - cultures. At this stage it is not known whether the pathway to 1 is blocked at an earlier stage by the absence of specific enzymes or whether the capability is present but repressed in some manner.

Prior to my work [1] on the intact conversion of doubly radiolabelled 5 into 1, two biosynthetic routes were proposed for the conversion of 5 into 1 (Fig. 5). First one, path d was suggested by Potier [13] on the

Table I

Incorporation of anhydrovinblastine into vinblastine by cell-free extracts[a] of <u>C.roseus.</u>

Expt.	Precursor	AVLB ^3H	fe(D.P.M.) ^{14}C	$^3H/^{14}C$ ratio	VLB isolated (d.m.p.)[e] ^3H	^{14}C	$^3H/^{14}C$ ratio	incor
1	[21'α-^{3}H],methyl-^{14}C AVLB	$5.03x10^6$	$7.37x10^5$	6.83=0.05	$8.59x10^4$	$1.23x10^4$	6.98=0.24	1.67
2	[21'α-^{3}H] methyl-^{14}C AVLB	$5.70x10^6$	$5.70x10^6$	8.25=0.05	$3.69x10^4$	$4.56x10^3$	8.10=0.30	0.8
3[b]	[21 α-^{3}H][d] AVLB	$8.24x10^7$	-	-	$1.32x10^4$	-	-	0.02

a. Cell-free extracts were prepared as described (A.I.Scott and S.L.Lee, J.Am.Chem.Soc., 1975, <u>97</u>, 6906). Approximately 10 g fresh leaves/10 ml of 0.05 M tris-maleate buffer (pH 7.0) were used in expts 1 and 3. The concentration of plant material was halved in expt 2.

b. Extract boiled for 5 min. prior to feeding. c. Specific activities 4.01 mCi $mmol^{-1}$ for ^3H; 0.22 mCi $mmol^{-1}$ for ^{14}C. d. Specific activity 38.25 mCi $mmol^{-1}$. e. At the end of each expt vinblastine sulphate (8-10 mg) was added, the extract was adjusted to pH 11 (NH_4OH) and extracted (CH_2Cl_2); AVLB and VLB were separated by t.l.c., VLB was further purified by h.p.l.c. and crystallised to constant activity as its 0.5 MeOH: 0.5 Et_2O solvate. f. No adjustment was made for recovered precursor.

Table-II

Expt.	21' - ^{3}H AVLB administered		VLB isolated[a]		
	dpm	dpm/mg	dpm[b]	dpm/mg	Incorporation %
1	6.60×10^6	1.12×10^7	3.79×10^4	4062[d]	0.57
2[c]	6.21×10^6		0.98×10^3	105[d]	0.02
3.	2.91×10^7	1.34×10^8	1.41×10^5	6060[e]	0.49
4[c]	2.92×10^7		3.73×10^3	160[e]	0.01

(a) as 1/2MeOH ½Et_2O solvate; (b) mean of constant activity samples each counted to a 20 error of 3; (c) extracts boiled for 15 minutes prior to incubation; (d) 10 mg of VLB H_2SO_4 added as carrier; (e) 25 mg of VLB H_2SO_4 as carrier.

basis of synthetic analogy, since 20'-deoxyleurosidine 9 was found to be incorporated [14] into 1 by C.roseus plant. Another route (path e, Fig. 5) which was equally compatible with the experimental evidence available at that time was the existence of a biogenetic gride [14] in which conjugated immonium salt derived from a Polonovski type reaction of 7 and 6 a is in equilibrium with the 1,2-reduction product 5 and the 1,4-reduction product 10. Hydration of 10 could then give rise to 1. A feature common to both of the suggested pathways from 5 to 1 is the loss of one of the 21' hydrogens of 5, by a process which could be mediated by the biological equivalent of Polonovski elimination of the corresponding N-oxide [1]. If this were the case, then loss of the 21'α-hydrogen should be expected, as only this hydrogen can adopt an antiperiplanar orientation relative to the oxygen of 5-N-oxide. To test the possible intervention of an N-oxide (albeit indirectly) it was required to prepare 5 stereospecifically labelled with ^{3}H in the 21'α position. Inspection of models of the coupling reaction product (immonium salt) indicated that the steric congestion of the β-face of the dihydropyridinium ring might force hydride reduction to occur predominantly from the less hindered α-face. Treatment of the

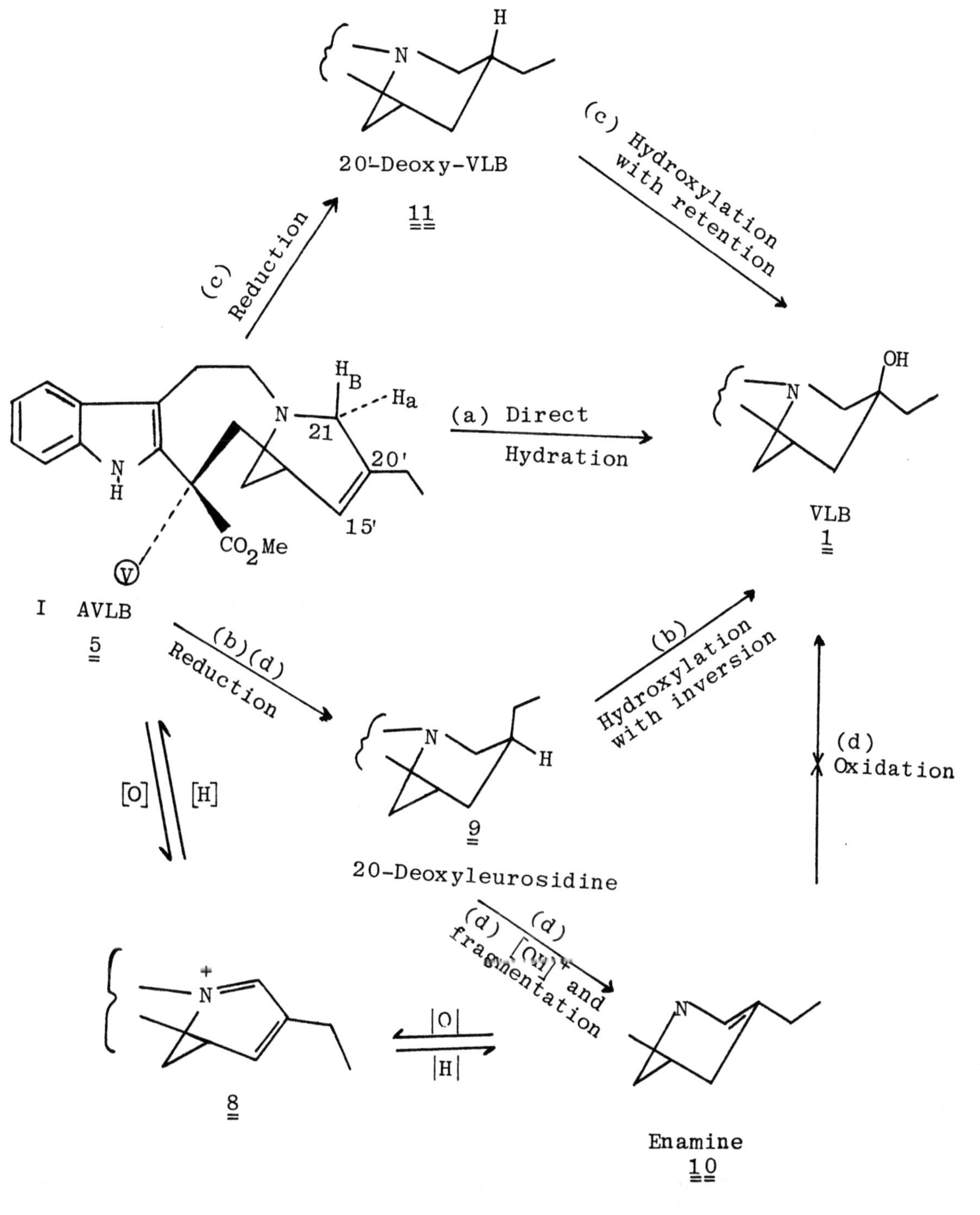

H
N
20'-Deoxy-VLB
11
(c) Reduction
(c) Hydroxylation with retention
OH
N
VLB
1
H
B
Ha
N
21
20'
15'
N-H
CO2Me
V
(a) Direct Hydration
I AVLB
5
(b)(d) Reduction
(b) Hydroxylation with inversion
N
H
9
20-Deoxyleurosidine
(d) Oxidation
[O]
[H]
(d) [OH]+ and fragmentation
(d)
N
N
8
Enamine
10

Fig. 5

immonium salt with $NaB(^2H)_4$ in methanol afforded monodeuterio-5, the ^1H n.m.r. spectrum (360 MHz) of which showed absence of the 21'α H doublet at δ 3.52 and the collapse of the 21' β-H doublet at δ 3.27 (J = 16 MHz) to a singlet. 21'α - ^3H 5 was prepared in a similar manner using $NaB\,^3H_4$.

Administration of 21'α- ^3H - methyl - ^{14}C - 5 ($R^5 = {}^{14}CH^*_3$) to cell free extracts of mature C.roseus leaves followed by isolation afforded 1 with no significant change in $^3H/^{14}C$ ratio (experiments I and II, Table I), showing that the 21' α-^3H was retained in the transformation.

There are several pathways which are compatible with these results, indicating the possibility that different paths suggested by earlier workers may not be interconnected. These possible pathways are shown in Fig.5, which are: (i) direct hydration of 5 (path a), (ii) reduction of 5 to 9 and hydroxylation with inversion of configuration (path b), (iii) reduction of 5 to 20' - deoxyvinblastine 11 and subsequent hydroxylation with retention of configuration (path c).

There is a possibility that the mechanism of this enzymic process is different in the two systems, i.e. in the whole plant and cell culture. After elucidation of the regulatory mechanism of this biosynthetic process, it will be possible to isolated and purify the enzyme or enzyme systems responsible for this process. Enzymes which are 100% efficient and stereo-specific will then lead to the production of VLB in larger amounts, thus solving the problems faced by pharmaceutical chemists today.

---

*Methyl - ^{14}C vindoline ($R^5 = {}^{14}CH_3$) was prepared by treatment of vindolic acid with ^{14}C diazomethane, coupling[5)] of the resultant methyl - ^{14}C vindoline ($R^5 = {}^{14}CH_3$) with catharanthine-N_b oxide and reduction of the product with $NaBH_4$

References

1. Baxter, R.L., Hasan, M., Mackenzie, N.E., and Scott, A.I., **J.Chem.Soc.,Chem.Commun.,** 791-793 (1982)

2. McLauchlan, W.R., Hasan, M., and Baxter, R.L., and Scott, A.I. **Tetrahedron**, Vol. 39, No.22, 3777-3780 (1983)

3. Scott, A.I., Gueritte, F., and Lee, S.L., **J.Am.Chem.Soc.**, 100, 6253 (1978)

4. Scott, A.I., Lee, S.L., Culver, M.G. Wan, W., Hirata, T., Gueritte, F.Baxter, R.L., Nordlov, H., Dorschel, C.A. Mizukami, H., and McKenzie, N.E., **Heterocycles**, 15, 1257 (1981)

5. Potier, P., Langlois, N., Langlois Y., and Guiritte, F., **J.Am.Chem.Soc.**, 98, 7017 (1976)

6. a) Baxter, R.L., Dorschel, C.A.; S-L. Lee, and A.I.Scott, **J.Chem.Soc.,Chem.Commun.**, 257 (1979)

   b)Stuart K.L., Kutney, J.P.Honda, T., and Worth, B.R., **Heterocycles**, 9, 1391 (1978).

7. Scott, A.I., Lee, S-L., Culver, M.G., Wan, W., Hirata, T.,Gueritte, F., Baxter, R.L., Nordlov, H., Dorschel,C.A.,Mizukami, H., and Mackenzic, N.E., **Heterocycles**, Vol. 15, No.2, 1257 (1981).

8. Scott, A.I., and Lee, S-L., **J.Am.Chem.Soc.**, 97, 6906 (1975)

9. Scott.A.I., Mizukami, H.Hirata, T. and Lee, S-L., **Phytochemistry**, 19, 488 (1980)

10. Langlois, N., Potier, P., **J.Chem.Soc., Chem.Commun.**, 102 (1978)

11. Scott, A.I., Lee, S-L., **J.Am.Chem.Soc.**, 97, 6906 (1975)

12. Kutney, J.P., Choi, L.S.L., Kolodziejczyk, P., Sleigh, S.K., Stuart, K.L., Worth, B.R., Kurz, W.G.W., Chatson, K.B., and Constabel, F. **Helv.Chim.Acta.**, 64, 1837 (1981)

13. Mangeney, P., Andriamialisoa, R.Z., Langlois,N., Langlois, Y., and Potier, P., **J.Am.Chem.Soc.**, 101, 2243 (1979)

14. Gueritte, F., Bac., N.V., Langlois, Y., and Potier, P., **J.Chem.Soc., Chem.Commun.**, 452 (1980).

Chemistry of Pakistani Artemisias

Syed Khaqan Hasan

Pharmaceutical and Fine Chemicals Research Division, P. C. S. I. R. Laboratories, Karachi-39, Pakistan

Artemisia (Family Compositae) is a large genus of small herbs and its 280 species are found in Northern hemisphere in arid regions [1]. Artemisias are bitter aromatic herbs or low shrubs often much divided leaves and inconspicuous flowers borne on numerous small heads.

About thirty species of Artemisias have been reported to grow naturally in Pakistan [2]. Very little work on chemical constituents of these plants has been done in Pakistan although research work has been reported on ten of the species, existing also in Pakistan, from Japan, China, Russia and other countries. No work is reported on the remaining species occurring in our Northern areas therefore Pakistani Artemisias present a wide scope of work.

These herbs are of economic importance to Pakistan as some of these are of medicinal value and yield essential oils while few are useful as fodder. Some of these species contain santonin, an anthelmintic drug [3]. Several santonin containing species of Artemisia like A.maritima Linn., A.brevifolia Wall., A.herba-alba Asso, A.kurramensis Qazilbash occur in Pakistan [2,4,5] but these different species appear to be only varieties of A.maritima L. instead of being separate species [2,4].

Artemisia maritima Linn. is found in Kurram Valley, Gilgit Agency, Kaghan Valley, Chitral, Dir Swat, Khyber Agency, Indus, Kohistan and

Natural Product Chemistry
Ed. by Atta-ur-Rahman

large barren areas of Baluchistan. Santonin is extracted at Kurram Chemical Company, Rawalpindin, one of the largest manufacturers of this drug in the world [6], from Artemisia, mainly collected near Parachinar in Kurram Valley.

Artemisia absinthium L. is used as tonic and digestive and dried leaves and flowering tops of the plant have medicinal properties [7]. Artemisia annua, a Chinese traditional herb, contains the antimalarial constituent, quinghaosu [8]. Artemisia elegantissima Pamp. is useful in treatment of jaundice while A. maritima L. is anthelmintic. Dried young plants of A. scoparia Waldst. and Kit are used in Ing-Chen Hao, a Chinese drug for treatment of jaundice [9].

A new oral hypoglycaemic agent, vulgarin, has been reported as the main constituent [10] of A vulgaris L. which appears to possess important medicinal properties. Its leaves are given in Indian medicine for nervous and spasmodic affections and it is said to be antiseptic, expectorant and anthelmintic [11].

The main use of Artemisia plants, particularly of A. maritima L. in Pakistan has been for the extraction of santonin, an anthelmintic drug for human and veterinary use. At one time, Pakistan was earning about a million dollars yearly from the export of this drug. Lately, export of this drug has considerably decreased on account of the competition from much cheaper and equally effective synthetic anthelmintics. It has resulted in a considerable loss of foreign exchange earnings, coupled with the loss of means of livelihood to local population, as these plants occur in comparatively underdeveloped areas of the country. It is therefore necessary to find alternate or new uses for these plants to keep the collection of these herbs profitable for the people of our Northern areas.

With this view, a project was started to study the chemical constituents of various Artemisia species occurring in Pakistan and to assess the potential use of these plants and their constituents. Some work, carried out under this project, is described here.

Various types of chemical constituents, occurring in Artemisia species, can be classified into essential oils, sesquiterpene lactones, flavones, coumarins, furans, acetylenes and miscellaneous compounds. Only some of the components of Artemisia oils, and some sesquiterpene lactones and coumarins of Artemisias will be described in this review.

1. ESSENTIAL OILS

Presence of essential oils appears to be a characteristic of these herbs. A first time visitors to Parachinar cannot miss the fragrant smell of Artemisia in the fields around the city.

Study of the chemical constituents of various essential oil from A. absinthium L. [12-15], A. annua L. [16-19], A. macrocephala [20], A. karramensis [21,22], A. maritima L. [11], A. pallens Wall teste Pamp. [23] A.sacroroum Ledeb. [11] and A. vulgaris L. [24,25], A. herba-alba [26] Asso show that these oils contain thujone (1), cineole (2) and camphor (3) as the main constituents. Oil of A. pallens Wall teste Pamp.,

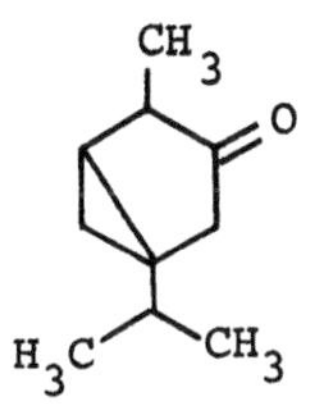

(1)

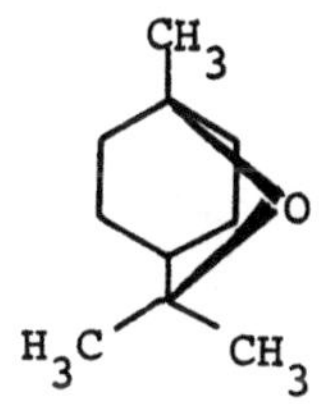

(2)

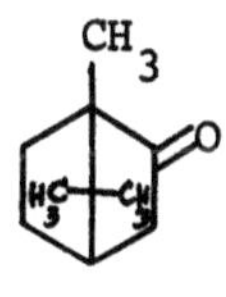

(3)

A. absinthium L., and A. dracunculus L are used in perfumery industry and these oils appear to be comparable to Artemisia oil from A. maritima L. in Pakistan in their chemical composition. Artemisia oil is obtained as by-product of santonin manufacture in Pakistan.

Palmitic (33.90%), arachidic (26.2%) and linoleic (27.5%) were the major fatty acids in the petroleum ether extract of A. absinthium L. [14].

Four non-head-to-tail monoterpene alcohols, comprising 30% of the oil, were identified in the essential of A.herba-alba Asso growing in Negev desert. These monoterpene alcohols, relatively rare in nature, were artemisia alcohol (4), santolina alcohol (5), yomogi alcohol (6) and lyratol (7) [27]. The essential oil of A. herba-alba Asso is antibacterial and this activity appears to be mainly due to santolina alcohol (5) [28].

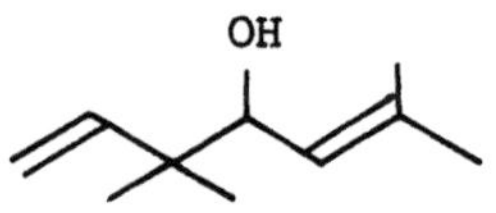

(4)

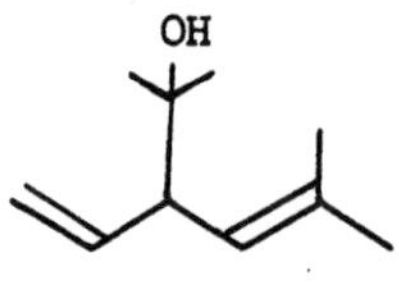

(5)

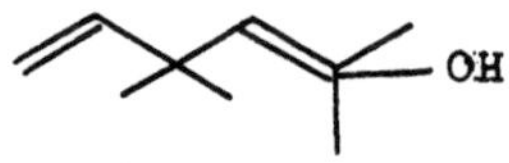

(6)

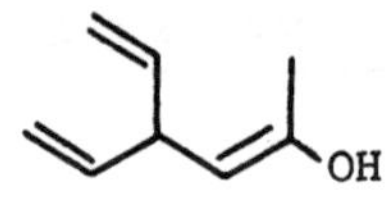

(7)

Several other interesting compounds occur in Artemisia oils. Artemisia (8) and isoartemisia ketones (9) occur in the essential oil of A. annua L. [17,18] , Essential oil from A. macrocephala Jacq ex Boss. contains chamazulene [20] (10) which is an anti-inflammatory and antipyretic agent [29]. Essential oil of A. capillaris Thunb contains β-pinene, scopoletin, esculetin 6,7-dimethyl ether, capillin (11), capillene (12) and capillone (13) [30]. Vulgerol (14) was isolated from A. vulgaris L. together with camphor, β-thujone and myrcene [31].

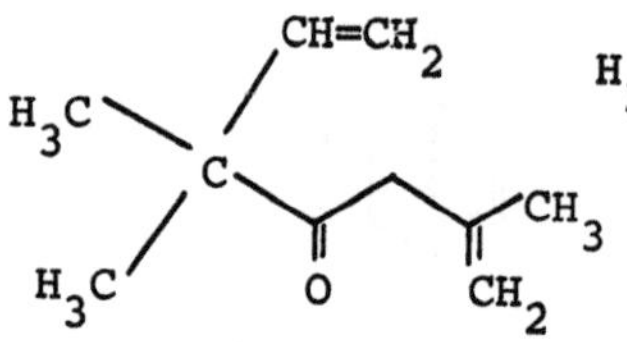

(8)

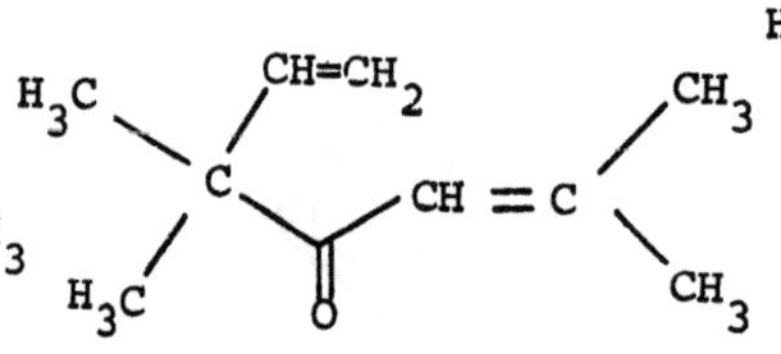

(9)

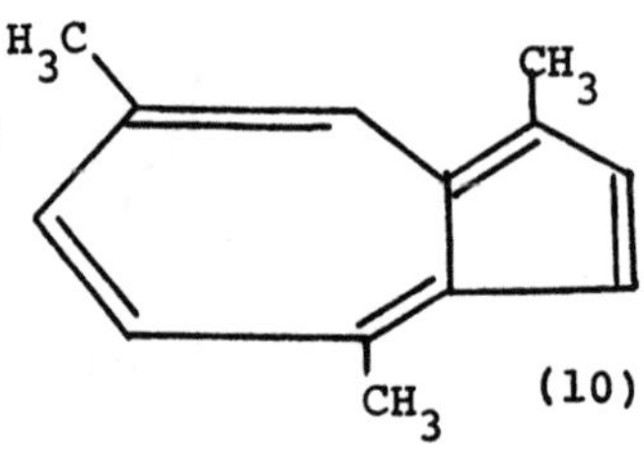

(10)

$$C_6H_5-\overset{\overset{O}{\|}}{C}-C{=}C-C{=}C-CH_3$$

(11)

$$C_6H_5CH_2CH{=}CH-C{=}C-CH_3$$

(12)

$$C_6H_5-\overset{\overset{O}{\|}}{C}-CH{=}CH-CH{=}CH-CH_3$$

(13)

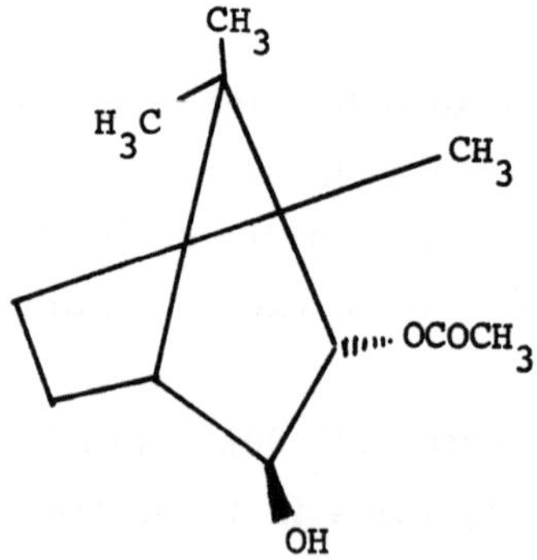

(14)

2. SESQUITERPENE LACTONES

Another characteristic of Artemisia plants, appears to be the presence of sesquiterpene lactones. α-Santonin (15) is present in various varieties of A. maritima L. like A. brevifolia [32] A. herba-albea [33] and A.kurramensis [34]. Various strains of A. kurramensis Qazilbash have been reported to yield both α-(15) and β-santonine (16) [21].

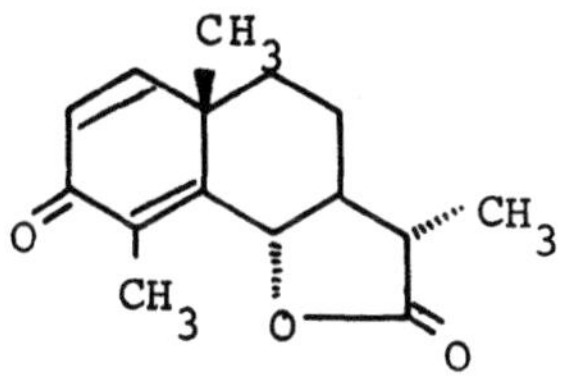

(15)

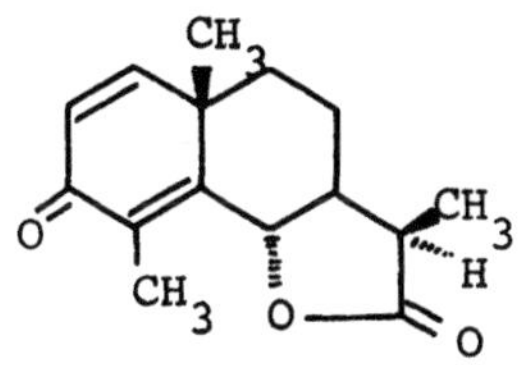

(16)

Artemisia growing near Parachinar City in Kurram Valley is specially rich in santonin content. Parachinar is situated in a remote Northern Area of Pakistan near Afghan border. Santonin content is high during month of September therefore we have collected plants from various villages around Parachinar during September for santonin analysis.

The reported methods for determination of santonin in Artemisia herb appear to be tedious and involve gravimetric determination by precipitation with barium hydroxide [34,35], chromatography [36] and colorimetric [37] methods. A very saimple procedure has been developed by us for gravimetric determination of santonin in the herb.

Dried aerial parts of the herb (15 g), after removing stems were extracted in a thimble in a soxhlet apparatus. Evaporation of the extract gave a dark green residue which afforded santonin. The santonin content, shown in Table-1 of various samples collected from Parachinar area, varied from 1.27 to 2.75% [38]. Samples from Calmi Village and Parachinar City appeared to be more rich in santonin.

Table 1 Santonin content of samples from Parachinar area

S.No.	Area of Collection	Santonin (% w/w)
1.	Militia Garden (Parachinar)	
2.	Karakhela	1.28
3.	Kachkina	.127
4.	Sahra Kili	1.6
5.	Lalmi	2.75
6.	Burki	2.26

Other sesquiterpene lactones, similar in structure to santonin, like torrentin (17) [39] from A. herba-alba Asso, artabin (18) [40,41] and (19) [42,43] from A. absinthium L. have been reported. Vulgarin (20) a new oral hypoglycaemic agent, is the main constituent of A.vulgaris [10].

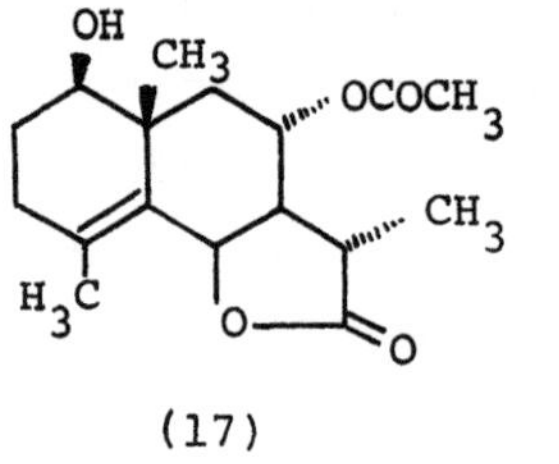

(17)

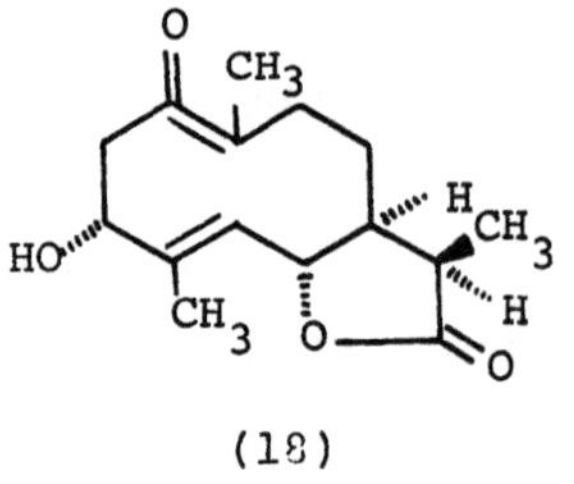

(18)

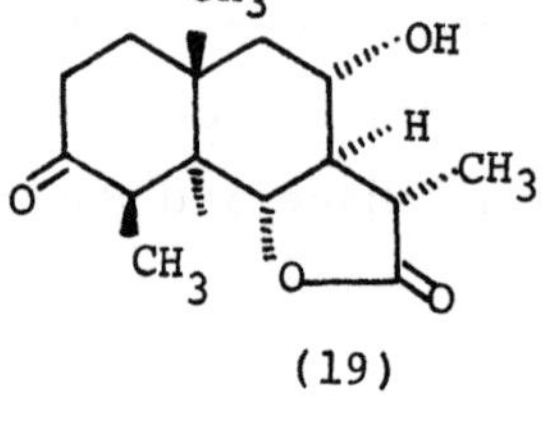

(19)

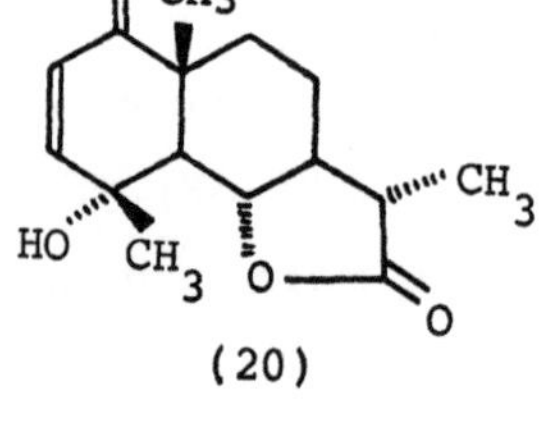

(20)

It is known that α-methylene-γ -lactone moiety shows cytotoxic and antitumor activities and santonin was recently converted into (+) tuberiferine (21) and (+) artecalin (22) [44], derivatives containing -methylene--lactone moiety.

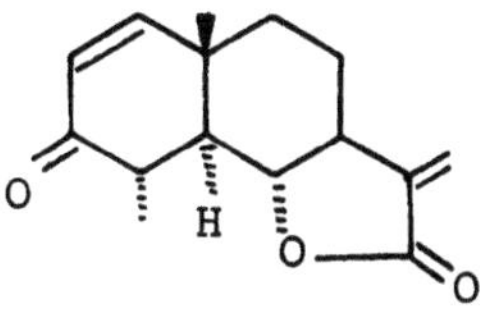

(21)

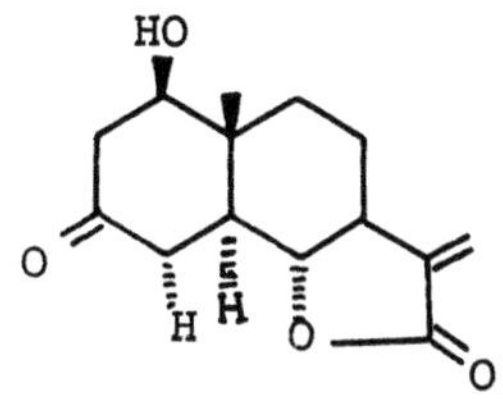

(22)

Aziridine compounds also show antitumor activity and compounds like 2,4,6-tris(l-aziridinyl)-S-triazine (TEM) (23) [46], tris (l-aziridinyl) phosphine oxide (TEPA) (24) [45] and mitomycin-C (25) [46], containing aziridine ring, are tumor inhibiting. Aziridine compounds are also insect chemosterilants [47].

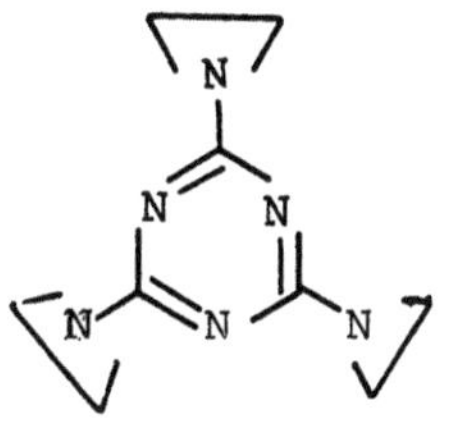

(23)

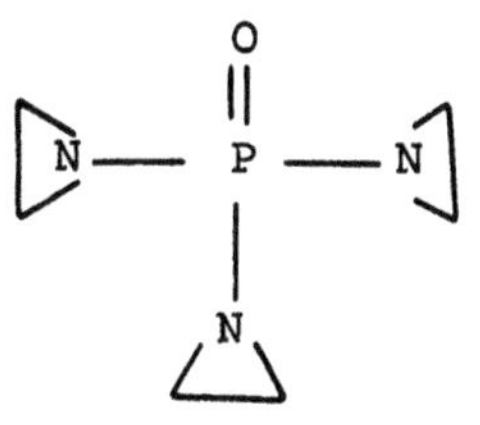

(24)

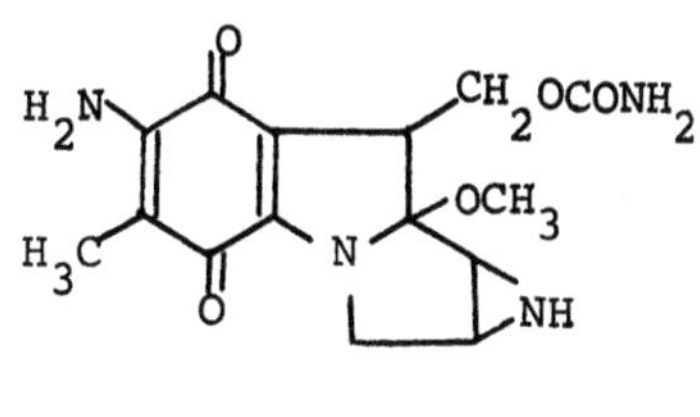

(25)

Aziridine derivatives can be obtained by several procedures [48-51] (Fig-1) like reactions of α, β-dihaloketones with primary amines to give ethyleneimine ketones (2-acyl-aziridines) via an intermediate β-amino-β-haloketone by reacting alkyl halides with aziridines and addition of aziridines at low temperature to carbonyl compounds to obtain moderately stable α-hydroxyalkyl derivatives which could react further with aziridine to afford diaziridinyl derivatives.

It was therefore thought that it would be of interest to introduce aziridine ring into santonin molecule to obtain possible tumor inhibiting compounds. Aziridine appeared to react partially with santonin at room

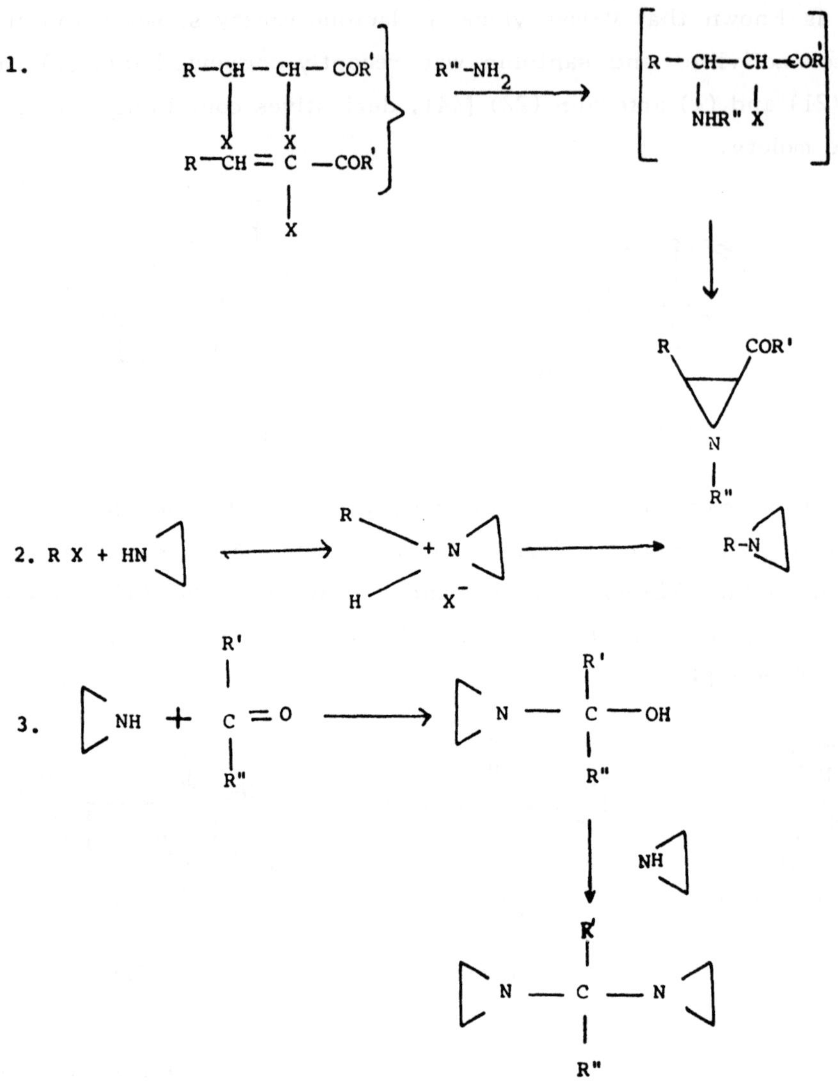

Fig. 1. Some reactions to form aziridine derivatives

temperature affording a polymeric material. Refluxing of santonin in aziridine also gave a mixture of products which did not separate on chromatography. One of these products appeared to be aziridinium salt of santoninic acid but was difficult to characterize. No reaction appeared to occur between santonin and aziridine in presence of base like sodium carbonate, triethylamine and sodium metal.

In order to prepare 2-halosantonins, halogenation of santonin was also studied. Wedekind [52-54] reported formation of dichlorosantonin (26) on direct chlorination of santonin in chloroform at room temperature. Monochlorosantonin (27) was obtained on treatment of 1,2-dichlorosantonin with aniline in ethanolic solution [54].

We have reported chlorination of santonin in these conditions and, interestingly, isolated monochlorosantonin (27) on passing chlorine gas in direct sun light at room temperature in chloroform solution of santonin followed by evaporation of the solvent and chromatography of the residue on neutral alumina. On repeating the reaction at 0°, 1,2-dichlorosantonin (26) could be isolated directly on recrystallization of the residue from chloroform petroleum ether, obtained by evaporation of the reaction mixture in vacuo. It appeared that elimination of hydrogen chloride from 1,2-dichloro santonin is very facile and it was established later that

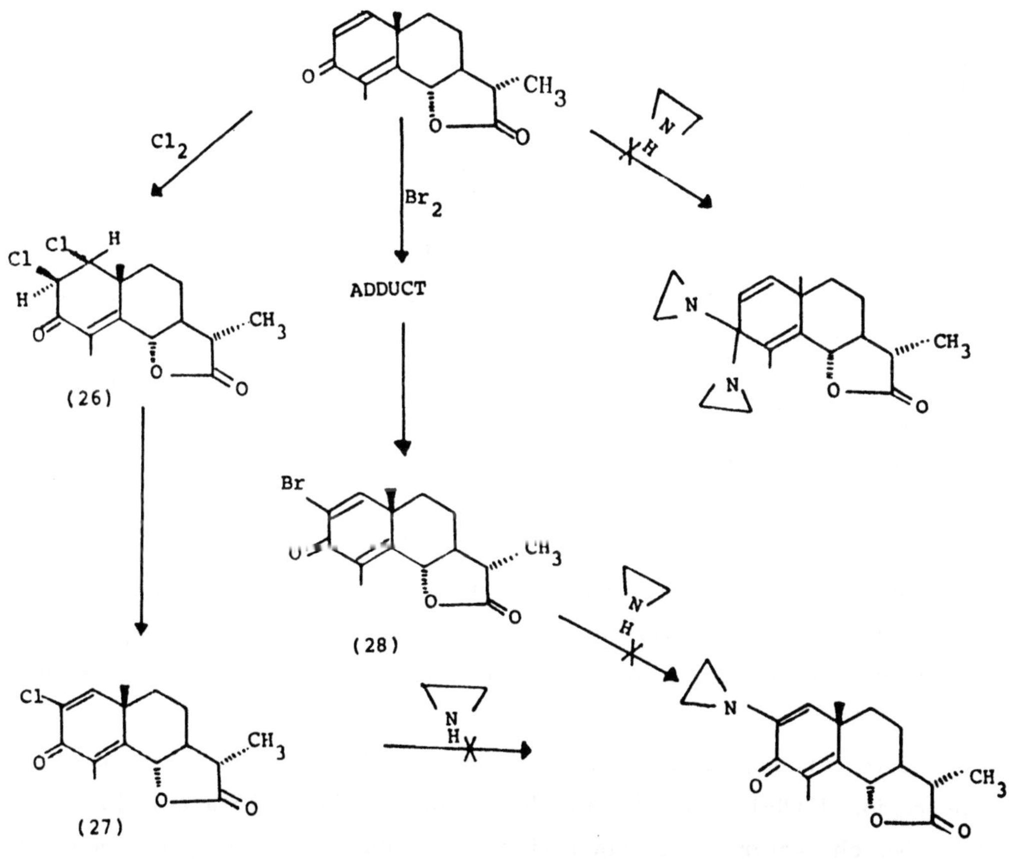

Fig. 2. Halogenation of santonin and attampted addition of aziridine to santonin

chromatography of 1,2-dichlorosantonin on neutral alumina led to elimination resulting in 2-chlorosantonin. It was suggested that 1,2-dichlorosantonin is 1β, 2α -dichlorosantonin [55]. An adduct appears to be formed on bromination of santonin in acetic acid [56,57] affording 2-bromosantonin (28) on treatment with aniline. This adduct was difficult to characterize but appeared to be dibromosantoninoxonium bromide. An adduct of santonin was also formed on reacting santonin with iodine in presence of 25% hydroiodic acid [52]. The adduct appeared to be less stable and lost iodine on recrystallization or on keeping at room temperature.

2-Bromosantonin did not react with aziridine in presence of triethylamine and appeared to form polymeric products on reacting in presence of sodium carbonate and sodium metal. 2-Bromo and 2-chlorosantonins remained unreacted in presence of benzylamine and cyclohexylamine.

3.COUMARINS

Artemisia elegantissima Pamp. occurs in Islamabad, Chitral and Kashmir areas of Pakistan. As no work has ever been reported on this species, therefore this herb was collected from Islamabad to study its chemical constituents. Extraction of the aerial parts of the herb with chloroform and concentration of the extract yielded a colourless crystalline material, m.p. 138-140°C, which appeared to be the main product from the plant. High resolution mass spectrum of the compound gave the molecular ion formual as $C_{11}H_{10}O_4$ which was also indicated by its elemental analysis.

The compound burnt with a sooty flame and showed very strong absorption in the carbonyl region in its infrared spectrum. It was neutral in character. It dissolved in aqueous alkali and was reprecipitated on acidification indicating it to be an aromatic lactone, possible a coumarine.

The carbonyl frequency in coumarins [58a] (α -pyrones)occurs in the region 1700-1750 cm^{-1} and absorptions at 1722, 1610 and 1552 cm^{-1} are also characteristic of coumarins [59]. The infrared spectrum of the isolated compound shows strong carbonyl absorption at 1710 cm^{-1} and absorptions at 1610 and 1560 cm^{-1} further indicated it to possess a coumarin nucleus.

The n.m.r. spectrum of the compound shows two singlets (3 protons each) at δ 3.9 and 3.92 corresponding to the two methoxyl groups and two doublets at δ 6.27 and 7.62, each corresponding to one proton, with the same coupling constant of 9.5 Hz. In addition, two singlets, each corresponding to one proton, appear in the aromatic region at δ 6.7 and 6.75. Appearance of a pair of doublets with a coupling constant of 9.5 Hz, centred at δ 6.1-6.4 and δ 7.5-8.3 in the n.m.r. spectrum (recorded in deuterochloroform) of a product strongly indicates a coumarin unsubstituted in the pyrone ring [58b,60]. These characteristic signals arise from hydrogens attached to C-3 and C-4 respectively and absorptions in the rangeδ 7.5 to 8.3 occurs due to C-4 hydrogen. Doublets observed at 6.27 andδ7.62 in the isolated compound appear to be due to C-3 and C-4 hydrogen of a coumarin structure. Lack of the significant splitting of the two aromatic protons arising near δ 7.0 indicates presence of the hydrogens at the para position in the aromatic ring at C-5 and C-8, and is characteristic of a 6,7-disubstituted coumarin [59]. The two singlets thus at δ 6.7 and 6.75 in the isolated compound appear to be due to C-5 and C-8 hydrogens of a coumarin nucleus. Aromatic methoxyl groups normally resonate in the range δ 3.8-4.4 in substituted coumarins therefore two singlets arising at δ 3.9 and 3.92 in the NMR spectrum of the isolated compound appear to be due to the aromatic methoxyl groups. The NMR spectrum thus indicated strongly 6,7-dimethoxycoumarin to be the structure of the isolated compound.

The presence of 6,7-methoxycoumarin chromophore moiety is indicated by u.v. absorption [59] near 227, 253, 297 and 327 nm. Scopoletin (6-methoxy-7-hydroxycoumarin) (34) and isoscopoletin (35) (6-hydroxy-7-methoxycoumarin), show absorptions at 228, 252, 260, 295, 344 nm and 229, 253, 260, 295, 350 nm respectively [58c]. The U.V. spectra of these compounds are very similar to that of the isolated compound which shows absorptions at 220, 225, 250, 272, 295, 310 and 340 nm. It is known that there is a great similarity in the u.v. spectra between the hydroxyl derivatives and their ethers [61] and 6,7-dimethoxycoumarin (33) shows a more complicated u.v. spectrum than of coumarin [62]. The u.v. spectrum of the compound thus also strongly favours 6,7-dimethoxycoumarin (33) structure.

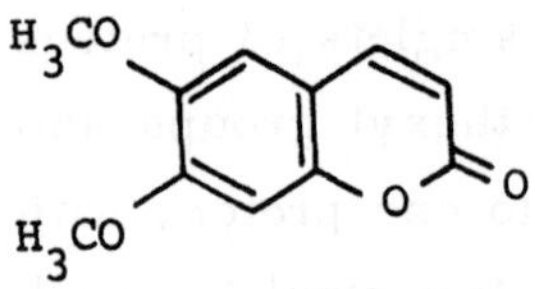

(33)

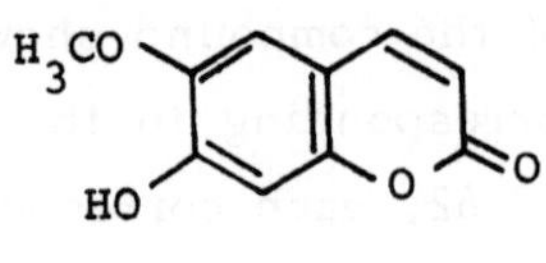

(34)

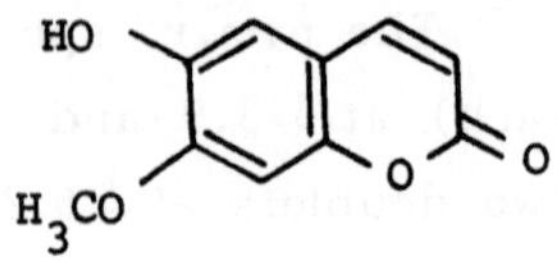

(35)

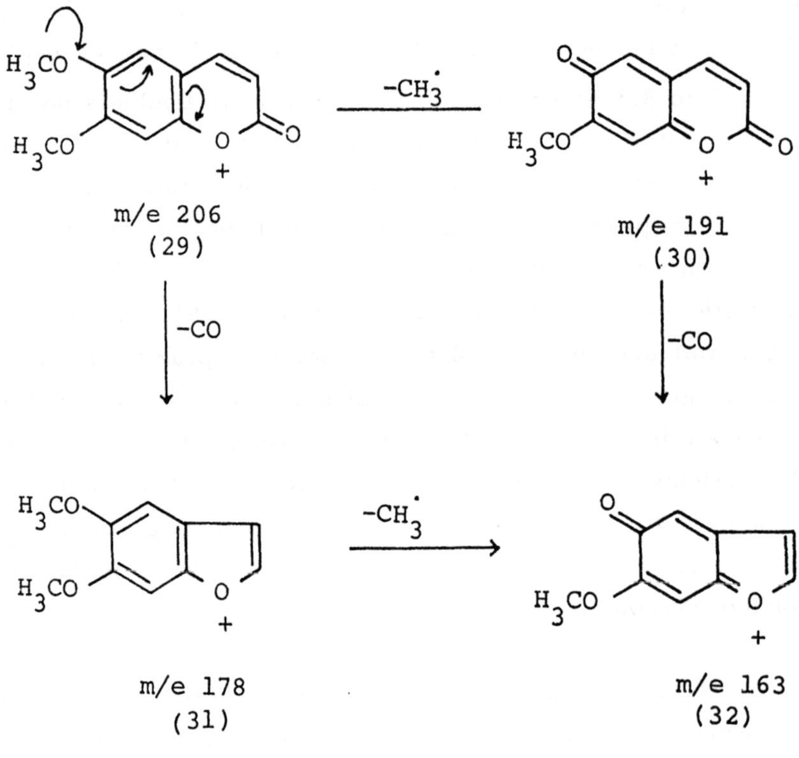

Fig. 3. Fragment ions from 6,7-dimethoxycoumarin

It is known that compounds derived from coumarin lose oxygen as carbon monoxide under electron impact and usually form stable benzofuran ions [63]. Djerassi [64] reported the mass spectrum of 6,7-dimethoxycoumarin and observed the molecular ion peak to be the most intense (100%) and the other ions appearing at m/e 191, m/e 178 and m/e 163 and interpreted these ions in terms of structures (29)-(32).

We have obtained the high resolution mass spectrum of the compounds, isolated from A.elegantissima Pamp., and the observed spectrum also shows molecular ion peak to be the most intense (100%) and is the same as that reported earlier by Djeressi [64] for 6,7-dimethoxycoumarin. The molecular weights observed at m/e 206.0587 (M^+,100%), m/e 191.0343 (M^+-CH_3,46.76%), m/e 178.0627 (M^+-CO, 18.19%) and 163.0392 (M^+-(CH+CH_3), 30.45%) corresponding to the structures $C_{11}H_{10}H_4$, $C_{10}H_7O_4$,$C_{10}H_{10}O_3$ and $C_9H_7O_3$ respectively. The high resolution mass spectrum thus confirms the loss of methyl radical from molecular ion (29) to give p-quinonoid system (30) which could loss CO to generate (32) which could also be obtained from ion (31) by loss of methyl groups. The ion (31) is generated from molecular ion by loss of carbon monoxide. Djerassi had shown earlier [64] from deuterium labelling of C-7 methoxyl group of 6,7-dimethoxycoumarin that 6-methoxy group is lost preferentially from molecular ion and the high resolution mass spectrum, observed by us, confirms further the interpretation of the observed peaks in terms of the structures (29) to (32). Spectral evidence thus indicates clearly that the compound isolated from A.elegantissima pamp. of Islamabad area is 6,7-dimethoxy-coumarin (33).

6,7-Dimethoxycoumarin (33) scoparone [65], esculetin dimethyl ether [66], 6,7-dimethylesculetin [67[also occurs in A.scoparia [65,68-71] and A.capillaris which are components of the Chinese drugs [72]. Ing-Chen-hao and Inchinko-to respectively, which are used for treatment of jaundice in Chinese medicine. 6,7-Dimethoxycoumarin, a component of these drugs, increased the bile secretion and appeared to be responsible for the therapeutic activity of the above crude Chinese drugs [72].

6,7-Dimethoxycoumarin (33) also occurs in A.campestris, A.dracunculoides Persh, A.dracunculus L., A.parviflora Roxb. and A.tridentata var vaseyana (Rydb.) Bectle [58d].

Recently, lot of interest has been shown in the pharmacological activity of 6,7-dimethoxycoumarin. Besides possessing choleretic activity [72,73], it also shows hypotensive [73-76] activity but its anthelmintic action is weak [77]. It also shows anti-inflammatory and analgesic properties in addition to choleretic activity [73].

Scopoletin (34) (6-methoxy-7-hydroxycoumarin was identified in A.parviflora Roxlb [78]. A.japonica Thunb [79] and A.persica Boiss [80] while esculetin (35) (6,7-dihydjroxycoumarin) occurs in A.parviflora Robx [78].

It is hoped that this short talk would have indicated to you that Pakistani Artemisia herbs are useful medicinal plants and contain chemical constituents of potential medicinal uses.

Acknowledgement

Pakistan Science Foundation is gratefully acknowledged for providing a research grant (No. PSF/RES/S/CSIR/CHEM(105) under which this work was carried out. Collaboration of Dr. Yusuf Ahmad and Miss Nuzhat K.Sherwani, is also thankfully acknowledged.

References

1. **The Wealth of India**, C.S.I.R, India Publication, Delhi, 1948, p. 120.

2. R.R.Stewart,**An Annotated Catalogue of the Vernacular Plants of West Pakistan and Kashmir**, Fakhri Printing Press, Karachi (1972) p. 714.

3. **Mono and Sesquiterpenoids**, P. de Mayo, Interscience Publishers Inc., New York, 1959, p. 216.

4. I.I.Chowdhry, **Economic Botany** 9, 224 (1955)

5. Sher Ahmad Khan, Government of Pakistan,Department of Agricutre Report, Bureau of Agricultural Information, West Pakistan (1966)

6. **Mfg.Chemist**, 34, 64 (1963)

7. V.L.Kretovich, Yu S.Rall and L.A.Trisvyatskiv, **Biokhimiya**, 11, 493 (1946); **Chem.Abstr.**, 41, 250le (1947)

8. Tu,You-You, Mi Mu-Yun, Zhong Yu-rong, Li Lan-na, Cui Shu-Lian, Zhang Mu-qun, Wang Ziu-Zhen, Ji Zheng and Liang Xiao-Tian, **Planta Medica**, 44, 143 (1982)

9. Jun-sheng Hu, Pao-Chen Li and Mei Chen, **Yao Hsueh Pao**, 12, 289 (1965); **Chem.Abstr.**, 63, 6021b (1965)

10. C.Rodriguez de Vera, S.Peran Mesa, A.M.de Diego and A.Villar del Fresno, **Bull.Chem.Farm.**, 115, 445, 451 (1976); **Chem.Abstr.**, 85, 171817 f (1976)

11. R.N.Chopra, **Indigenous Drugs of India**, U.N.Dhar and Sons Private Ltd., Calcutta-12 (1958), p.65.

12. F.W.Bieloszabska and W.Sawicke, **Farm.Polska**, 22, 278 (1966); **Chem.Abstr.**, 65, 8661d (1966)

13. V.K.Kaul,S.S.Nigam and A.K.Banerjee, **Indian Perfum**, 23, 1 (1979); **Chem.Abstr.**, 92, 116240g (1980)

14. M.D.Sayed, A.M.El-Shamy, F.M.Soliman and A.O.El-Shabrawy, **Bull.Fac.Pharm.**.(Cairo Univ.) 16, 85 (1977); **Chem.Abstr.**, 92, 90953q (1980)

15. V.Paolini and R.Romonaco, **Univ.Rome.Atti accad.**, 23, 123 (1914); **Chem.Abstr.**, 9, 1323[6] (1915)

16. Y.Tucakov, Perfumery Essentl.Oil Record, 46, 75 (1955); **Chem.Abstr.**, 49, 11245a (1955)

17. Y.Asahina and S.Takagi, **J.Pharm.Soc.Japan**, 837 (1920); **Chem.Abstr.**, 15, 1301 (1921)

18. L.Ruzicka, T.Reichstein and R.Pulver, **Helv.Chim.Acta.**, 19, 646 (1936)

19. Tu You-You, Ni Mi-Hu, Zhong Yu-rong, Li Lan-na, Cui Shu-lian, Zhang Mu-qun, Wang Ziu-zhen, Liang Xiaetion, **Yaoxue Xuebao**, 16, 366 (1981); **Chem.Abstr.**, 97, 52497q (1982)

20. T.P.Berezovskaya, V.V.Dudko, E.A.Serykh, and R.G.Usynina, **Rastit Rsesur.**, 12, 565 (1976); **Chem.Abstr**, 86, 40142e (1977)

21. T.Kuroda, **Yakugaku Zassi**, 82, 179 (1962); **Chem.Abstr.**, 57 2329h (1962)

22. M.I.Goryaev, G.K.Kruglykhina, P.P.Polyakov and I.M.Shabanov, Trudy, Inst., **Khim Nauk Akad.Nauk Kazakh,S.S.R.**, 4, 97 (1959); **Chem.Abstr.**, 54, 11384a (1960)

23. B.C.Gulati, M.N.A.Khan, **Parfum.Flavor**, 5, 23 (1980); **Chem.Abstr.**, 93, 225488k (1980)

24. M.Nakao and C.Shibue, J.Pharm.Soc.Japan, 636 (1924); **Chem.Abstr.**, 19, 2258[9] (1925)

25. I.S.Kohlmuenzer, **Dissertation Pharm.**, 12, 143 (1960); **Chem.Abstr.**, 58, 10509d (1963)

26. B.Benjilali and H.Richard, Rev.Ital.Essenze Profumi, Piante Off, Aromat. Syndets, Saponi, Cosmet Aerosols, 62, 69 (1980); **Chem.Abstr.**, 93, 22621a (1980)

27. R.Segal, A.Bruer and I.Feurstein, **Phytochemistry**, 19, 2761 (1980); **Chem.Abstr.**, 94, 171014c (1981)

28. J.Yashpe, R.Segal, A.Breuer and G.Erdreich-Naftali, **J.Pharm.Sci.**, 68, 924 (1979)

29. Paul G.Stecher (Editor). **The Merck Index.** Merck and Co.Inc., Rahway, N.J.(U.S.A.), VIII Ed. (1968), p.227.

30. Sze, Chin-Quey and Y.S.Chen, **S.African** 7900, 375 (1980); **Chem.Abstr.**, 93 120256e (1980)

31. G.M.Nano, C.Bicchi, C.Frattini and M.Gallino, **Chim.Ind.**(Milan), 59, 123 (1977); **Chem.Abstr.**, 87, 135977r (1977)

32. H.G.Greenish and C.E.Pearson, **Pharm.J..**, 106, 2 (1921); **Chem.Abstr.**, 15, 924[7] (1921)

33. S.M.Khafagy, A.S.Gharbo and T.M.Sarg, **Planta Medica.**, 20, 90 (1971); **Chem.Abstr**, 75, 85176h (1971)

34. M.Yamagishi, I.Nakamura and K.Nakajima, **J.Pharm.Soc.Japan**, 74, 1354 (1954); **Chem.Abstr.**, 49, 4940i (1955)

35. N.A.Qazilbash, **Pak.J.Sci.Res.**, 8, 78 (1956); **Chem.Abstr.**, 50, 17315b (1956)

36. T.Kuroda, **Yakugaku Zasshi**, 82, 164 (1962); **Chem.Abstr.**, 57, 2329e (1962)

37. R.A.Khan and H.Mohiuddin, **J.Pharm.Pharmacol.**, 12, 544 (1960); **Chem.Abstr.**, 55, 3002f (1961)

38. Y.Ahmad, S.K.Hasan and N.K.Sherwani, **Pak.J.Sci.Ind.Res.**, 26, 50 (1963)

39. J.D.Gomis, J.A.Marco, J.R.P. Llinares, J.S.Parareda, J.M.Sandra and E.Seoane, **Phytochemistry**, 18, 1523 (1979); **Chem.Abstr.**, 92, 181379y (1980)

40. I.S.Akhmedov, S.Z.Kasymov and G.P.Sidyakin, **Khim.Prir.Soedin**, 6, 622 (1970);**Chem..**, 74, 50523w (1971)

41. I.S.Akhmedov, S.Z.Kashymov and G.P.Sidyakin, **Khim.Prir.Soedin**, 6, 691 (1970); **Chem.Abstr.**, 74, 112239u (1971)

42. I.S.Akhmedov, S.Z.Kasymov and G.P.Sidyakin, **Khim.Prir.Soedin.**, 245 (1972); **Chem.Abstr.**, 77, 85567e (1972)

43. S.K.Zakirov, S.Z.Kasymov and G.P.Sidyakin, **Khim.Prir.Soedin.**, 548 (1976); **Chem.Abstr.**,86, 16805c (1977)

44. K.Yamakawa, K.Nishitani and T.Tominaga, **Tet Letters**, 2829 (1975).

45. J.A.Montgomery, **Progress in Drug Research**, Vol.8, Birkhauser Verlag, Basel, 1965, p.442.

46. E.J.Kirsch. **Mitomycins. Antibiotics** II, Edited by D.Gottlieb and P.D.Shaw, Spinger Verlag Publishers, New York, 1967, p.66.

47. C.N.Smith, G.C.La Brecque and A.B.Borkovec.**Ann.Rev.of Entomology**, 9,269 (1964)

48. . E.H.Rodd (Ed.), **Chemistry of Carbon Compounds.** Vol. IV A, Elsevier Publishing Co., London (1957) p.14.

49. R.C.Elderfield (Ed.), **Heterocyclic compounds**. John Wiley and Sons Inc., London, Che. 2, p.61.

50. E.Miller (Ed.)., **Methoden der Organischen Chemie (Houben-Weyl)**, Vol. XI/2, George Thiem Verlag Stuttgart (1958), p.223.

51. A.Weissberger (ed.), Part I, **Heterocyclic compounds with three or four Membered Rings.** Interscience Publishers, London (1964), p.524.

52. E.Wedekind and A.Koch, **Ber.**, 38, 429 (1905)

53. E.Wedekind and A.Koch, **Ber.**, 38, 1845 (1905)

54. E.Wedekind and K.Tettweiler, **Ber** 64, 387 (1931)

55. Takayangi Hireoki, Ogura Haroue and Litaka Yaichi, **Chem.Pharm. Bull.**, 22, 1433 (1974)

56. J.Klein, **Ber.**, 40, 939 (1907)

57. E.Wedekind, **Ber** 41, 364 (1908)

58. R.D.H.Murry, J.Mendez and S.A.Brown, **The Natural Coumarins** (John Wiley and Sons Ltd., New York, 1982), (a) p.34 (b) p.35 (c) p. 30 (d) p.482.

59. J.P.Kutney, A.K.Verma and R.H.Young, **Tetrahedron**, 28, 5091 (1971)

60. B.E.Nielsen and J.Lemmich, **Acta.Chem.Scand.**, 19, 601 (1965)

61. R.H.Goodwin and B.M.Pollock, **Arch.Biochem.Biophys.**, 49, 1 (1954); **Chem.Abstr.**, 48, 6827z (1954)

62. T.Nakabayashi, T.Tokoroyama, H.Miyazaki and S.Irono, **J.Pharm. Soc.Japan.**, 73, 669 (1953); **Chem.Abstr.**, 47 10348d (1953)

63. C.S.Barnes and J.L.Occolowitz, **Aust.J.Chem.**, 17, 975 (1964); **Chem.Abstr.**, 61, 12774h (1964)

64. R.H.Shapiro and C.Djerassi, **J.Org.Chem.**, 30, 955 (1965); **Chem.Abstr.**, 62, 13016d (1965)

65. R.S.Thakur, M.P.Jain, P.R.Rao, **Res.Ind.**, 20, 129 (1975); **Chem.Absr.**, 85, 142951q (1976).

66. S.Sera and C.Shibuye, **J.Agr.Chem.Soc.**, Japan, 6, 600 (1930); **Chem.Abstr.**, 24, 5742[9] (1930)

67. Y.Johji, F.Youichi, S.Tokounosuke, F.Hajime, **Shoyakugaku Zasshi**, 35, 108 (1981); **Chem.Abstr.**, 95, 197103b (1981)

68. Jun-Shen Hu, Pao-Chen Li and Mei Chen, **Yao Hsueh Hsueh, Pao**, 12, 289 (1965); **Chem.Abstr.**, 63, 6021b (1965).

69. G.Singh, G.V.Nair and K.P.Agarwal, **Chemistry and Industry**, 1294 (1954); **Chem.Abstr.**, 49, 13236 d (1955)

70. M.Tomova, **formatsiya**, 14, 18 (1964); **Chem.Abstr.**, 61, 11959e (1964).

71. M.Stefanovic, K.Krstic and S.Mladenovic, **Phytochemistry**, 12, 2996 (1973); **Chem.Abstr.**, 81, 1278f (1974)

72. A.Masaki, S.Hiroshi and H.Masatoshi, **Yakugaku Zasshi**, 96, 147 (1976); **Chem.Abstr.**, 85, 56660n (1976)

73. Y.Joji, M.Hisashi, S.Tokunosuke, M.Hiroyuki, F.Hajime, **Yakugaku Zasshi**, 102, 285 (1982); **Chem.Abstr.**, 96, 210458p (1982)

74. K.S.Jamwal, M.L.Sharma, N.Chandhoke, B.J.R.Ghatak, **Indian J.Med.Res.**, 60, 763 (1972); **Chem.Abstr.**, 77, 160212 (1972)

75. U.Zutshi, P.G.Rao, A.Soni, G.K.Atal, **Indian.J.Expl.Biology**, 16, 836 (1978); **Chem.Abstr.**, 89, 173347j (1978)

76. R.S.Thakur, S.C.Bagadia and M.L.Sharma, **Experientia**, 34, 158 (1978); **Chem.Abstr.**, 88, 163896c (1978)

77. K.Ishifuku, H.Sakurai, H.Okomoto and S.Sato, **J.Pharm.Soc.Japan.**, 73, 332 (1953); **Chem.Abstr.**, 48, 2695e (1954)

78. M.P.Jain, R.S.Thakur and P.R.Rao, **Curr.Sci.**, 45, 640 (1976); **Chem.Abstr.**, 85, 139765b (1976).

79. O.A.Konovalova, K.S. Rybalko and A.I.Shreter, **Khim.Prir.Soedin.**, 97 (1976); **Chem.Abstr.**, 85, 30626f (1976).

80. K.S.Rybalko, I.A.Gubanov and M.I.Vlasov, **Med.Prom.SSSR**, 18, 19 (1964); **Chem.Abstr.**, 61, 1710b (1964).

Biogenetic Aspects of Sesquiterpene Lactone Chemistry

Werner Herz

Department of Chemistry, The Florida State University, Tallahassee, FL 32306, USA

Distribution of Sesquiterpene Lactones in the Plant Kingdom

A. γ-Lactones formed by oxidation of "head" methyl Groups

Compositae	ca. 875 taxa
Umbelliferae	16
Lauraceae	2
Burseraceae	1
Chloranthaceae	2
Magnoliaceae	10
Cupressaceae	1
Hepaticae	12

B. Other Lactones scattered in Amaranthaceae, Aristolochiaceae, Cannellaceae, Lauraceae

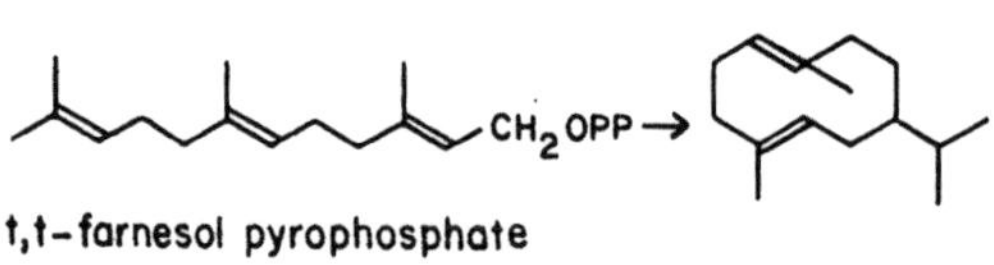

t,t-farnesol pyrophosphate

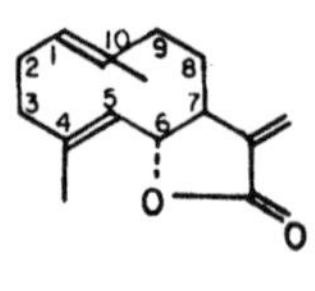

1 costunolide

Scheme I

Common lactones formed by oxidation of terminal methyl

Natural Product Chemistry
Ed. by Atta-ur-Rahman

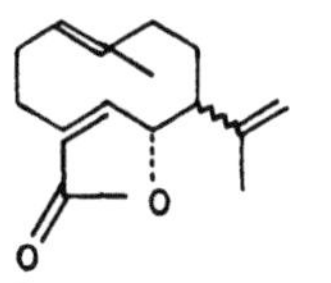
2 aristolactone

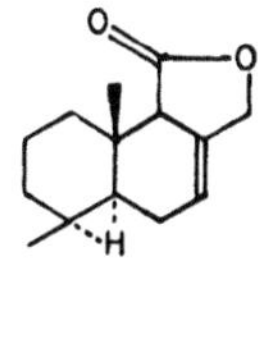
3 drimenin

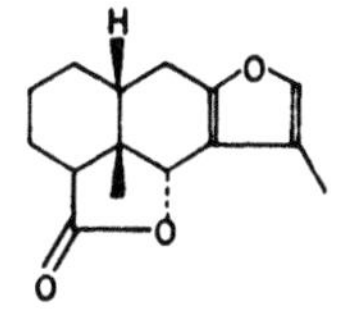
4

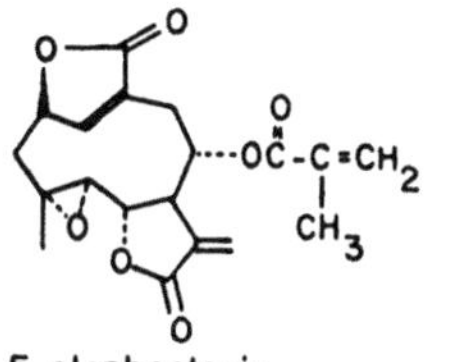
5 elephantopin

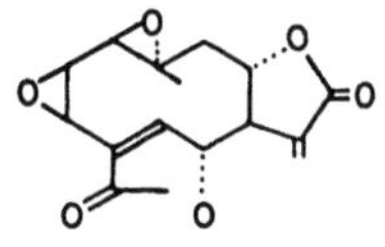
6 mikanolide

Rarer lactones formed by oxidation of non-terminal methyl or of both terminal and non-terminal methyls

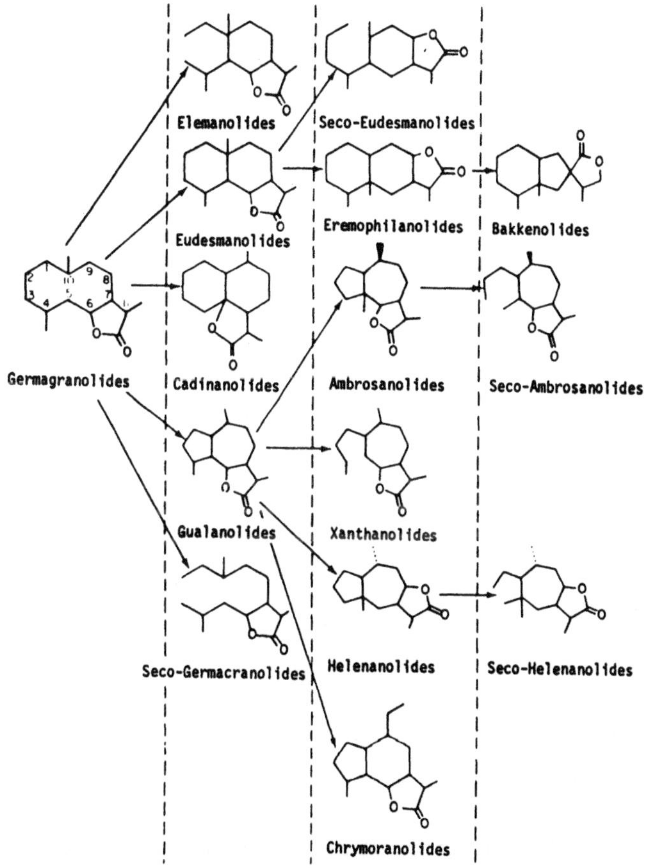

Arrangement of lactones by "biogenetic complexity" of carbon skeleton

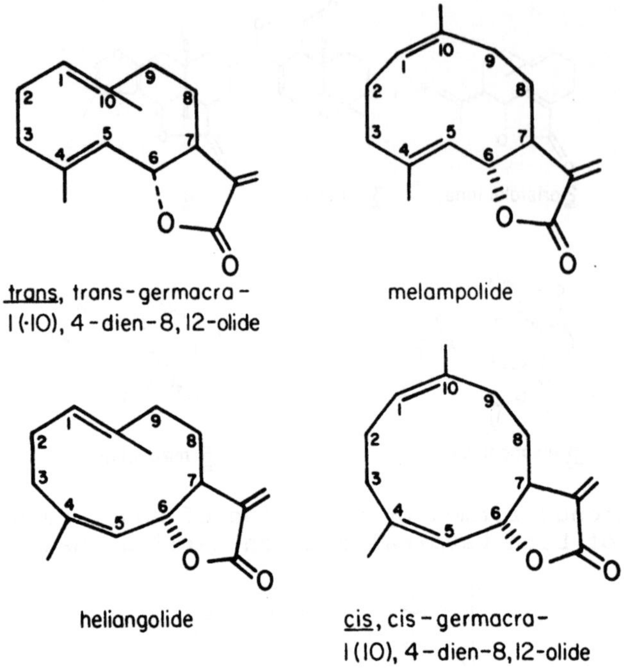

CONFIGURATIONAL ISOMERS

Possible geometrical 1,5-germacradienolide isomers

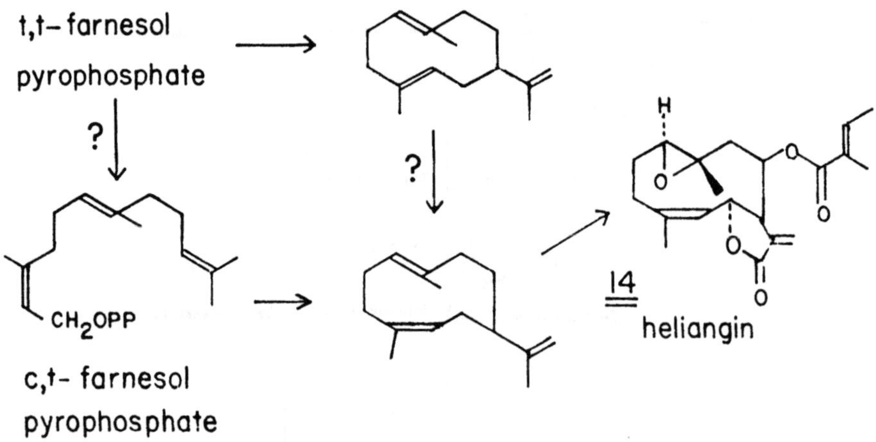

Possible routes to heliangolides

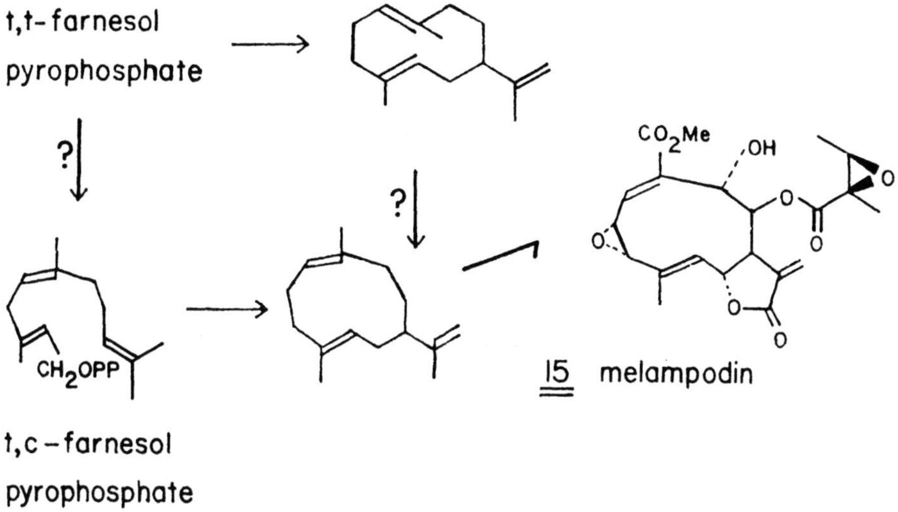

Possible routes to heliangolides

Melampolide formation

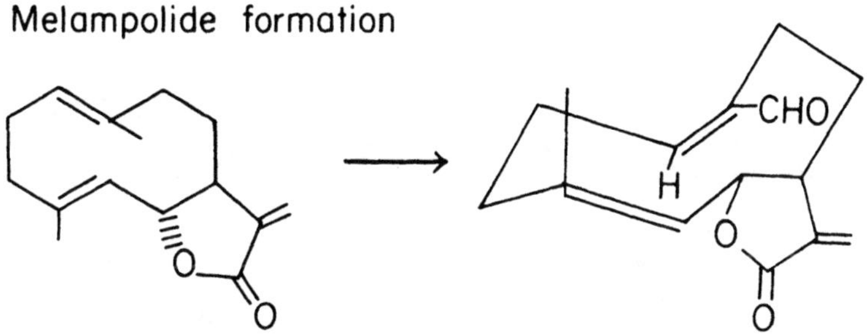

S. Krishnan et al. J. Indian Chem. Soc. 55, 1142 (1978)

Laboratory analogy (I)

Melampolide formation

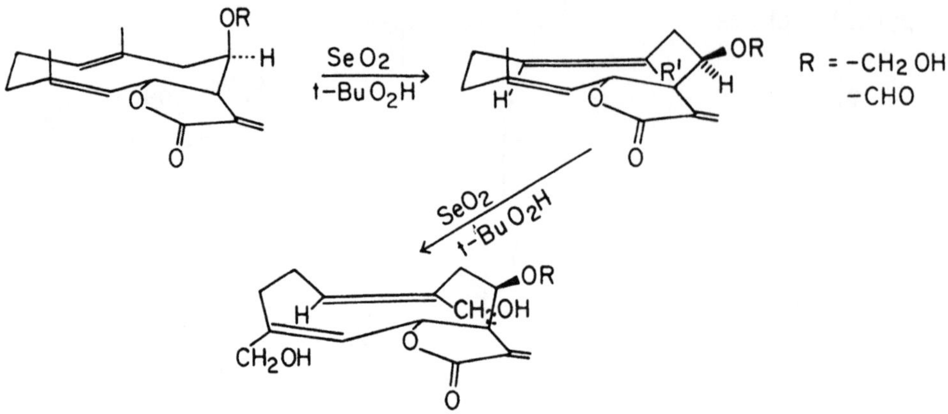

Haruna and Ho, J. Chem. Soc. Chem. Commun. 1981, 483.

Laboratory analogy (II)

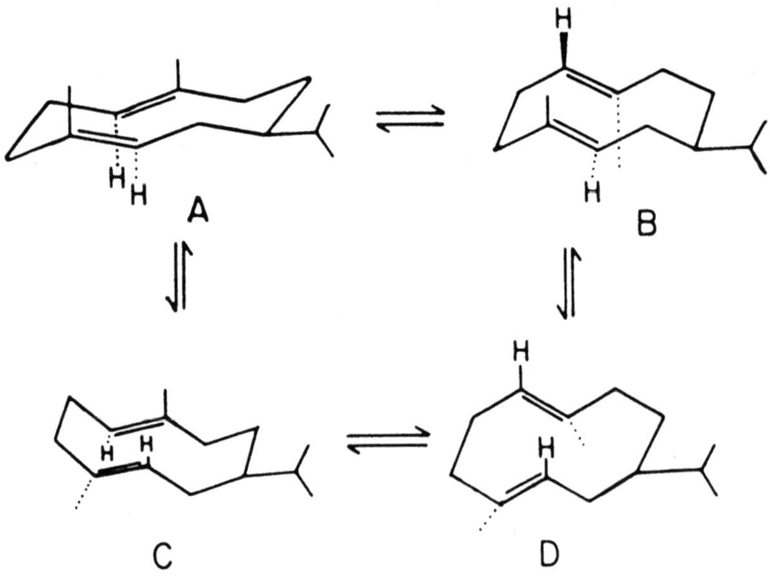

Possible conformations of (E,E)-1,5-germacradienes
(chair and boat only)

Germacradiene – 6βH, 12 olides

Preferred conformation (solid and solution)

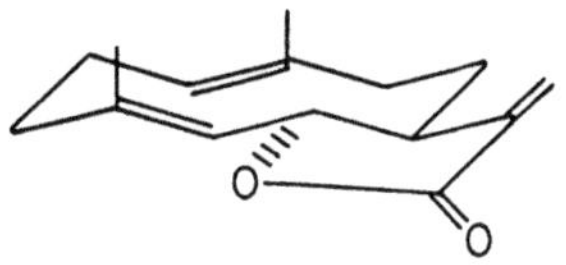

Germacradiene – 8βH,12–olides

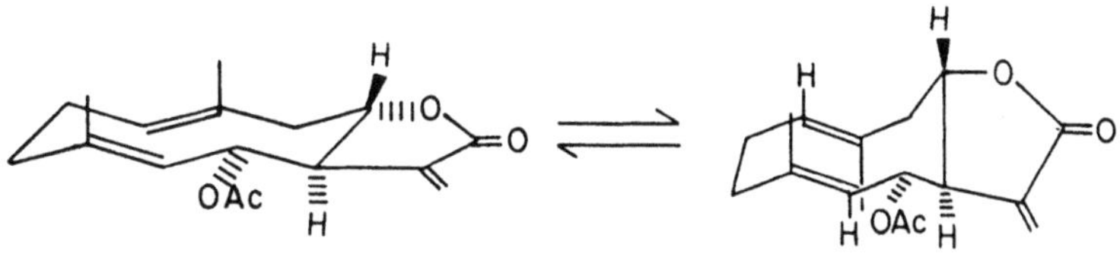

major laurenobiolide (solution) minor

Preferred conformations of (E,E)-1,5-germacradien-6βH,12- and 8βH,12-olides

PREFERRED CONFORMATIONS

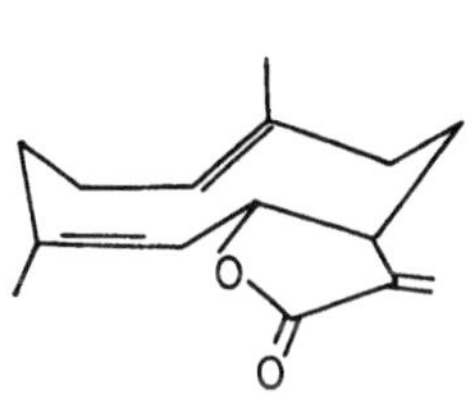

heliangolide

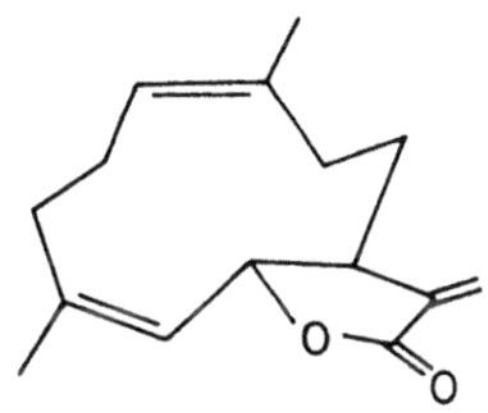

cis, cis-germacranolide

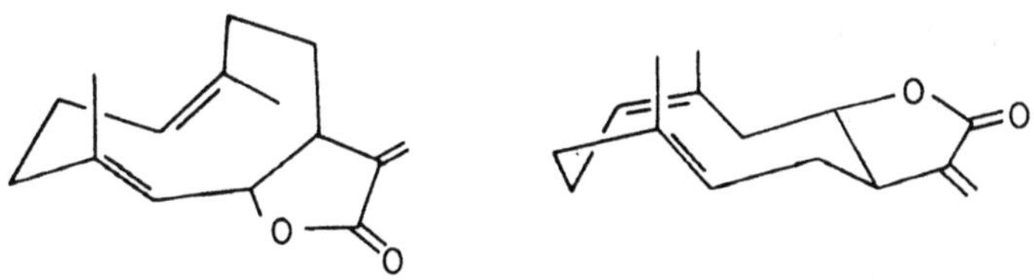

melampolide

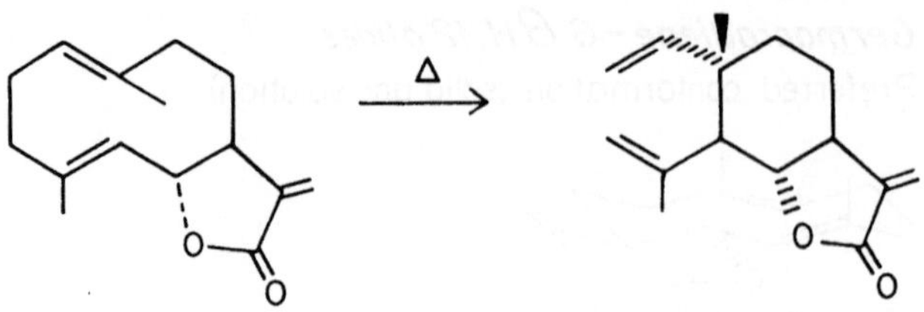

costunolide saussurea lactone

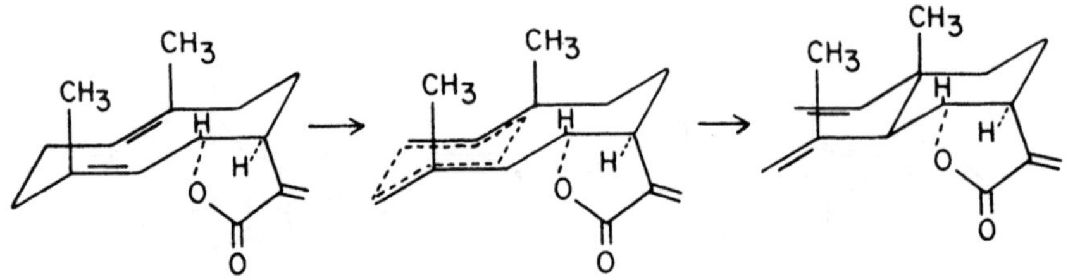

Cope rearrangement of (E,E)-1,5-germacradien-6βH,12-olides (crown conformation) ⟶ 5αH,10β-methylelemadienolides

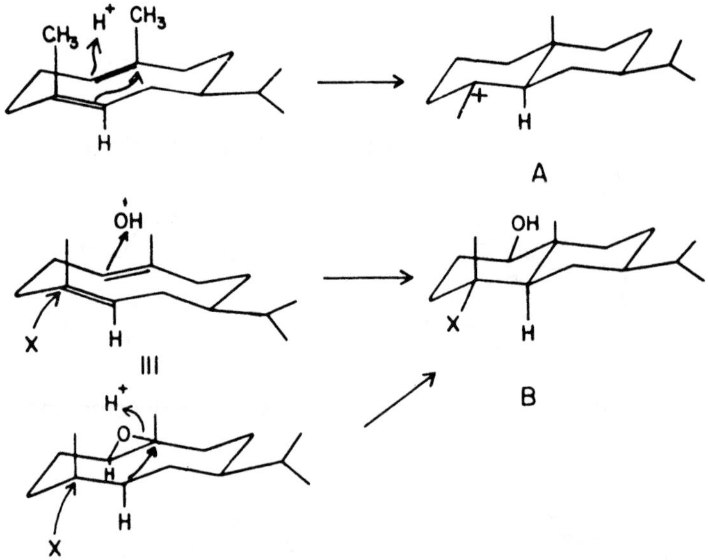

Markovnikov-orientated cyclization of (E,E)-1,5-germacradien-6βH,12-olides (crown) ⟶ 5αH,10β-methyleudesmanes

Preferred Conformations of some other 1,5-germacradienolides

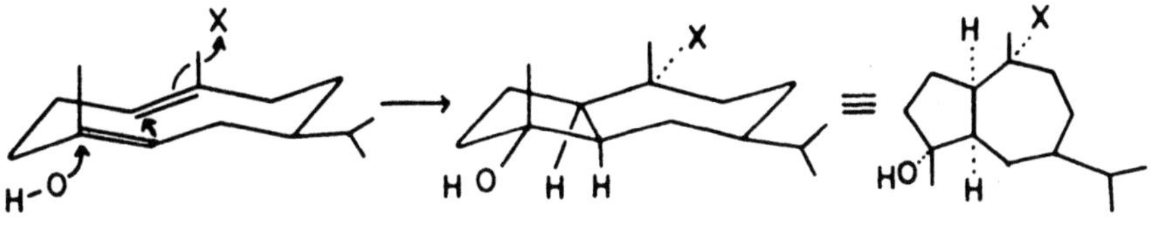

A

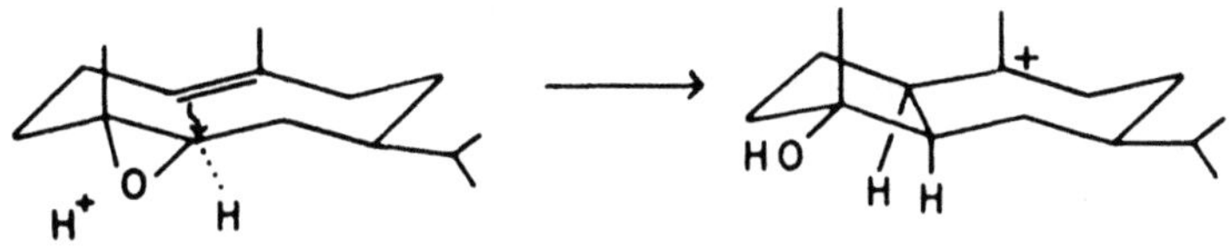

B

Anti-Markovnikov orientated cyclization of (E,E)-1,5-germacra-dien- 6βH,12-olides (crown) ⟶ 1αH,5αH-guianes

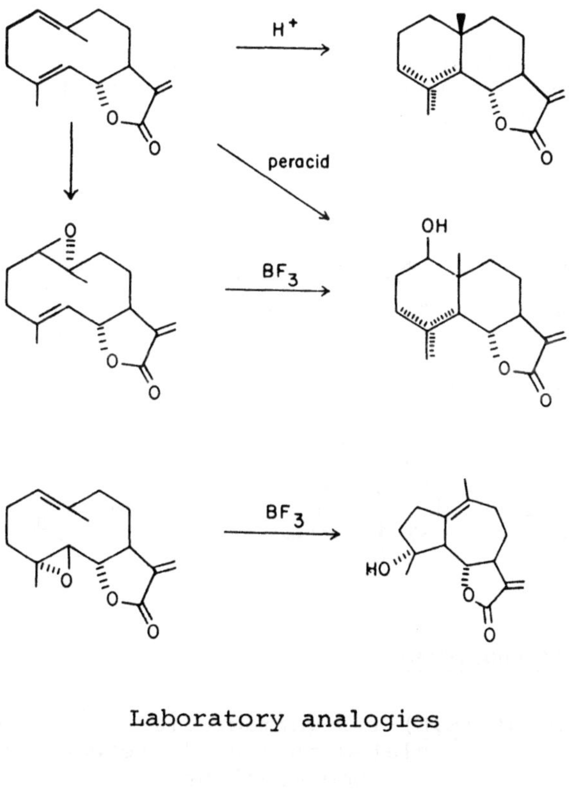

Laboratory analogies

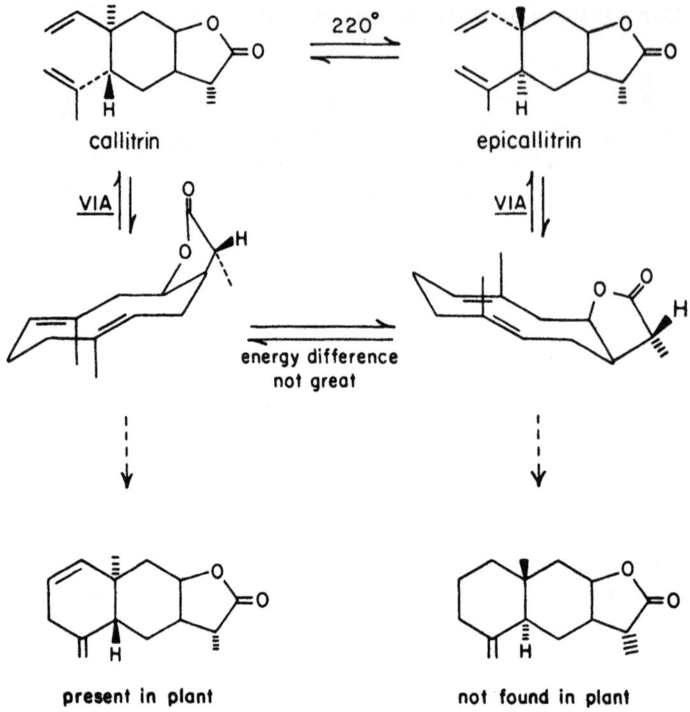

"Anti"-stereochemistry of elemadien-8αH,12-olides and eudesman-8αH,12-olides in Callitris related to conformation of hypothetical intermediates

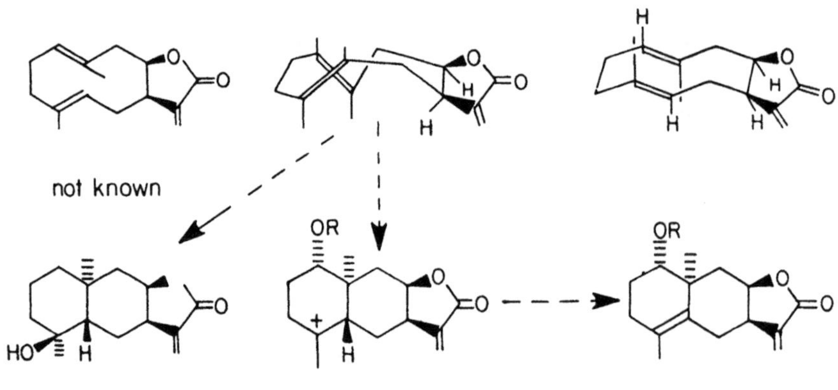

Steiractinia, Wedelia, Aspilia

Conformation of (E,E)-1,5-germacradien-8αH,12-olide in Steiractinia etc. related to "anti" stereochemistry of eudesmanolide

Montanoa, Ursinia species (1979, 1980)

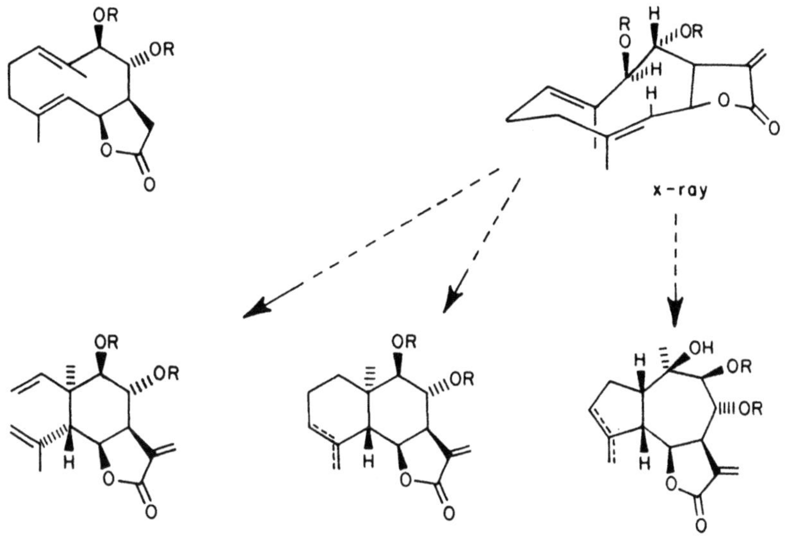

Conformation of (E,E)-1,5-germacradien-6αH,12-olides in Montanoa related to "anti" stereochemistry of hypothetical elemadienolides, eudesmanolides, guaianolides

Pagolettia senegalense (Bohlmann et al. 1983)

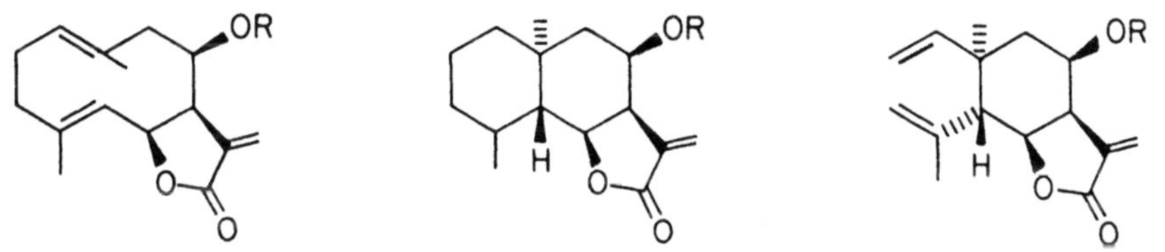

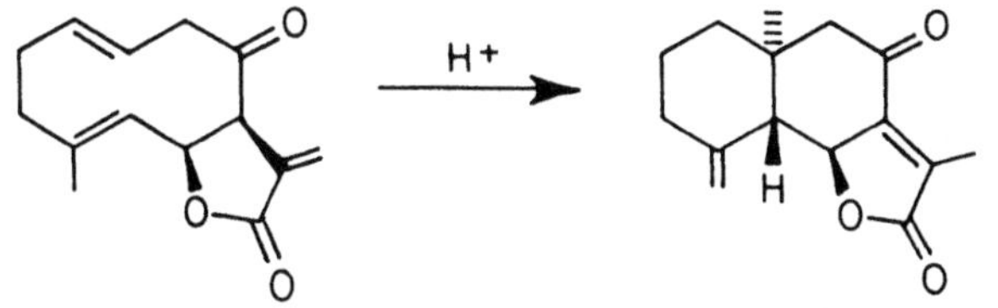

6αH,12-olides in Pegolettia and laboratory analogy

Laserpitium siler

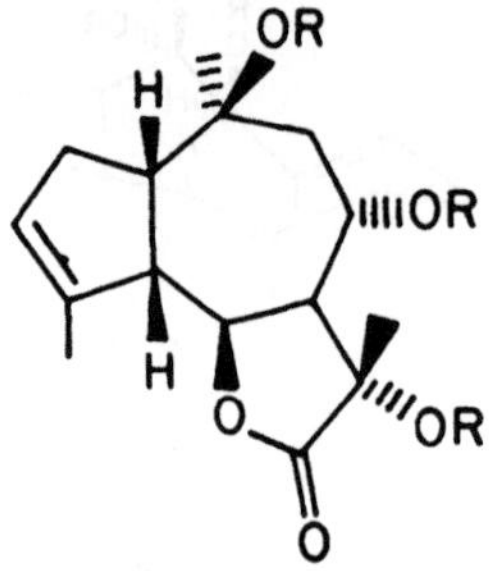

Holub et al. 1978, 1981

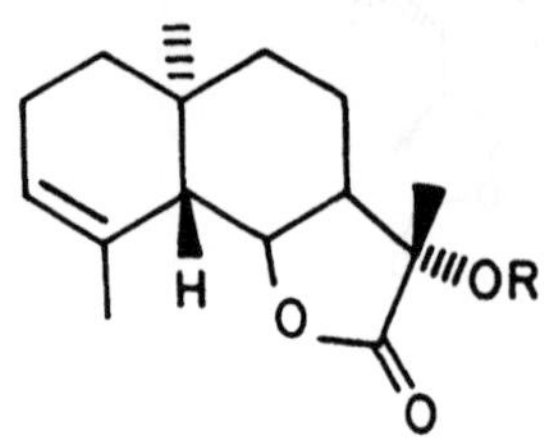

Holub et al. 1982

"Anti"-guaian- and eudesman-6αH,12-olides in Laserpitium

Schkuhria schkuhrioides (Valley of Mexico)

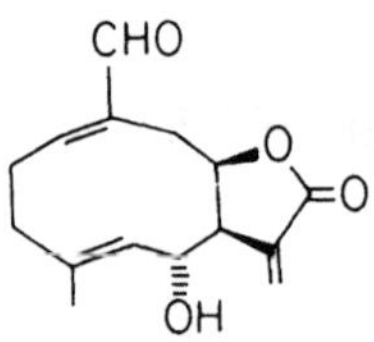

schkuhriolide

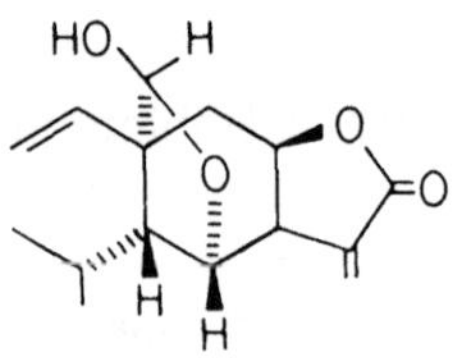

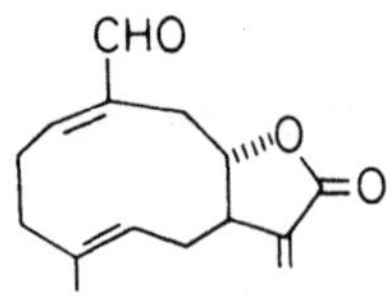

frutescin

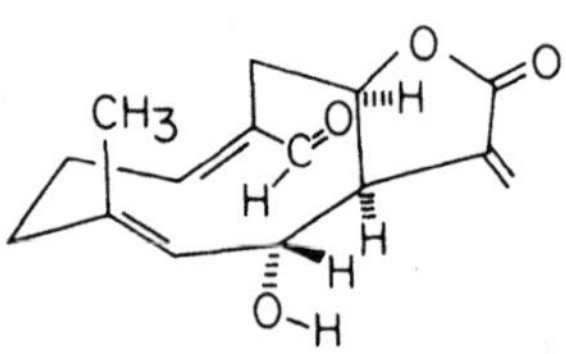

by X-ray

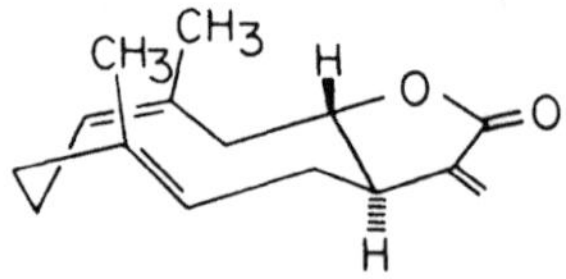

by X-ray

Melampolides and elemanolides in Schkuhria

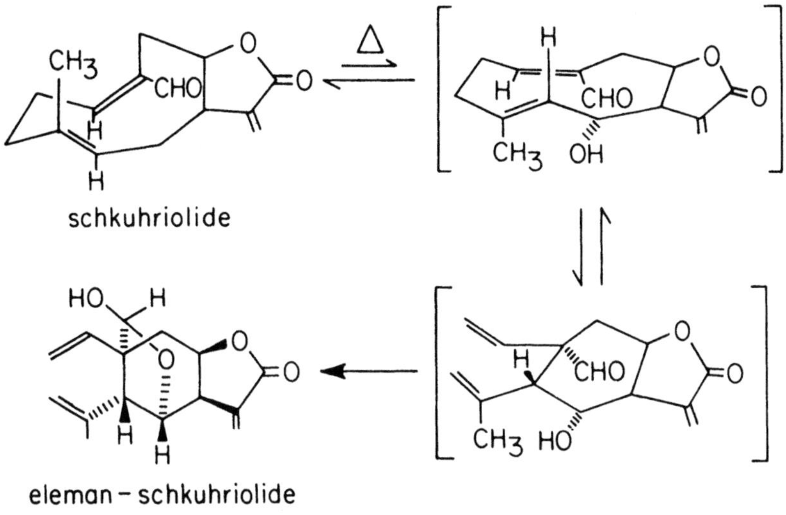

Cope rearrangement of schkuhriolide

eremophilanolides

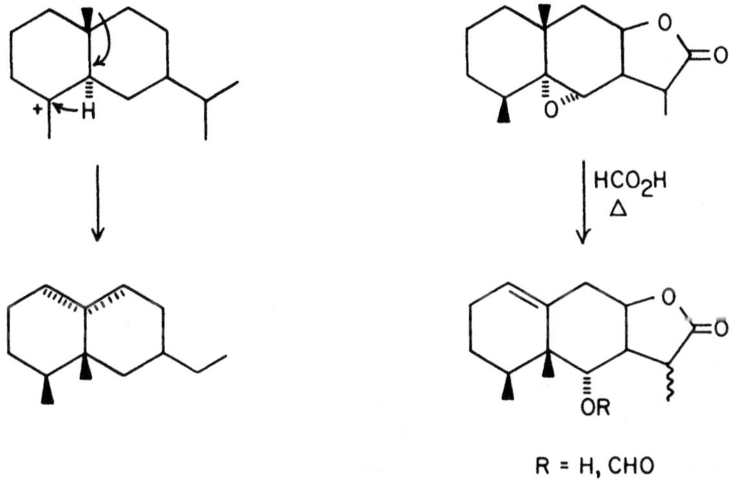

Eremophilanolides - hypothesis and laboratory analogy

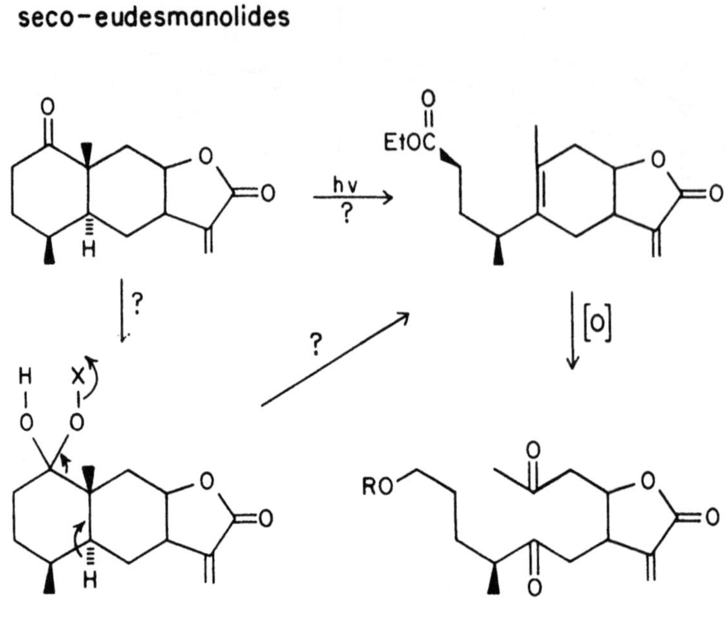

Seco-eudesmanolides I - hypothesis

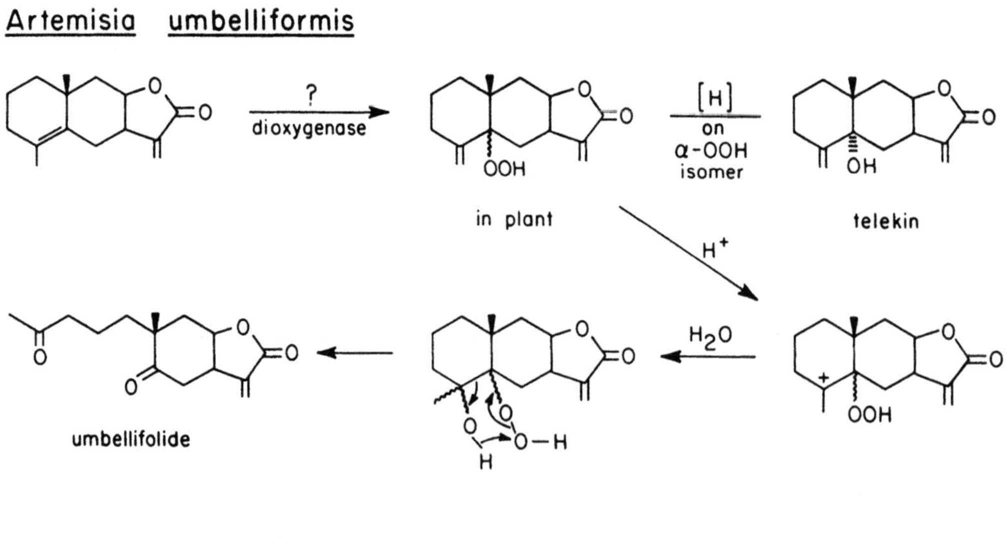

Seco-eudesmanolides II - hypothesis

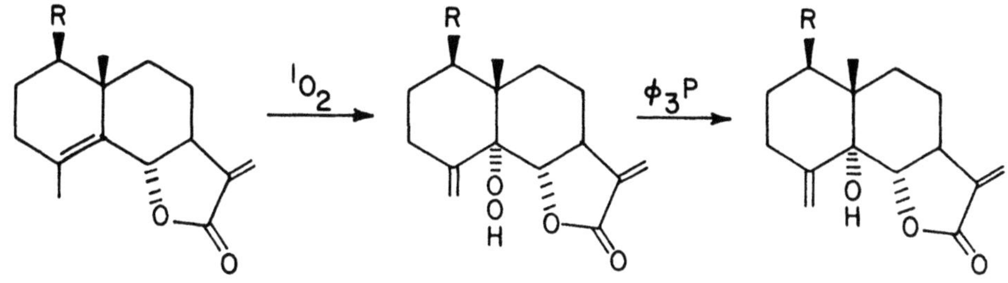

R = H

R = OH magnolialide

R = H arbusculin

R = OH artemin

El-Feraly et al. J.C.S. Perkin I, 1983, 355

Laboratory analogy for initial step - $^1O_2$ as a model for dioxygenase

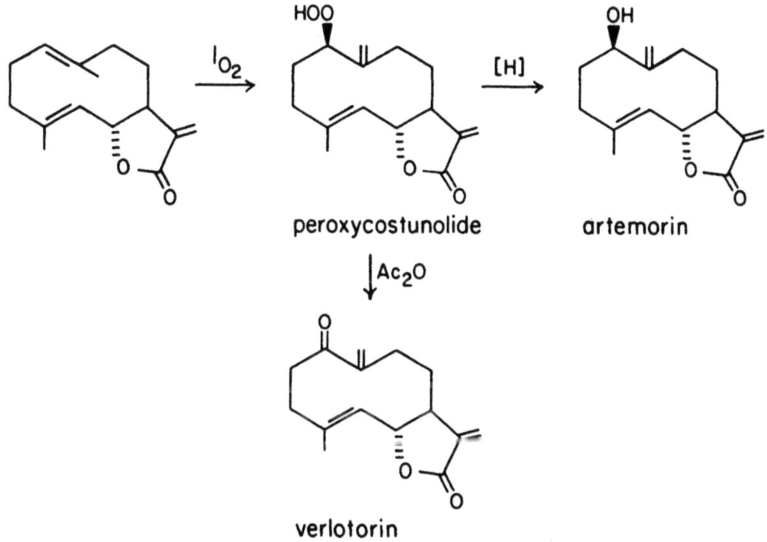

El-Feraly et al, J. Org. Chem. 44, 3953 (1979)

Another laboratory analogy using $^1O_2$

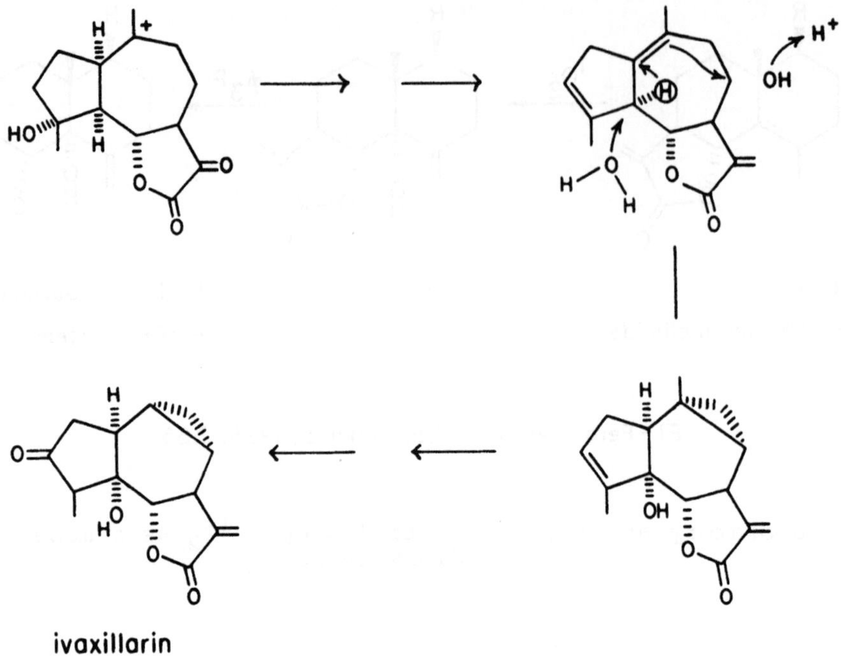

Ivaxillarin-type guaianolides - hypothesis

xanthanolides

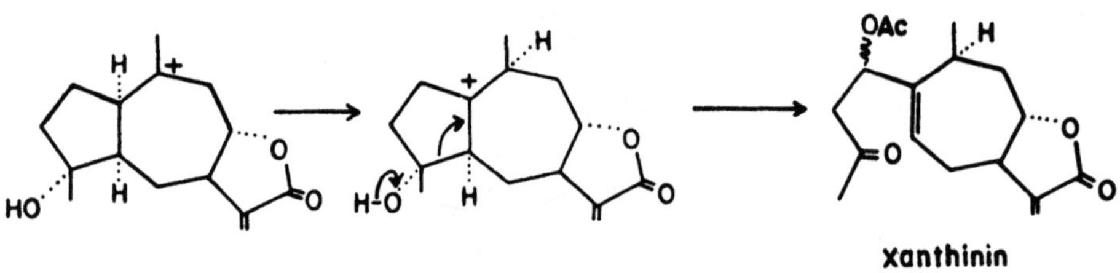

Xanthanolides (seco-guaianolides) - hypothesis

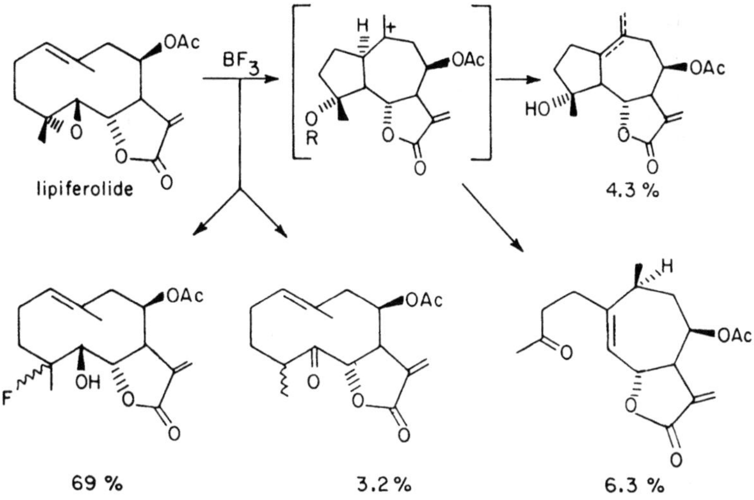

Wilton and Doskotch, J. Org. Chem. 1983

Xanthanolides - laboratory analogy

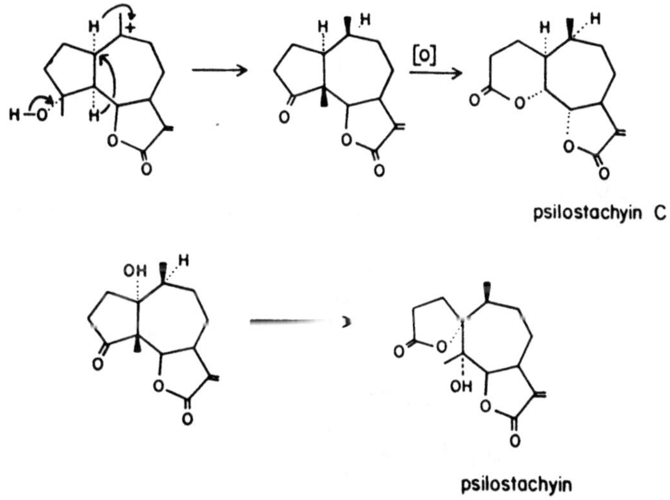

Ambrosanolides and seco - ambrosanolides

HELENANOLIDES, SECOHELENANOLIDES

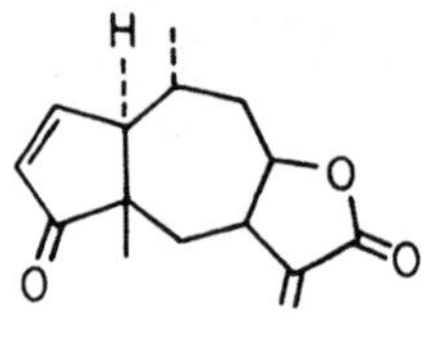

helenalin

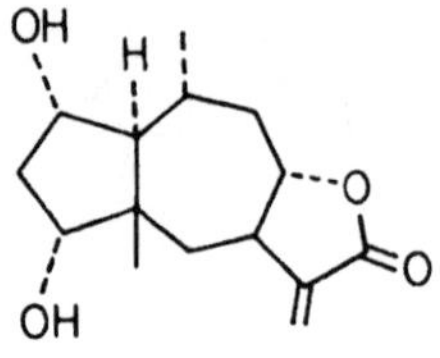

pulchellin

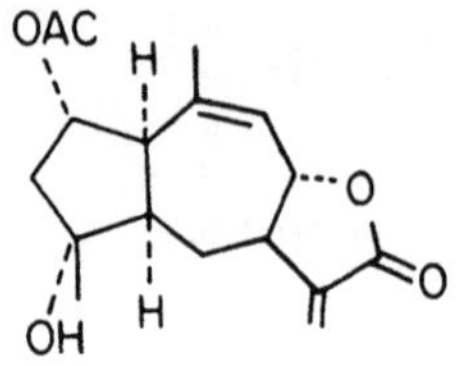

gaillardin

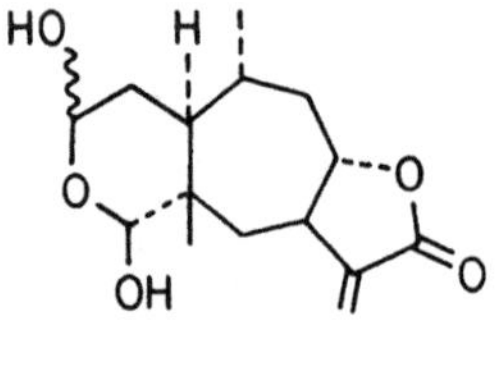

hymenovin

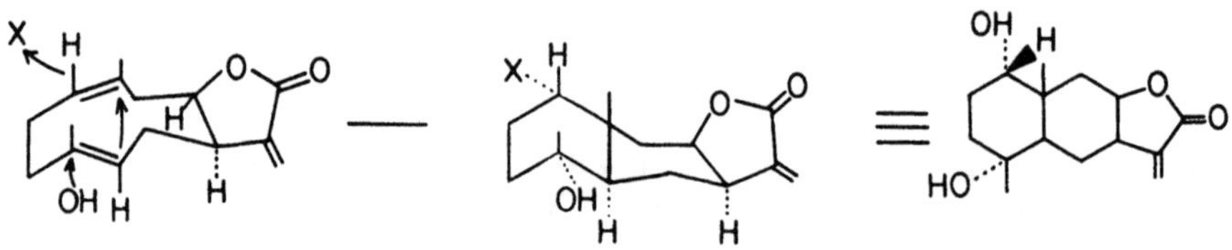

microcephalin

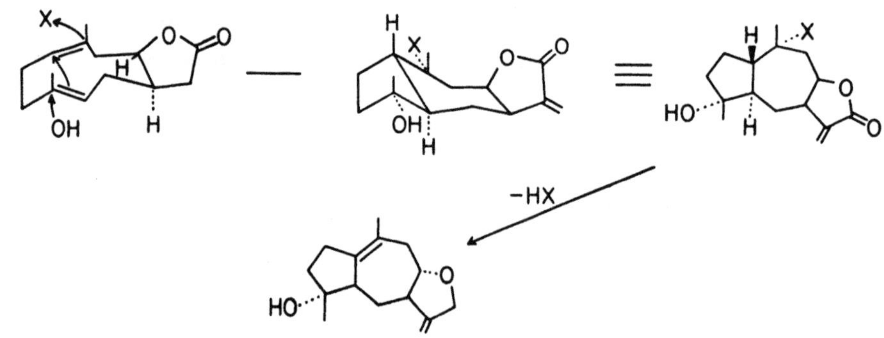

pseudoivalin

Melampolides - precursors for microcephalin and 1βH,5αH-guaianolides?

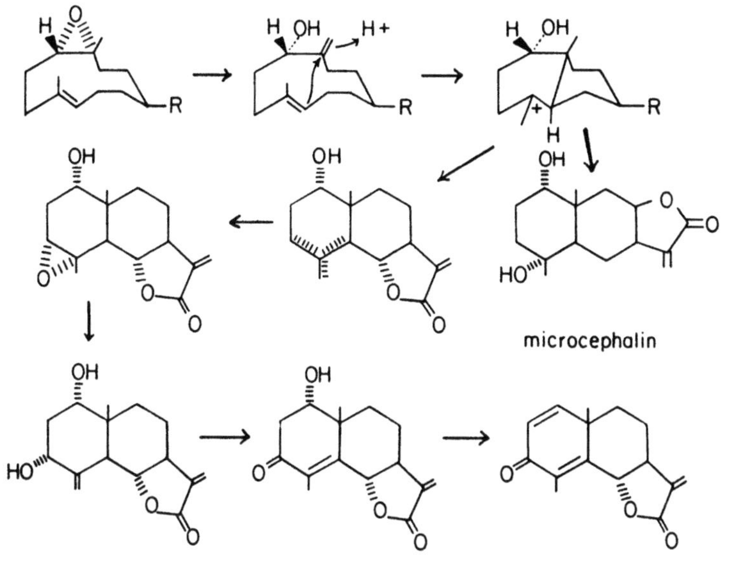

Alternative formulation of melampolide precursor

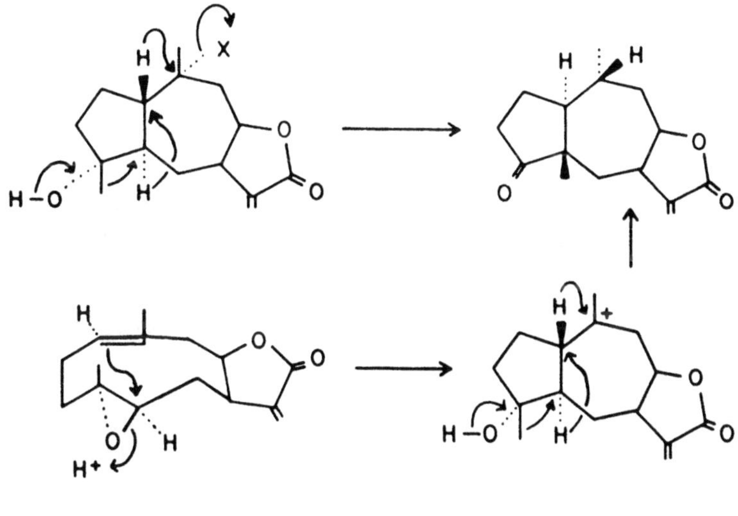

1βH,5αH - guaianolides - precursors of helenanolides?

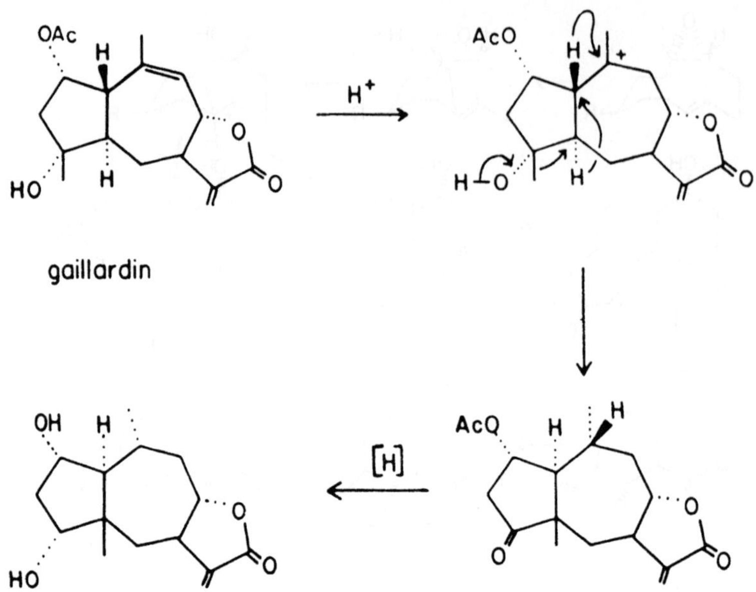

Indirect evidence from Gaillardia

Baileyea multiradiata Harv.

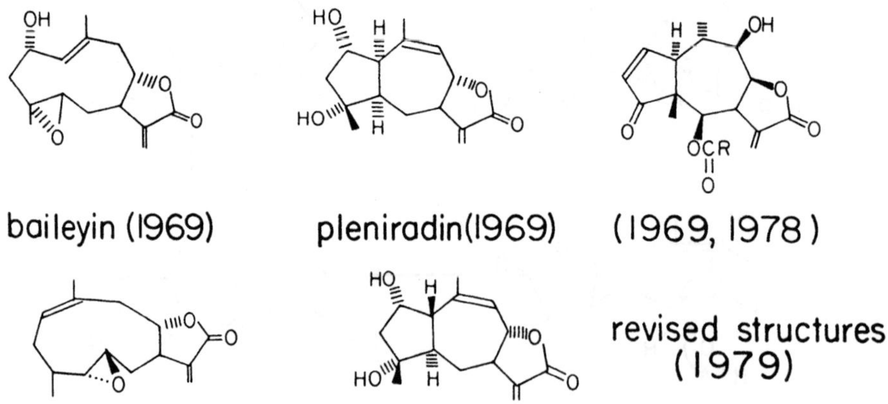

Indirect evidence from Baileya

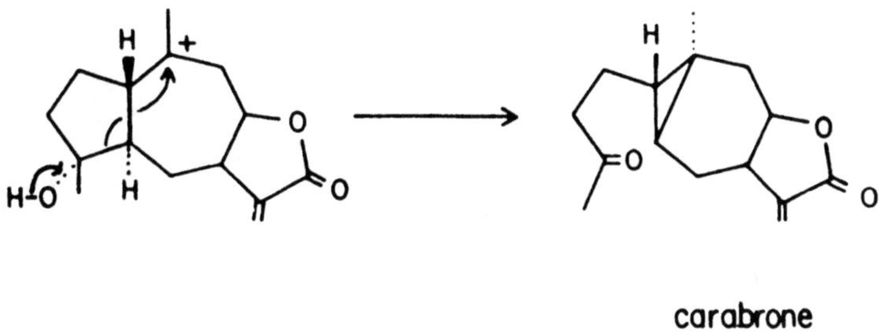

1βH,5αH- guaianolide - precursor of carabrone?

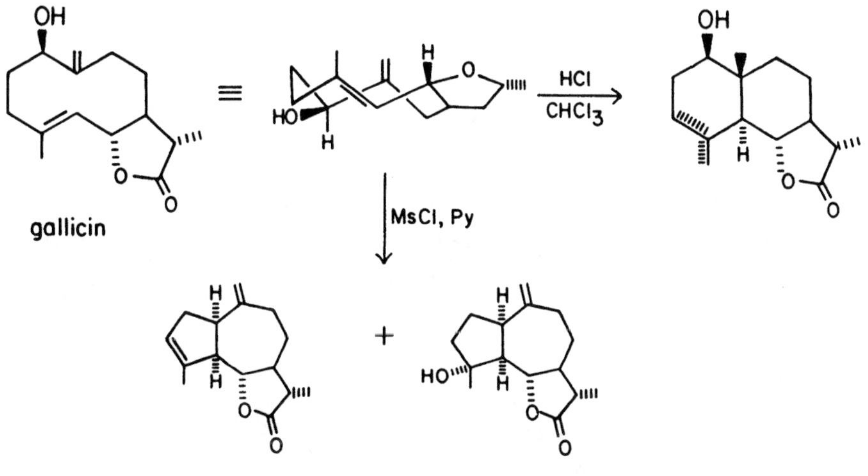

Laboratory analogy I (normal eudesmanolides and
1αH,5αH - guaianolides)

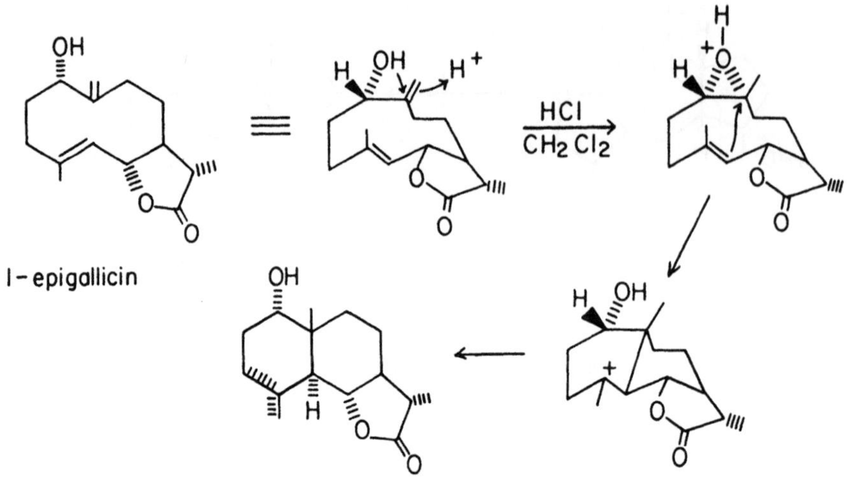

Laboratory analogy II (microcephalin)

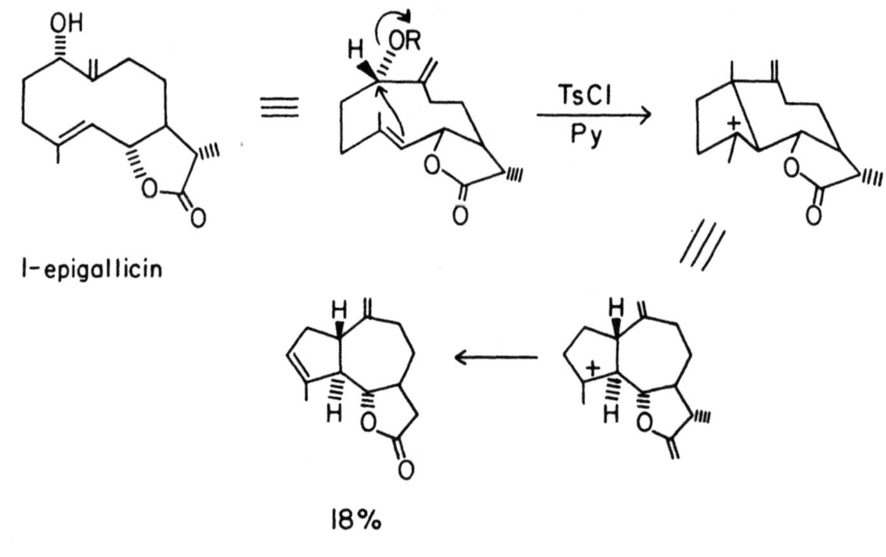

Laboratory analogy III (1βH,5αH-guaianolide)

Synthetic Study of Natural Products Via Three-Membered Heterocycles

Tetsuji Kametani

Institute of Medicinal Chemistry, Hoshi University, Ebara 2-4-41, Shinagawa-ku, Tokyo 142, Japan

1 Introduction

The outstanding characteristic of the three-membered hetero rings is their high reactivity to a wide variety of reagents because of the favorable release of strain energy involved. For this reason, these heterocycles may be converted to a wide variety of functionalized compounds by nucleophilic or electrophilic ring openings. In this symposium, I will mention syntheses of alkaloids, terpenes and β-lactam antibiotics, in which the reaction of a tentatively formed or an easily available three-membered heterocycles becomes the key step in a formation of the basic skeleton of natural products.

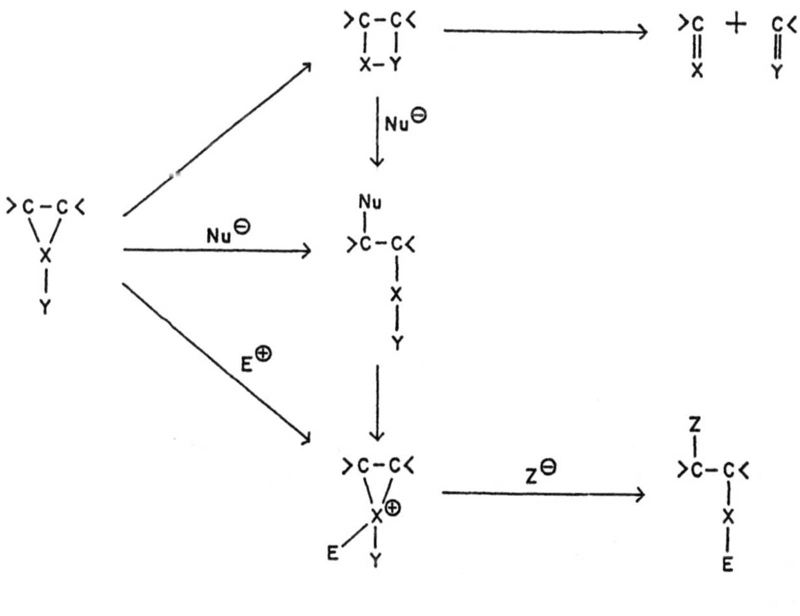

Natural Product Chemistry
Ed. by Atta-ur-Rahman

The three-membered rings discussed in this symposium are as follows.

1. Oxirane ring 

2. Thiirane ring 

3. Selenirane ring 

4. Aziridine ring 

2 Synthesis Via Oxirane Intermediate —— Biogenetic-type Synthesis of Camptothecin

Indoles are cleaved at the C_2-C_3 bond by an action of singlet oxygen through a dioxetane derived from an initially formed perepoxide as shown below.

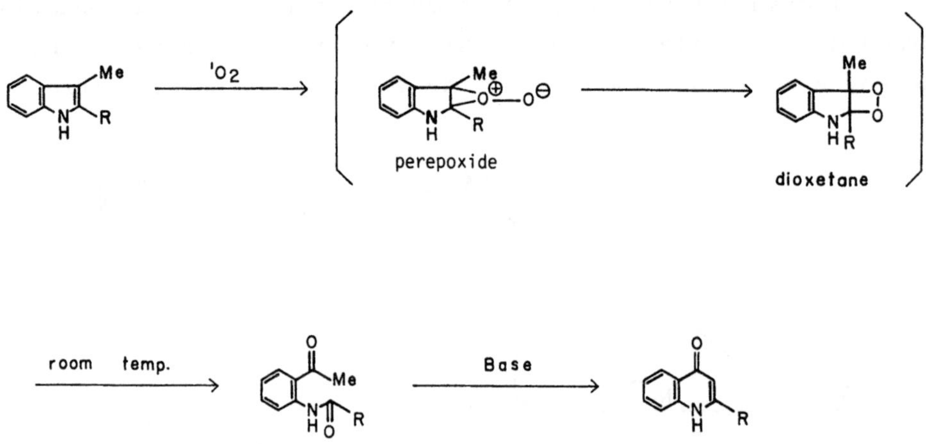

This reaction forms one of the key steps in a biogenesis of the antitumor alkaloids, camptothecin (1) and its 10-hydroxy derivative (2), isolated from <u>Camptotheca</u> species in China; thus these alkaloids are biosynthesized by an oxidative transformation of strictosamide (3), derived from tryptophan and secologanin via isovincoside, into a quinoline derivative, followed by an aromatization and a hydroxylation.

Biogenesis of Camptothecin

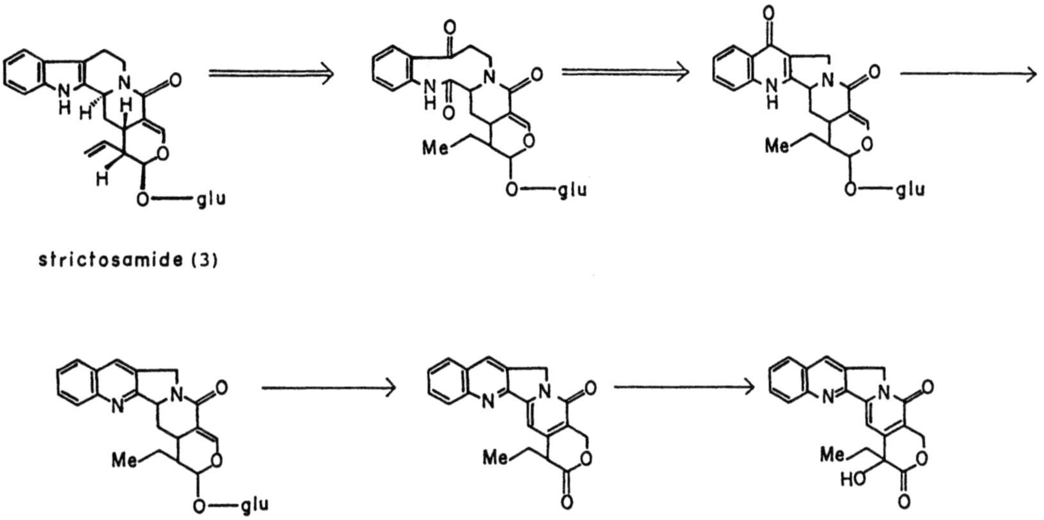

strictosamide (3)

camptothecin (1)

We have examined a total synthesis[1] of camptothecin (1) and 10-hydroxycamptothecin (2) along the biogenetic route and succeeded in a synthesis of both alkaloids in the racemic form as follows. Condensation of 3,4-dihydro-1-methyl-β-carbolines (4 and 5) with a mixture of methyl γ,γ-di-*tert*-butoxycarbonyl-α-methoxycarbonylcrotonate and *tert*-butyl α-*tert*-butoxycarbonyl-γ,γ-dimethoxycarbonylcrotonate, followed by sodium borohydride reduction gave the indolo[*a*]quinolizin-4-ones (6 and 7), a biogenetic equivalent of strictosamide (3).

Photo-oxygenation of 6 and 7 in the presence of Rose bengal in methanol and a subsequent treatment with sodium hydrogen carbonate produced the indolizino[1,2-*b*]quinolones (8 and 9), which were converted to the pyridones (10 and 11) by a reaction with thionyl chloride and then a catalytic hydrogenation. Treatment of pyridones (10 and 11) with DDQ, followed by a successive reduction with DIBAL and sodium borohydride and then hydrolysis with TFA afforded the dehydroxyethylcamptothecins (12 and 13), which were converted into camptothecin (1) and 10-hydroxycamptothecin (2) by application of the established method.

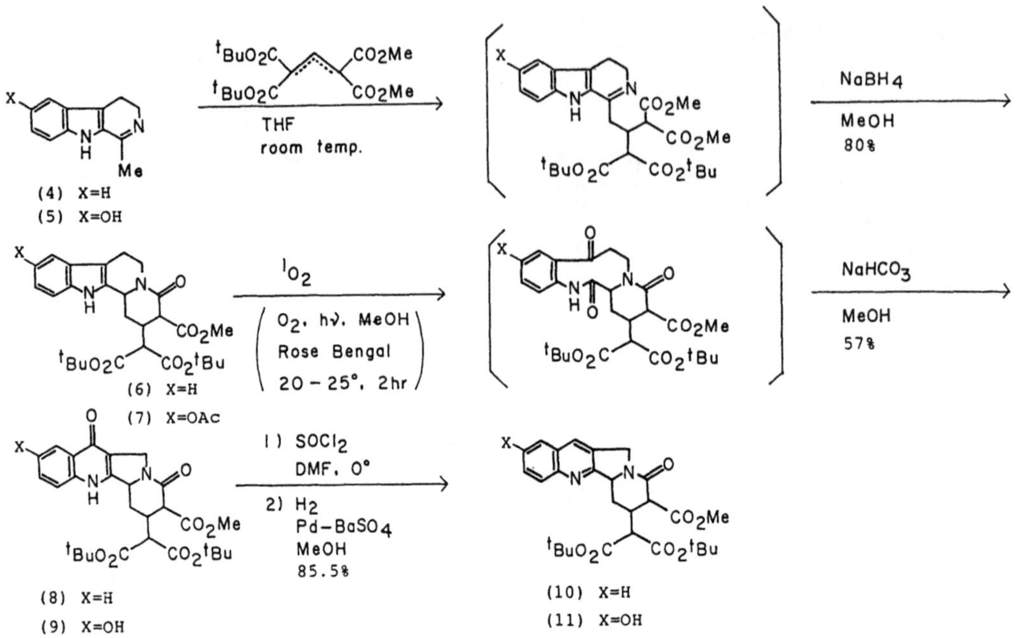

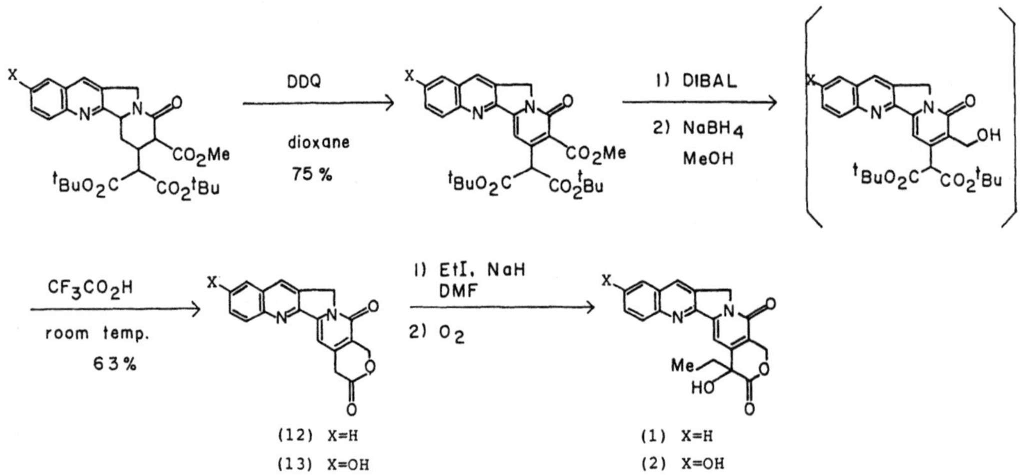

The above synthesis, however, has some demerit from synthetic point of view; thus, the overall yield to gain (±)-camptothecin was not satisfactory because of a poor result for the introduction of the ethyl group at C_{20} position.
Therefore more efficient synthesis of the indoloquinolizidine derivative having all carbon atoms required for the synthesis of camptothecin was further investigated[2].

Reaction of 1-methyl-3,4-dihydro-β-carboline (4) with pyrone (14) caused a double annelation to produce the pentacyclic compound (15), which was converted to the quinolone (16) by photo-oxygenation followed by basic treatment. After transformation of 16 to the quinoline (17), the epoxyetheno-bridge was cleaved by the action with trifluoroacetic acid in the presence of ethanedithiol. Desulfurization of 18 and the subsequent oxidation with DDQ furnished the acetate (19) which had been correlated with (±)-camptothecin (1) in two steps.

Modified Synthesis of Camptothecin

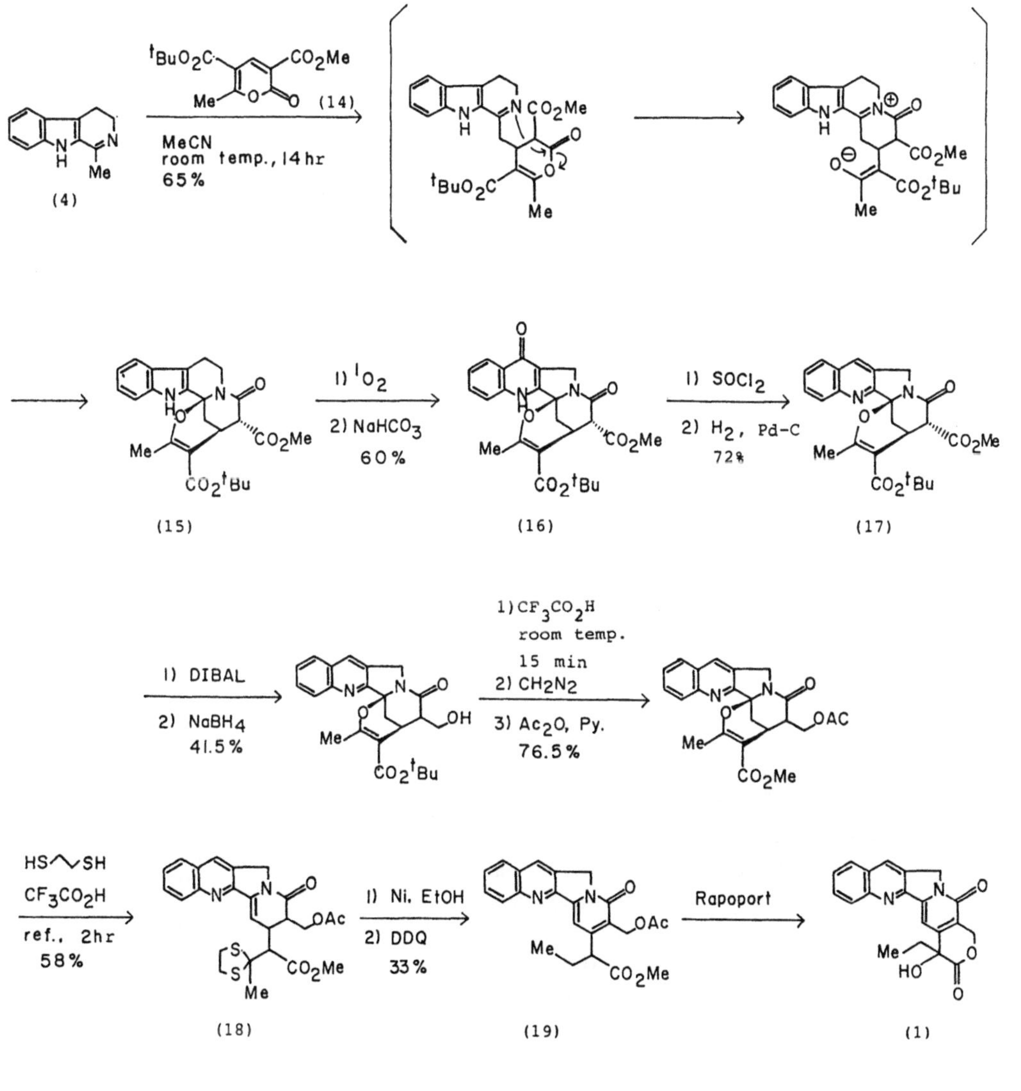

The reaction of indole with singlet oxygen in the presence of a nucleophile has been known to proceed as follows.

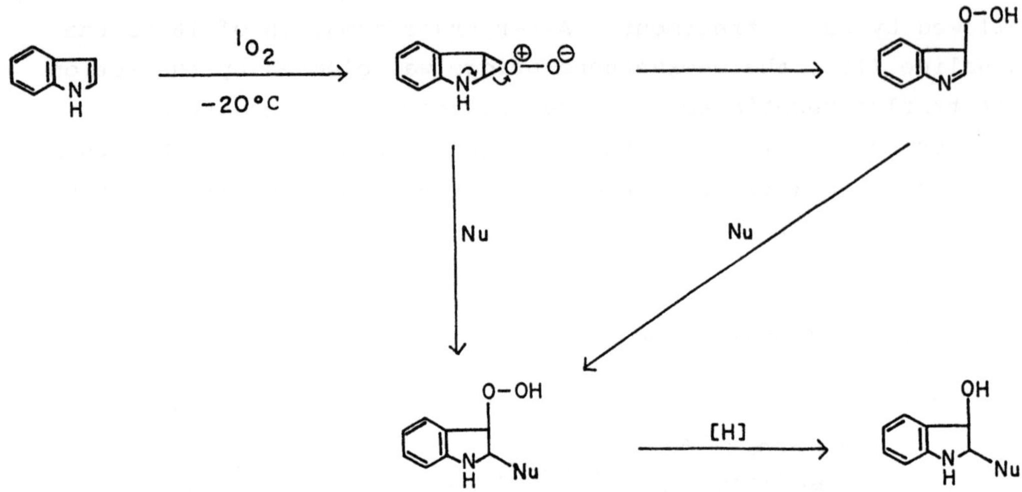

Application of this reaction provided the total synthesis of brevianamide E (20)[3].

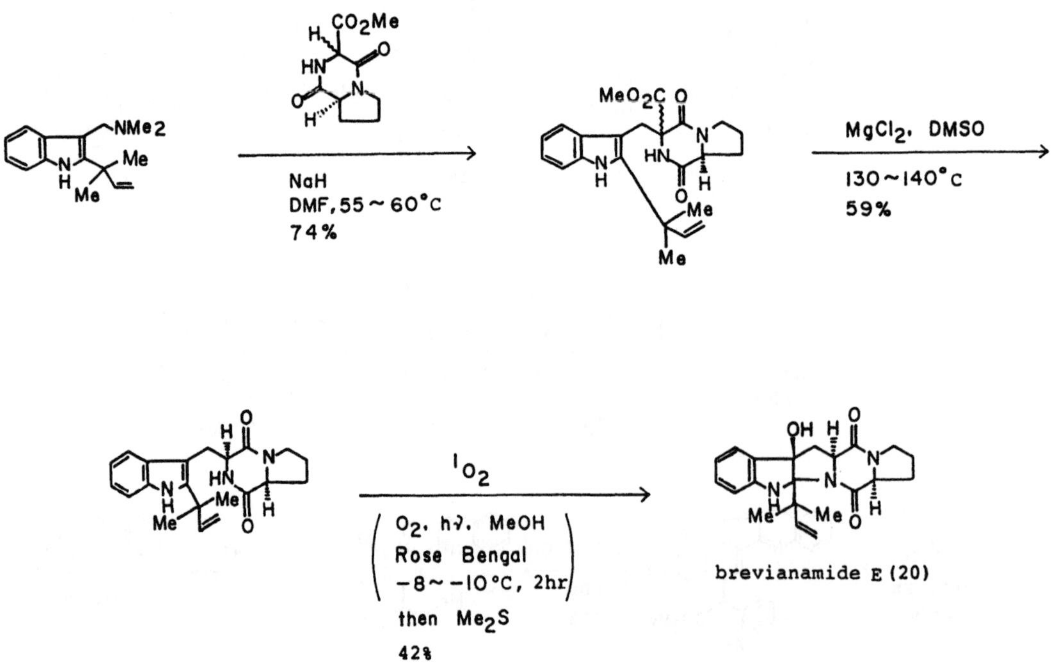

brevianamide E (20)

3 Synthesis Via Episulfide Intermediate (Sulfenocycloamination)

Total Synthesis of Necine Bases

Recently we developed a new monocyclic β-lactam synthesis by an addition of benzenesulfenyl chloride to α,β-unsaturated amides followed by a base treatment in the presence of a phase transfer catalyst as follows[4].

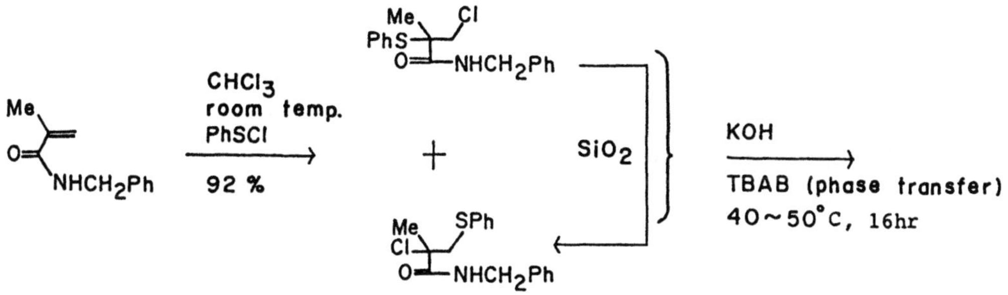

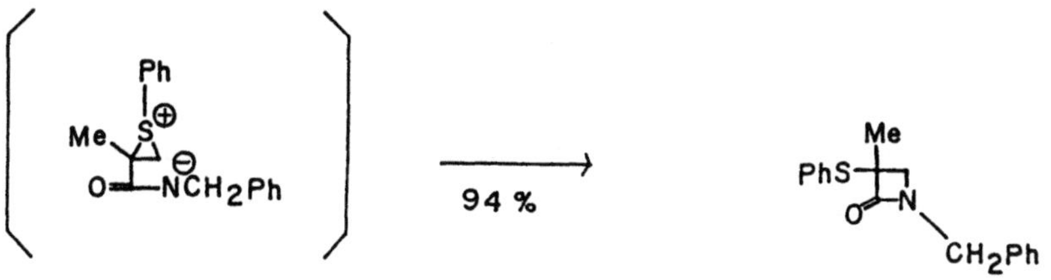

This type reaction is called "Sulfenocycloamination", and has been applied to a formal synthesis of nocardicinic acid[5].

Thus, the optically active α,β-unsaturated amide (20) was treated with phenylsulfenyl chloride and subsequently sodium hydride to give the azetidinone (21), whose oxidative elimination reaction afforded the exo-methylene lactam (22).

Since the lactam (22) had been already converted to nocardicinic acid, this synthesis provided a formal total synthesis of nocardicinic acid.

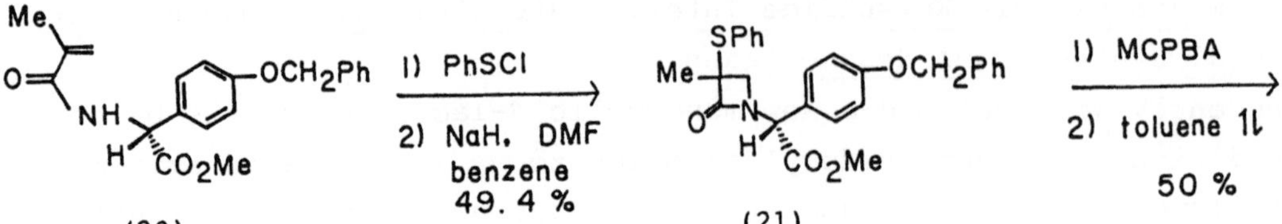

(20) (21)

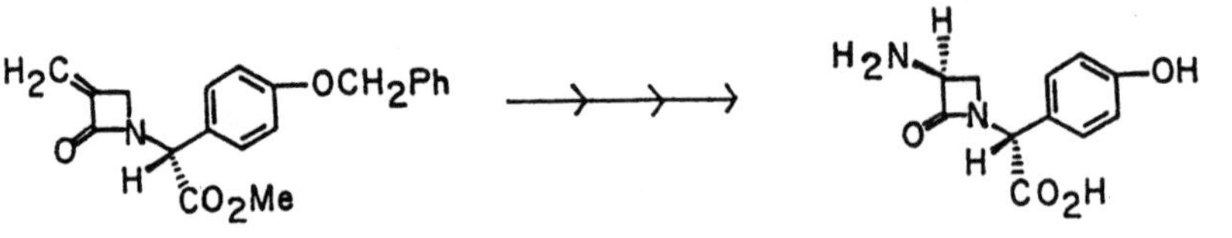

$[\alpha]_D$ -16.2°

(22)

nocardicinic acid

Furthermore, application of this synthetic strategy leads to the stereoselective synthesis of necine bases, retronecine and turneforcidine[6]. A readily available 3-pyrrolidinone (23) was treated with the cis-butene (24) in the presence of p-toluenesulfonic acid and sodium sulfate to give the compound (25), whose reduction with sodium borohydride and benzylation as usual afforded the carbamate (26). After hydrolysis of 23 with potassium hydroxide the resulting amine (27) was applied to the sulfenocycloamination. Namely, treatment of the hydrochloride of 24 with benzenesulfenyl chloride in methylene chloride at 0°C followed by the ring closure with potassium carbonate in acetonitrile in the presence of sodium iodide produced the sulfide (28) in 71.9 % yield from 24. The hydrochloride of 25 was oxidised with m-chloroperbenzoic acid to afford the sulfoxide, which was then subjected to the syn-elimination reaction by refluxing in xylene giving the olefin (29). Subsequently, the olefin (29) was debenzylated with lithium in liquid ammonia to furnish (±)-retronecine (30). Desulfurisation of 25 with Raney nickel followed by debenzylation gave (±)-turneforcidine (31).

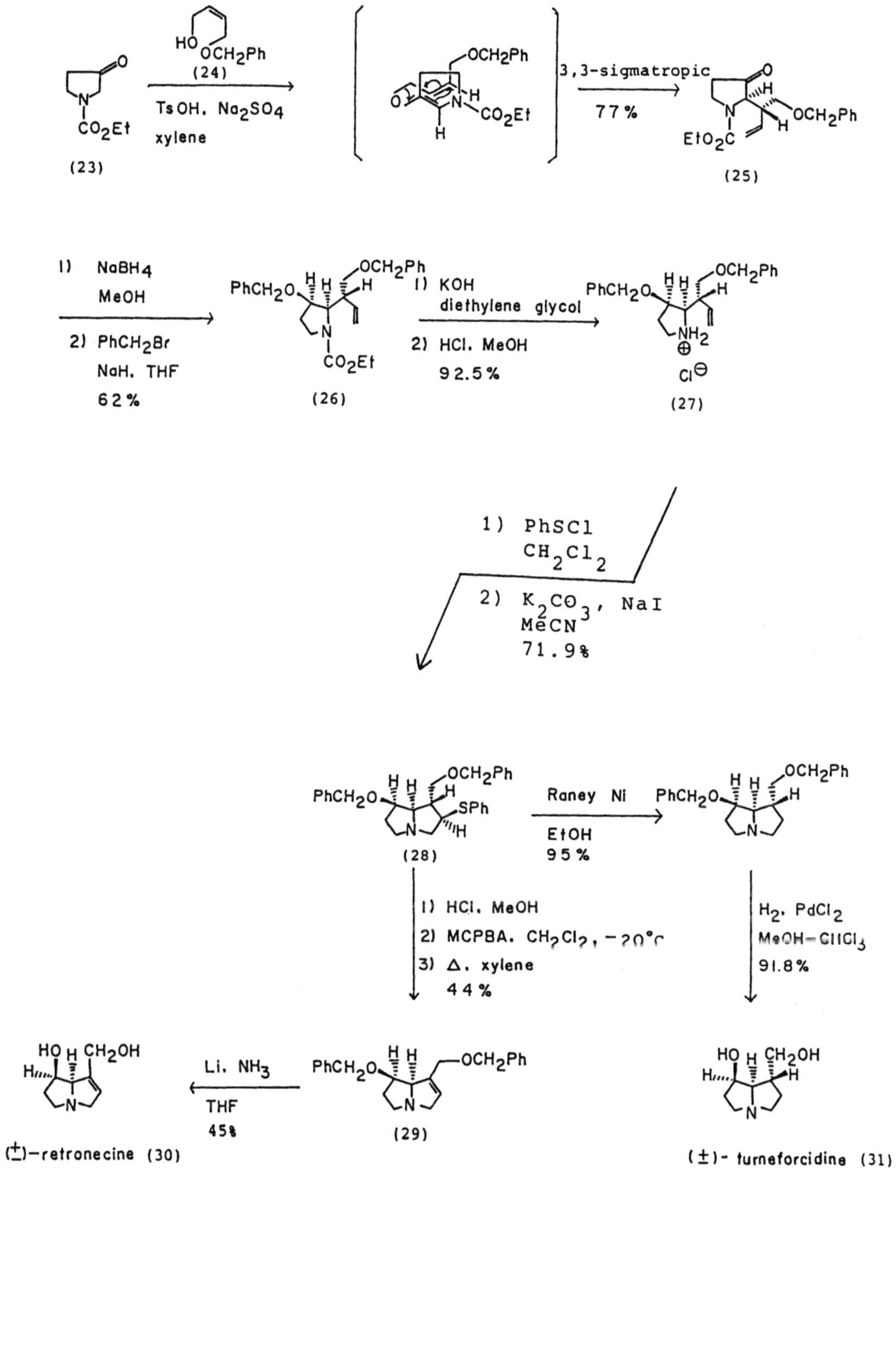

HO
OCH2Ph
(24)
TsOH. Na2SO4
xylene
(23)
3,3-sigmatropic
77%
(25)
1) NaBH4
MeOH
2) PhCH2Br
NaH. THF
62%
(26)
1) KOH
diethylene glycol
2) HCl. MeOH
92.5%
(27)
1) PhSCl
CH2Cl2
2) K2CO3, NaI
MeCN
71.9%
(28)
Raney Ni
EtOH
95%
1) HCl. MeOH
2) MCPBA. CH2Cl2, −20°C
3) Δ. xylene
44%
H2, PdCl2
MeOH−CHCl3
91.8%
Li. NH3
THF
45%
(29)
(±)-retronecine (30)
(±)- turneforcidine (31)

4 Synthesis of Mono-terpenes

Phenylselenenyl group mediated olefin cyclization leads to the formation of a new carbon-carbon bond and this reaction was successfully applied for the synthesis of mono-terpenes such as safranal, p-menthane derivatives, and caparrapi oxide as follows.

Synthesis of Safranal[7]: Treatment of geranylacetate (32) with phenylselenenyl chloride in acetic acid afforded the addition product (33), whose cation cyclization gave rise to the expected acetate (34). Oxidative elimination of the selenide group, followed by dehydration and oxidation, leads to the synthesis of safranal (35).

Total Synthesis of Safranal

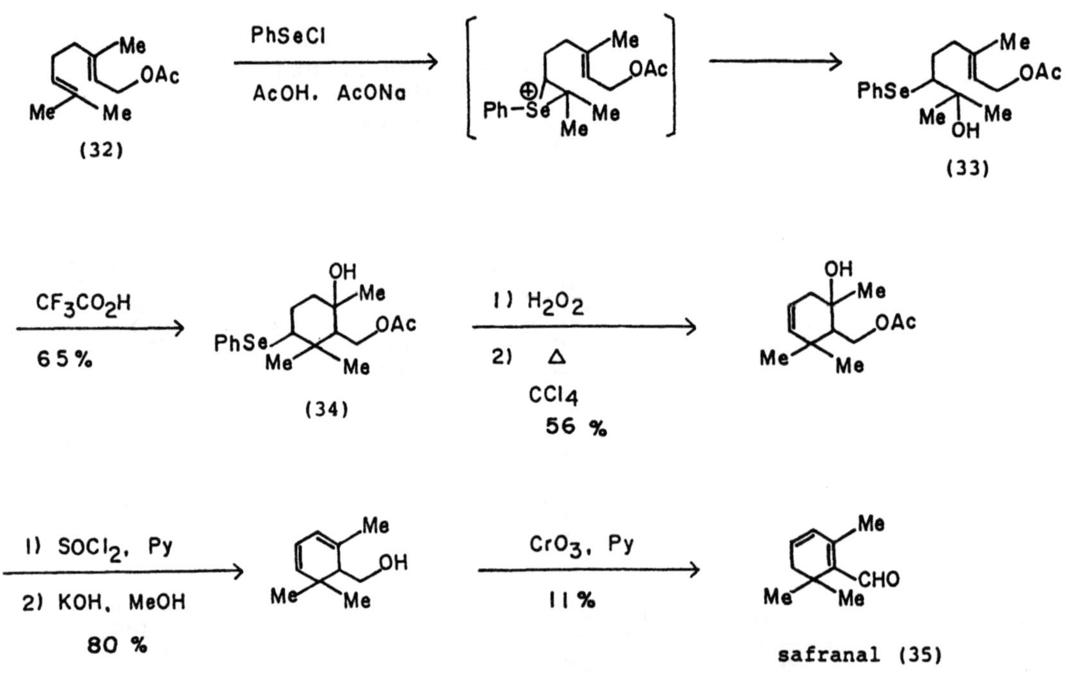

Synthesis of p-Menthane Derivative[8]: p-Menthane derivatives are also prepared by the application of the above olefinic cyclization.

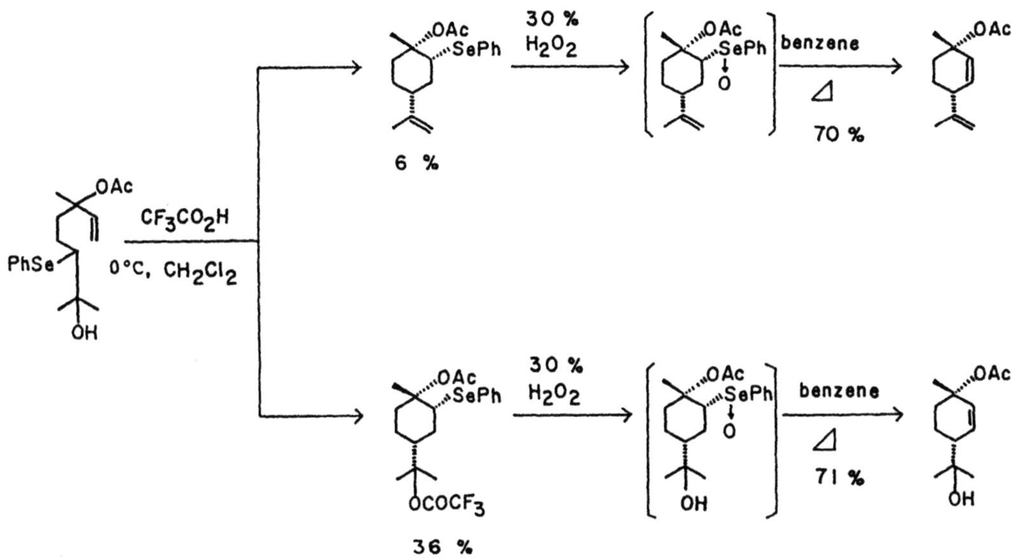

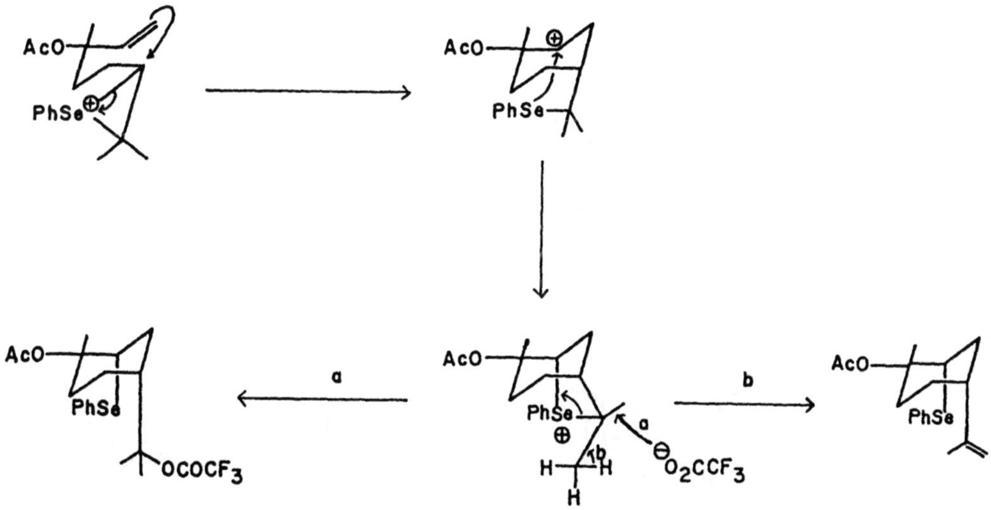

Synthesis of Caparrapi Oxide[9]: Synthesis of caparrapi oxide (36) from nerolidol involved the similar olefin cyclization as a key reaction as follows.

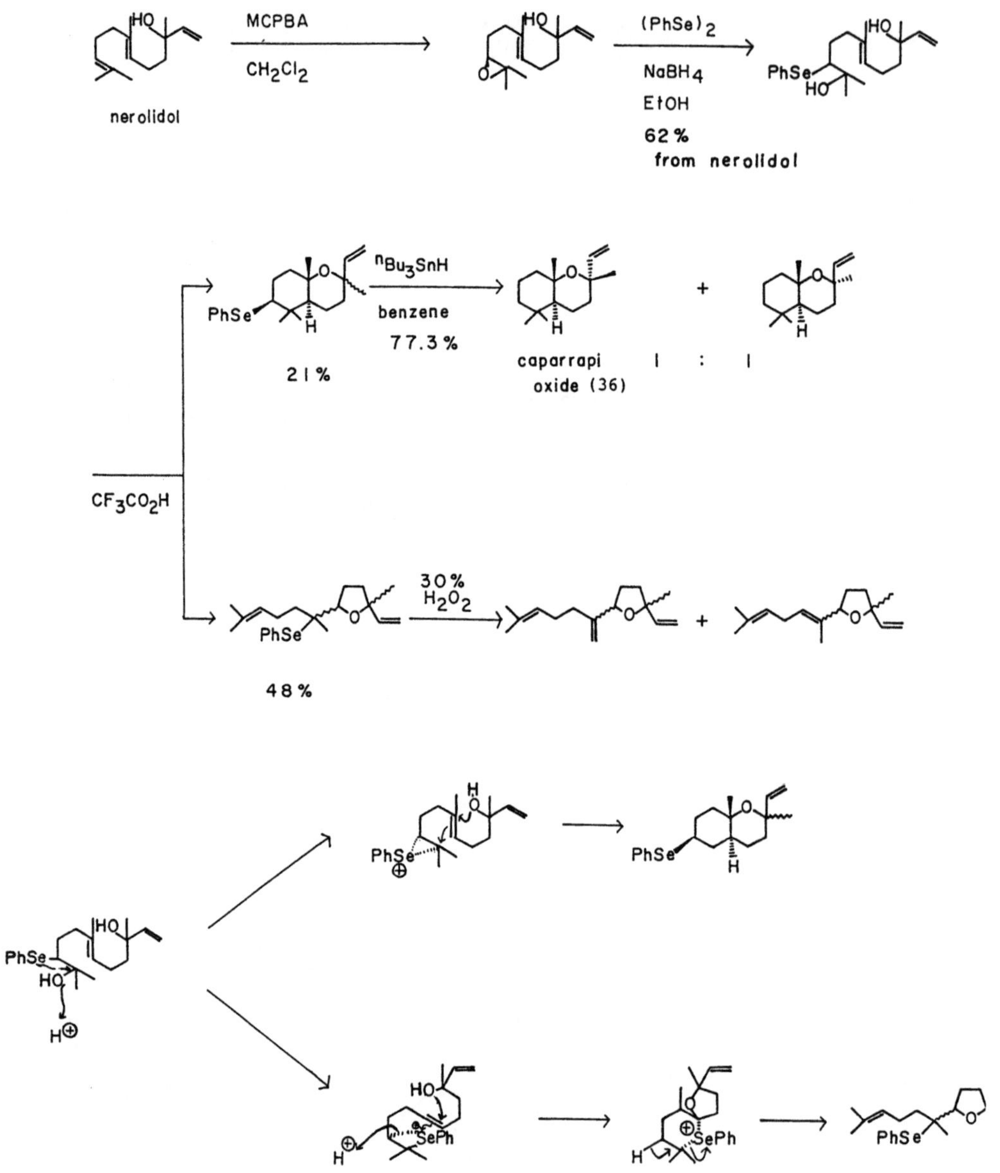

5 Synthesis of Isoquinoline Alkaloids Via Aziridines

a) Employing diazo-compounds: The 3,4-dihydroisoquinoline derivative (37), derived from the corresponding deoxybenzoin, was reacted with diazomethane to give the aziridinium salt (38) which without isolation was treated with 1 % hydrochloric acid in methanol to afford the isopavine alkaloid, reframidine (39)[10].

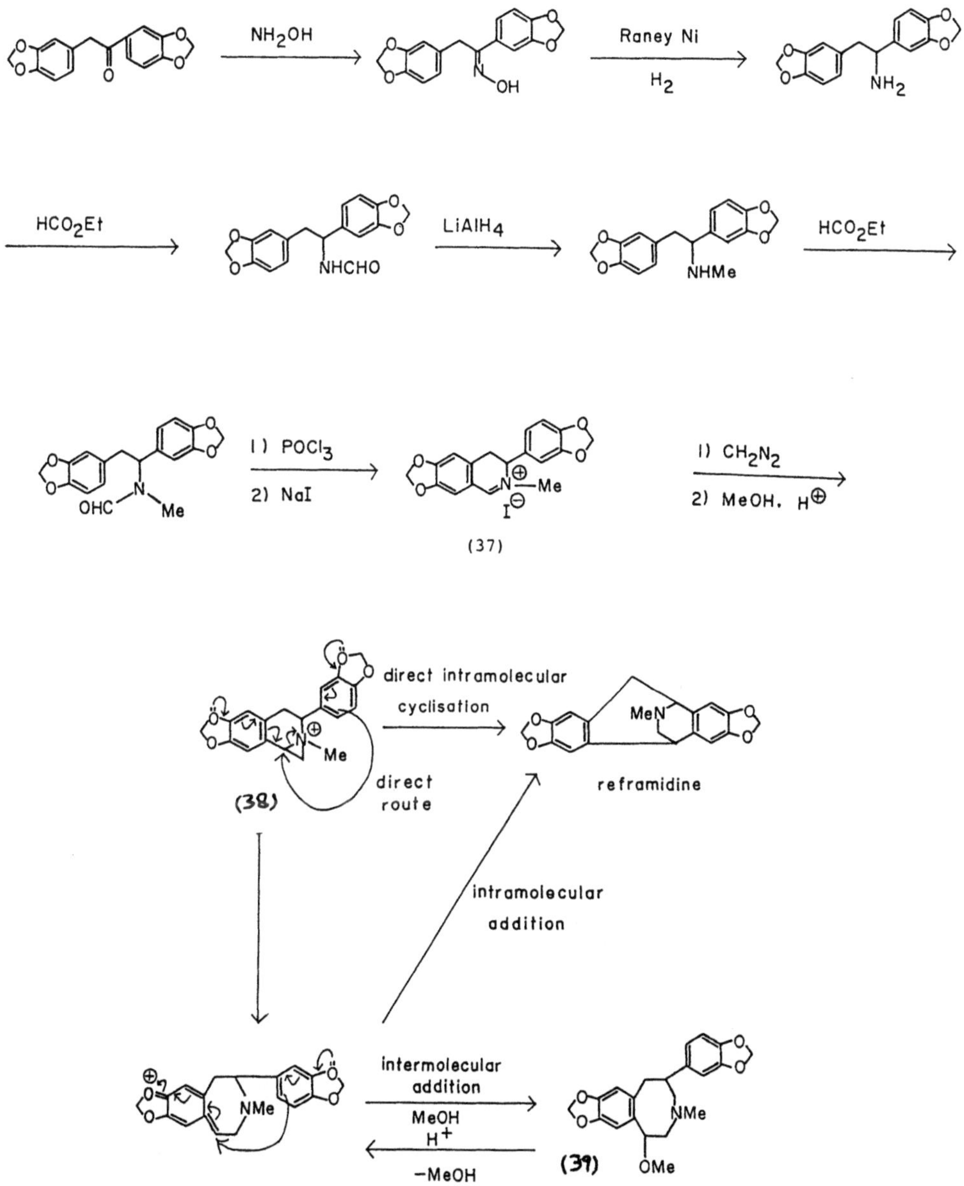

Application of this type of reaction employing a phenyldiazomethane (40) leads to the synthesis of phthalidyl alkaloids, cordrastine (41) and hydrastine (42)[11].

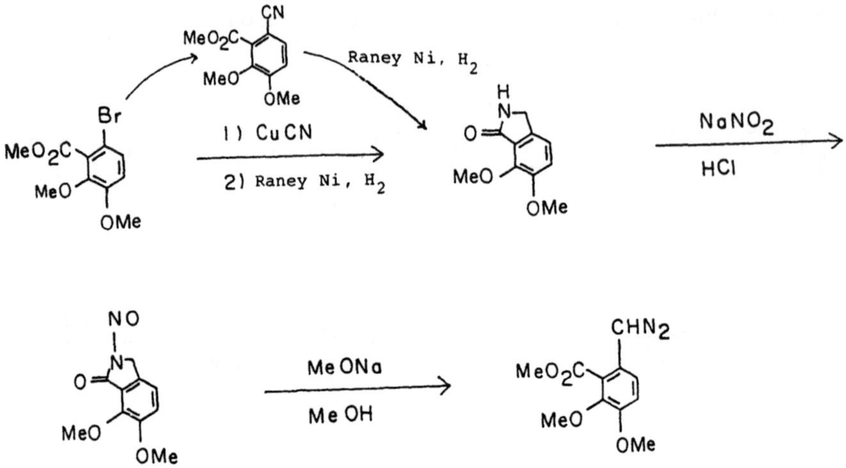

(40)

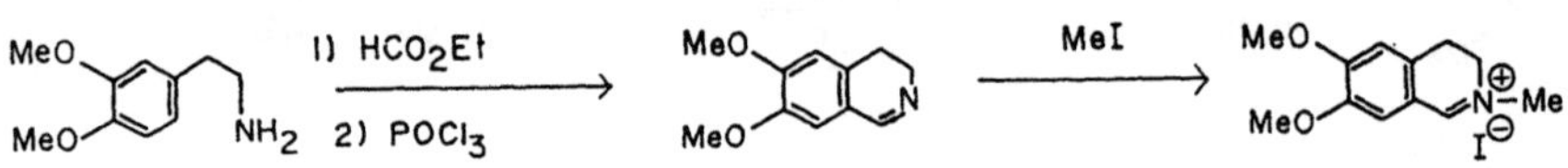

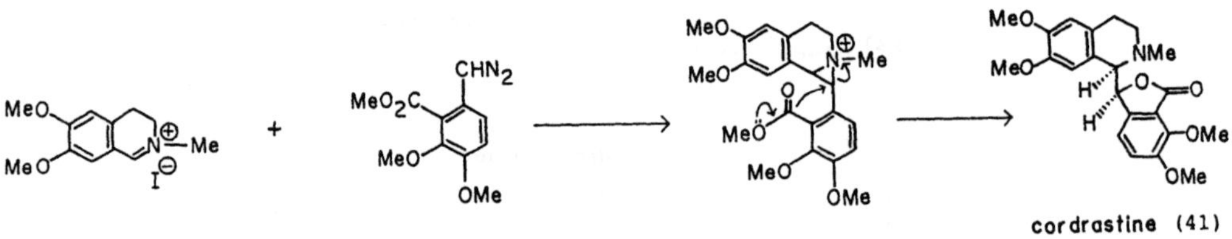

cordrastine (41)

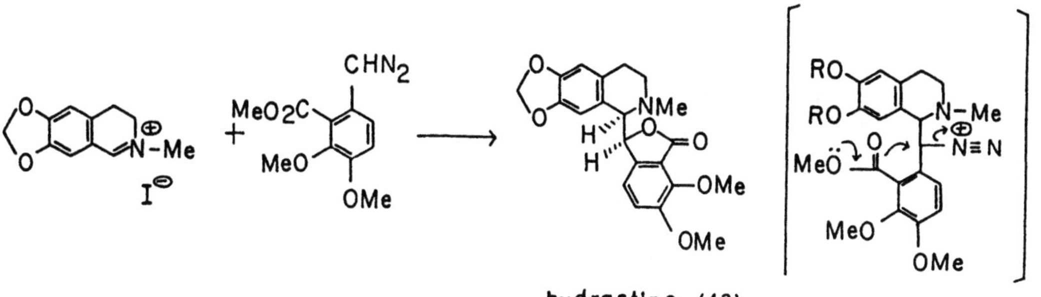

hydrastine (42)

4-Arylisoquinoline alkaloid, cherylline (44) was also synthesized[12] by treatment of the quaternary salt (43) with diazomethane, followed by acid treatment.

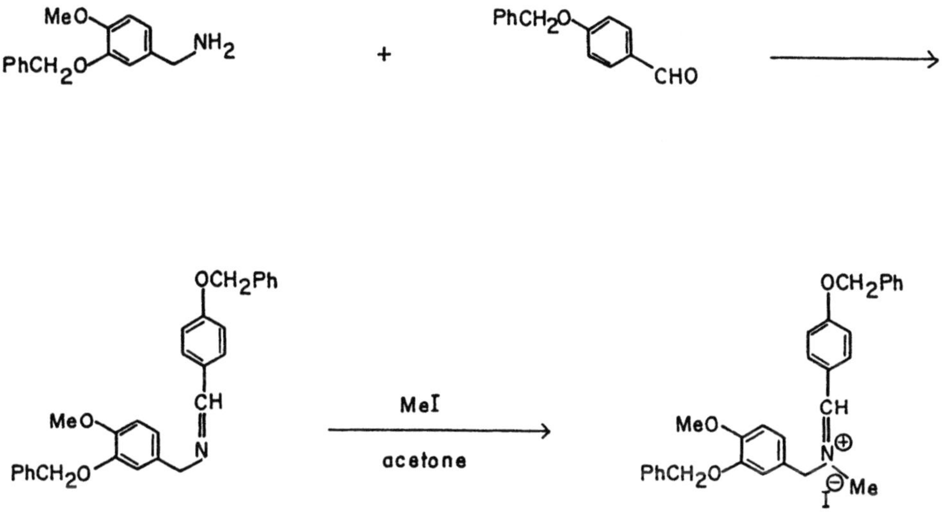

(43)

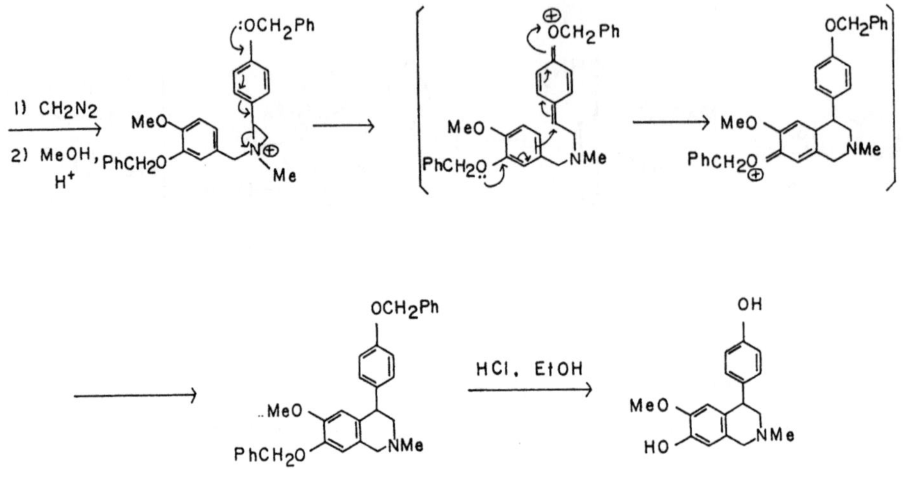

cherylline (44)

b) <u>Emploing ethyleneimine:</u> 2-Benzylisoquinoline alkaloids, sendaverine (45) and corgoine (46) were synthesized[13] from N-benzylaziridine utilizing a ring-opening reaction of the quaternary salt as a key step.

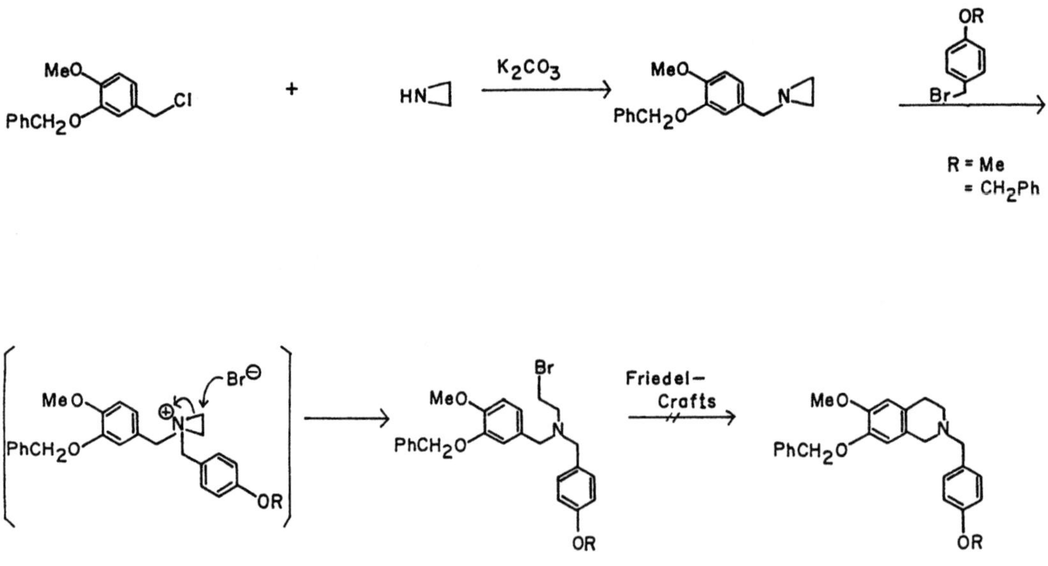

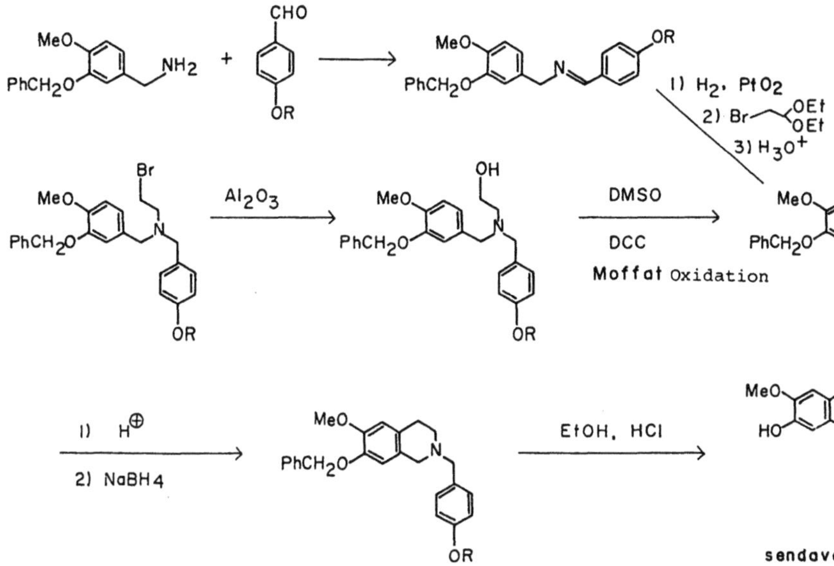

sendaverine (R = Me) (45)
corgoine (R=H) (46)

Simple synthesis of γ-lycorane (47) was carried out as follows[14].

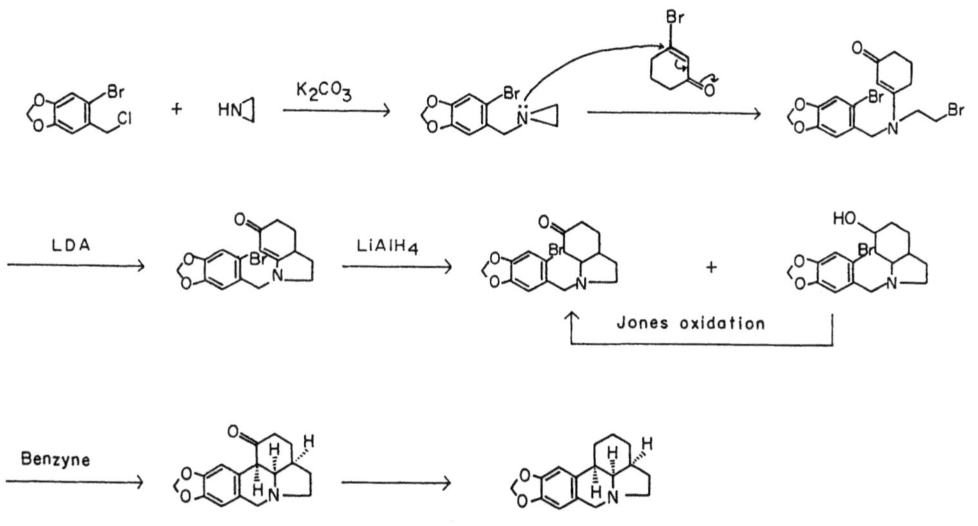

α-dihydrocaranone

γ-lycorane (47)

Similary, this reaction was applied to the synthesis of erythrinan skeleton (48)[15].

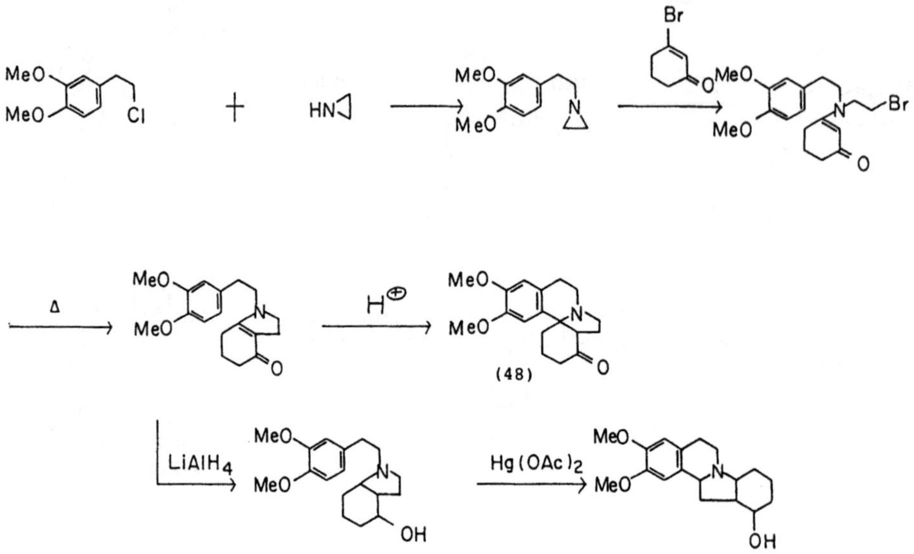

Further application of this reaction provided an alternative synthesis of reframidine (39)[16].

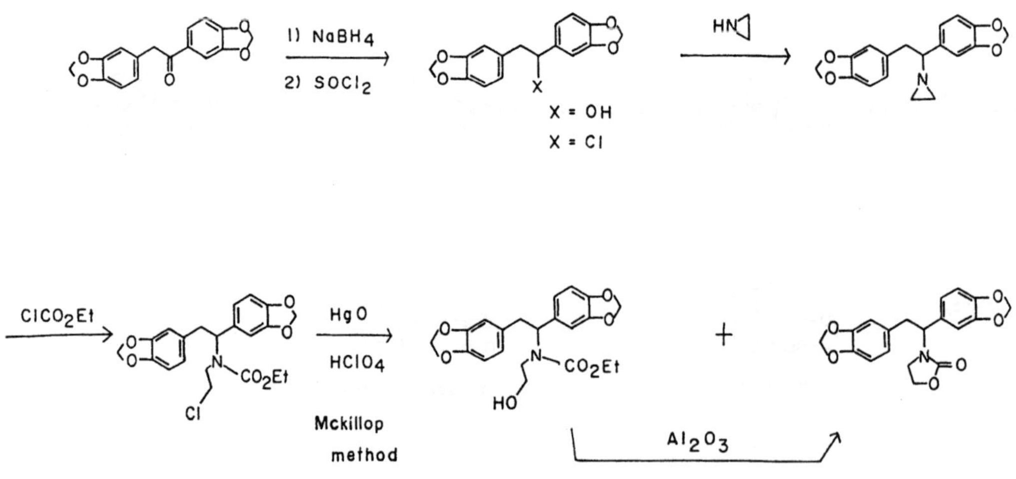

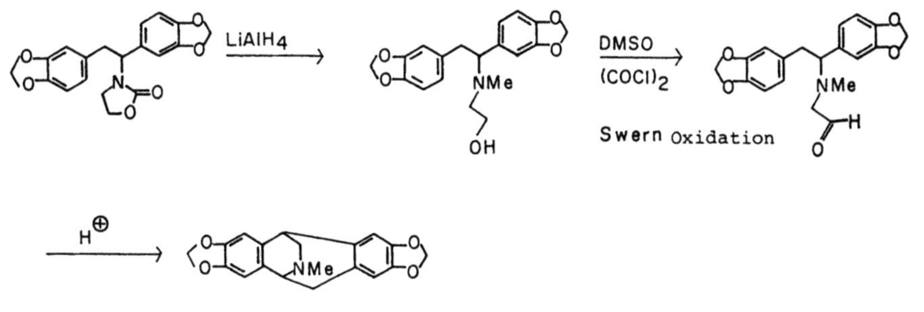

reframidine (39)

The pyrrolizidine alkaloids, trachelanthamidine (49) and supinidine (50) are also synthesized as follows[17].

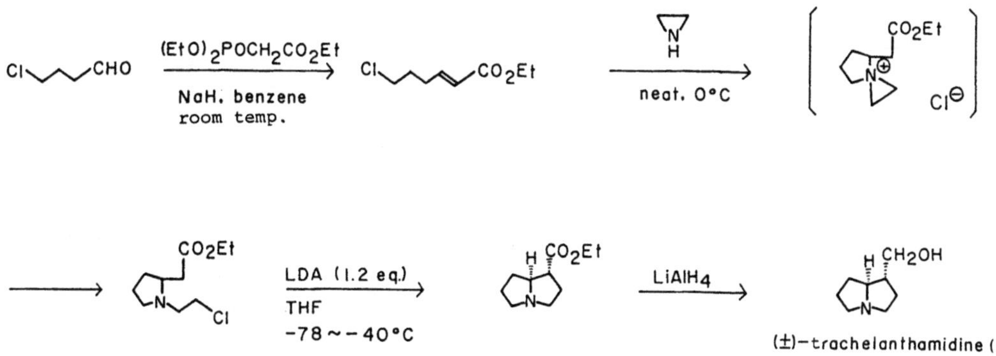

(±)-trachelanthamidine (49)

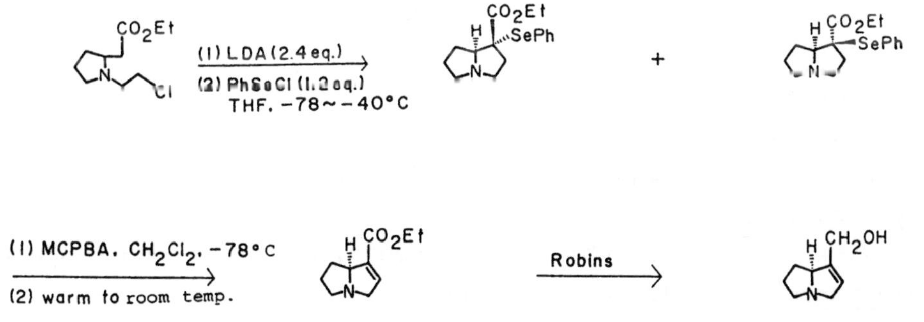

(±)-supinidine (50)

References

1. T. Kametani, T. Ohsawa, and M. Ihara, J. Chem. Soc. Perkin I, 1981, 1563.
2. M. Ihara, K. Noguchi, T. Ohsawa, K. Fukumoto, and T. Kametani, J. Org. Chem., 1983, 48, 3150.
3. T. Kametani, N. Kanaya, and M. Ihara, J. Am. Chem. Soc., 1980, 102, 3974.
4. M. Ihara, K. Fukumoto, and T. Kametani, Heterocycles, 1982, 19, 1435.
5. M. Ihara, Y. Haga, M. Yonekura, T. Ohsawa, K. Fukumoto, and T. Kametani, J. Am. Chem. Soc., 1983, 105, 7345.
6. T. Ohsawa, M. Ihara, K. Fukumoto, and T. Kametani, J. Org. Chem., 1983, 48, 3644.
7. T. Kametani, K. Suzuki, H. Kurobe, and H. Nemoto, Chem. and Pharm. Bull. (Japan), 1981, 29, 105.
8. T. Kametani, H. Kurobe, and H. Nemoto, J. Chem. Soc. Perkin I, 1981, 756.
9. T. Kametani, H. Kurobe, H. Nemoto, and K. Fukumoto, J. Chem. Soc. Perkin I, 1982, 1085.
10. T. Kametani, S. Hirata, and K. Ogasawara, J. Chem. Soc. Perkin I, 1973, 1466.
11. T. Kametani, T. Honda, H. Inoue, and K. Fukumoto, J. Chem. Soc. Perkin I, 1976, 1221.
12. T. Kametani, K. Higashiyama, T. Honda, and H. Otomasu, J. Chem. Soc. Perkin I, 1982, 2935.
13. H. Otomasu, K. Higashiyama, T. Honda, and T. Kametani, J. Chem. Soc. Perkin I, 1982, 2399.
14. K. Higashiyama, T. Honda, H. Otomasu, and T. Kametani, Planta medica, 1983, 268.
15. T. Kametani, K. Higashiyama, T. Honda, and H. Otomasu, Heterocycles, in press.
16. T. Kametani, K. Higashiyama, T. Honda, and H. Otomasu, Chem. and Pharm. Bull (Japan), in press.
17. T. Kametani, K. Hiagshiyama, H. Otomasu, and T. Honda, Heterocycles, in press.

Studies in the Alkaloids of *Rauwolfia caffra* Sonder

Mohammad Ataullah Khan
Department of Chemistry, University of Karachi, Karachi-32, Pakistan

Abstract: Ajmaline, ajmalicine, diacetyl ajmaline, 17-O - acetyl ajmaline, perakine, two phytosterols and four new indolenine alkaloids (raucaffricine, raucaffriline, raucaffinoline and raucaffridin) have been isolated from the roots and root bark of Rauwolfia caffra Sonder. The structure of raucaffricine, raucaffriline and raucaffrinoline have been elucidated mainly on the basis of spectroscopic studies.

Rauwolfia caffra Sonder, a 15-18 meter high tree having white flowers and simple 6 mm. long fruits, is known to be widely distributed in Central, East and South Africa [1]. It has also been grown successfully under an experimental plantation project sponsored by the Pakistan Council of Scientific and Industrial Research in Bangla Desh (formerly East Pakistan). Botanically R.caffra and R.natalensis Sond. are taken as nonspecific under the name R.caffra Sond. whereas R.inebrians. K.Schum is regarded as a synonym. Because of the widespread distribution of the species in Africa it seems likely that the synonyms are even more extensive. R.obliquinervis Stapf., R.coetzei Stapf., R.welwitachii Stapf and R.ochrosiodes, K.Schum are only doubtfully distinct from R.caffra Sonder.

Among certain South African tribes the plant is regarded as insecticidal [2] and is employed against maggots in wounds when applied as poultice to infected parts. Brayant [3] refers to its use as a blood purifier and

Natural Product Chemistry
Ed. by Atta-ur-Rahman

in the treatment of measles, witicaria and other rashes. The bitter bark is said to be strongly purgative in action. It is also employed to relieve griping pains in the stomach and other abdominal troubles. The name "Quinine Tree" (applied to the tree formerly known as R.natalensis) indicates a belief in the febrifugal properties of the bark [4] but evidence on this point is conflicting. Lewin states that preparations of the bark may find occasional use as an arrow-poison supplement [5]. Rindle and Groenewood [6] in 1932, isolated two uncrystallised alkaloidal fractions from R.natalensis (R.caffra) one of which gave the blood-red colour reaction with nitric acid, characteristic of dihydroindole derivatives. In the same year Koepfli [7] reported the isolation of a crystalline alkaloid from the bark of R.caffra to which he assigned the molecular formula $C_{20}H_{26}O_3N_2$, on the basis of the analysis of the base and its halogen salts.

This base was quite distinct from ajmaline, giving indications of a quarternary character, and was named rauwolfine. He further reported the isolation of two insufficiently characterized crystalline bases A and B, the former of which has been recorded as showing m.p. 294°C (uncorr). Schuler and Warren [8] in 1956 reported the isolation of ajmaline and reserpine from R.natalensis (R.caffra). whereas Paris and Dilleman [9] published the isolation of reserpine and rescinnamine in 1957 from the same plant.

Following the studies in the isolation of alkaloidal complexes from the fresh undried roots of R.serpentina through a technique of dialysis with organic solvents, it was considered of interest to re-investigate the constituents of R.caffra using freshly harvested, undried plant material. As a result of studies in the roots and root bark of the plant, obtained through air transport (to keep them as fresh as possible) from an experimental plantation in East Pakistan (Bangla Desh) and also from South Africa, the compounds shown in Fig. 1 have been isolated.

Out of the alkaloids mentioned in Fig.1 only ajmaline has been reported by Schuler and Warren. Reserpine also mentioned by these authors could not be isolated in the present study. It may further be noted that the

1.	Raucaffrine (identified as perakine)	$C_{21}H_{22}O_3N_2$	186-9° m.p.
2	Raucaffridine	$C_{21}H_{24}O_3N_3$	221°(decomp)
3.	Raucaffricine	$C_{27}H_{32}O_8N_2H_2O$	220°(decomp)
4.	Ajmalicine	$C_{21}H_{24}O_3N_2\frac{1}{2}H_2O$	253°
5.	Ajmaline	$C_{20}H_{26}O_2N_2$	158°
6.	Raucaffriline	$C_{21}H_{22}O_3N_2$	200-201°
7.	Raucaffrinoline	$C_{21}H_{24}O_3N_2$	235-36°
8.	Diacetyl ajmaline	$C_{24}H_{30}O_4N_2$	187-89°
9.	17-O-Acetyl ajmaline	$C_{22}H_{28}O_3N_2$	215°
10.	Serposterol	$C_{30}H_{48}O_2$	160°
11.	Caffrosterol (An uncharacterised-sterol)	$C_{20}H_{34}O_2$	297°(decom)

Fig 1

base, raucaffridine, was obtained only once in the first working of the sample from South Africa, and that it was not possible to isolate Koepfli's rauwolfine from either of the two samples.

The separation of raucaffricine and raucaffriline was mainly based on differences in the strength of these bases and the solubility of their hydrochloride. While the weak bases, ajmalicine and raucaffriline were obtained from the ethyl acetate soluble fraction of the total alcoholic extractive, the other crystalline bases like raucaffricine, raucaffrin (perakine) and ajmaline were obtained from the water soluble hydrochlorides of the aqueous solution of the total extractive. A little amount of raucaffricine was also obtained from the water-insoluble hydrochlorides, the latter forming about 50 percent of the total alkaloids.

On a careful check up of the analytical data, the melting points, colour reactions and other physical properties, it appears that raucaffricine, raucaffriline and raucaffrinoline are not identical with any of the known bases and form an addition to the already long list of alkaloids isolated from the various Rauwolfia species. Although ajmalicine m.p. 250-252°C. and tetrahydro-alstonine m.p.228-230°C have the same molecular formula ($C_{21}H_{24}O_3N_2$) as raucaffridine the melting points of the former bases differ markedly from the latter (221°C.) (decomp.). The infra red spectrum shows that raucaffridine has not twin peaks at 6 μ region, characteristic of the unsaturated ester grouping ($CH_3OOC-\overset{|}{C}=$) which is present in ajmalicine and tetrahydro-alstonine. The relevant properties of the known alkaloids which might be considered as possibly identical with raucaffridine are tabulated in Fig.2 for comparison.

Raucaffricine seems to be the major alkaloid of Rauwolfia caffra root bark (yield* 0.17%) whereas raucaffrine (perakine) is the major alkaloid of Rauwolfia caffra thin roots (yield 0.11%).

Raucaffricine which analyses for $C_{27}H_{32}O_8N_2$ is a monoacid base and is tertiary in character, forming a crystalline monomethiodide. Its I.R. spectrum indicated that it is a poly-hydroxy base, and the formation of tetra-acetyl and tetrabenzoyl derivatives showed that it contains four hydroxyl groups. The presence of an acetyl group was shown by IR spectrum and microanalytical values of one acetyl in the base and five in its tetra acetyl derivative. It reduces Fehling's and ammonical silver nitrate solutions, indicative of a possible aldehyde grouping, but this could not be confirmed through IR or the formation of oxime and phenylhydrazone derivatives. The alkaloid, therefore, appeared to be a glycoside and was hydrolyzed by heating at 100°C with 1N HCl for ½ hr. This hydrolysis yielded two components, one of which was identified as D(+)-galactose through paper chromatography as detailed below.Hydrochloric acid was removed from the hydrolysate and the residue dried. It was taken up in water and run on Whatman paper No.1. Development by benzene-n-butanol -pyridine-water (1:5:3:3) and spraying of the chromatogram with a satura-

---

*All the yields have been calculated on dry weight basis.

Name	Empirical formula	m.p. (°C)	
Raucaffridine	$C_{21}H_{24}O_3N_2$	221°(dec.)	-
Ajmalicine		250-25°	
-yohimbine		258-59°	
Raubasine(Alkaloid II)	$C_{21}H_{24}O_3N_2$	247-48°	-61°($CHCl_3$)
Alkaloid F		253-54°	-45°
Tetrahydroserpentine			
Corynanthine (Rauhimbine)	$C_{21}H_{26}O_3N_2$	218-225°	-82°(py)
Isorauhimbine	$C_{21}H_{26}O_3N_2$	225-228°	-104°(py)
Yohimbine	$C_{21}H_{26}O_3N_2$	234-236°	+76°(py)
3-epi- α-yohimbine	$C_{21}H_{26}O_3N_2$	125-28° 181-183°	-96°(py)
Serpine	$C_{21}H_{26}O_3N_2$	213°	+70°(py)
Rauwolscine	$C_{21}H_{26}O_3N_2$	239-41°	-15°(py)
β-yohimbine	$C_{21}H_{26}O_3N_2$	246-49°	-48°(py)
Serpentidine	$C_{21}H_{22}O_3N_2$	262-63°	-
Tetrahydroalstonine	$C_{21}H_{24}O_3N_2$	228-230°	

Fig.2

ted solution of aniline oxalate [10] revealed the sugar as D(+)-galactose which was confirmed by preparation of its osazone.

As regards the aglycone, the UV spectrum of raucaffricine exhibits absorption in the regions known for indolenine alkaloids: λ_{max} 219.5, 258nm λ_{min} 236 nm [11]. It was proposed earlier that the easy conversion of raucaffricine to indole bases under mild alkaline conditions, similar to such conversion of perakine (I) [12] suggests that it is an indolenine derivative. Thus we recorded the NMR spectrum (60 MHz) in pyridine and observed a close similarly to that of vomilenine (II) [13].

(I)

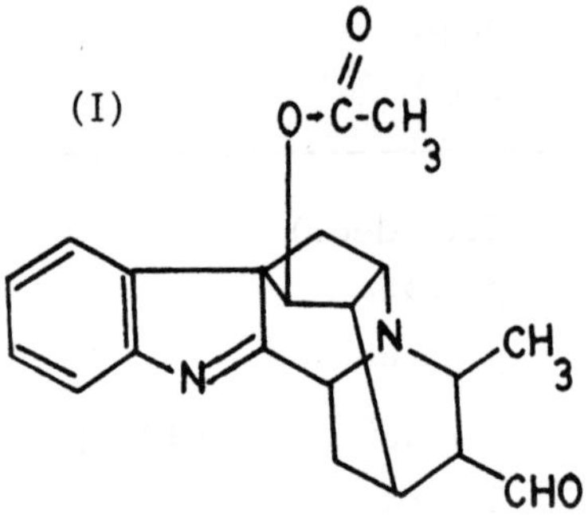

(II)

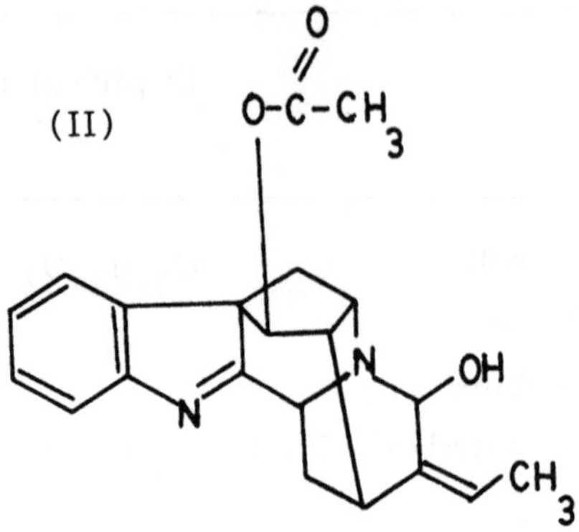

Attention is specially drawn to the peaks for the ethylidine group (1.56 δ, 3H, d, J 6.7 Hz; 5.98 δ, 1H, q, J 7 Hz) the acetyl group (2.23 , δ 3H, s,), the single hydrogens at C-15 (2.80, δ, 1H, m), C-5(3.33, δ, 1H, q) and at C-3(4.31 δ, 1H). The peak centred at δ 5.00 corresponds to the peak in the vomilenine spectrum [6] at exactly the same position. A noteworthy difference in the two spectra is for the anomeric hydrogen atom which appears as a broad doublet at δ 5.42 [14].

For confirming that the aglycone was indeed vomilenine, the hydrochloride of the base that was obtained from an hydrolysis experiment with 1N HCl was compared with that obtained from a known solution of vomilenine similarly refluxed in the same solvent. Both showed identical IR and UV spectra and gave the same R_f values on TLC. The free bases were again isolated therefrom and compared by IR and UV spectra and TLC which were found identical. The UV spectrum showed typical absorptions for the indole structure indicating that the base had changed completely from the indolenine to the indole type on heating with the acid as may be expected [12].

For finding out the nature of the linkage at the anomeric carbon an attempt was made to hydrolyse the alkaloid with sweet almond emulsin which contains β-galactosidase, but, to a lesser extend α-galactosidase. As it is not hydrolysed to any appreciable extent it is possibly an α-galactoside.

In order to establish that raucaffricine was indeed an α -galactoside of vomilenine its enzymic partial synthesis (vomilenine has not been synthesized so far) was attempted with the latter and D(+)-galactose (large excess) in presence of fresh brewer's yeast. After intermittent shaking of the mixture for 10 days at 25°C, TLC (butanol-acetic acid-water, 4:1:1) revealed the presence of raucaffricine (compared with a known sample run alongside).

On the basis of the above findings structure III was proposed for raucaffricine.

(III)

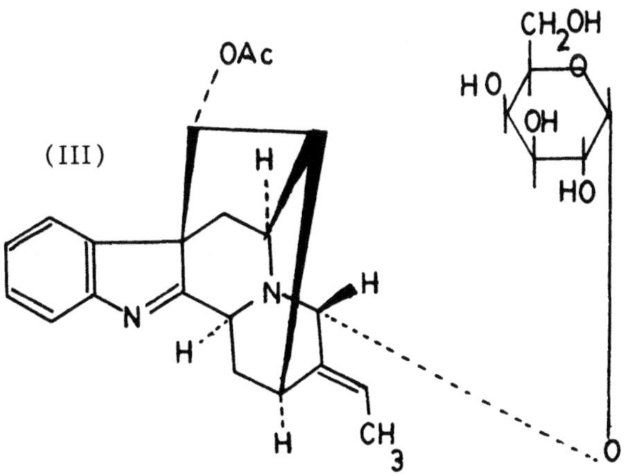

Molecular models of raucaffricine also indicate that the preferred glycosidic linkage at the anomeric C atom is α , and that C-21 oxygen of the vomilenine moiety is axial, (however for vomilenine as the free base, being a carbinolamine, equilibration at that junction is possible). Similar steric considerations imply that the geometry of the C-18 methyl attached to the ethylidine group is possibly that of the sarpagine alkaloids (IV) [15].

(IV)

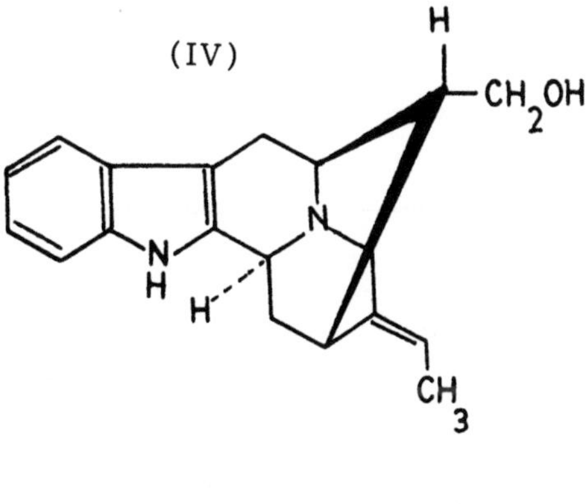

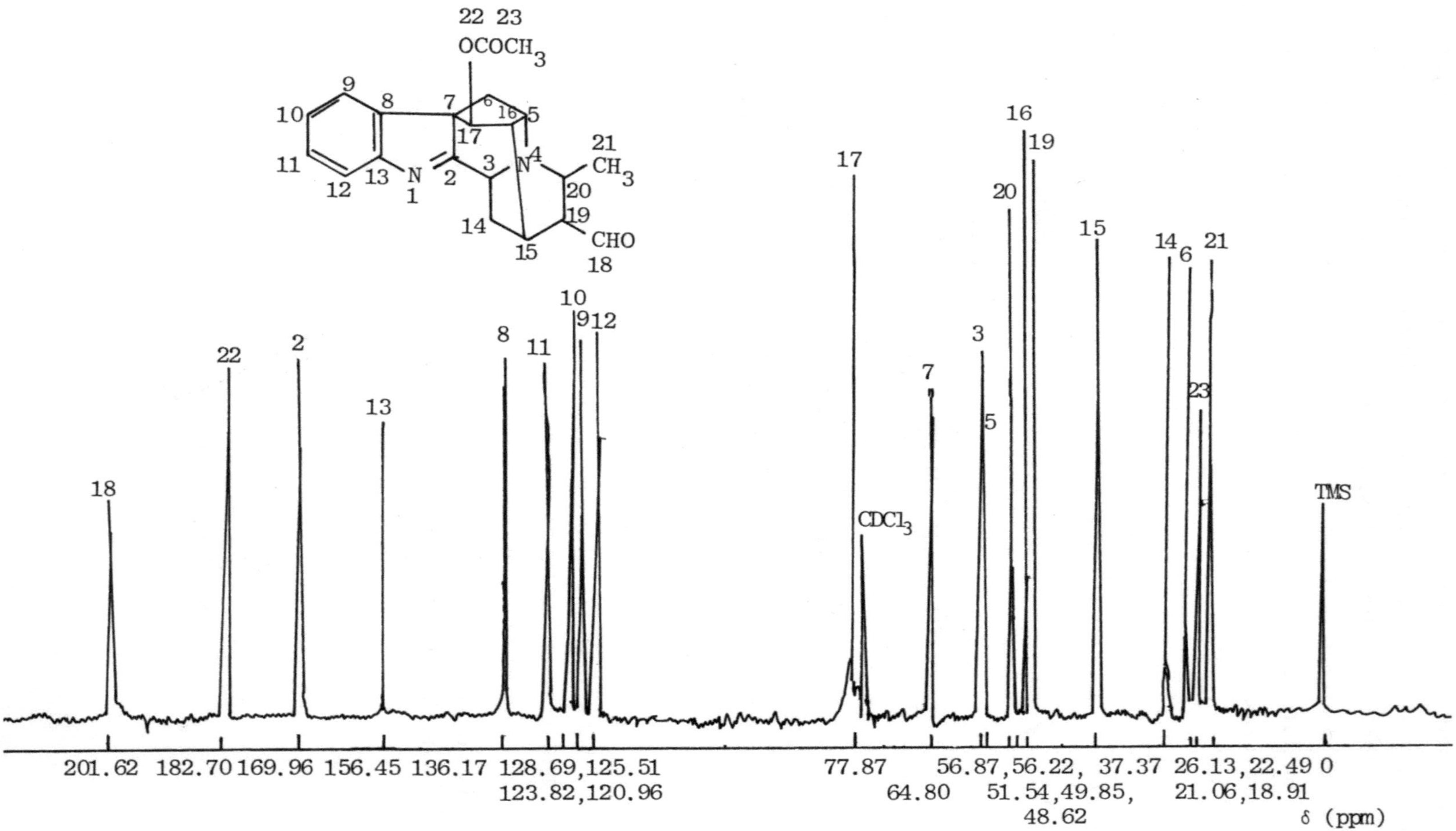

Fig.3. PFT ^{13}C (^{1}H) NMR spectrum of perakine, 22.63MHz, 180mg/1.5ml $CDCl_3$

In order to confirm the various structural features of raucaffricine, its ^{13}C NMR spectrum was studied [18]. The pulse Fourier transform ^{13}C NMR spectra of perakine and raucaffricine are shown in Figs. 3 and 4.

The total signal identification is based on the application of general chemical shift rules and on the correlation of the ^{13}C NMR spectra of perakine, raucaffricine, quinuclidine, 2,3-dimethyl-indole and indole Fig.5.

The resonances of the benzene residue in indolenine occur, with the exception of C-13, in similar shift ranges as those of benzene itself. The C-N residue of the indolenine ring can be considered as a heteroanalogue carbonyl group. Therefore one expects the resonance of C-2 to occur in the range of carbonyl carbons. However, as the nitrogen atom has a less electron withdrawing property, compared to the oxygen, the signal is shifted upfield compared to those of carbonyl carbons. Compared to catharanthine the resonance of C-17 shows a strong downfield shift as it is attached to an ester group. Correlation of the spectra of perakine and raucaffricine with quinuclidine leads to the signal identification for the carbons 3,5 and 20. The low field resonance in the spectrum of perakine is unambiguously assigned to the aldehyde carbon; the resonances at highest field in the spectrum of perakine are caused by the methyl carbons 21 and 23.

^{13}C NMR of raucaffricine is further correlated to that of the ^{13}C NMR spectra of different galactoses Fig.6. From the position of the resonance of the anomeric carbons of β-D-galactopyranosides and α-D-galactopyranosides it must be concluded that in raucaffricine and α-D-galactopyranoside residue is attached to the aglycone.

Ajmalicine (V) ($C_{21}H_{24}O_3N_2$) m.p. 253° (decomp.) was obtained as prismatic rods from petroleum ether insoluble part of the ethyl acetate soluble fraction of the total extractive. Its identity was established through micro-analysis, T.L.C., mixed m.p. determination and superimposibility of IR spectrum with an authentic sample.

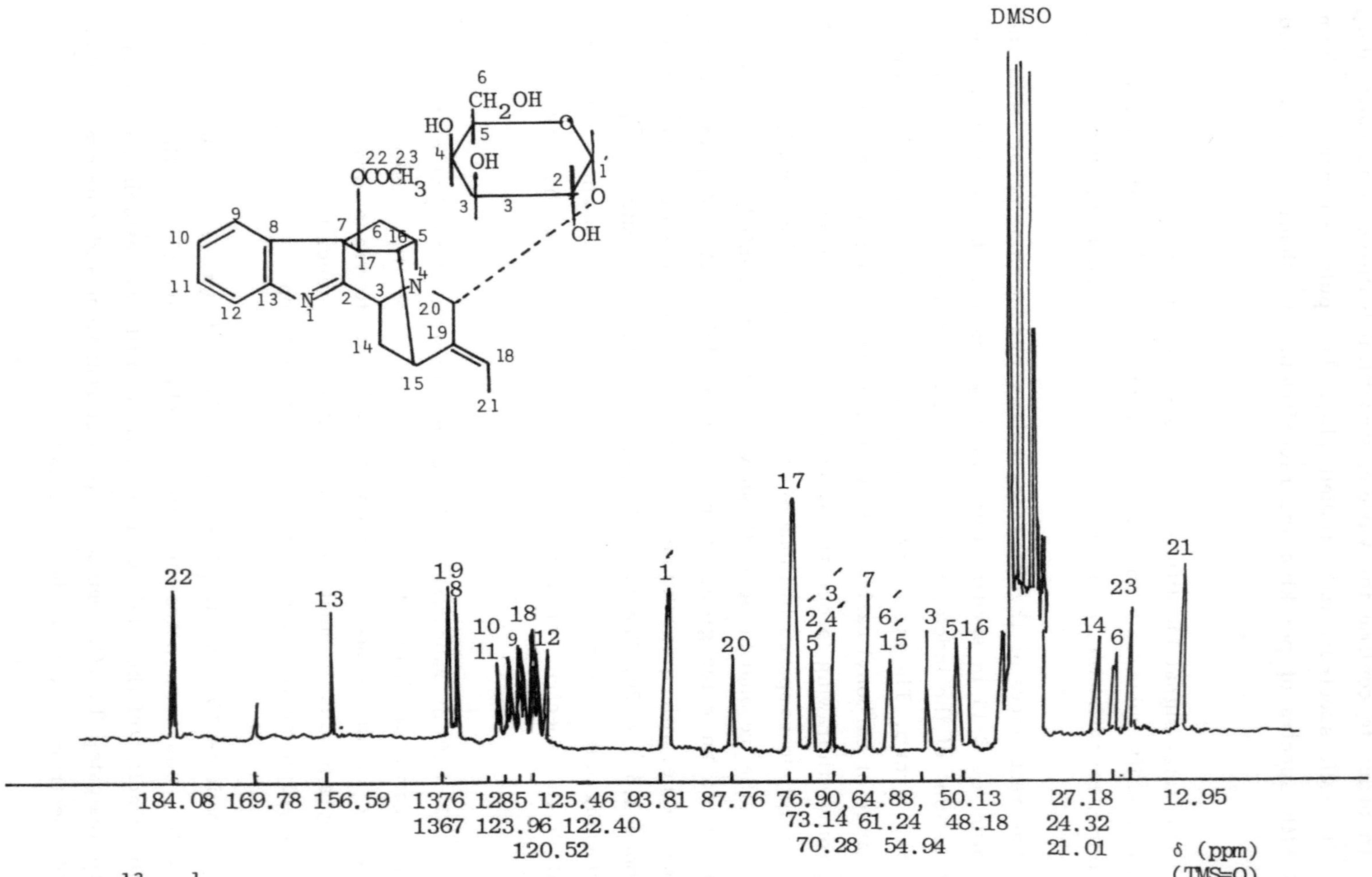

Fig.4. ^{13}C (^{1}H) NMR spectrum of raucaffricine, 22.63 MHz, 180mg/15ml DMSO

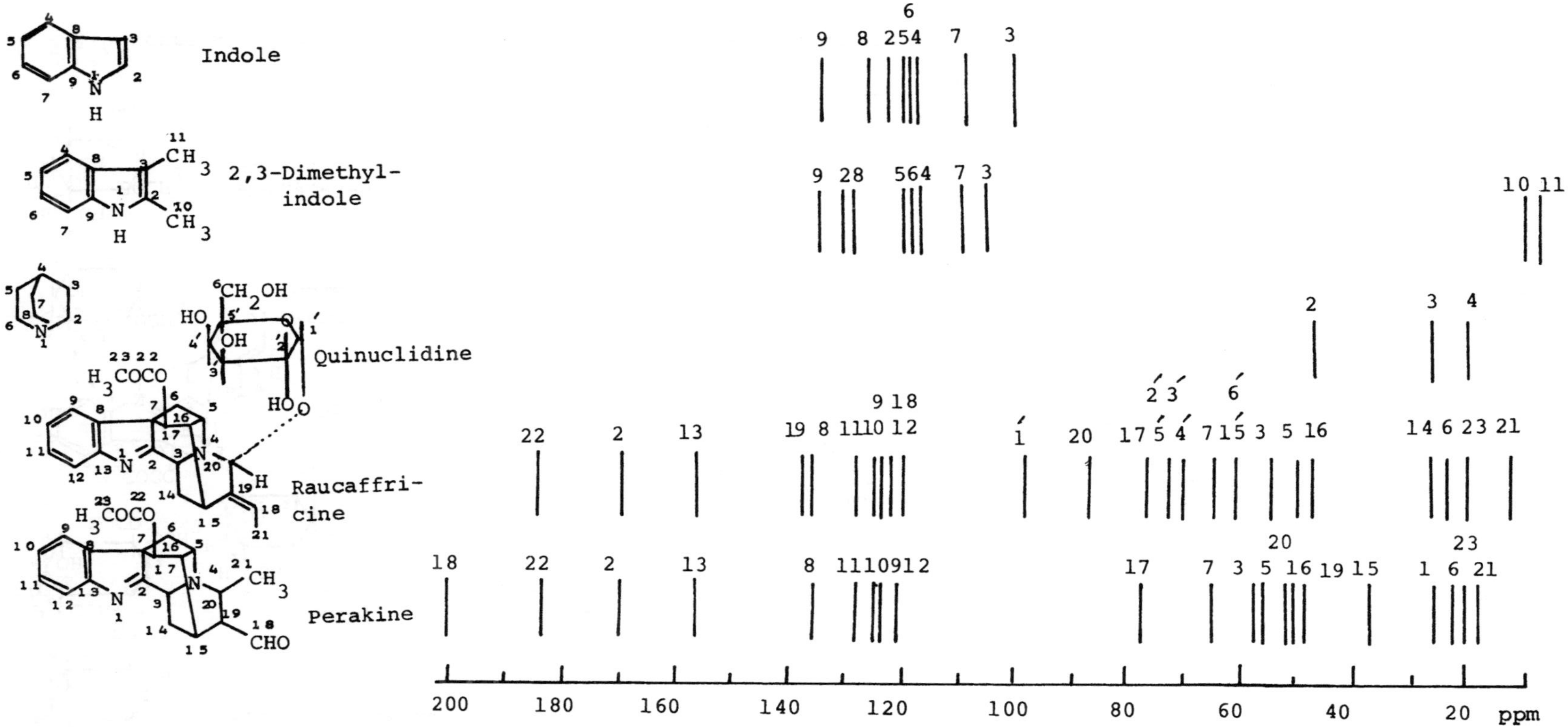

Fig.5. Correlation of ^{13}C-NMR spectra of perakine, raucaffricine and quinuclidine with indoles

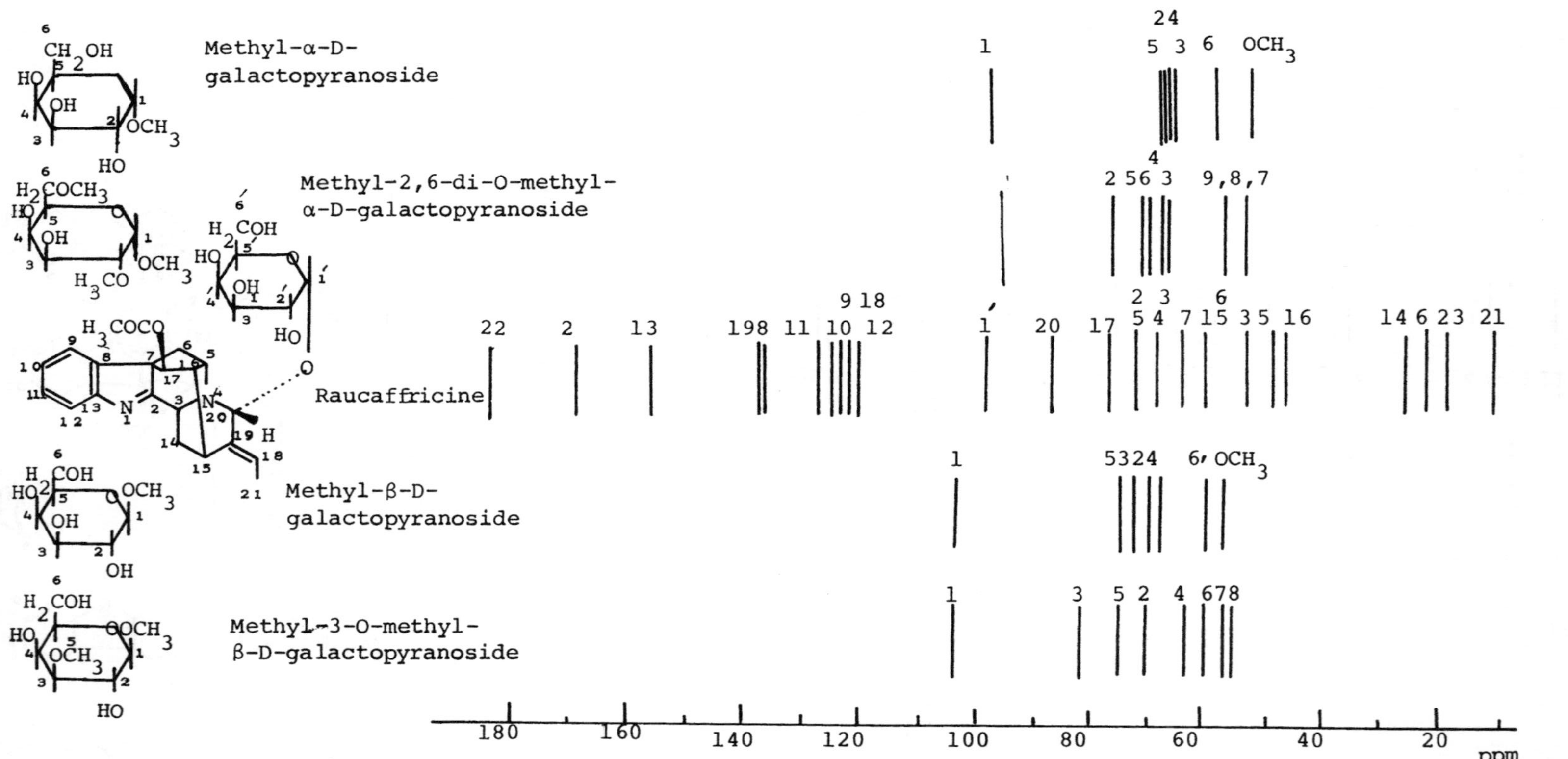

Fig.6. Correlation of ^{13}C-NMR spectra of different galactoses and raucaffricine

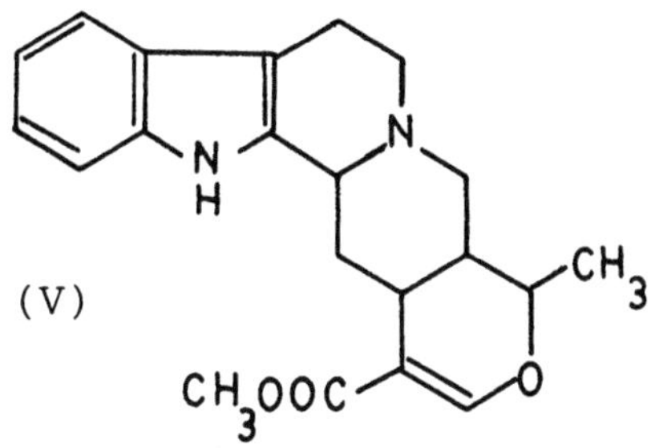

(V)

Raucaffriline, the third minor alkaloid, was isolated from the mother liquor of ajmalicine in etherial solution by extracting it out with dilute acetic acid and subjecting the base liberated from the acidic layer to column chromatography over neutral alumina. It crystallised from ethyl acetate in cubes m.p. 200-201° and analysed for $C_{21}H_{22}O_3N_2$ with two-CH_3, one active H and one-O-CO-CH_3 group. Its ultraviolet absorption spectrum in ethanol (λ_{max}, 250 mμ, 255 mμ, 259 mμ , 261 mμ , 265 mμ , and 268 mμ and λ_{min}. 220 mμ and 226 mμ) indicated that it also belongs to the indolenine type of alkaloids. Its IR spectrum showed a strong peak at 1740 cm^{-1} indicating the presence of an ester (> CO) group and is probably -$OCOCH_3$. The IR and UV spectra of raucaffriline is similar but not superimposable to that of vomelinine which indicated it to be a stereoisomer of vomilenine.

The fourth subsidiary alkaloid was isolated from the basic fraction of the alcoholic extractive of fresh undried root bark of the plant. Carefully avoiding any contact with strong alkali and acids or excessive heat, the alcoholic extractive was worked up in the following way: The alcoholic extracts were freed of the solvent in vacuo, the residue thus obtained was digested with hot water and distributed into aqueous and ethyl acetate layers. A small quantity of non-basic material did not go into either of these solvents and was neglected. The well cooled aqueous fraction was saturated with sodium chloride, and the water-insoluble hydrochloride formed was filtered out. The filterate was basified with 10% ammonia and repeatedly extracted with ethyl acetate. The ethyl acetate solution was washed with water, dried over anhydrous sodium sulphate and filtered. On keeping overnight, the solution gave a crystallisate which on recrystallisation from a mixture of moist ethyl acetate and methanol (5:1) yielded raucaffricine [16,17,18]. The mother liquor of raucaffricine was freed of the solvent in vacuo and the residue was exhaustively extracted with hot benzene. The benzene solution of the base was passed through a column of alumina (Aluminium oxide 'Woelm' activity grade I, M. Woelm-Eschweger; Fabrik Chemisch-Pharmazeutischer Preparate) and eluted with benzene which yielded perakine [19] m.p. 186-189°C. The column was subsequently eluted with a mixture of benzene and ethyl acetate (1:1) which yielded a glassy mass on removal of the solvent in vacuo. It crystallized from methylene dichloride in prismatic rods melting at 236°C with decomposition

	Raucaffrinoline(δ)	Perakine(δ)	N-Demethyl-Δ′ deoxy-ajmaline-O-acetate (δ)
1. Aromatic (4 protons)	Multiplet 7.33-7.73	Multiplet- 7.20-7.75	Multiplet 7.06-7.84
2. $C_{(3)}$-H	Distorted quartet centred at 4.20(J=4,8)	Distorted quartet centred at 4.20(J=4.1,8.4)	Quartet centred at 4.20(J=3.4,8.8)
3. $C_{(17)}$-H	Singlet 5.12	Singlet 4.97	Doublet centred at 5.00(J=1.8)
4. Me of O-Ac(3 protons)	Singlet 2.2	Singlet 2.17	Singlet 2.15
5. C-Me (3 protons)	Doublet centred at 1.26 (J=7)	Doublet centred at 1.29(J=6.27)	-
6. $C_{(18)}$-H aldehydic proton	-	Singlet 9.93	-

The NMR-data were obtained on deuterochloroform solutions with a varian model HR 60 spectrometer at 60 mc/sec with tetramethylsilane acting as internal standard. Chemical shifts are expressed in δ and spin-spin coupling (J) constants in cycles/second.

Fig.7

and analyzed for $C_{21}H_{24}O_3N_2$. Found after drying to constant weight at room temperature over P_2O_5 in vacuo: C, 71.25; H, 6.73; O, 14.03; N, 8.20% and MW, 352 (mass spectrum). Calculated for $C_{21}H_{24}O_3N_2$: C, 71.57; H, 6.86; O, 13.62; N, 7.95% and MW, 352. It showed UV-absorption spectrum of indolenine derivatives (λ_{max}. 265, 227, 220 nm, λ_{min}. 230 nm). Its IR spectrum shows the presence of an alcoholic-OH (3500 cm^{-1})

and a single peak at 1725 cm^{-1} for CO against two peaks in the case of perakine accounting for $-OCOCH_3$ and $-CHO$, respectively.

Comparison of proton magnetic spectra of raucaffrinoline, perakine [12] (I) and N α-demethyl-Δ' 'deoxyajmaline-O-acetate [20] (VI) revealed a close similarity in the vital regions of their spectra (Fig.7).

(VI)

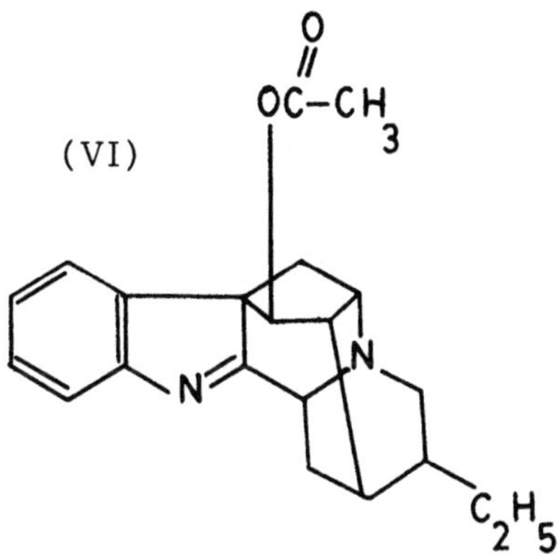

The multiplet corresponding to the aromatic protons of raucaffrinoline and perakine was found almost identical in their height, shape and number of peaks. Likewise the signal associated with the protons of $C_{(3)}$-H and $C_{(17)}$-H appeared as distorted quartet centered at δ 4.20 and singlet at δ 5.12 and δ 4.97 in the NMR-spectra of raucaffrinoline and perakine, respectively. Similarly the signals at δ 2.20 and δ 5.17 (singlet) and δ 1.26, J=7, and δ 1.29, J=6.27 (doublet centered), accounting for the acetyl ester methyl and C-methyl protons are almost the same in their chemical shifts and coupling constants in their spectra. Furthermore, in contrast to perakine which showed an aldehydic proton at low field, δ 9.93, it showed a singlet at δ 3.4 which may be accounted for by an alcoholic proton. Raucaffrinoline may therefore be a derivative of raucaffrine, the aldehyde group of which has been reduced to an alcoholic group.

This inference was further substantiated through mass spectral studies of raucaffrinoline. The mass spectrum showed the principal peak at m/e = 352 confirming the molecular formula $C_{21}H_{24}O_3N_2$. The peak at 43 may be attributed to the acetyl ion (CH_3CO^+) whereas the peaks at 168, 293, 309, 321 and 335 could be accounted for by the presence of the following moieties in the raucaffrinoline molecule, respectively.

A small peak at 115 appears to be due to the indolenine moity or its rearrangement product.

Considering the results of microanalysis IR-, UV-, nuclear magnetic resonance, mass spectra and the probable genetic relationship with the other alkaloids of Rauwolfia caffra Sond., especially with perakine, the following tentative structure is proposed for raucaffrinoline (VII).

(VII)

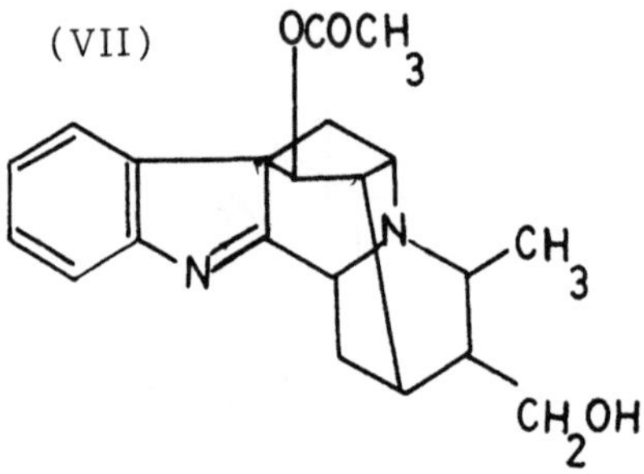

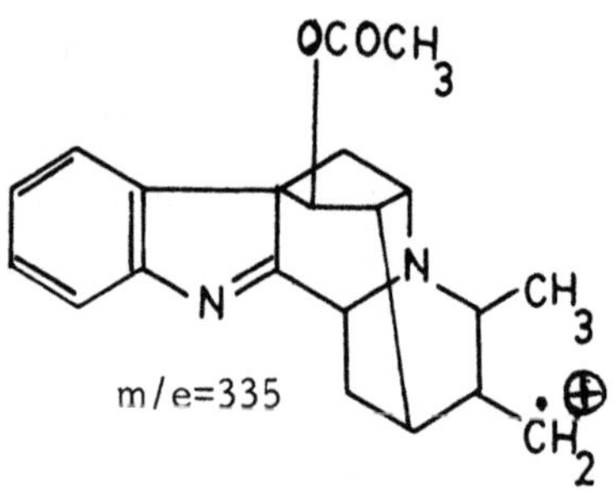

m/e=335

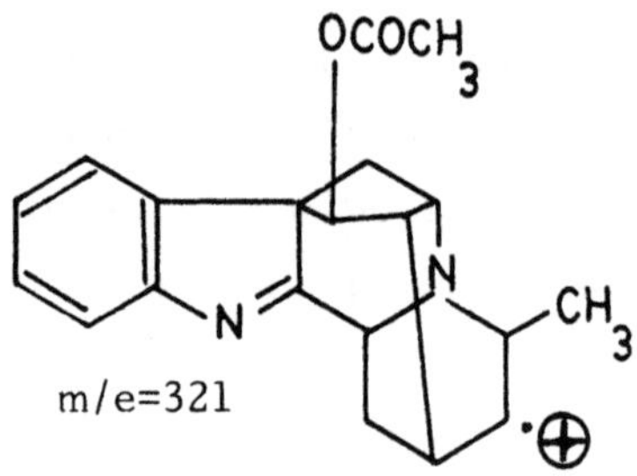

m/e=321

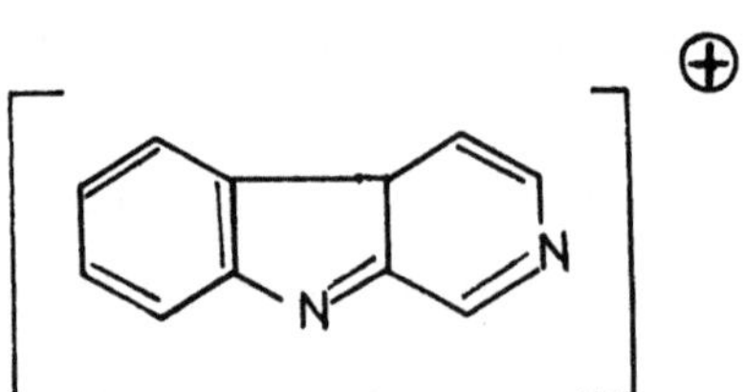

m/e=169

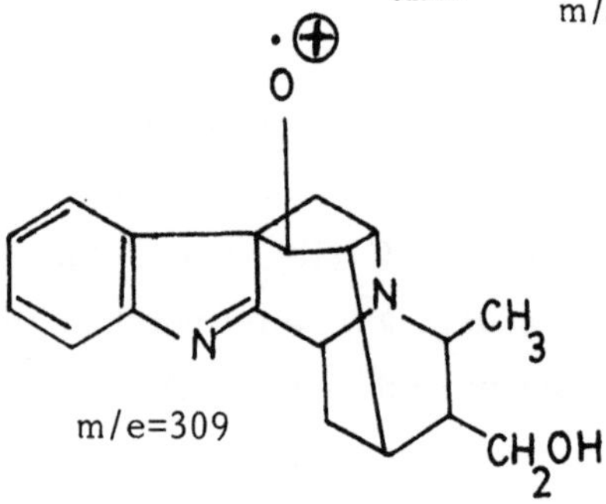

m/e=309

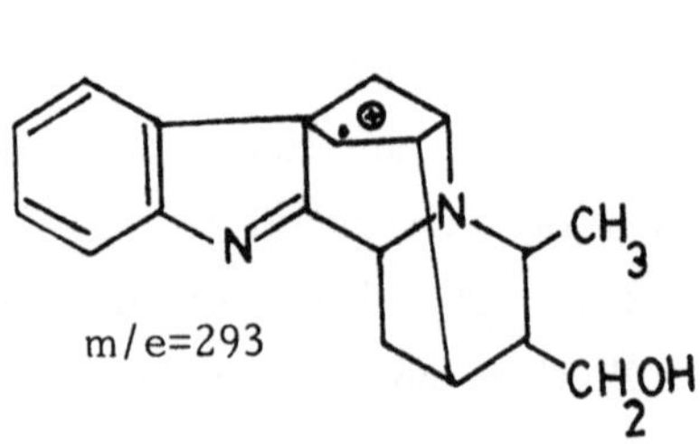

m/e=293

The fifth and sixth subsidiary alkaloids were isolated from the column chromatography when it was further eluted from ethyle acetate. They were identified as diacetyl and 17-O-acetyl ajmaline respectively with the help of microanalysis CO-tlc, IR, NMR, and mass spectral studies.

References

1. Colonial plants and Animal products (Her Majesty's Stationery office, London 1955), Vol. 1, 5, page-7.

2. **Ibid** Vol. 5, page-13.

3. Bryant, "Zulu Medicine Men"., Annals of the Natal Government Museum, Vol. (ii), pt. 1, July, 1909.

4. Juritz, The Urgency of a Definite Forward Movement in the study of the Active principles of South African Medical Record, 13th Nov. 1915.

5. Same as 2.

6. M.Rindle and P.W.G.Groenewood, **Trans.Roy.Soc.S.Africa**, 21, 55, (1932)

7. J.B.Koepfli, **J.Am.Chem.Soc.**, 54, 2412 (1932).

8. O.G.Schuller and E.C.Warren, **J.Chem.Soc.**, 215 (1956).

9. Paris,R.A. and Dillemann, G.(1957), Fr. 15,310.

10. B.D.E.Gaillard, **Nature**, 171, 1160 (1953).

11. N.H.Khan, M.Ataullah Khan and S.Siddiqui, **Pakistan J.Sci.Ind. Res.**, 8, 23 (1964).

12. P.R.Ulshafer, M.F.Bartlett, L.Dorfman, M.A. Gillen, E.Schlittler and Ernest Wenkert, **Tetrahedron Letters**, 11, 363 (1961)

13. W.I.Taylor, A.J. Fray and A.Hofmann, **Helv.Chim.Acta.**, 45, 611 (1962).

14. M.Sharma, M.G.Kelly and Sr.M.A.Podczasy, **Tetrahedron Letters**, 56, 4951 (1969).

15. J.G.Nouls, P.Wollast, J.C.Breakman, G.Van Binst, J.Pecher and R.H.Martin, **Tetrahedron Letters**, 23, 2731 (1968).

16. S.Siddiqui, N.H.Khan and M.Ataullah Khan, **Pak.J.Sci.Ind.Res.**, 9, 210 (1966).

17. M.Ataullah Khan and A.M.Ahsan, **Tetrahedron Letters**, 39, 5137 (1970).

18. M.Ataullah Khan, H.Horn and W.Voelter, **Z.Naturforschung**, 37b, 494 (1982).

19. A.K.Kiang and A.S.S.Wan, **J.Chem.Soc.**, (1960), 1394.

20. M.F.Bartlett, **J.Chem.Soc.**, 84, 624 (1962).

The Oxidative Metabolism of the Aromatic A-Ring of Steroidal Estrogens

Philip W. Le Quesne[1] and Robert H. Purdy[2]

[1]Department of Chemistry, Northeastern University, Boston, MA 02115, USA
[2]Department of Organic Chemistry, Southwest Foundation for Biomedical Research, San Antonio, TX 78284, USA

Abstract

The pathways of oxidative metabolism of the aromatic A-ring of steroidal estrogens such as estradiol involve initial hydroxylation reactions at C-2 and C-4. Estrogen 2-/4-hydroxylase activities, which catalyze the formation of the resulting catechol estrogen products, are found in microsomes from a variety of mammalian tissues. The relatively high rate of catechol estrogen formation in the liver has generally been regarded as a deactivating reaction because the catechol products were considered to be biologically inert. Recently, however, there has been a resurgence of interest in 2- and 4-hydroxy-estradiol due to their potential biological activities when they are formed within tissues, since these catechols are further metabolized extremely rapidly when they are administered to animals or humans. It has also become evident from differences between some of the biological activities of 2- and 4-hydroxy-estrogens that it is no longer appropriate to consider catechol estrogens as a single class of compounds.

There is relatively little information on the mechanism of microsomal estrogen 2-/4-hydroxylase activities. In this paper it is proposed that the

Natural Product Chemistry
Ed. by Atta-ur-Rahman

active intermediates are stabilized tautomers of dienol epoxides, related in structure to some of the intermediates known to be involved in the conversion of carcinogenic polycyclic aromatic hydrocarbons into their reactive ultimate carcinogens, the dihydrodiol epoxides. The chemistry of estrogen-derived dienol epoxides will be discussed and syntheses described. Biochemical results will be provided which implicate one of these dienol epoxide taumomers as an intermediate in the conversion of estradiol to 2-hydroxyestradiol.

Introduction

Biochemical investigations since 1957 have provided increasing evidence[1] that the natural human estrogenic hormone estradiol 1 is metabolized in large part to 2-hydroxyestradiol 2, and in much smaller amount to 4-hydroxyestradiol 3.

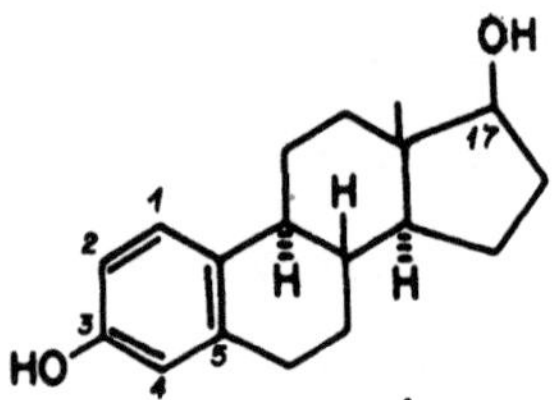

1

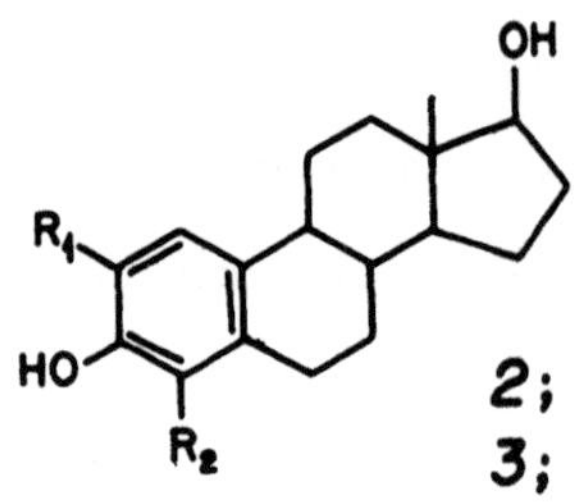

2; R_1 = OH, R_2 = H
3; R_1 = H, R_2 = OH

These catechols are rapidly deactivated by conversion to their 2- and 4-monomethyl ethers, respectively, by catechol-O-methyltransferase activity which is particularly high in liver and red blood cells.[2] The catechols and their monomethyl ethers are mainly excreted as glucuronides or sulfates.[2] The formation of catechol estrogens has been demonstrated to occur in a wide variety of normal and neoplastic tissues in experimental animals or humans.[2] It is now apparent that the catechol estrogens cannot be regarded merely as weak estrogens. The rapid metabolism of catechol estrogens in the circulation makes it difficult to interpret responses obtained following systemic administration of the parent estrogens compared to their catechols. Using in vitro conditions, where metabolism of the catechol estrogens is minimized, it has been clearly shown that the catechols themselves bind to estrogen receptors isolated from estrogen-responsive target tissues,[3] and may produce effects not observed with the parent estrogen.[2]

The formation of catechol estrogens is catalyzed by a family of membrane-bound enzymes, termed cytochrome P-450's or NADPH-dependent monooxygenases.[4] The key to an understanding of the mutagenic and carcinogenic potential of a polycyclic aromatic hydrocarbon such as benzo[a]pyrene 4 is its metabolic activation by cytochrome P-450 mixed function oxidases to a dihydrodiol epoxide 5, which is an ultimate carcinogenic metabolite that reacts with DNA in cells susceptible to carcinogenic transformation.[5]

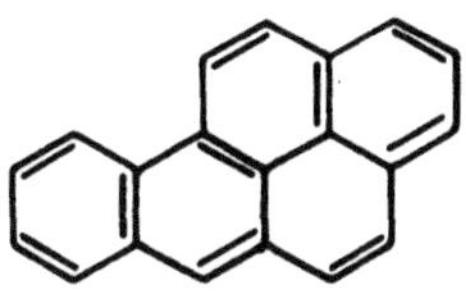

4

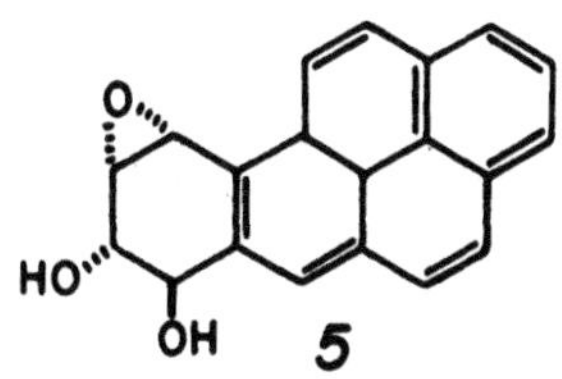

5

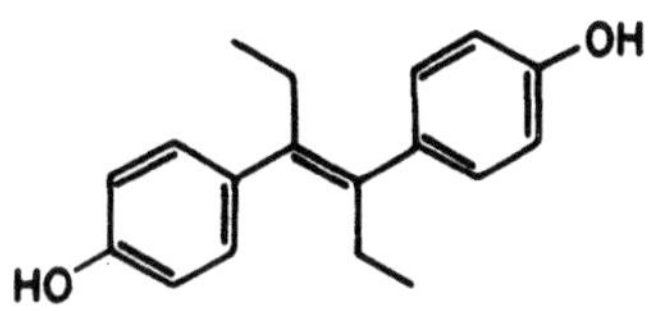

6

In contrast to tumor formation induced by benzo[a]pyrene, the occurrence of estrogen-induced carcinogenesis is still the subject of considerable debate. The synthetic estrogen diethylstilbestrol 6 (DES) has been classified as a carcinogen by the International Agency for Research on Cancer on the basis of a positive association between exposure and the occurrence of various cancers in humans, squirrel monkeys and rodents.[6] The continous administration of ethynylestradiol or other estrogens produces a wide variety of cancers in experimental animals.[7] In humans, the estrogen-responsive breast and endometrium are not believed to develop cancer unless they are stimulated by estrogens, which also promote the growth of these tissues.[8] It is therefore not surprising that estrogens are usually referred to as tumor promoters[9] rather than as inducers of cellular transformation.

Numerous attempts to apply the tester strains of the Ames bacterial assay system[10] to the detection of the mutagenic activity of estrogens in the presence of various metabolic activating systems have yet to succeed.[10,11] A weak mutagenic response to DES has been reported in a strain of yeast,[12] in Drosophila after larval feeding,[13] and in a mouse lymphoma cell system.[14] DES has been conclusively shown to induce the neoplastic transformation of Syrian hamster embryo fibroblasts.[15] After demonstrating that DES also induced the neoplastic transformation of a subclone of Balb/c 3T3 cells[16] that is particularly sensitive to transformation by carcinogenic polycyclic aromatic hydrocarbons,[17] we have directed our further efforts toward a study of the mechanism involved in the transformation induced by steroidal estrogens in this mammalian cell model system.

Results and Discussion

a) General Considerations

As outlined above, carcinogenic polycyclic aromatic hydrocarbons such as benzo[a]pyrene 4 are not active as such, but require cellular metabolic activation to reactive epoxides such as 5, which can then react with DNA to form DNA-adducts. Microsomal cytochrome P-450-mediated mono-oxygenases are principally responsible for catalyzing the formation of different phenols of benzo[a]pyrene *via* arene oxides.[18] In an earlier general paper[19] we drew attention to the possible wide occurrence of *benzene* oxide intermediates in phenol biosynthesis as a result of the generality on NIH-shift-type mechanisms, and suggested that these intermediates may have biological effects in and of themselves.

In 1980 we proposed [20] that conversion of estradiol 1 to the catechol estrogen 3 is accomplished as shown in Scheme I. The putative 4,5-arene oxide

(II) of estradiol metabolism can either stabilize as its keto-tautomer (III) or be converted to 4-hydroxyestradiol (IV) either by spontaneous aromatization

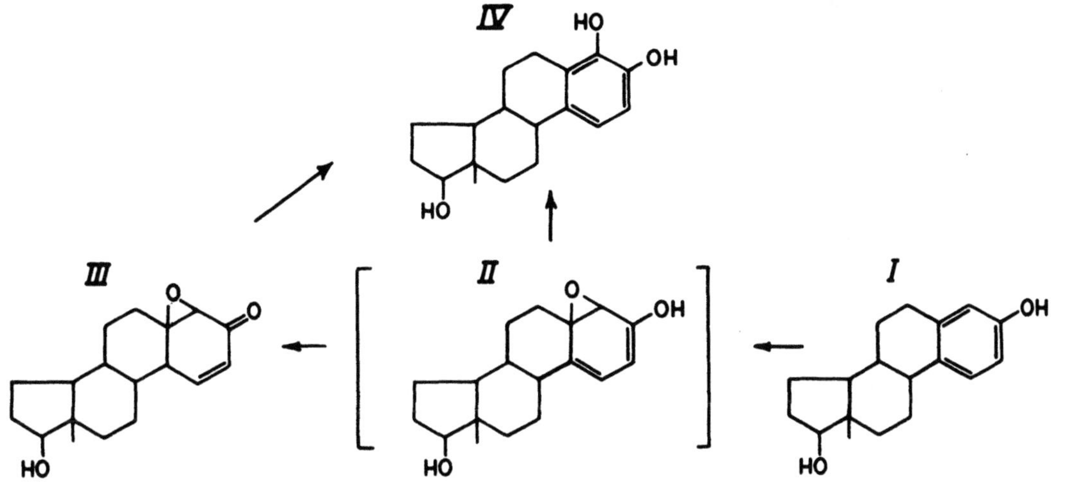

Scheme I. Postulated metabolic pathways for the formation of 4-hydroxyestradiol (IV) from estradiol (I). The 4α,5α-arene oxide intermediate (II) can spontaneously aromatize to (IV) or stabilize as the epoxyenone (III).

(acid-or base-catalyzed) or by enzymatic catalysis. Therefore, for conversion of estradiol 1 into the two catechol estrogens 2 and 3, two regio-isomeric dienol epoxides 7 and 8 can be evisaged.

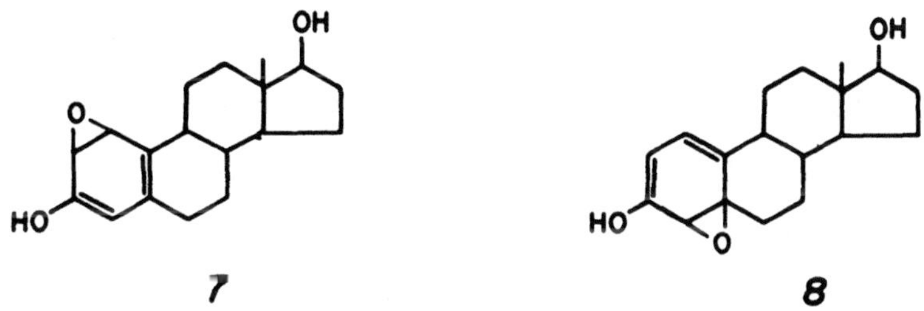

The chirality introduced by epoxidation means that two diasteroomers of each epoxide regioisomer must be considered. Once tautomerism of the dienol to the α,β- unsaturated enone takes place an additional chiral center, that at C-10, is created, which leads to doubling of the number of potential intermediates. We have so far considered the four intermediates 9 through 12 (and their derived 17-acetates) in our work. The C-10 epimers 13 through 16 are targets of ongoing synthetic and biological investigation.

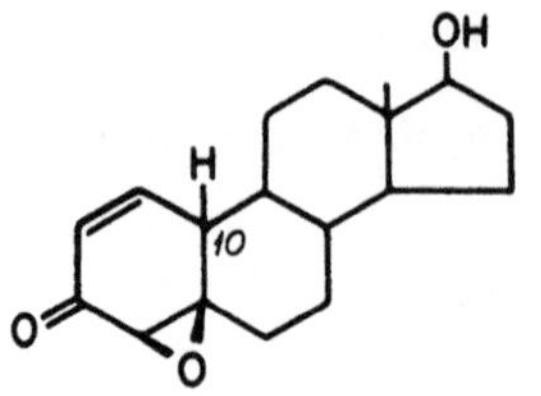

9; 10β-H
13; 10α-H

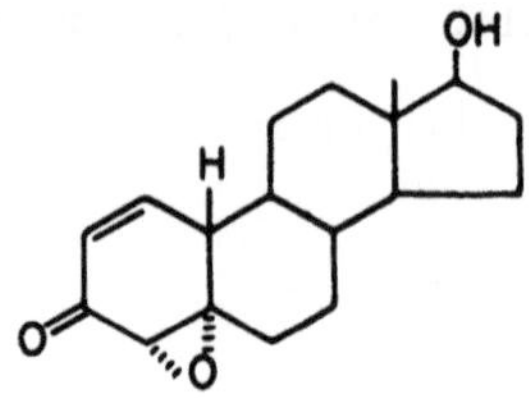

10; 10β-H
14; 10α-H

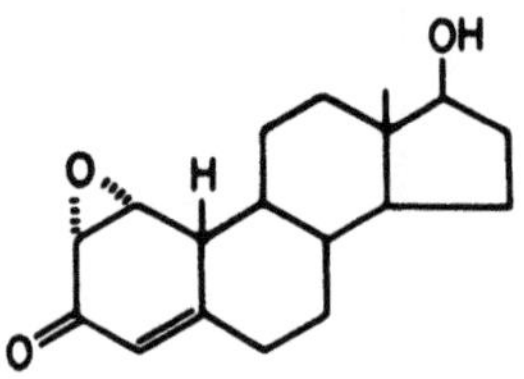

11; 10β-H
15; 10α-H

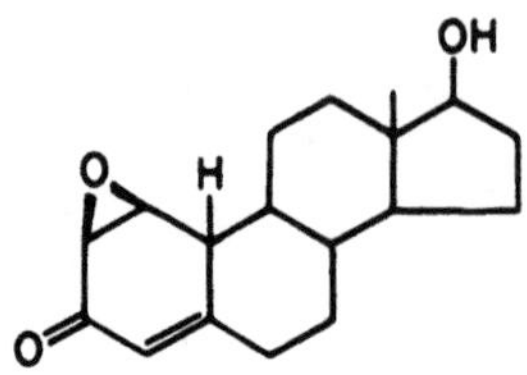

12; 10β-H
16; 10α-H

b) Synthesis of Potential Intermediates of Estrogen Metabolism

The syntheses of compounds 7 through 10 have been accomplished by a variety of routes[21] of which the following summary is representative of our established procedures. The 4β,5β-epoxy-17β-hydroxyestr-1-en-3-one was synthesized from 19-nortestosterone by the route outlined in Scheme II. The 4α,5α-isomer was prepared from a previously known epoxy-17-acetate[20] as shown in Scheme III. The 1α,2α-epoxy-17β-hydroxyestr-4-en-3-one was synthesized by the

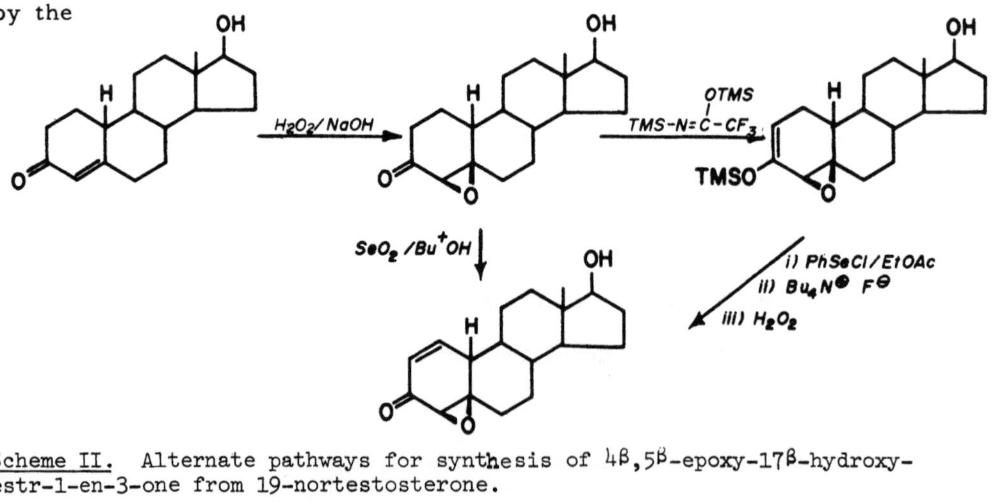

Scheme II. Alternate pathways for synthesis of 4β,5β-epoxy-17β-hydroxyestr-1-en-3-one from 19-nortestosterone.

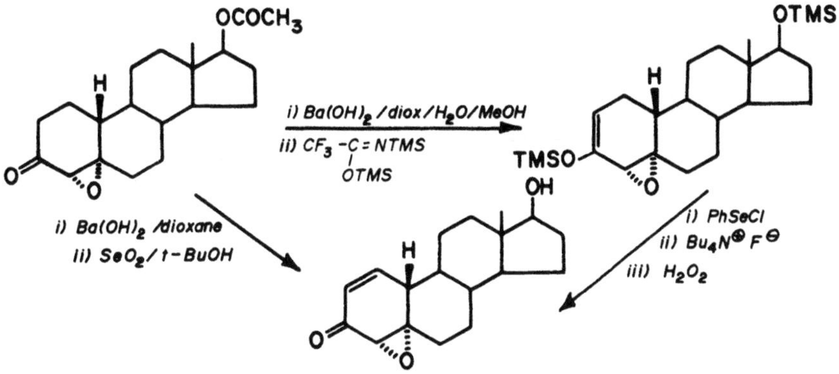

Scheme III. Alternate pathways for the synthesis of 4α,5α-epoxy-17β-hydroxy-estr-1-en-3-one 10.

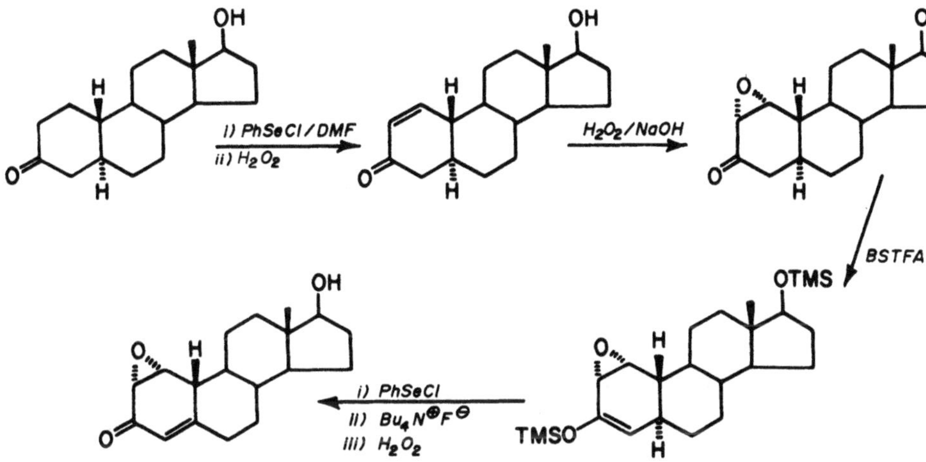

Scheme IV. Sequence of reactions used to prepare 1α,2α-epoxy-17β-hydroxy-estr-4-en-3-one 11 from 17β-hydroxy-5α-estran-3-one.

sequence shown by scheme IV, starting with the product of Birch reduction of 19-nortestosterone. The 1β,2β-epoxy isomer was prepared from 17β-acetoxy-1β,2β-epoxy-5α-estran-3-one, as shown in Scheme V. Because of its relative instability, this last epoxide has yet to be obtained in sufficient

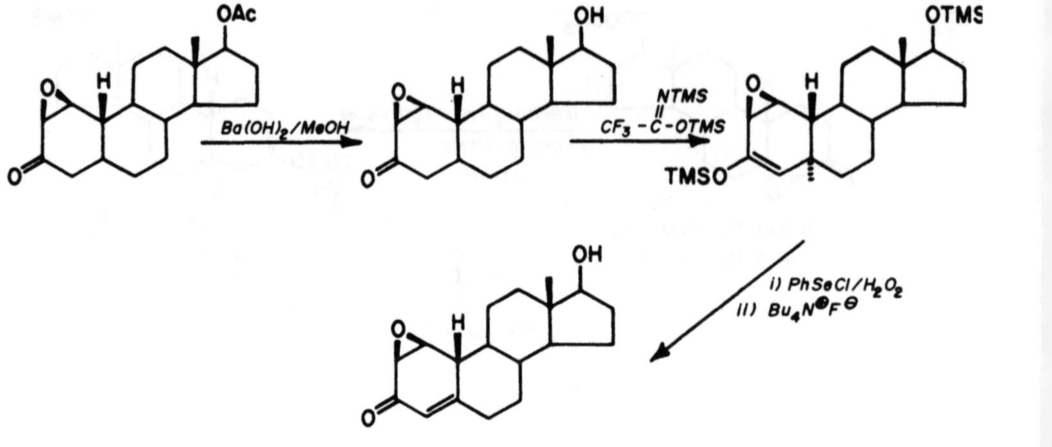

Scheme V. Sequence of reactions used to prepare 1β,2β-epoxy-17β-hydroxyestr-4-en-3-one 12 from 17β-acetoxy-1β,2β-epoxy-5α-estran-3-one.

yield for rigorous evaluation, but work along these lines is in progress. We have been able to obtain this 1β,2β-epoxide as its 17-acetate for use in isolation studies described in (d). With the exception of the 1β,2β-epoxy-17β-alcohol 12, these epoxides have been rigorously purified by HPLC to remove organoselenium impurities before final characterization as pure crystalline materials. All four of the corresponding acetates have been similarly characterized.

c) Chemical Transformation Studies

We have recently described[16] in some detail one way in which chemical transformation of Balb/c 3T3 cells by estrogens and their metabolites can be measured. When the epoxyenones (7, 8, and 9) were assayed for their ability to induce the transformation of Balb/c 3T3 cells, it was found that all three produced a significant transformation frequency which was three times greater than that of the control. The 4α,5α-epoxide 8 was the most toxic in terms of cell survival after 72 hr. treatment and also produced the highest calculated transformation frequency. These date support the hypothesis that one or more of these intermediates may be responsible for the neoplastic transformation of these cells by estradiol.

d) Attempted Isolation of Activated Intermediates of Estradiol Metabolism

Initial efforts to trap the putative intermediates of estradiol metabolism have been performed with microsomes prepared from a cultured line of MCF-7 human mammary cancer cells whose growth is known to be stimulated by 2-hydroxyestradiol.[22] Microsomes from samples of normal breast tissue were reported to be devoid of estrogen 2-hydroxylase activity.[23] Microsomes from MCF-7 cells were incubated with 5 μM [2,4,6,7-^{3}H(N)]-estradiol, 0.5 μM NADPH, and 5 μM magnesium chloride in 10 mM HEPES pH 7.5 buffer with the unlabeled compounds 7, 8, or 9 to serve as trapping agents and in the presence of the epoxide hydratase inhibitor 1,2-epoxy-3,3,3-trichloropropane. A radioactive product was formed which after isolation and acetylation let to a purified component having identical characteristics in high performance liquid chromatography with that of the 17-acetate of compound 10. This is strongly suggestive evidence for the validity of our hypothesis that an epoxyenone is an intermediate, at least in these human breast cancer cells, in the converstion of estradiol into 2-hydroxyestradiol.

e) Preliminary Studies with Fluorinated Estrogens

In the study of the biological action and metabolism of aromatic drugs, a time-honored method for probing the mechanism of action has been to substitute fluorine for key aromatic or aliphatic hydrogen atoms. Although the carbon-fluorine bond is approximately the same length as the carbon-hydrogen bond, its electronic characteristics and the electronic effects exerted by the fluorine atom upon the reactivity of adjacent atomic sites are profoundly different. With this background in hand, we investigated the use of 2-fluoroestradiol 17 as a potential inhibitor of the 2-hydroxylation process in the above MCF-7 system. It was reasoned that the microsomes would probably be unable to convert the fluorinated estrogen into a catechol and that 17 might, perhaps, bind tightly to enzymes involved in the reaction sequences of

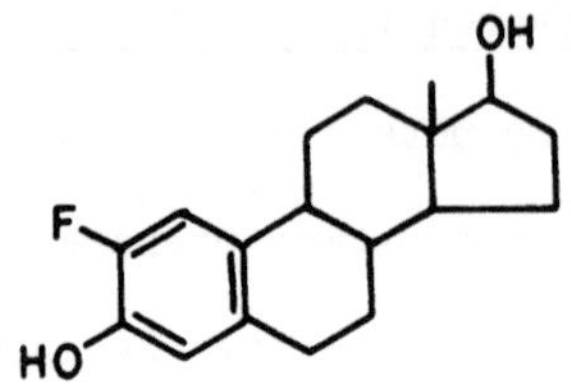

17

catechol estrogen formation. With this in mind an experiment was designed to determine if 2-fluoroestradiol is reacted upon in any way by the MCF-7 microsomal enzyme system.

The results clearly show that the fluorinated estrogen undergoes a smooth conversion to 2-hydroxyestradiol. This process formally must involve a reduction step somewhere in the conversion, and is reminiscent of an earlier report[24] of conversion of 4-fluorophenylalanine to tyrosine by enzymic hydroxylation systems. Clearly we are not able to utilize the fluorine substituent as an inert marker in studies of this kind. Further, the ability of a number of different microsomal systems to effect catechol formation from 2-fluoroestradiol and 2,4-difluoroestradiol is also noteworthy.

Acknowledgments

We thank the Council for Tobacco Research for support at Northeastern University (Grant 1271 M) and the National Cancer Institute, DHSS, for support at the Southwest Foundation (Grant CA 24629).

We are very grateful to our colleagues Samy Abdel-Baky, Adhyam V. Durga, Joseph W. Goldzieher, Connie K. Durocher, Perry H. Moore, Jr., and Johng S. Rhim for their valued collaboration in this work.

References

1. Fishman, J. (1983): Aromatic hydroxylation of estrogens. Ann. Rev. Physiol., 45, 61-72.

2. MacLusky, N.J., Naftolin, F., Krey, L.C. and Franks, S. (1981): The catechol estrogens. J. Steroid Biochem., 15, 111-124.

3. Merriam, G.R., MacLusky, N.J., Picard, M.K. and Naftolin, F. (1980): Comparative properties of the catechol estrogens, I: Methylation by catechol-O-methyltransferase and binding to cytosol estrogen receptors. Steroids, 36, 1-11.

4. Poth, M.A., Axelrod, J. and Hoffman, A.R. (1983): Estrogen-2-hydroxylase: developmental, hormonal, and pharmacological studies using a radioenzymatic method. In: Catechol Estrogens, edited by G.R. Merriam and M.B. Lipsett, pp. 19-30, Raven Press, New York.

5. Brown, H.S. Jeffrey, A.M. and Weinstein, I.B. (1979): Formation of DNA adducts in $10T\frac{1}{2}$ mouse embryo fibroblasts incubated with benzo[a]pyrene or dihydrodiol oxide derivatives. Cancer Res., 39, 1673-1677.

6. Tomatis, L., Agthe, C., Bartsch, H., Huff, J., Montesano, R., Saracci, R., Walker, E. and Wilbourn, J. (1978): Evaluation of the carcinogenicity of chemicals: A review of the monograph program of the International Agency for Research on Cancer. (1971 to 1977). Cancer Res., 38, 877-885.

7. Jull, J.W. (1976): Endocrine aspects of carinogenesis. ACS Monogr., 173, 52-82.

8. Lipsett, M.B. (1979): Interaction of drugs, hormones, and nutrition in the causes of cancer. Cancer, 43, 1967-1981.

9. Lipsett, M.B. (1983): Hormones, medications, and cancer. Cancer, 51, 2426-2l

10. McCann, J. and Ames, B.N. (1976): Detection of carcinogens as mutagens in the Salmonella/microsome test: Assay of 300 chemicals: Discussion. Proc. Natl. Acad. Sci. U.S.A., 73, 950-954.

11. Glatt, H.R., Metzler, M., and Oesch, F. (1979): Diethylstilbestrol and 11 derivatives. A mutagenicity study with Salmonella typhimurium. Mutat. Res. 67, 113-121.

12. Mehta, R.D. and von Borstel, R.C. (1982): Genetic activity of diethylstilbestrol in Saccharomyces cerevisiae. Enhancement of mutagenicity by oxidizin agents. Mutat. Res., 92, 49-61.

13. Creus, A., Xamena, N. and Marcos, R. (1983): Positive response of diethylsti bestrol in the sex-linked recessive lethal assay in Drosophila after larval feeding. Mutat. Res. 122, 309-313.

14. Clive, D., Johnson, K.O., Spector, J.F.S., Batson, A.G. and Brown M.M.M. (1979 Validation and characterization of the L5178Y/TK$^{+/-}$ mouse lymphoma mutagen as system. Mutat. Res., 59, 61-108.

15. Barrett, J.C., Wong, A. and McLachlan, J.A. (1981): Diethylstilbestrol induces neoplastic transformation without measurable gene mutation of two loci. Science, 212, 1402-1404.

16. Purdy, R.H., Goldzieher, J.W., Le Quesne, P.W., Abdel-Baky, S., Durocher, C.K., Moore, P.H., Jr. and Rhim, J.S. (1983): Active intermediates and carcinogenesis. In: Catechol Estrogens, edited by G.R. Merriam and M.B. Lipsett, pp. 123-140, Raven Press, New York.

17. Kakunaga, T., Lo, K.-Y., Leavitt, J. Ikenaga, M. (1980): Relationship between transformation and mutation in mammalian cells. Jerusalem Symp. Quantum Chem. Biochem., 13, 527-541.

18. Holder, G., Yagi, H., Dansett, P., Jerina, D.M., Levin, W., Lu, A.Y.H. and Conney, A.H. (1974): Effects of inducers and epoxide hydrase on the metabolism of benzo-[a]pyrene by liver microsomes and a reconstituted system: Analysis by high pressure liquid chromatography. Proc. Natl. Acad. Sci. U.S.A., 71, 4356-4360.

19. Soloway, A.H. and Le Quesne, P.W. (1980): Potential endogenous mutagens/carcinoge J. Theoret. Biol., 85, 153-163.

20. Le Quesne, P.W., Durga, A.V., Subramanyam, V., Soloway, A.H., Hart, R.W. and Purdy, R.H. (1980): Biomimetic synthesis of catechol estrogens: Potentially mutagenic arene oxide intermediates in estrogen metabolism. J. Med. Chem., 23, 239-240.

21. Abdel-Baky, S. (1983): Ph.D. Thesis, Department of Chemistry, Northeastern Univesity, Boston, MA.

22. Paul, S.M., Hoffman, A.R., and Axelrod, J. (1980): Catechol estrogens: Synthesis and metabolism in brain and other endocrine tissues. Front. Neuroendocrinol., 6, 203-217.

23. Paul, S.M., Purdy, R.H., Hoffman, A.R. and Axelrod, J. (1983): Radioenzymatic assays for catechol estrogens. In: Catechol Estrogens, edited by G.R. Merriam and M.B. Lipsett, pp. 83-90, Raven Press, New York.

24. Kaufman, S. (1961): The enzymic conversion of 4-fluorophenylalanine to tyrosine. Biochim. Biophys. Acta , 51, 619-621.

Pyranoid Chiral Building Blocks: Preparation from Monosaccharides and Utilization for Natural Product Synthesis

Frieder W. Lichtenthaler

Institut für Organische Chemie, Technische Hochschule Darmstadt, D-6100 Darmstadt, FRG

Over the past decade the use of monosaccharides as chiral substrates for the construction of enantiomerically pure, non-carbohydrate natural products has become increasingly popular[2)] and continues to provide challenge and stimulation in ever new facets. The attractiveness of sugars as a starting material for preparative synthetic purposes arises from the fact, that they are an exceedingly cheap source of enantiomerically pure compounds: D—glucose, D-fructose and D-mannitol, for example, are not only less expensive than other Nature-derived chiral educts such as amino and hydroxy acids, but they are, most notably, cheaper than standard organic solvents (Table 1). In addition, they are available in open-chain, furanoid or pyranoid form, since the methodology for fixing either one of these tautomer is well known and usually achievable in one, preparatively satisfactory step.

However, despite of the unique synthetic potential inherent particularly in furanoid and pyranoid tautomers of monosaccharides, their *direct* use as the starting material for natural product synthesis is encumbered by a number of obstacles: they are overfunctionalized by similar or identical groups, they have considerably more chirality centres than required for a non-carbohydrate target molecule, and they lack suitable functional groups, such as olefinic and/or carbonyl unsaturation to which the preparative armoury of modern organic chemistry may directly be applied.

Natural Product Chemistry
Ed. by Atta-ur-Rahman

Table 1. Cost of Nature-derived chiral substrates used enantiomerically pure educts for organic preparative purposes[3)]

AMINOACIDS / HYDROXYACIDS		DM/kg	MONOSACCHARIDES		DM/kg
L - Glutamic Acid	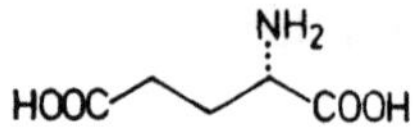	34.-	D - Glucose	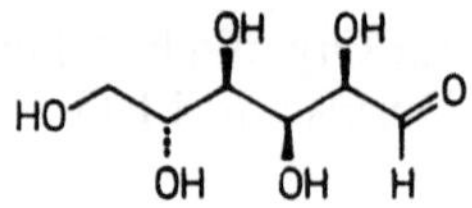	12.-
L - Leucine	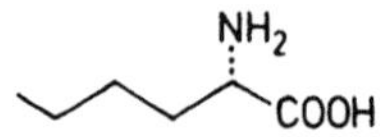	175.-	D - Fructose	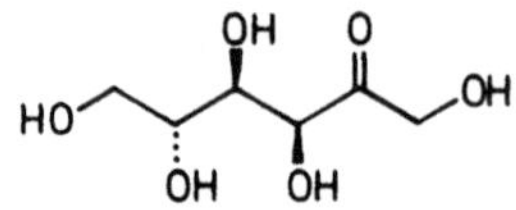	18.-
L - Valine	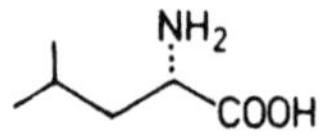	310.-	D - Mannitol	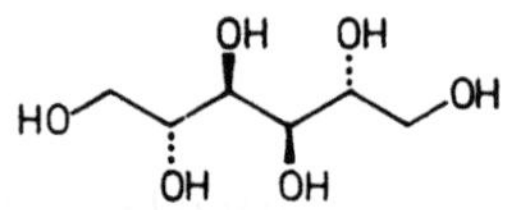	23.-
L - Proline	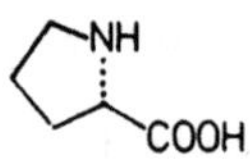	580.-	D - Galactose	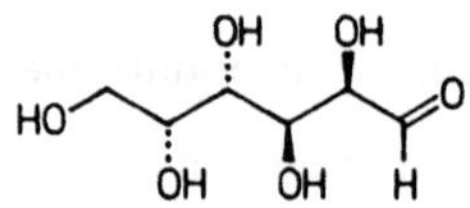	80.-
(S)-Lactic Acid	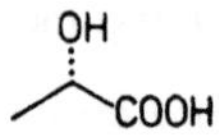	30.-	D - Mannose	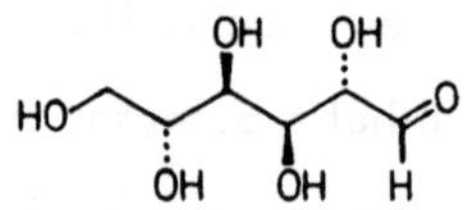	440.-
(S)-Malic Acid	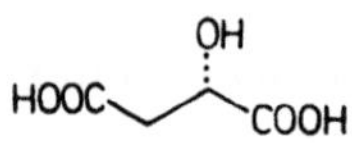	370.-	L - Arabinose	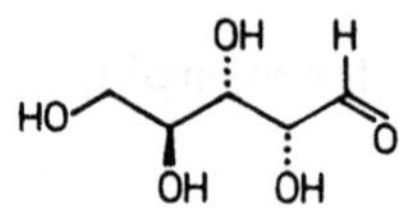	368.-
(R,R)-Tartaric Acid	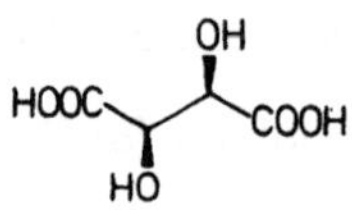	27.-	L - Rhamnose	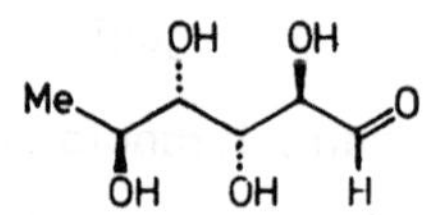	1190.-
(S,S)-Tartaric Acid	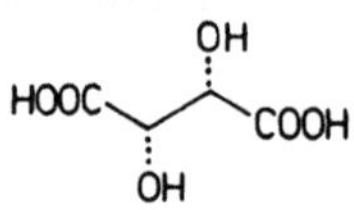	1010.-	Chloroform		25.-
			Toluol		25.-
			Nitromethane		47.-

MONOSACCHARIDES AS CHIRAL EDUCTS

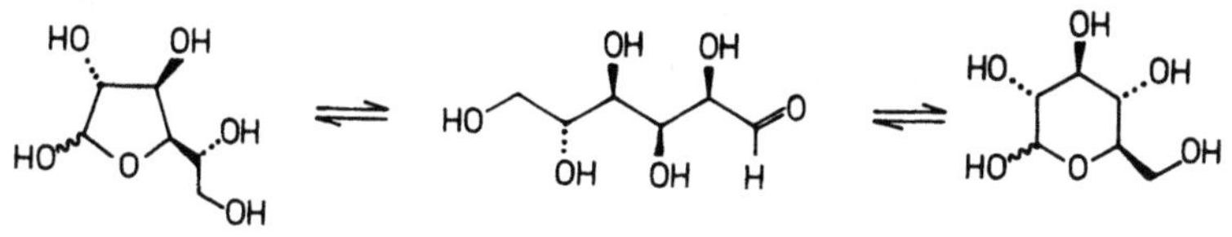

D-GLUCOSE

SHORTCOMINGS: overfunctionalized with hydroxyl groups
too many chirality centres
lack of C=C and C=O functionalities

This untoward situation has launched considerable efforts to reduce the number of chiral centres as well as of hydroxyl groups. One major approach involves the shortening or bisecting of the carbon skeleton as exemplified by the long known[4] conversion of D-mannitol into the 2,3-O-isopropylidene derivative of (R)-glyceraldehyde. However, aside of giving away half of the synthetic potential inherent in a six-carbon chain, incorporation of this simple three-carbon building block into the target molecule is usually effected via Wittig or aldol type methodology[5] which requires at least slightly basic conditions that in fact readily cause racemization next to the aldehyde function, thus seriously endangering the chiral integrity of the product.

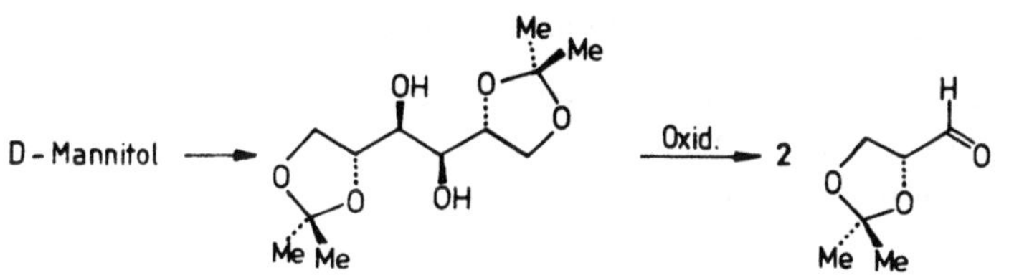

(R)-Glyceraldehyde

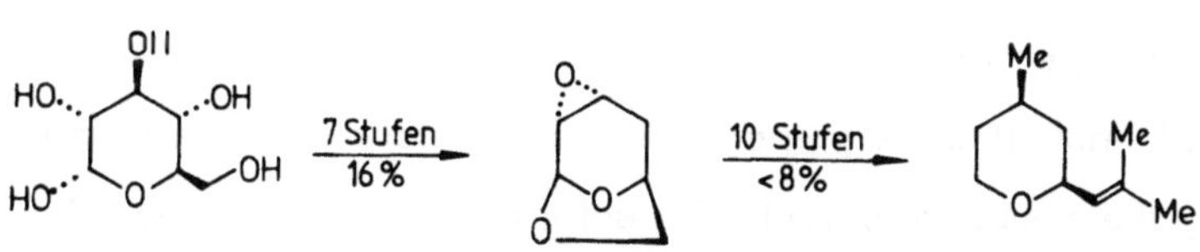

α-D-Glucose

cis-Rose oxide

Another approach pursued in a large number of cases comprises the gradual, step-by-step carving out of a target molecule segment from a monosaccharide, quasi an "exfoliation" that for every case requires a special, tailor-made reaction sequence. Whilst conceptually interesting and intellectually challenging, the majority of these undertakings have resulted in exceptionally long, low-overall-yield syntheses of limited general or practical value, eliciting — with some justification unfortunately —

quite sarcastic comments particularly from non carbohydrate chemists[6]. One such example is the elaboration of *cis* – rose oxide from D-glucose in 17 steps with an overall yield below 1%[7], which was hailed by its authors as "the first example of chemical transformation of a carbohydrate into an optically active perfume"[7]. Many other, markedly more ambitious total syntheses, such as the 23-step conversion of glucose into spectinomycin (overall yield: 2%)[8] or the assembly of the erythronolide A framework from D-glucose[9] are similarly encumbered with too many steps and impracticably low overall yields.

The alternate to these approaches — shortening or bisecting the carbon chain on one hand, and stepwise "exfoliation" of a given target molecule in a special tailor-made reaction sequence on the other — consists in the *utilization of open-chain, furanoid or pyranoid chiral, enantiomerically pure building blocks* which fulfill most, if not all, of the following attributes:

- only one or two chiral centres with suitable synthon functionalities
- broad applicability of methodologies effecting C extension or C branching with foreseeably high stereoselectivities
- crystalline compounds, stable enough to be stored
- ready accessibility from monosaccharides in four, at the most five, preparatively satisfactory, large scale adaptable steps, such that overall yields of 50% are attainable

Realization of most, or all of these requirements for a useful highly versatile chiral building block necessitates careful consideration of the stereochemical intricacies inherent in the respective monosaccharide educt, because only then 4-5 step reaction sequences will result that are preparatively satisfactory and of practical utility. Accordingly, the number of potentially useful chiral building blocks is large — as e.g. those outlined in the following scheme — but not limitless, since restriction to the preparatively "makable" must have priority over lengthy tailor-made "exfoliation" procedures.

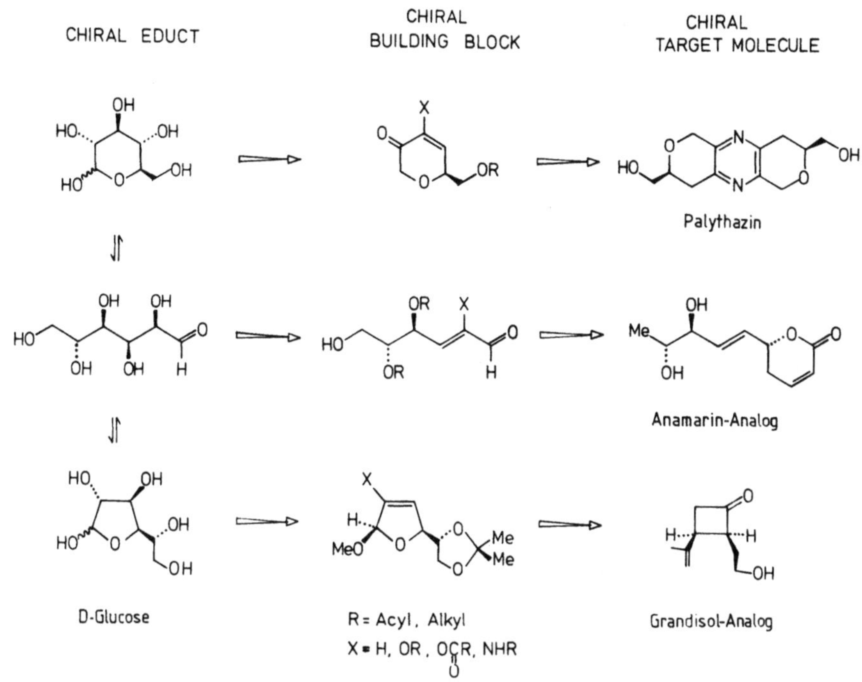

Methodical entries into monosaccharides fixed in pyranoid, open-chain or furanoid form are well known and go back to the beginning of preparative carbohydrate chemistry, notable examples being mercaptalization to the acyclic dithio acetals[10], isopropylidenation to furanoid systems — as e.g. the preparation of "diacetone glucose" by E. Fischer 100 years ago[11] — or the generation of glycal[12] or hydroxyglycal esters[13] in three-step procedures, which may readily be combined into one continuous operation.

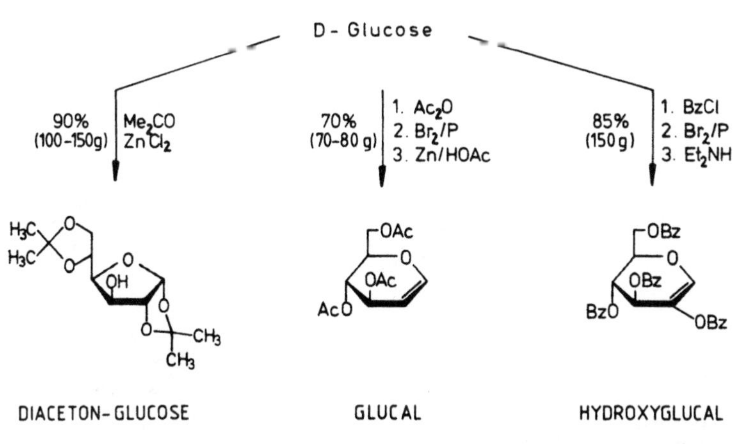

The yields obtainable are excellent, rendering these products as chiral starting materials of which it is worthwhile to further develop their chemistry systematically for attaining chiral building blocks with a wide range of use. The account given in the sequel will concentrate on such objectives, as well as the utilization of the building blocks obtained for the synthesis of carbohydrate and non-carbohydrate natural products.

Preparation of some Novel Pyranoid Chiral Building Blocks

Conversion of monosaccharides into dihydropyranones with two vicinal chirality centres left, may be accomplished via several preparatively useful procedures. Endiolones of D-*erythro*-configuration such as 2 and 4 are particularly readily accessible: D-glucose is convertible into 2 in a 4-step-procedure with a 60% overall yield, involving acetonation, oxidation (→1), acid removal of isopropylidene groups and benzoylation, the slightly basic conditions in the last step (pyridine) already sufficing to elicit β-elimination of benzoic acid[14]. To convert maltose in the corresponding disaccharide endiolone 4 requires only two steps even, since low temperature benzoylation affords the heptabenzoate 3[15], subsequent DMSO/acetic anhydride oxidation — not unexpectedly[16] — effecting both, oxidation of the 3 OH and β-elimination[17].

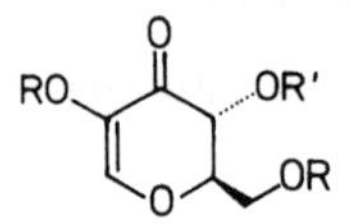

PYRANOID ENEDIOLONES

D-erythro
(R,R)

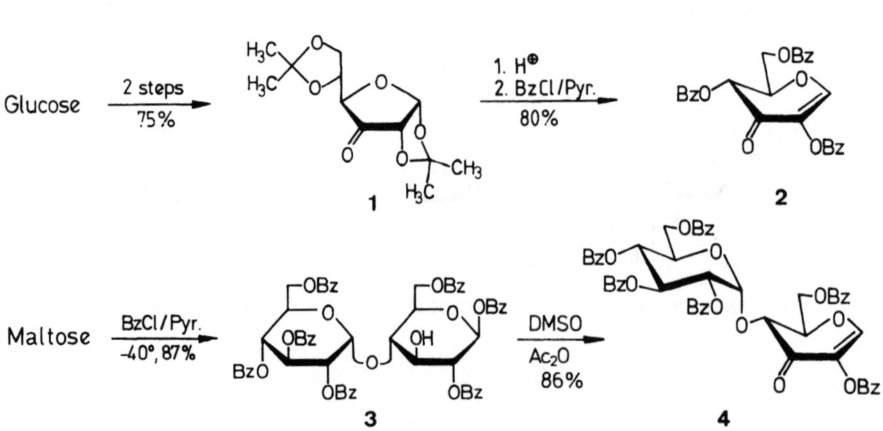

Whilst peroxidation of glycal and hydroxyglycal esters yields normal products of oxirane ring opening or of acyloxy group migration[18] the presence of a Lewis acid such BF_3 shifts the reaction into an entirely different direction. Initiated by a BF_3-induced removal of the allylic acyloxy function to form the allylcarboxonium ion 6a ↔ 6b, the m-chloroperbenzoic acid (MCPBA) attacks solely at C-1 as expected from a "hard" nucleophile; the resulting intermediate 7 then undergoes fragmentation as indicated by the arrows to yield enlactones or enollactones of type 8[19]. This most readily effected conversion (15 – 30 min at ambient temperature) provides a series of hitherto only with difficulty accessible dihydropyran-2-ones in excellent yields (cf. table 2), as e.g. the acetate of the naturally occuring osmundalactone from L-rhamnose via di-O-acetyl-L-rhamnal (entry 4 in table).

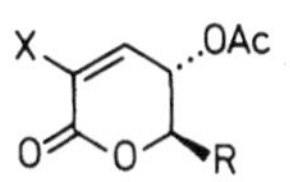

ENLACTONES (X=H)
ENOLLACTONES (X=OAc)

(R = H, CH_3, CH_2OAc)

BF_3-catalyzed Peroxidation of Glycal Esters

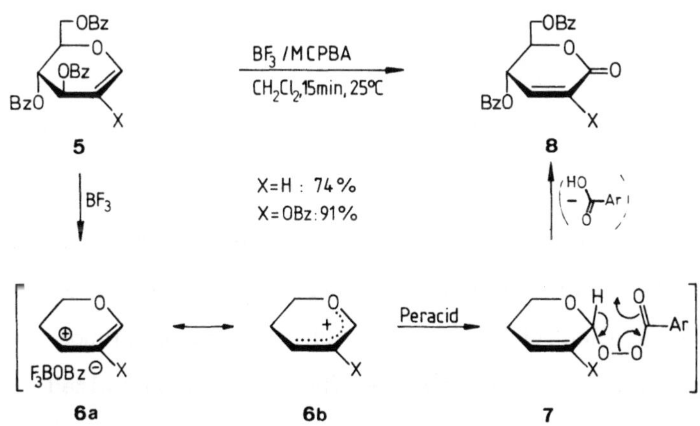

Hydroxylaminolysis of alkyl or aryl esters is a standard procedure for the preparation of hydroxyamic acids[20]. Enol esters react similarly, whereby the ketone liberated (I→II→III) is further transformed into the respective oxime IV. Application of the

Table 2. BF_3-catalyzed Oxidation of Glycal Esters with m-Chloroperbenzoic Acid[19)]

Educt		Product		Yield	m.p. (°C) b.p. (°C/mm)	$[\alpha]_D$ $CHCl_3$
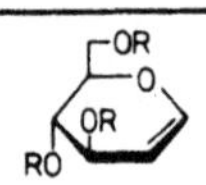	R = Ac R = Bz	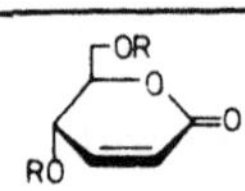	R = Ac R = Bz	69 74	135/0.1 syrup	+128 +195
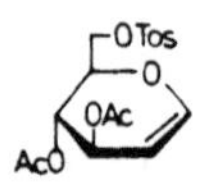		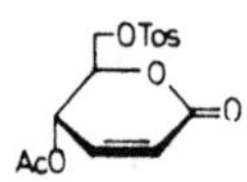		78	92-93	+ 92
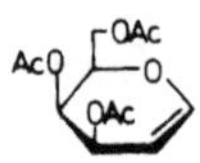		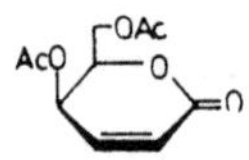		81	135/0.1	-347
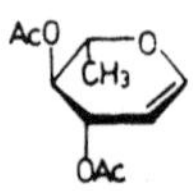		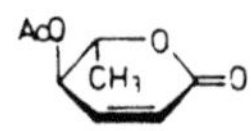		89	110/0.5	-179
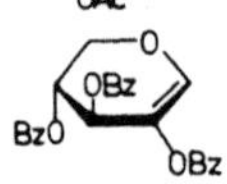		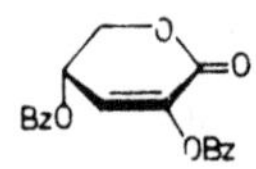		78	134-135	+ 79
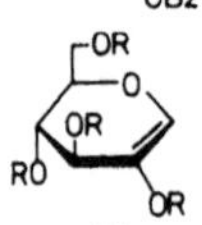	R = Ac R = Bz	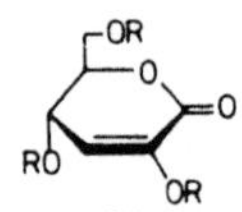	R = Ac R = Bz	67 91	110/0.1 110-111	+109 +104
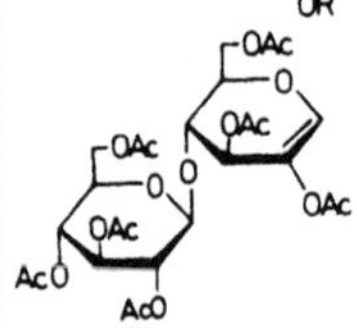		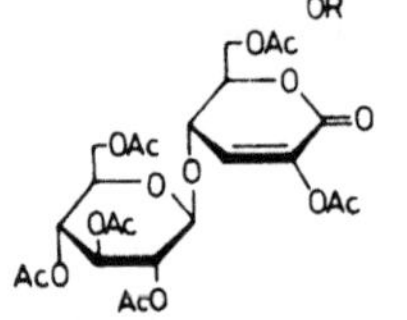		87	130-131	+ 28

procedure I→IV to hydroxyglycal esters proved to be particularly advantageous since the respective ketoximes of 1,5-anhydro-ketoses — hitherto inaccessible — are obtained in excellent yields[21)] without affecting normal primary or secondary ester functions elsewhere in the molecule. The examples summarized in Table 3 amply illustrate the preparative feasibility and the apparent generality of this selective enol ester hydroxylaminolysis[21)].

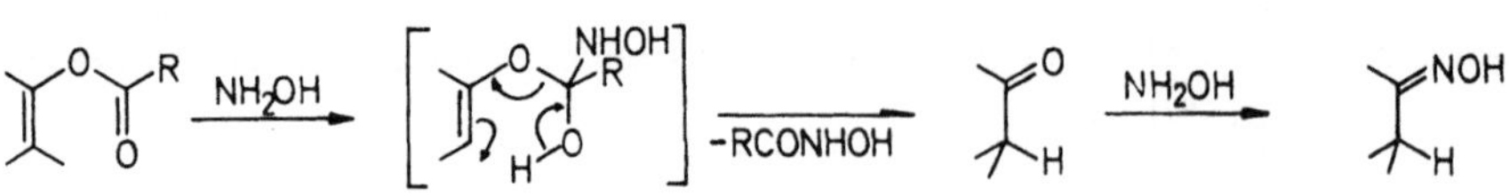

I II III IV

Table 3. Selective Deacylation of pyranoid Enol Esters with Hydroxylamine (e.g. $NH_2OH \cdot HCl$/pyridine, 25°C, 12 h)[21]

Educt		Product		Yield	m.p.(°C)	$[\alpha]_D$ in $CHCl_3$ (c, °C)
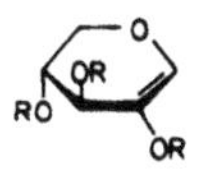	R=Ac R=Bz	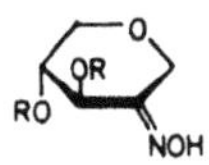	R=Ac R=Bz	76 79	103 - 104 198 - 200	-58.3 (0.3, 21) -86.5 (0.6, 21)
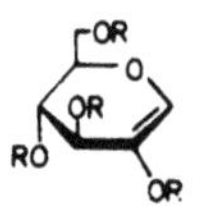	R=Ac R=Bz	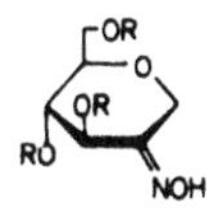	R=Ac R=Bz	86 93	89 - 90 176 - 177	-39.0 (0.4, 21) -52.9 (0.3, 21)
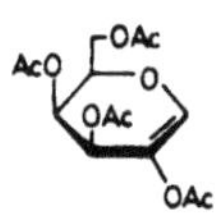		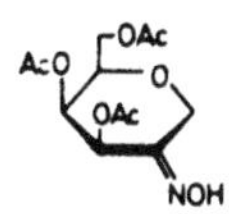		74	155	-41.2 (0.5, 22)
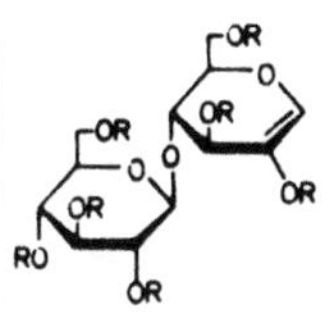	R=Ac R=Bz	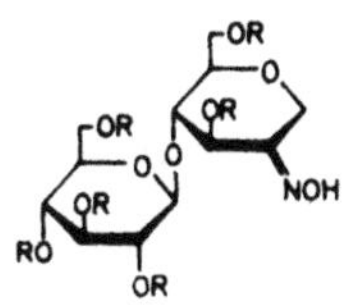	R=Ac R=Bz	89 76	125 - 127 amorph	-22.3 (0.3, 21) +12.3 (0.6, 22)
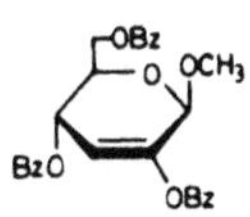		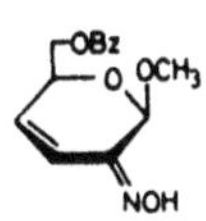		89	syrup	-148 (0.3, 22)

The acylated ketoximes, thus obtainable, are crystalline, stable intermediates; they may be subjected to Zemplén conditions (sodium methoxide/methanol) to smoothly afford the respective 1,5-anhydroketose oximes (e.g. the oxime of 10), from which the parent ketoses as, for example, 1,5-anhydro-D-fructose (10) may be generated by deoximination[22]; a direct conversion of type 9→10 is infeasable, since the basic de-O-acetylation conditions induce β-eliminations (e.g. 12→13) and benzilic acid type rearrangements, resulting in a whole series of products.

Deoximination of the oximes, however, may readily be accomplished, the conversion 11→12, e.g. being feasible in 88% yield[22,23]. Elimination of benzoic acid from the 3,4-positions is as easily effected (stirring with sodium acetate in acetone for 30 min at 25°C, 92% isolated), to afford — in an overall yield from 9 of 75% — the pyranoid enolone ester 13[23], a crystalline, stable, highly versatile chiral building block (vide infra for its use in the synthesis of *S,S*-palythazine).

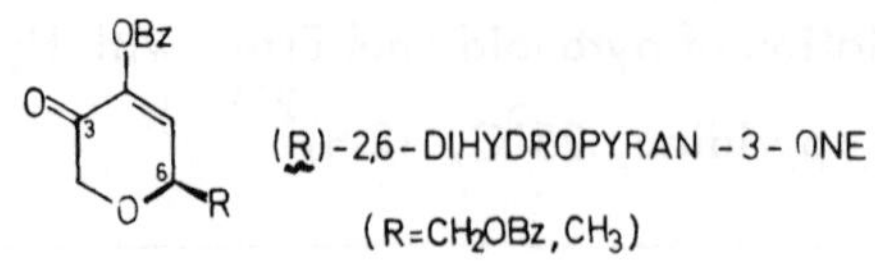

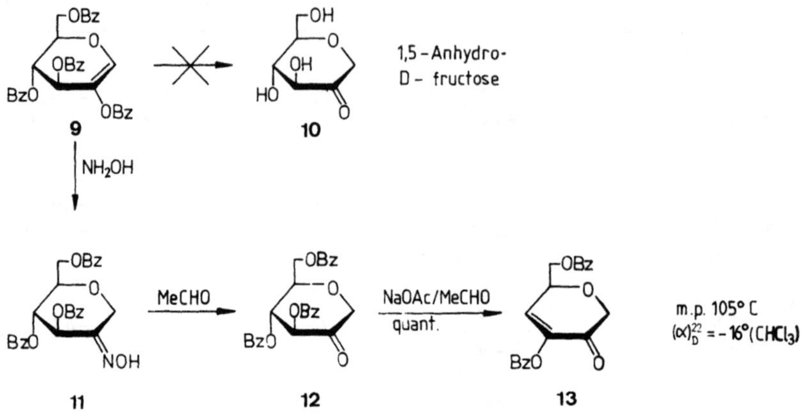

Oximes of type 11 and their parent ketones 12 are of preparative interest also, albeit for the synthesis of oligosaccharides. Due to a push-pull substitution pattern at the proanomeric centre — the ring oxygen pushes electrons whilst the oximino and keto group at C-2 are strongly electron withdrawing — a hydrogen atom at C-1 is most readily removed under free radical conditions. Accordingly, photobromination with N-bromosuccinimide or, even bromine, is effected with particular ease, affording glycosyl bromides or their oximino analogs in high yields[24,25].

FUNCTIONALIZATION AT PROANOMERIC CENTRES
BY PHOTOBROMINATION

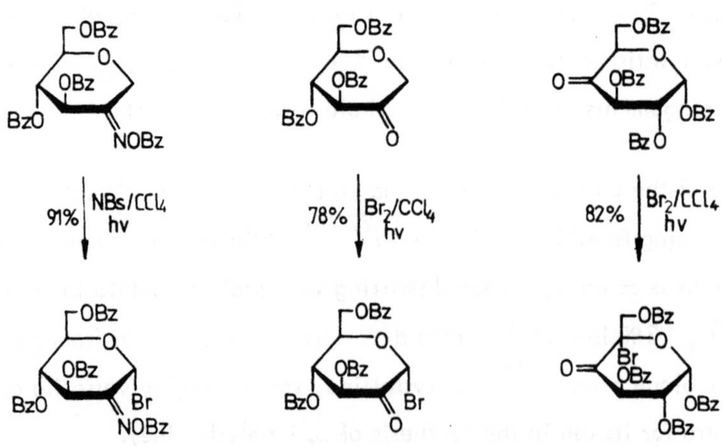

The availability of such α-bromides in stable, storable form provides an efficient access to oligosaccharides containing aminosugars (e.g. β-D-mannosamine or α-D-glucosamine) or β-D-mannose units by simple displacement of the bromine via β- or α-selective Königs-Knorr-glycosylation, reduction, and deblocking. This approach has close analogies to the "nitroso halogenide procedure"[26], yet holds the additional advantage that it is readily applicable to disaccharide-derived hydroxyglycal esters, e.g. 14 → 15, thus providing disaccharide building blocks for the construction of lactosamine-, maltosamine-, and cellobiosamine-containing oligosaccharides in a most simple manner[27]. On the other hand, 2-oxoglycosyl bromides are highly useful synthetic intermediates for the construction of dioxane-anellated pyrano-dioxane systems via alcoholysis with bi-functional alcohols such as glycol or cyclohexane-1,2-diol[28].

SYNTHESIS OF AMINOSUGAR-CONTAINING OLIGOSACCHARIDES

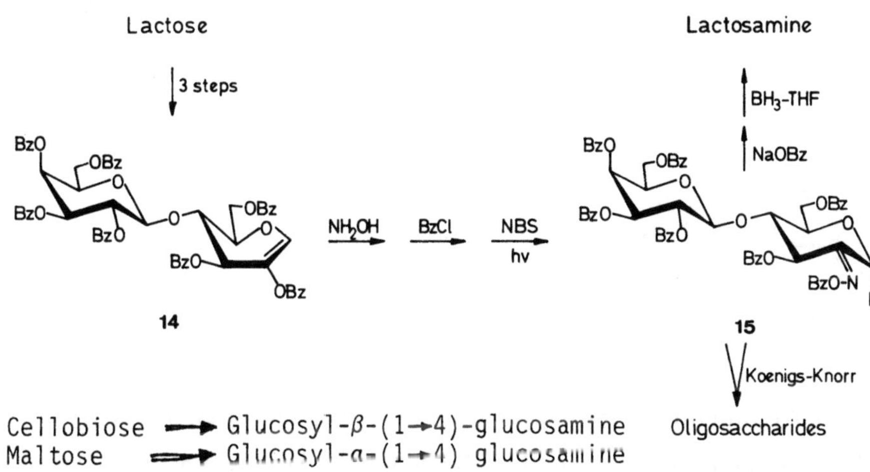

In the dihydropyranone 13, the enolone ester functional unit is flanked by one asymmetric carbon atom. Analogous systems in which the enolone structural element is framed by chiral centres — they are apt to determine even more strongly the stereochemistry of addition reactions — are as readily accessible from hydroxyglycal esters, namely by a chlorination-hydrolysis-elimination sequence[16,29]. In the case of 9, e.g. low temperature-chlorination affords a mixture of α-D-*manno*-dichloride 16 and the benzoxonium salt 17, which only at room temperature is converted into the β-D-*gluco*-dichloride 18.

Hence, when a low temperature-chlorination of 9 is followed by hydrolysis with water, a mixture of 16 and the hexosulose 20 is obtained, which on brief warming with moist sodium hydrogen carbonate yields enolone 21, readily separable from 16 by fractional crystallisation. All these steps can be performed in one continous operation providing 21 in an overall yield of 69 %[29,30].

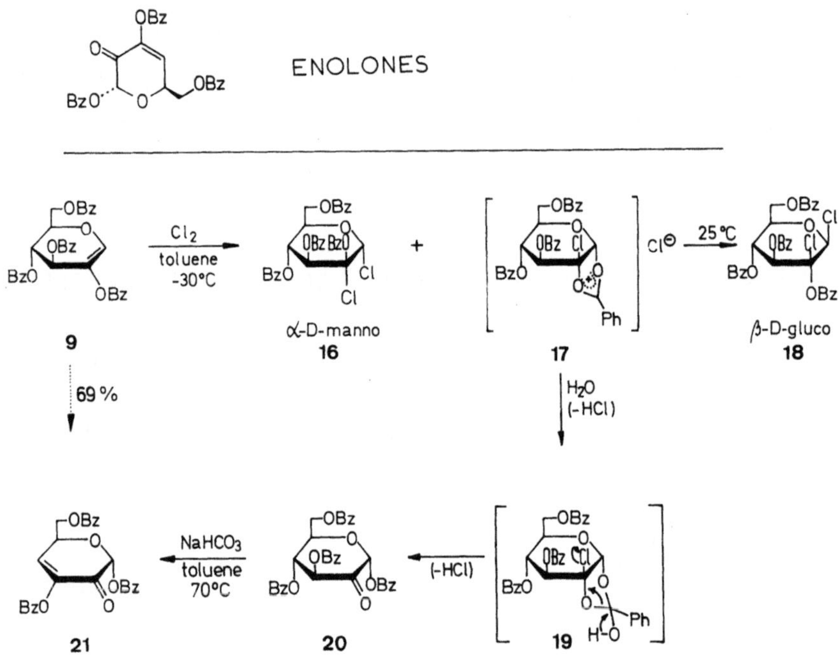

Being remarkably insensitive towards acidic conditions, enolones of type 21 are readily convertible into the respective 1-halides, as shown by the essential quantitative conversion 21→22 on treatment with HBr/acetic acid[29]. Alcoholysis of such enolone α-D-bromides may be performed with high β-selectivity when employing sodium hydrogen carbonate or silver carbonate as acid scavenger, providing the β-glycosides in yields well over 75%[29,3 Similarly, when subjected to glycolysis with ethylene glycol in the presence of Ag_2CO_3, the bromide 22 smoothly elaborates the pyrano[2,3-b]dioxane system 25, the β-selective glycosylation obviously being followed by cyclization to the hemiketal and subsequent 3-O→2-O-benzoyl migration (arrows in 24) with liberation of the C-3 carbonyl function[31] Thus, from the stage of a hydroxyglycal ester of type 9, a simple four-step sequence (i.e.

9→21→22→25) is developed to dioxane-anellated glycosides of the spectinomycin and uscharidine type (cf. below).

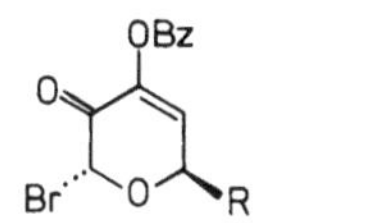

($R = CH_2OBz, CH_3$)

ENOLONE CONVERSIONS

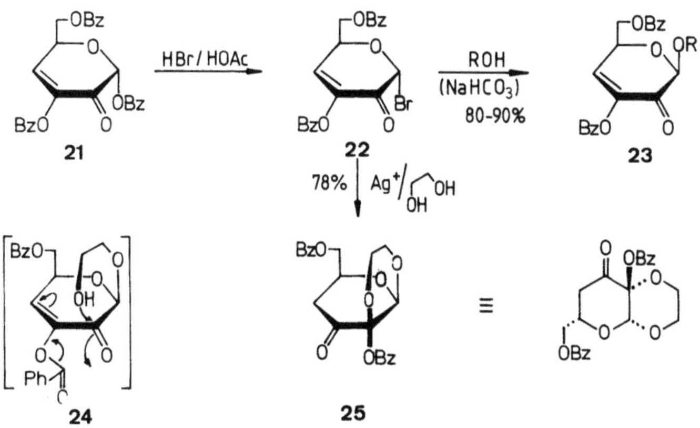

The stereoselectivities attainable with β-D-enolone esters on addition reactions are noteworthy; hydride addition as executed with $NaBH_4$, already give the 4-deoxyglycoside of β-D-*lyxo*-configuration (27) in respectable stereochemical preference[32], which when increasing the steric demand of the nucleophile by using L-Selectride (lithium tri-sec-butyl-borohydride) proceeds in an essentially stereospecific manner[30]. The less reactive zinc tetrahydroborate allows the saturation of only the free carbonyl function, as evidenced by the conversion 26→28, which may be mechanistically understood[32] as proceeding via hydride addition to the C-2 carbonyl, subsequent benzoyl group migration from O-3 to O-2 to form a zinc enolate complex that is stable towards further hydride addition; on aqueous workup, then, the 3-ulose is generated, which most readily undergoes β-elimination to give the enediolone 28.

C-Branching also may readily be effected with high stereoselectivities, lithium dialkyl cuprates resulting in conjugate addition as e.g. 26→29, whilst Grignard or alkyl lithium reagents give 1,2-addition products with retention of the enol ester function[33].

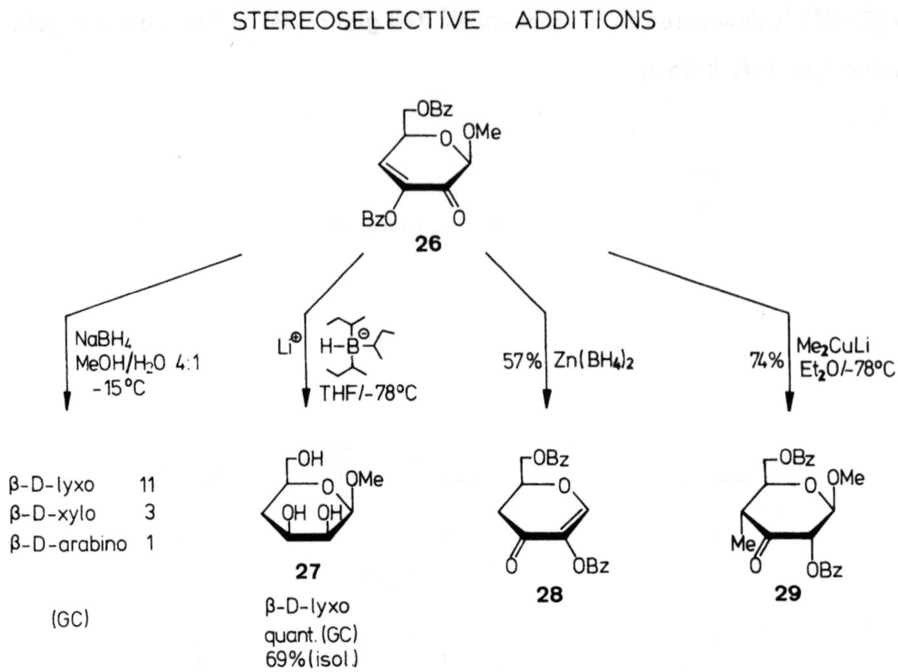

Aside these highly stereoselective addition reactions, enolone esters of type 26 may undergo a most intriguing, essentially stereospecific rearrangement reaction under very mild basic conditions[17]. Stirring at room temperature with 1.5 molar equivalents of an alcohol in DMSO, acetonitril or butanone and K_2CO_3 or Bu_4NOAc, 26 is converted into dihydropyranones of structure 30 in high yields[17].

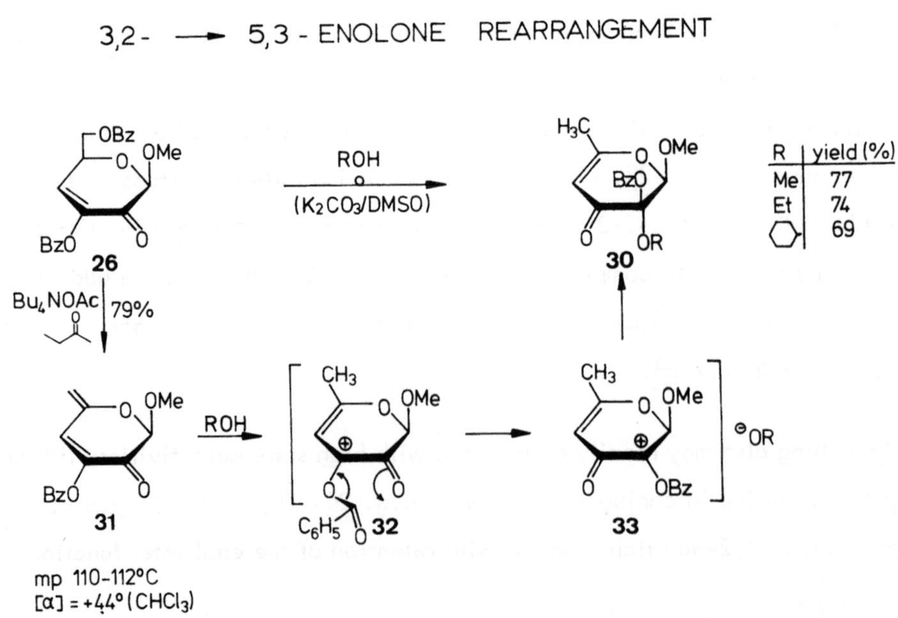

The compounds are crystalline and exhibit high negative rotations (-260° and more). Their structure readily followed from spectral data, whereas configurational proof for the tertiary centre was more laborious. It was delineated from stereochemical considerations, particularly from the fact that the corresponding α-D-enolone (i.e. the anomer of 26), when subjected to these conditions, gave the optical antipodes of 30 as evidenced by identical melting points, yet opposite rotations[17]. Absolute configurational proof was provided by an X-ray crystallographic structure determination of the bis-methoxy derivative (cf. Fig.), clearly showing the nearly antiparallel arrangement of the two methoxy groups.

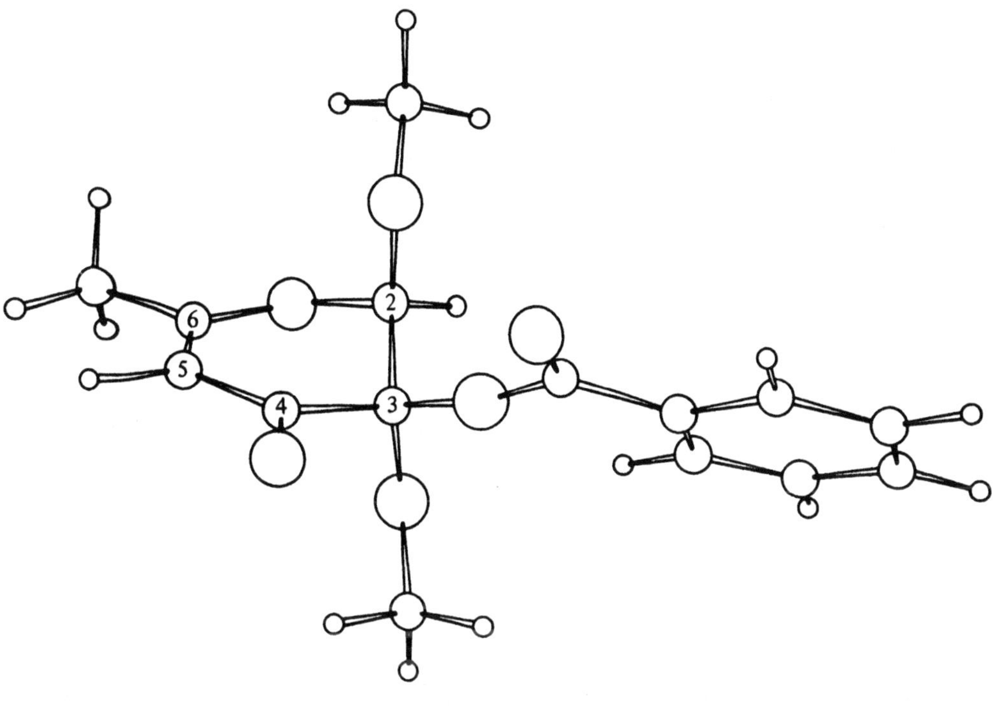

Fig. 1. X ray stereoview of (2*R*,3*R*)-3-benzoyloxy-2,3-dimethoxy-6-methyl-2H-pyran-4(3H)-one (30, R = Me); the dihedral angle O2−C2/C3−O3 is -172.3° [34]

The conversion 26 → 30 is obviously initiated by 5,6-elimination of benzoic acid to give the dienolone 31 — its isolation in crystalline form substantiates this strongly — is followed by protonation of the exocyclic double bond and benzoyl migration, and is concluded by addition of the nucleophile ($RO^{\ominus}$) from the side opposite to the anomeric substituent.

The utility of enantiomerically pure 6-carbon-building blocks with only one chiral centre, such as 13, 41, 47 and 50, has been amply demonstrated. Another chiral building block particularly useful for the synthesis of a variety of macrodialide-type natural products are the 5*S*-hexenolactone 54 — contained[53] in the red berries of mountain ash (*Sorbus aucuparia*), hence named parasorbic acid — and its 5*R*-enantiomer.

The 5*S*-version is most readily prepared from L-rhamnose via the known di-O-acetyl-rhamnal (55) and the 2,3-unsaturated ethyl glycoside 56, which on standard tosylation (tosyl chloride / pyridine) affords the 4-chloro-compound 57 rather than the 4-O-tosylate of 56, since the chloride ions formed upon tosylation suffice to displace the tosyloxy function at room temperature already; the subsequent dehalogenation (57→58) and BF_3-catalyzed peroxidation smoothly gives *S*-parasorbic acid 54, the first synthesis of this simple compound in enantiomerically pure form[54].

54

(S)-(+)-PARASORBIC ACID
(mountain ash berries)

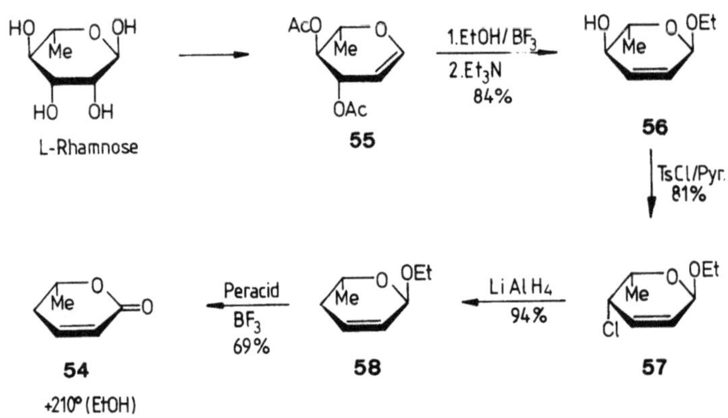

Starting from 6-deoxy-D-glucose (D-rhamnose), an analogous set of reactions as readily leads to the 5*R*-enantiomer 59 as a key intermediate. Hydrolysis affords the 5*R*-hydroxy hexenal (61) or, when preceeded by hydrogenation, the 5*R*-hydroxyhexanal (60), whilst peroxidation yields *R*-parasorbic acid (62)[55] — all enantiomerically pure 6-carbon building blocks that represent major segments of macrodiolides: half of diplodialide- and phorocantholide-type pheromones, the entire right half of colletodiol or the C11–C16-section of, e.g., carbonolide. Efforts towards the total syntheses of these compounds via C-extension and lactonization are presently being underway in this laboratory[56,57].

APPROACH TO MARCODIOLIDES

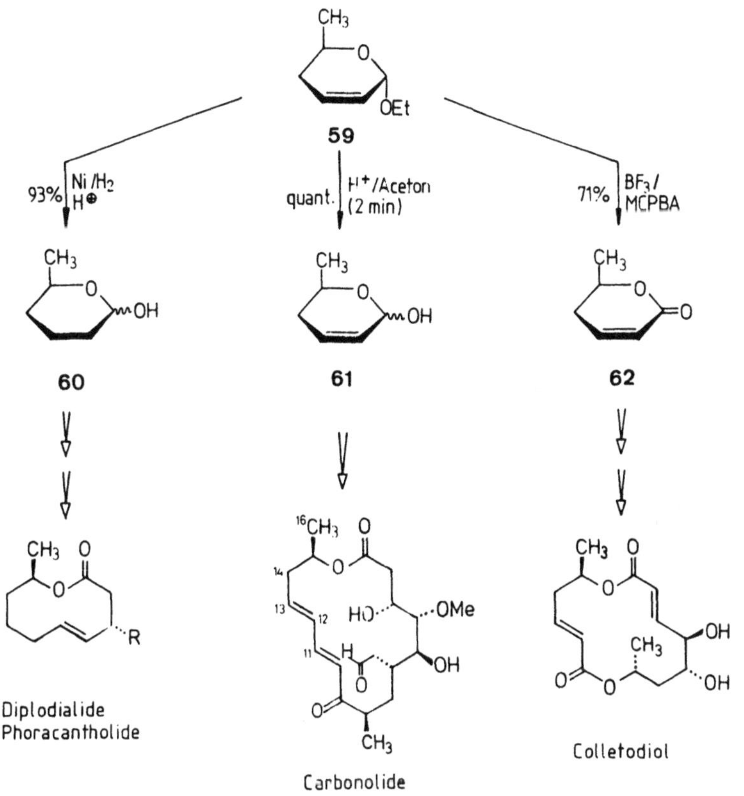

Finally, I wish to give special acknowledge to the talented students and postdoctoral coworkers who have carried out the experiments upon which this account is based: Dr. Eckehard CUNY, Dr. Pan JARGLIS, Walter HEMPE, Dr. Eisuke KAJI, Franz-Dietrich KLINGLER, Peter KÖHLER, and Sabine WEPREK. Support for this research was provided by grants from the Deutsche Forschungsgemeinschaft and the Fonds der Chemischen Industrie.

References and Notes

1. First communication of the new series "Sugar-derived Chiral Building Blocks".— Part II: F.W. Lichtenthaler, F.-D. Klingler, and P. Jarglis, *Carbohydr. Res.*, (1984), to be published.
2. B. Fraser-Reid and R.C. Anderson, *Progr. Chem. Org. Nat. Prod.*, 39, 1 (1980); A. Vasella, *Modern Synthetic Methods*, Vol. 2, pp. 173 - 267, (R. Scheffold, ed.), O. Salle Verlag, Frankfurt, Germany (1980); S. Hanessian, Total Synthesis of Natural Products, Pergamon Press, Oxford, 1983.
3. Prices listed are the lowest found in the most recent fine chemicals catalogues of Merck-Schuchardt (Darmstadt), Aldrich-Chemie (Steinheim), Fluka (Neu-Ulm) or Sigma Chemie (Taufkirchen). The DM (Deutsche Mark) price quoted for D-glucose corresponds to about 50 Rupees/kg; in ton quantitites the kg price is considerably lower (~ DM 1.50, i.e. about 7 Rupees).
4. H.O.L. Fischer and E. Baer, *Helv. Chim. Acta*, 17, 622 (1984); *J. Biol. Chem.*, 128, 463 (1939).
5. For a recent overview see: J. Mulzer, *Nachr. Chem. Techn. Lab.*, 32, 146 (1984).
6. Examples: "plastic surgery on sugars", "academic sport of carbohydrate maniacs"; the author, though, prefers more elaborate literary formulations, e.g. that of John of Gaunt, Duke of Lancaster: "Things sweet to taste prove in digestion sour" [W. Shakespeare, *King Richard II*, 1, 3 (1593)].
7. T. Ogawa, N. Takasaka, and M. Matsui, *Carbohydr. Res.*, 60, C4 (1978); no yield was given for the last two steps.
8. S. Hanessian and R. Roy, *J. Am. Chem. Soc.*, 101, 5839 (1979).
9. S. Hanessian and G. Rancourt, *Pure Appl. Chem.*, 49, 1201 (1977); *Can. J. Chem.*, 56, 1843 (1978).
10. E. Fischer, *Ber. Dtsch. Chem. Ges.*, 27, 673 (1984); J.D. Wander and D. Horton, *Adv. Carbohydr. Chem. Biochem.*, 32, 15 (1976).
11. E. Fischer, *Ber. Dtsch. Chem. Ges.*, 28, 1145 (1885); O.T. Schmidt, *Methods Carbohydr. Chem.*, 2, 318 (1963).
12. E. Fischer and K. Zach, *Sitzungsber. Kgl. Preuss. Akad. Wiss.*, 16, 311 (1913); *Ber. Dtsch. Chem. Ges.*, 47, 196 (1914); W. Roth and W. Pigman, *Methods Carbohydr. Chem.*, 2, 405 (1963).
13. K. Maurer, *Ber. Dtsch. Chem. Ges.*, 62, 332 (1929); 66, 955 (1933); R.J. Ferrier and N. Prasad, *J. Chem. Soc. C*, 1969, 570.

14. H. Hartmann, Diploma Thesis, Technische Hochschule Darmstadt, 1984; to be published.

15. K. Takeo and S. Okano, *Carbohydr. Res.*, 59, 379 (1977).

16. F.W. Lichtenthaler, *Pure Appl. Chem.*, 50, 1343 (1978).

17. F.W. Lichtenthaler, S. Nishiyama, and P. Jarglis, *Angew. Chem.*, 91, 1001 (1979); *Angew. Chem. Int. Ed. Engl.*, 18, 936 (1979).

18. F.W. Lichtenthaler and P. Jarglis, *Chem. Ber.*, 113, 489 (1980).

19. P. Jarglis and F.W. Lichtenthaler, *Tetrahedron Lett.*, 23, 3781 (1982).

20. S.R. Sandler and W. Karo, *Organic Functional Group Preparations*, Vol. III, 406 ff., Academic Press, New York, 1972.

21. F.W. Lichtenthaler and P. Jarglis, *Tetrahedron Lett.*, 21, 1425 (1980).

22. F.W. Lichtenthaler, E.S.H. El Ashry, and V.H. Göckel, *Tetrahedron Lett.*, 21, 1429 (1980).

23. P. Jarglis and F.W. Lichtenthaler, *Angew. Chem.*, 94, 140 (1982); *Angew. Chem. Int. Ed. Engl.*, 21, 141 (1982); *Angew. Chem. Suppl.*, 1982, 175.

24. F.W. Lichtenthaler and P. Jarglis, *Angew. Chem.*, 94, 643 (1982); *Angew. Chem. Int. Ed. Engl.*, 21, 625 (1982); *Angew. Chem. Suppl.*, 1982, 1449.

25. F.W. Lichtenthaler, P. Jarglis, and W. Hempe, *Liebigs Ann. Chem.*, 1983, 1959.

26. R.U. Lemieux, Y. Ito, K. James, and T.L. Nagabhushan, *Can. J. Chem.*, 51, 7, 42, 48 (1973).

27. E. Kaji and F.W. Lichtenthaler, unpublished results.

28. F.W. Lichtenthaler, E. Cuny, and S. Weprek, *Angew. Chem.*, 95, 906 (1983); *Angew. Chem. Int. Ed. Engl.*, 22, 891 (1983).

29. F.W. Lichtenthaler and U. Kraska, *Carbohydr. Res.*, 58, 363 (1977).

30. P. Köhler, *Doctoral Dissertation*, Technische Hochschule Darmstadt, to be submitted.

31. F.W. Lichtenthaler, A. Löhe, and E. Cuny, *Liebigs Ann. Chem.*, 1983, 1973.

32. F.W. Lichtenthaler, U. Kraska, and S. Ogawa, *Tetrahedron Lett.*, 1978, 1323.

33. P. Köhler, *Diploma Thesis*, Technische Hochschule Darmstadt, 1982.

34. H.J. Lindner and F.W. Lichtenthaler, to be published.

35. F.W. Lichtenthaler, *Methods Carbohydr. Chem.*, 6, 348 (1972).

36. F.W. Lichtenthaler, S. Ogawa, and P. Heidel, *Chem. Ber.*, 110, 3324 (1977).

37. W. Hempe, *Doctoral Dissertation*, Technische Hochschule Darmstadt, 1984.

38. E.J. Reist, R.R. Spencer, D.F. Calkins, B.R. Baker, and L. Goodman, *J. Org. Chem.*, 30, 2312 (1965).

39. G. Hesse, F. Reicheneder, and H. Eysenbach, *Liebigs Ann. Chem.*, 537, 67 (1939); F. Brüschweiler, K. Stöckel, and T. Reichstein, *Helv. Chim. Acta*, 52, 2086, 2276 (1969).
40. G. Hesse, H. Hertel, and K. Mix, *Liebigs Ann. Chem.*, 625, 174 (1959).
41. H.T.A. Cheung, T.R. Watson, J.N. Seiber, and C. Nelson, *J. Chem. Soc., Perkin Trans. 1*, 1980 (2169).
42. M. Hoeksema, A.D. Argoudelis, and P.F. Wiley, *J. Am. Chem. Soc.*, 85, 2652 (1963); W. Rosenbrook, Jr., *J. Antibiot. Sect.*, B2, S-211 (1979).
43. E. Cuny and F.W. Lichtenthaler, to be published.
44. F.W. Lichtenthaler, H. Leinert, and T. Suami, *Chem. Ber.*, 100, 2383 (1967).
45) D.R. White, R.D. Birkenmeyer, R.C. Thomas, S-A. Mizsak, and V.H. Wiley, *Tetrahedron Lett.*, 1979, 2737.
46. J.C. Sowden, *Methods Carbohydr. Chem.*, 1, 132 (1962).
47. R.E. Moore, and G. Bartolini, *J. Am. Chem. Soc.* 103, 2491 (1981); D. Uemura, K. Ueda, Y. Hirata, H. Naoke, T. Iwashita, *Tetrahedron Lett.*, 22, 2781 (1981); Y. Kishi, *Current Trends in Organic Synthesis*, (H. Nozaki, ed.), Pergamon Press, Oxford-New York, 1982, p. 115 ff.
48. D. Uemura, Y. Toya, I. Watanabe, and Y. Hirata, *Chem. Lett.*, 1979, 1481.
49. H. Gutknecht, *Ber. Dtsch. Chem. Ges.*, 12, 2290 (1879); Y.T. Pratt, *Heterocyclic Compounds* (R.C. Elderfield, ed.), Vol. 6, p. 385 ff., Wiley, New York 1957.
50. F.W. Lichtenthaler and P. Jarglis, XI. Internat. Carbohydr. Symposium, Vancouver, August 23, 1982, Abstract I-10, and unpublished results.
51. Grateful acknowledgement is made to Professors Hirata and Uemura for kindly providing samples of *palythoa*-derived 45 and 46.
52. G. Liebezeit and R. Dawson, *Kontakte (Merck)*, 1982, 19; *Marine Chem.*, 10, 55 (1980).
53. R. Kuhn and D. Jerchel, *Ber. Dtsch. Chem. Ges.*, 76, 413 (1943); R. Kuhn and K. Kum, *Chem. Ber.*, 95, 2009 (1962); R. Lukes, J. Jarý, and J. Nemec, *Collect. Czechoslov. Chem. Commun.*, 1962, 735.
54. F.W. Lichtenthaler and F.-D. Klingler, *Carbohydr. Res.*, (1983), in press.
55. F.-D. Klingler, *Doctoral Dissertation*, Technische Hochschule Darmstadt, to be submitted.
56. K.H. Neff, *Diplomarbeit*, Technische Hochschule Darmstadt, 1984.
57. K. Lorenz, *Diplomarbeit*, Technische Hochschule Darmstadt, 1984.

Gastrin and Cholecystokinin

An Arduous Task for the Peptide Chemist

Luis Moroder and Erich Wünsch

Max-Planck-Institut für Biochemie, Abteilung Peptidchemie,
D-8033 Martinsried bei München, FRG

Summary

Both during synthesis and storage as well as handling of gastrin- and cholecystokinin-peptides serious difficulties related to a sequence-dependent high reactivity of various side chain functions were encountered. Thus, methods were devised to bypass or at least control side reactions along the synthetic routes, and analogs were designed to enhance the stability of the peptide factors with concomitant retainment of full hormonal potency. The access of these gastrin- and cholecystokinin-peptides allowed intensive physiological, biological and conformational studies aimed at a better understanding of the mechanism of action of this class of hormones. Additional attention was paid to particular derivatives well suited to improve the immunochemical methods as needed for a deeper insight into the physiological significance of the remarkable heterogeneity of these gut hormones.

Abbreviations: For amino acids and amino acid derivatives standard abbreviations recommended by the JUPAC-IUB commission on biochemical nomenclature were used.
Mox = methoxinine (oxa-analog of methionine); DCC = dicylohexylcarbodiimide; HONSu = N-hydroxysuccinimide; HOBt = 1-hydroxybenzotriazole; DMF = dimethylformamide; NMP = N-methylpyrrolidone; MEI = 2-morpholinoethylisocyanide; CCK = cholecystokinin-pancreozymin; HG = human gastrin, whereby the additional number indicates the chain-length of the gastrin peptide.
tlc = thin-layer chromatography; hplc = high performance liquid chromatography.

Natural Product Chemistry
Ed. by Atta-ur-Rahman

The main structural characteristics of the gastrin-cholecystokinin family of peptide hormones are: 1) the common C-terminal tetrapeptide amide sequence Trp-Met-Asp-Phe-NH_2 which already in early structure-function studies on gastrin was recognized as the minimum fragment with physiological activity although at low potency, and thus as a kind of active center of the hormone (1); 2) the tyrosine-O-sulfate residue which for cholecystokinin (CCK) represents an essential structural feature for full hormonal expression, whereas for the gastrins its role has not yet been identified; they are isolated in the nonsulfated form I as well as in the sulfated form II (2,3) and no significant differences in their physiological actions have been detected so far; 3) the complex size-heterogeneity as resulting from the posttranslation processing of their precursor mole-

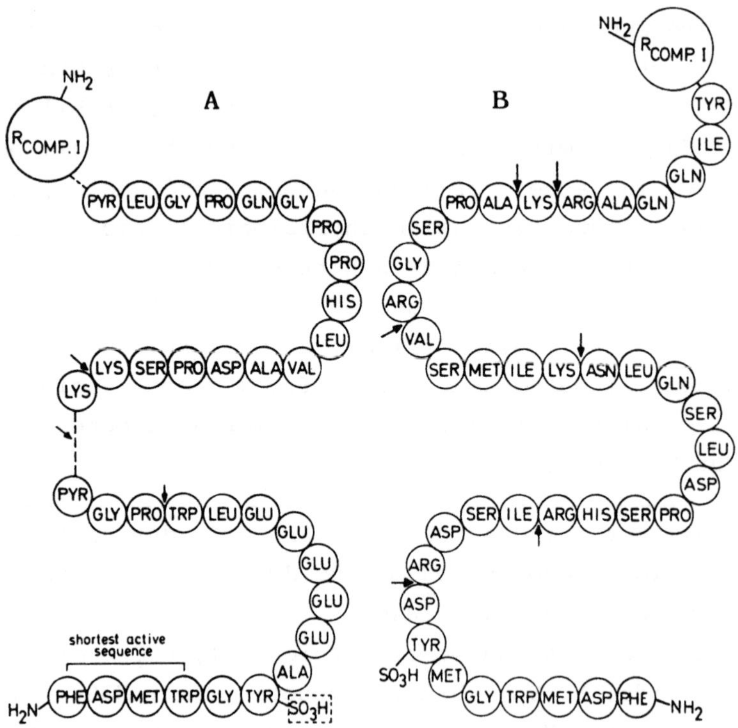

Fig. 1. Primary structures of A) human gastrins (HG) and B) porcine cholecystokinins. Arrows indicate enzymatic cleavages of the prohormonal molecules with production of gastrins and cholecystokinins of decreasing chain-length. The C-terminal amides result from enzymatic processing of glycine-extended forms

cules (Fig. 1) (4). To better understand the structural bases for the biological actions of this class of hormones and to possibly identify the physiological significance of the observed heterogeneity both in respect to sulfation and extension at the N-termini, extensive synthetic studies have been performed in the last years in our laboratory (5 - 15). Aim of these intensive efforts was i) to obtain the various components of this class of hormones at the highest possible degree of purity for structure-function studies ii) to identify fully active analogs stable on storage and handling as needed for physiological and clinical studies, and finally iii) to synthetize derivatives well suited for an efficient improvement of the immunological methods necessary for a reliable quantitative and selective evaluation of the different components both when circulating in the plasma and in the tissue of their biosynthesis. Our contributions in the field are reviewed in the following:

Synthesis of Gastrins

The main difficulties encountered in the synthesis of gastrin and CCK related peptides derive from those structural elements which are essential for the biological activity and thus, directly involved in the mechanism of action of this class of hormones, i.e. from the thioether function of methionine, the indole side-chain function of tryptophan and the tyrosine-O-sulfate moiety. This sequence-dependent enhanced reactivity particularly of the methionine and tryptophan residue towards oxidants and electrophiles as well as the instability of the sulfate-ester function has represented for years a challenge for the peptide chemist.

Oxidation of methionine: The pronounced tendency of the methionine residues in the gastrins and CCK-peptides to oxidation to the corresponding S-oxide derivatives with concomitant loss of activity (16 - 17) implies careful avoidance of traces of oxidants, e.g. peroxides and heavy metals along the various steps of synthesis and purification as well as of air-oxygen to prevent desactivation. Even if reduction of S-oxides with thiols is possible (18), handling of these peptides for in vitro and in vivo assays becomes delicate.

Additionally the lability of the tyrosine-O-sulfate ester prevents an "a posteriori reduction" of S-oxide derivatives by thiols (15). Basing on first indications of Morley (17), Kenner et al. (19) and Wünsch et al. (5 - 6) both independently revealed the possibility to replace the methionine residue in the gastrins by the branched aliphatic side-chain of leucine with retainment of full biological activity. Subsequent substitutions of methionine by norleucine or its oxa-analogue methoxinine proved to be equally efficient (20). The related analogs were equipotent to the parent methionine- hormone in stimulating gastric acid secretion.

Oxidative degradation of tryptophan: Since the first studies on tryptophan-containing peptides it is known that the indole function undergoes facile oxidation with production of usually colored complex mixtures consisting of components which are not yet studied in detail. This oxidative degradation occurs in pronounced manner in acid solutions, e.g. as needed in acidolytic deprotection steps. It has been, however, observed that addition of phenols to free the acid mixtures from halogens as well as operating in inert atmosphere such as under argon or adding antioxidants, e.g. diethylphosphite or preferentially thiols, constitute efficients methods to prevent or at least to minimize the oxidative destruction of tryptophan residues.

Alkylation of tryptophan: A more serious side reaction at the level of tryptophan residues was detected in our laboratory during the synthetic work on gastrins, i.e. the alkylation of the indole ring as generated in the acidolytic removal of protecting groups. First indications of a possible chemical modification of the indole group derived from the synthetic studies on glucagon (21) and then, particularly on 15-leucine-human little gastrin-I (6).
An exact identification of the side products and thus, of the nature of the side reaction which had occurred, was difficult because of the presence of two tryptophan residues in the little gastrin molecule. We then succeeded to identify this side reaction during the synthesis of a leucine-analogue of

minigastrin-I (10) initially proposed to correspond to the C-terminal tridecapeptide amide of little gastrin-I (22).

Upon removal of the protecting groups on tert-butanol basis by exposure to trifluoroacetic acid and partition chromatography of the resulting crude product on Sephadex G-25, the main side fraction was identified as 10-N^{in}-tert-butyl-tryptophan-11-leucine-minigastrin-I by mass- and nmr-spectroscopy of the C-terminal chymotryptic fragment. Subsequently in collaboration with Kisfaludy and associates, studies on model compounds allowed the characterization of the various alkylation products of the indole ring formed in the acidolytic cleavage of protecting groups on tert-butanol basis. Besides the N^{in}-tert-butyl-derivative, additionally mono-C-tert-butyl-, poly-C-tert-butyl- and poly-C-, N-tert-butyl-derivatives are produced whereby the extent of such alkylations and the distribution pattern of the various substitutions at the indole function were found to strongly depend on the acids used, on time and temperature (23). To possibly minimize this deleterious side reaction which causes remarkable loss of product, the use of scavangers, e.g. indole derivatives, ethers and thioethers as well as a series of thiols and their mixtures, have carefully been examined by extensive comparative studies. All efforts to entirely suppress indole alkylation failed, even if 1,2-ethanedithiole was found to efficiently reduce the rate of this side reaction as shown in Fig. 2 (11).

Fortunately various analytical methods e.g. ^{1}H-nmr, tlc, hplc, were found to be sufficiently sensitive to permit detection and quantitative monitoring of the extent of such a side reaction, thus remarkably facilitating the purification processes. In this context noteworthy is the observation that the rate of alkylation of the indole function strongly depends on the chemical environment of the tryptophan residue. While during the synthesis of gastrins and gastrin analogs as well as gastrin-related peptides tert-butyl-tryptophan-derivatives have been detected in extents up to 15 % even applying efficient scavangers (11), no alkylation

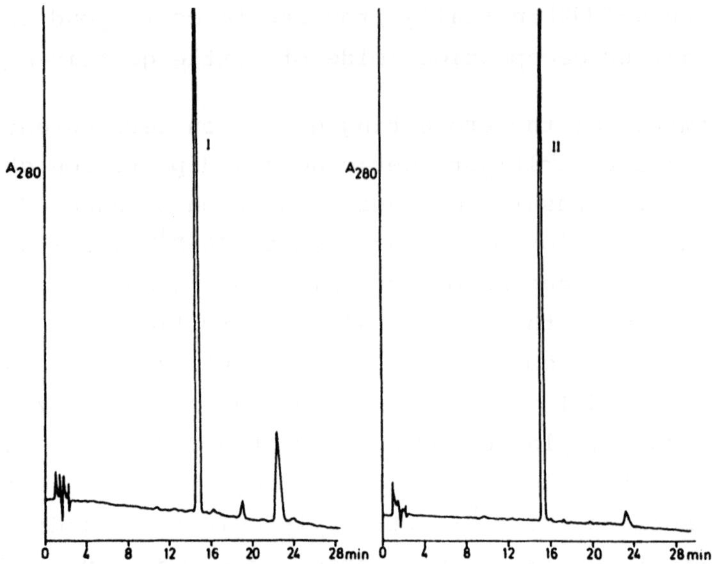

Fig. 2. Hplc of crude 1-des-tryptophan, 11-leucine-minigastrin-I as resulting from exposure of the fully protected tridecapeptide derivative Boc-Leu-[Glu(OBu^t)]$_5$-Ala-Tyr(Bu^t)-Gly-Trp-Leu-Asp(OBu^t)-Phe-NH$_2$ to I) trifluoroacetic acid/anisole, 10:1 and II) trifluoroacetic acid/anisole/1,2-ethanedithiol, 10:1:0.25. Chromatographic conditions: μ-Bondapak C 18 (30 x 0.4 cm); eluents: A) 0.1 M sodium phosphate, pH 5.4; B) acetonitrile/eluent A, 45:55; linear gradient of 35 % B to 80 % B in 30 min; flow rate: 1.7 ml/min

of the tryptophan residue was observed upon acidolytic removal of tert-butanol and 1-adamantanol derived protecting groups from fully protected somatostatin (24). Consequently, this sequence-dependent reactivity to electrophiles strongly questions the general efficacy of acidolytic deprotection procedures devised on the basis of model studies with simple tryptophan-compounds.

Racemization: An additional serious difficulty encountered in the synthesis of gastrins is related to the last fragment condensation step, i.e. assembly of the N-terminal pentapeptide Pyr-Gly-Pro-Trp-Leu-OH with H-[Glu(OBu^t)]$_5$-Ala-Tyr(Bu^t)-Gly-

Trp-X-Asp(OBu^t)-Phe-NH_2 via DCC/HONSu even using a 3-to 4-fold excess of the carboxyl component (5). The low acylation rate induced us to a detailed investigation (11) of various known coupling methods (Table I).

With benzotriazole as additive the acylation rate was sensibly enhanced; concomitantly quantitative racemization of the C-terminal leucine residue occurred (25). This coupling procedure proposed by König and Geiger (26) has widely been used and it is surprising that its negative side effects have not been recognized earlier. The danger of related remarkable racemization rates has been confirmed by Sieber et al. in the synthesis of insulin (27) and by us in the synthesis of secretin (28) and PHI (29). The experiments listed in Table I indicate as only alternative to the DCC/HONSu method (30) the use of 2-morpholinoethylisocyanide in presence of N-hydroxysuccinimide (31).
These above summarized meticulous studies on the observed side reactions as well as the experiences gained in parallel synthetic studies on other gastrointestinal hormones allowed an efficient feedback for an optimization of the synthetic routes to gastrins-I (11) (Fig. 3) and gastrin-I-related peptides (12) (Fig. 4). The desired peptides became accessible at a high degree of purity as judged by thin-layer chromatography under various conditions, high pressure liquid chromatography in optimized conditions, electrophoresis, uv- and nmr-spectroscopy, amino acid analyses of the acid and enzymatic hydrolysates as well as by chiral analysis according to the method of Bayer and associates (32).

Synthesis of tyrosine-O-sulfate peptides

As mentioned initially the closely related gut hormones CCK and gastrin-II share as common characteristic a tyrosine-O-sulfate residue, whereby for latter hormone the nonsulfated form-I has also been isolated. The question arises of whether circulating in plasma the gastrins are sulfated and desulfation takes place during isolation or of whether both the sulfated and nonsulfated gastrins constitute native components. To shed light on this undoubtedly important question synthetic

Table I

H-Pyr-Gly-Pro-Trp-Leu-OH [1-5] + H-[Glu(OtBu)]$_5$-Ala-Tyr(tBu)-Gly-Trp-Nle-Asp(OtBu)-Phe-NH$_2$ [6-17]

H-Pyr-Gly-Pro-Trp-Leu-[Glu(OtBu)]$_5$-Ala-Tyr(tBu)-Gly-Trp-Nle-Asp(OtBu)-Phe-NH$_2$ [1-17]

	ratio[mol/mol]				content in crude product of:	
	[1-5] / [6-17]	method	activation	reaction conditions	[6-17] %	D-Leu %
1)	3	DCC/HOBt (3 experiments)	-	4° C 24h;t_R 3d	10-15	50 ± 3
2)	3	DCC/HONSu (5 experiments)	-	4° C 24h;t_R 3d	15-25	0 ± 8
3)	3	MAI/HONSu (2 experiments)	90min;t_R	t_R 4d	10-15	0

Nps-Ala-Tyr(tBu)-Gly-Trp-OH + H-Met-Asp(OtB)-Phe-NH_2

↓ DCC /HONSu

Nps-Ala-Tyr(tBu)-Gly - Trp-Met-Asp(OtBu)-Phe-NH_2

↓ HCl/ 2-methylindole

Nps-[Glu(OtBu)]$_5$-OH + H-Ala-Tyr(tBu)-Gly-Trp-Met-Asp(OtBu)-Phe-NH_2

↓ DCC /HONSu

Nps-[Glu(OtBu)]$_5$-Ala-Tyr(tBu)-Gly-Trp-Met-Asp(OtBu)-Phe-NH_2

↓ HCl/ 2-methylindole

Pyr-Gly-Pro-Trp-Leu-OH + H-[Glu(OtB)]$_5$ - Ala-Tyr(tBu)-Gly-Trp-Met-Asp(OtBu)-Phe-NH_2

↓ DCC /HONSu

Pyr-Gly-Pro-Trp-Leu-[Glu(OtBu)$_5$ - Ala-Tyr(tBu)-Gly-Trp-Met-Asp(OtBu)-Phe-NH_2

↓ CF_3COOH / anisole/1,2-ethanedithiol

partition chromatography

Pyr-Gly-Pro-Trp-Leu-[Glu]$_5$-Ala-Tyr-Gly-Trp-Met-Asp-Phe-NH_2

Fig. 3. Synthetic route to human little gastrin-I

Z-Glu(OBu^t)-Ala-Tyr(Bu^t)-Gly-OH[10–13] + H-Trp-Nle-Asp(OBu^t)-Phe-NH$_2$ [Nle^{15}][14–17b]

↓ DCC/HONSu

Z-Glu(OBu^t)-Ala-Tyr(Bu^t)-Gly-Trp-Nle-Asp(OBu^t)-Phe-NH$_2$ [Nle^{15}][10–17a]

↓ H_2/Pd

Z-[Glu(OBu^t)]$_4$-OH[6–9] + H-Glu(OBu^t)-Ala-Tyr(Bu^t)-Gly-Trp-Nle-Asp(OBu^t)-Phe-NH$_2$ [Nle^{15}][10–17b]

↓ DCC/HONSu

Z-[Glu(OBu^t)]$_5$-Ala-Tyr(Bu^t)-Gly-Trp-Nle-Asp(OBu^t)-Phe-NH$_2$ [Nle^{15}][6–17a]

↓ H_2/Pd

Glu-Gly-Pro-Trp-Leu-OH[1–5] + H-[Glu(OBu^t)]$_5$-Ala-Tyr(Bu^t)-Gly-Trp-Nle-Asp(OBu^t)-Phe-NH$_2$ [Nle^{15}][6–17b]

↓ DCC/HONSu

Glu-Gly-Pro-Trp-Leu-[Glu(OBu^t)]$_5$-Ala-Tyr(Bu^t)-Gly-Trp-Nle-Asp(OBu^t)-Phe-NH$_2$ [Nle^{15}][1–17a]

↓ CF_3CO_2H/anisole/1,2-ethanedithiol
partition-chromatography

Glu-Gly-Pro-Trp-Leu-Glu-Glu-Glu-Glu-Glu-Ala-Tyr-Gly-Trp-Nle-Asp-Phe-NH$_2$ [Nle^{15}][1–17b]

Fig. 4. Synthetic route to 15-norleucine-human little gastrin-I. The identical synthetic scheme was followed for the preparation of the 15-leucine- and 15-methoxinine-analog

sulfated gastrins should be accessible. So far, the synthesis of sulfated gastrins has not been achieved; previous attempts in Kenner's laboratory (33) based on direct sulfation of gastrin using pyridine/SO_3 failed. Surprisingly more success was obtained with this method for the preparation of CCK-peptides upon optimization of the reaction conditions and extensive purification of the final compounds (34 - 35). Nonetheless several side reactions such as nuclear sulfonation of the tyrosine side-chain could not be prevented; the extent of sulfonation was found to strongly depend on sulfating agents used, on the operating conditions and on the chemical environment of the tyrosyl residue (36). The mentioned difficulties compelled us to devise a new synthetic route to tyrosine-O-sulfate-peptides based on the direct use of suitable derivatives of tyrosine-O-sulfate. This key intermediate was obtained by sulfation of Z-Tyr-OH with pyridine/SO_3 followed by its isolation as barium salt. The homogeneity of this starting material is easily assessed by chromatographic and spectroscopic methods: free tyrosine or sulfonated derivatives show red-shifted uv-maxima of strong intensities in basic solution. Equally sensitive is the detection of such contaminants by ^{1}H-nmr (13). The usefulness of N^{α}-acyl-tyrosine-O-sulfate as starting material in the synthesis of peptides was then investigated on the C-terminal sequence 24-33 of CCK (Fig. 5). For this purpose Fmoc-Tyr($SO_3Ba_{1/2}$)-OH was prepared and coupled to the C-terminal hexapeptide derivative as outlined in the synthetic scheme; subsequent removal of the N^{α}-protecting group with piperidine proceeded smoothly as well as the N-terminal elongation to the decapeptide derivative 24-33, which upon N^{α}-deprotection by treatment with piperidine represents a suitable key-fragment for the synthesis of higher molecular weight CCK-peptides. Removal of the residual acid-labile protecting groups by exposure to 90 % trifluoroacetic acid followed by chromatographic purification of the crude product led to the CCK-decapeptide in over 60 % yield as homogeneous material as judged by all the analytical assays performed (14). The sulfate-ester moiety was found to be surprisingly stable towards the acidolytic deprotection step, most probably because of the sequential proximity of the strong basic function of the arginine side-chain. In fact the sulfate ester function in tyrosine-O-sulfate and related peptides is hydrolysed under identical conditions usually to about 20 %. As expected the alkylation of the tryptophan residue and the high tendency of the methionine residues to oxidation were again observed. Unfortunately reduction of the S-oxides with thiols as well as partial suppression of indole-alkylation by thiols as scavangers is precluded by the facile hydrolysis of the sulfate ester under acidic conditions in the presence of thiols (15).

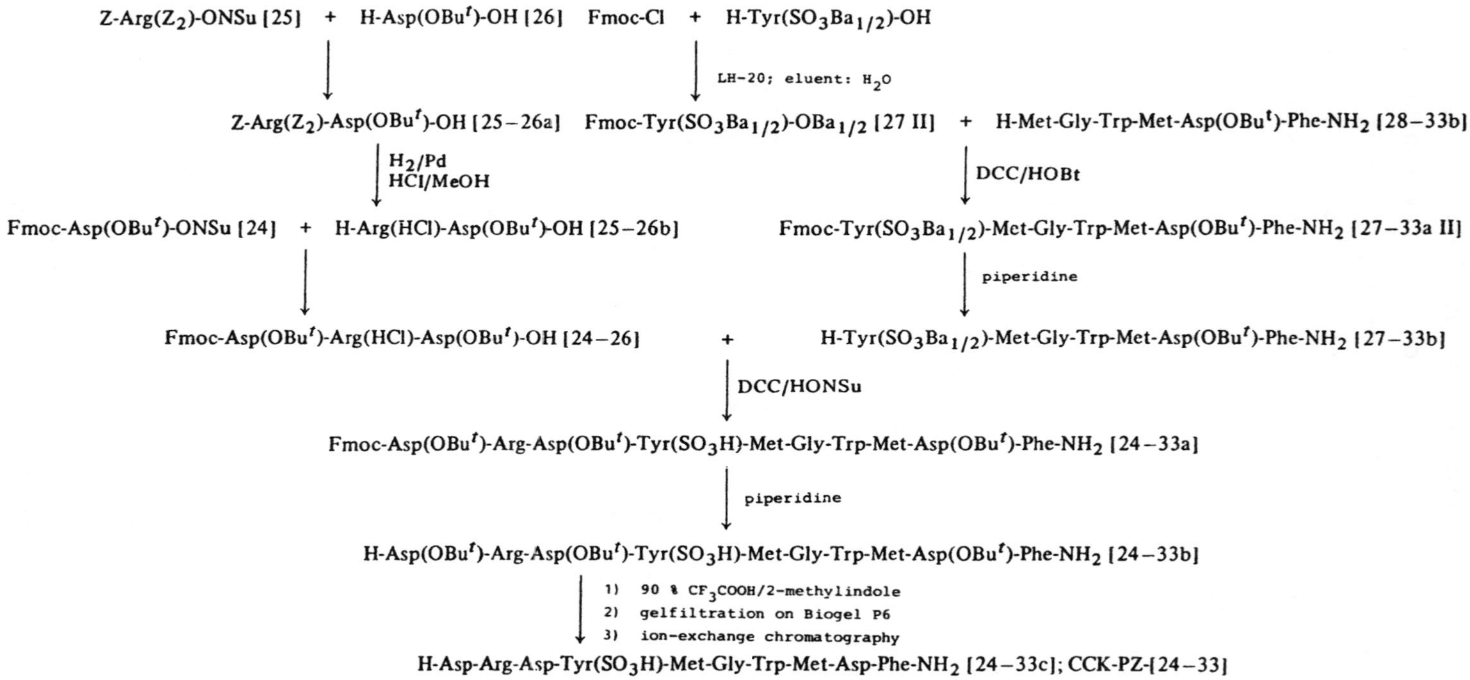

Fig. 5. Scheme of synthesis of CCK-[24-33]

Thus, as for the gastrins the search for stable analogs of CCK-peptides became important leading us to the design of two nonapeptides corresponding to the C-terminal CCK-sequence 25-33 in which methionine-28 was replaced by threonine - in analogy to caerulein, the CCK-active peptide of amphibian skin - and methionine-31 by leucine and norleucine, respectively. The syntheses of the two analogs were again successfully performed using N^{α}-acyl-tyrosine-O-sulfate as intermediate and the desired nonapeptides were obtained according to the scheme outlined in Fig. 6 as analytically homogeneous products in satisfactory yields (15). In vivo and in vitro assays revealed the norleucine-analog as active as CCK-8 and caerulein, whereas replacement of methionine by leucine was found to lower the hormonal potency by a factor of 10. These findings surprisingly differ from those related to the gastrins and may be explained by the need of a well balanced amphiphilicity which is better achieved with the combination threonine-norleucine according to the hydrophobic/hydrophilic increments of the various residues using the factors of Tanford (37). The observed high stability of Thr^{28}, Nle^{31}-CCK-[25-33] both in solution and on storage as lyophilized powder, resulting from the salt bridge between the O-sulfate- and guanido- function (15,16), its fast conversion in vitro to the corresponding CCK-octapeptide form (38) as well as its equipotency to CCK-8 and caerulein (15), revealed this analog as perfectly suited for physiological and clinical research.

The experiences gained in these synthetic studies on CCK-peptides have opened the way to the preparation of sulfated gastrin-peptides.

Immunochemical studies

The pronounced tendency of the methionine residue to oxidation and of tryptophan to oxidative degradation may lead to deleterious chemical modifications of the gastrin molecule during its labelling with radioactive iodine unter oxidative conditions as needed for the preparation of tracers.
These structural modifications of the C-terminal

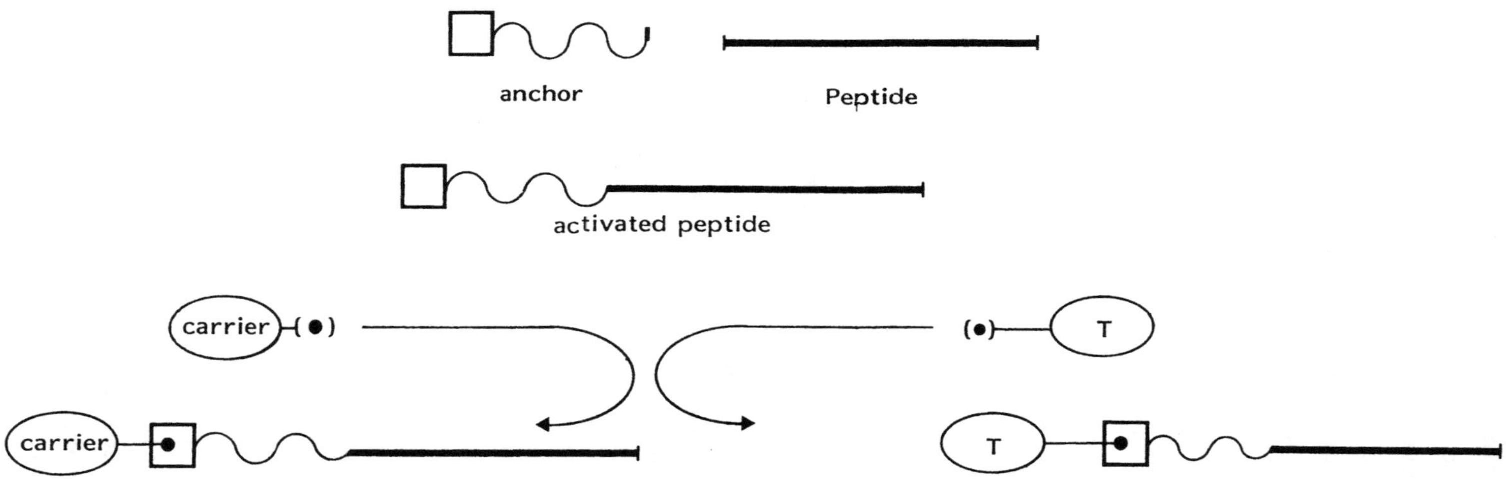

Fig. 7. Scheme for a selective synthesis of tracers and antigens of peptide factors

Mal-ß-Ala-ONSu + H-Gly-Pro-Trp-Leu-(Glu)$_5$-Ala-Tyr-Gly-Trp-Met-Asp-Phe-NH_2

↓ DMF / ACN

Mal-ß-Ala-Gly-Pro-Trp-Leu-(Glu)$_5$-Ala-Tyr-Gly-Trp-Met-Asp-Phe-NH_2

peptide content (M_r = 2138.4) : 90 %

hplc (μ-Bondapak C18): isocratic elution with acetonitrile/0.1 M sodium phosphate (pH 5.0), 28:72, flow rate 1.4 ml/min, 280 nm;

tlc: n-BuOH/AcOH/Pyridine/H_2O; 60:6:20:24

n-BuOH/AcOH/H_2O; 60:20:20

Fig. 8. Synthesis of maleoyl-ß-alanyl-human little gastrin-I-[2-17]

was found to be well suited for a smooth and clean preparation of radiolabelled tracers via its reaction with [^{125}I]-desamino-tyrosyl-cysteine, [^{3}H]-acetylcysteine or [^{35}S]-cysteine; the related derivatives were found to retain full immunoreactivity, e.g. Fig. 9. With this approach the problem related to the methionine and tryptophan residue would be resolved.
However, to bypass also radiolysis of the gastrin molecule and thus, the storage problem as well as waste-problems, an alternative fluorescence-immunoassay based on a fluorogenic enzyme substrate linked to the hormone molecule (Fig. 10) was elaborated. The fluorogenic tracer was used to develop a double-antibody enzyme-immunoassay for gastrin which showed to be sufficiently sensitive to determine normal gastrin levels in healthy humans (41).
The conventional conjugation procedures used in the preparation of antigens employ reagents, e.g. carbodiimides, which undoubtedly involve the most reactive functions along the hapten molecule, e.g. in the case of gastrins the various carboxyl functions in the central region of the molecule. This fact may be responsible for the observed high immunogenicity of the C-terminal

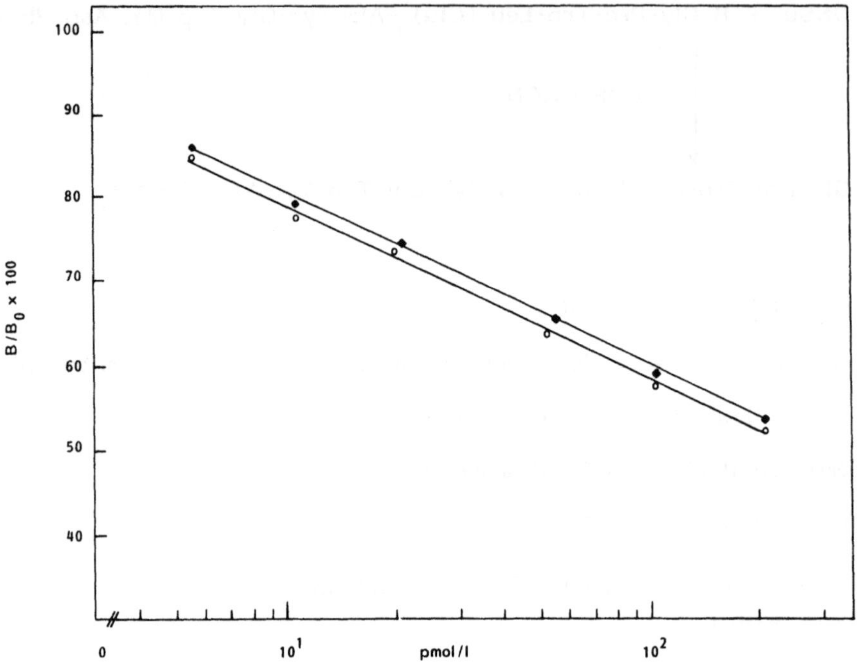

Fig. 9. Comparative immunoreactivities of HG-17 (o—o—o) and of the conjugate cysteine/Mal-ß-Ala-HG-[2-17] (●—●—●) in cross reactions versus ^{125}I-HG-[1-17] using antigastrin-antiserum No. 2127 of Rehfeld

tetrapeptide and thus, for the difficult production of gastrin- or CCK-specific antisera (42). On the assumption that selective linkage of the gastrin at its N-terminus to carriers with concomitant full exposure of the unmodified hapten molecule to the antigen recognition may lead to more selective and specific antisera, maleimido-ß-alanyl-human little gastrin-[2-17] was conjugated to reduced RNase A. The antisera raised in rabbits with this antigen were used to titrate the immunoreactivity of the gastrin sequence. As shown in Fig. 11, a sharp cooperative transition was observed upon incorporation of the third glutamic acid residue, while the C-terminal tetrapeptide is recognized to very low extents. A similar sigmoidal curve in function of chain-length of the gastrin was determined for the acid secretion potency (12) as shown in Fig. 12 and for the onset of the proposed "active" ß-structure as determined by circular dichroism measurements

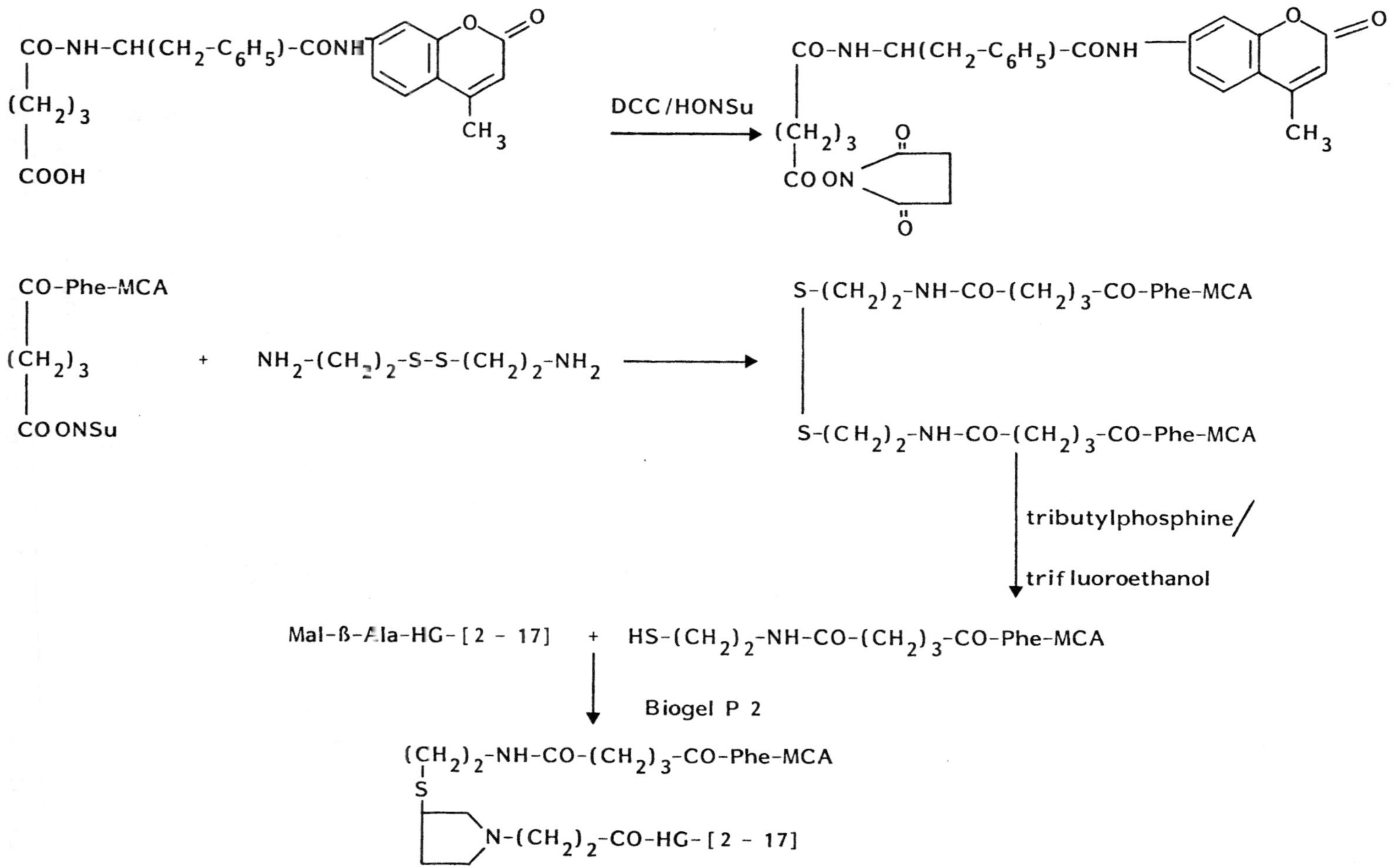

Fig. 10. Synthesis of a fluorogenic gastrin tracer as chymotrypsin substrate

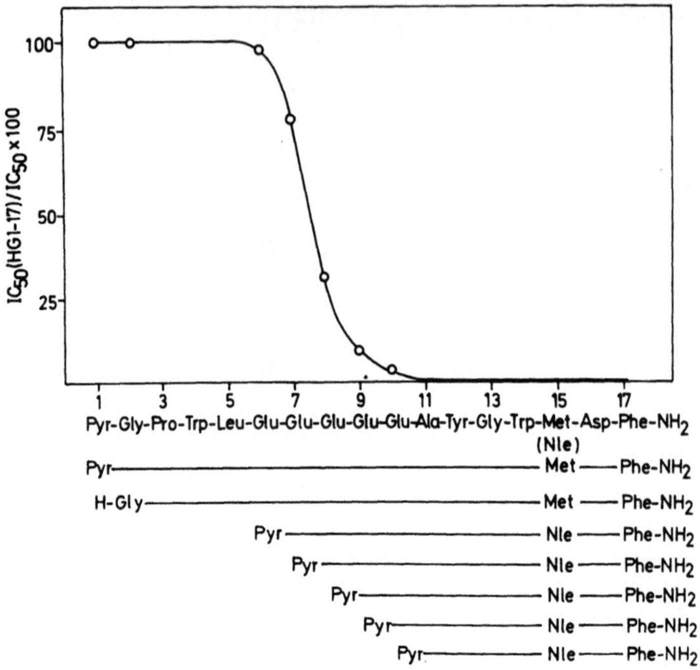

Fig. 11. Titration of the immunoreactivity of the human little-gastrin-I sequence

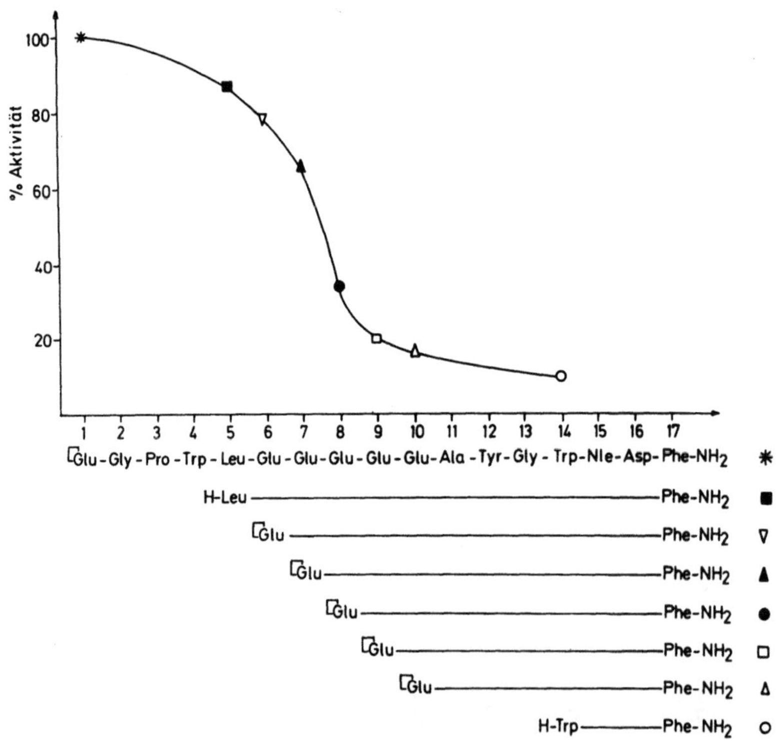

Fig. 12. Stimulation of gastric acid secretion in rats by gastrin-I-peptides expressed as percentage of the response to Nle^{15}-HG-17

of the gastrin peptides in trifluoroethanol (43 - 44). This parallel development of the conformational, biological and immunological properties in function of chain-length strongly suggests that the immunodeterminant expressed by this new approach may possibly be conformation-dependent. This should be equivalent to an enhancement of selectivity and specificity of the antisera. Further experiments related to both the gastrins and cholecystokinins are in progress to confirm these promising findings.

References

1. Tracy, H.J. and Gregory, R.A. (1964) Physiological properties of a series of synthetic peptides structurally related to gastrin I. Nature, 204, 935-938.

2. Gregory, H., Hardy, P.M., Jones, P.S., Kenner, G.W. and Sheppard, R.C. (1964) The antral hormon gastrin; structure of gastrin. Nature, 204, 931-933.

3. Gregory, R.A. and Tracy, H.J. (1966) Isolation of two gastrins from human antral mucosa. Nature, 209, 583.

4. For review see Mutt, V. (1983) Chemistry of the Gastrointestinal Hormones and Hormone-like Peptides and a Sketch of Their Physiology and Pharmacology. Vitamins and Hormones. Vol. 39, 231-427.

5. Wünsch, E. and Deimer, K.H. (1972) Zur Synthese des [15-Leucin]Human-Gastrin I. II. Mitteilung: Herstellung der Gesamtsequenz. Hoppe-Seyler's Z. Physiol.Chem. 353, 1255-1258.

6. Wünsch, E., Jaeger, E., Deffner, M. and Scharf, R. (1972) Zur Synthese des [15-Leucin]Human-Gastrins I. III. Mitteilung: Zur Reindarstellung des synthetischen Heptadecapeptidamids. Hoppe-Seyler's Z. Physiol.Chem. 353, 1716-1720.

7. Wünsch, E., Wendlberger, G., Hallet, A., Jaeger, E., Knof, S., Moroder, L., Scharf, R., Schmidt, I., Thamm, P. and Wilschowitz, L. (1977) Zur Totalsynthese des Human-Big-gastrins I und seines 32-Leucin-Analogons. Z. Naturforsch. 32c, 495-506.

24. Moroder, L., Gemeiner, M., Göhring, W., Jaeger, E., Thamm, P. and Wünsch, E. (1981) New Synthesis of Somatostatin According to the S-tert-Buthylthiocysteine Procedure. Biopolymers, 20, 17-37.

25. Wendlberger, G. (1976) Discussion. Peptides 1976, Proc. 14th Eur. Peptide Sympos. (Loffet, A., ed.) p. 60, Editions de l'Université de Bruxelles, Bruxelles.

26. König, W. and Geiger, R. (1970) Eine neue Methode zur Synthese von Peptiden: Aktivierung der Carboxylgruppe mit Dicyclohexylcarbodiimid unter Zusatz von 1-Hydroxy-benzotriazolen. Chem. Ber. 103, 788-798.

27. Sieber, P., Kamber, B., Hartmann, A., Jöhl, A., Riniker, B. and Rittel, W. (1977) Totalsynthese von Humaninsulin. IV. Beschreibung der Endstufen. Helv. Chim. Acta, 60, 27-37.

28. Wünsch, E. (1983) Peptide Factors as Pharmaceuticals: Criteria for Application. Biopolymers, 22, 493-505.

29. Moroder, L., Göhring, W., Lucietto, P., Musiol, J., Scharf, R., Thamm, P., Bovermann, G., Wünsch, E., Lundberg, J., Tatemoto, K. and Mutt, V. (1983) Synthesis of the Porcine Intestinal Peptide PHI and its 24-Glutamine Analogue. Hoppe-Seyler's Z. Physiol. Chem. 364, 1563 - 1584.

30. Weygand, F., Hoffmann, D. and Wünsch, E. (1966) Peptid-synthesen mit Dicyclohexylcarbodiimid unter Zusatz von N-Hydroxysuccinimid. Z. Naturforsch. 21b, 426-428.

31. Aigner, H., Koch, G. and Marquarding, D. (1982) Isocyanides as Activating Reagents in Peptide Synthesis. Chemistry of Peptides and Proteins, Proc. 3th USSR-FRG Symp. 1980 (Voelter, W., Wünsch, E., Ovchinnikov, Y. and Ivanov, V. eds.) pp. 207-216, Walter de Gruyter and Co., Berlin-New York.

32. Frank, H., Nicholson, G.J. and Bayer, E. (1977) Rapid Gaschromatographic Separation of Amino Acid Enantiomers with a Novel Chiral Stationary Phase. J. Chromatogr. Sci. 15, 174-176.

33. Beacham, J., Bentley, P.H., Kenner, G.W., Macleod, J.K., Mendive, J.J. and Sheppard, R.C. (1967) The structure and synthesis of human gastrin. J. Chem. Soc. (C), 2520-2529.

34. Anastasi, A., Bernardi, L., Bertaccini, G., Bosisio, G., de Castiglione, R. Erspamer, V., Goffredo, O. and Impicciatore, M. (1968) Synthetic Peptides Related to Caerulein. Experientia, 24, 771-773.

35. Ondetti, M.A., Pluščec, J., Sabo, E.F., Sheehan, J.T. and Williams, N. (1970) Synthesis of Cholecystokinin-Pancreozymin. I. The C-Terminal Dodecapeptide. J. Am. Chem. Soc. 92, 195-199.

36. Bodansky, M. (1977) Peptide Synthesis: An Undiminshed Challenge in Peptides. Proc. 5th. Am. Peptide Sympos. (Goodmann, M. and Meienhofer, J., eds). pp. 1-13, John Wiley and Sons, New York.

37. Tanford, C. (1962) Contribution of Hydrophobic Interactions to the Stability of the Globular Conformation of Proteins. J. Am. Chem. Soc. 84, 4240 - 4247.

38. Koulischer, D., Moroder, L., and Deschodt-Lanckmann, M. (1982) Degradation of cholecystokinin octapeptide, related fragments and analogs by human and rat plasma in vitro. Arch. Int. Physiol. Biochim. 90, B 42 - B 45.

39. Alexander, N.M. (1974) Oxidative Cleavage of Tryptophanyl Peptide Bonds during Chemical- and Peroxidase-Iodinations. J. Biol. Chem. 249, 1946-1952.

40. Mourier, G., Moroder, L. and Previero, A. (1984) Prevention of Tryptophan Oxidation during Iodination of Tyrosyl Residues in Peptides. Z. Naturforsch. 39b, in press.

41. Moroder, L., Nyfeler, R., Gemeiner, M., Kalbacher, H. and Wünsch, E. (1983) Immunoassays of Peptide Hormones and Their Chemical Aspects. Biopolymers, 22, 481-486.

42. Rehfeld, J.F., Larsson, L.I., Goltermann, V.R., Schwartz, T.W., Holst, J.J., Jensen, S.L. and Morley. J.S. (1980) Neural regulation of pancreatic hormone secretion by the C-terminal tetrapeptide of CCK. Nature, 284, 33-38.

43. Peggion, E., Jaeger, E., Knof, S., Moroder, L. and Wünsch, E. (1981) Conformational Aspects of Gastrin-Related Peptides: A Circular Dichroism Study. Biopolymers, 20, 633-652.

44. Peggion, E., Foffani, M.T., Wünsch, E., Moroder, L., Borin, G.and Goodmann, M. (1984) Conformational Properties of Gastrin Fragments of Increasing Chain Length. Biopolymers ; submitted for publication.

Glycoprotein Structure Function Relationship: Study of Epithelial Cell Secretion

Nasir-ud-Din, M. Saleh Ajaz, and Irfan-Ullah Khan
Department of Chemistry, University of Baluchistan, Quetta, Pakistan

Glycoproteins are polymers in which proteins and carbohydrates are covalently linked, and in which carbohydrates contribute to major biological functions. Glycoproteins are abundantly distributed in microorganisms, plants, animals and humans. There are sufficient indications to suggest that the carbohydrate moiety of the glycoprotein performs significant biological roles such as biological recognition. i.e., specification of blood type, control of the lifetime of glycoproteins in the circulatory system, stage specific differentiation, control of glycoprotein uptake by the cells, acceptors for carbohydrate or protein components and protection of exposed tissue from invasive organisms or molecules. This capability of the carbohydrate arises from the great variety of oligosaccharides that can be formed from a relatively small number of monosaccharides. Oligosaccharide chains can influence the ability of the protein moiety of the glycoprotein to fold properly. The transfer, cotranslational event, of oligosaccharide having the structure $(Glc)_3$ $(Man)_9(GlcNAc)_2$ to an aspargine residue of the protein during the biosynthesis in the endoplasmic reticulum may be essential for the protein to acquire correct tertiary structure. Inhibition studies of protein glycosylation support the concept that carbohydrate chains influence the conformation of proteins. The consequence of glycosy-

Abbreviations: N-Acetylglucosamine: GlcNAc; N-acetylgalactosamine: GalNAc: N-acetylgalactosaminitol: GalNAc-ol

Natural Product Chemistry
Ed. by Atta-ur-Rahman

lation for protein folding and assembly may be examplified by the synthesis of virus glycoprotein, the G protein of vesicular stomatitis virus (VSV), initially described by Gibson et al., [1]. The main molecular basis underlying the function of carbohydrates in biological recognition is the interaction of oligosaccharides with the carbohydrate binding proteins. The specific determinant formed by a saccharide sequence has affinity to complementary structures in other molecules. Furthermore cell surface or membrane glycoproteins possibly stabilize cell surface structures, particularly in eukaryotes. It has been proposed that the carbohydrate component of the major sialoglycoproteins of erythrocytes, such as glycophorin, interact to form a lattice over the internal cell membranes and thus provide the membrane with greater rigidity [2].

The carbohydrates in the glycoproteins range from less than one per cent, collagen, to eighty five per cent, soluble blood group substances, of the total macromolecule. Glycoproteins also differ in the number of carbohydrate chains present. This may vary from one, in ribonuclease B, to eight hundred, in submaxillary mucin. Furthermore the oligosaccharide chains in glycoprotein differ in the number of different constituting monosaccharides, ranging from one to a dozen or more.

Carbohydrate-peptide linkage

The oligosaccharide linkages in the glycoproteins can be classified according to the type of bond to the protein moiety as being N- or O-glycosidic. The N-glycosidically linked oligosaccharide chains are joined through N-acetylglucosamine to the amide nitrogen of asparagine in the protein. The O-glycosidic chains may be linked to the hydroxyl group of L-serine, L-threonine, L-hydroxylysine or L-hydroxyproline in the protein moiety. Glycoproteins from epithelial secretions mainly contain O-glycosidic type linkages between carbohydrate chains and protein backbone. Amongst the carbohydrate residues the joining site is provided by N-acetylgalactosamine whereas L-serine and L-threonine from the protein moiety is involved in the covalent linkage.

Epithelial secretions

Mucins, epithelial glycoproteins, are dominant components of bronchial, gastrointestinal and cervical secretions. Epithelial fluids are produced by a variety of cells such as goblet, columnar, mucous and serous cells. In systems like tracheobronchial and gastrointestinal, mucus is secreted by different types of cells, whereas in cervical system [3] and gall bladder [4], a component of the gastrointestinal system, mucins are produced only by columnar cells. In the case of tracheobronchial and gastrointestinal mucin fluids are secreted by a variety of cells, suggesting that the mucus may be a heterogeneous milieu of secreted components, including glycoproteins. In order to study structure function relationship cervical and gall baldder secretions provide excellent systems.

Cervical mucus is a complex mixture, the principal constituents of which are mucin type glycoproteins that share the chemical composition and physical behaviour with other epithelial secretions. As the glycoproteins in the mucus represent the major macromolecular components, it is reasonable to assert that they are responsible for its physiological and biophysical properties [5]. The secretion regulated during the menstrual cycle to influence sperm passage, its survival and nutrition. Treatment of cervical mucus in vitro with proteolytic enzymes enhanced sperm migration [6] and altered the biophysical behaviour of cervical mucus. In addition, treatment of the midcycle human cervical mucus with Pronase destroyed the protein backbone or bridges between the glycoprotein fibrils. The cyclic changes in mucus behavior are accompanied by alterations in carbohydrate composition [7,8] and in chemical structure [9]. The mechanism by which glycoproteins, specifically, and mucus, in general, function in above described processes is unknown. In order to assess their role the structural features of a main glycoprotein from midcycle mucus was investigated, and the structures of neutral oligosaccharides are discussed. Understanding of chemical structures and their relation to physiological functions will provide a biochemical and immunological procedure to regulate glycoprotein functions.

TABLE II

COMPOSITION OF GLYCOPROTEIN SEPARATED ON DEAE-CELLULOSE

Carbohydrate	Fraction-A %	Fraction-A Molar Ratio[1]	Fraction-B %	Fraction-B Molar Ratio[1]
Fucose	8	0.8	7	0.9
Galactose	25	2.2	18	2.2
N-Acetylglucosamine	14	1.0	10	1.0
N-Acetylgalactosamine	22	1.6	12	1.2
N-Acetylneuraminic acid	17	0.6	11	0.6
Sulfate	1.4	0.2	3.8	0.9

Amino acids[2]	Fraction A: Original glycoprotein	Fraction A: Difference after treatment with alkali	Fraction B
Aspartic acid	74		68
Threonine	234	- 148	264
Serine	140	- 62	124
Glutamic acid	58		51
Proline	48		70
Glycine	70		69
Alanine	90	+ 49	99
α-Aminobutyric acid		+ 120	
Cysteine	16		9
Valine	51		48
Isoleucine	48		39
Leucine	51		58
Tyrosine	20		18
Phenylalanine	30		20
Lysine	33		30
Histidine	14		12
Arginine	23		21

1 Molar ratio relative to N-acetylglucosamine

2 Residues per 1000 residues

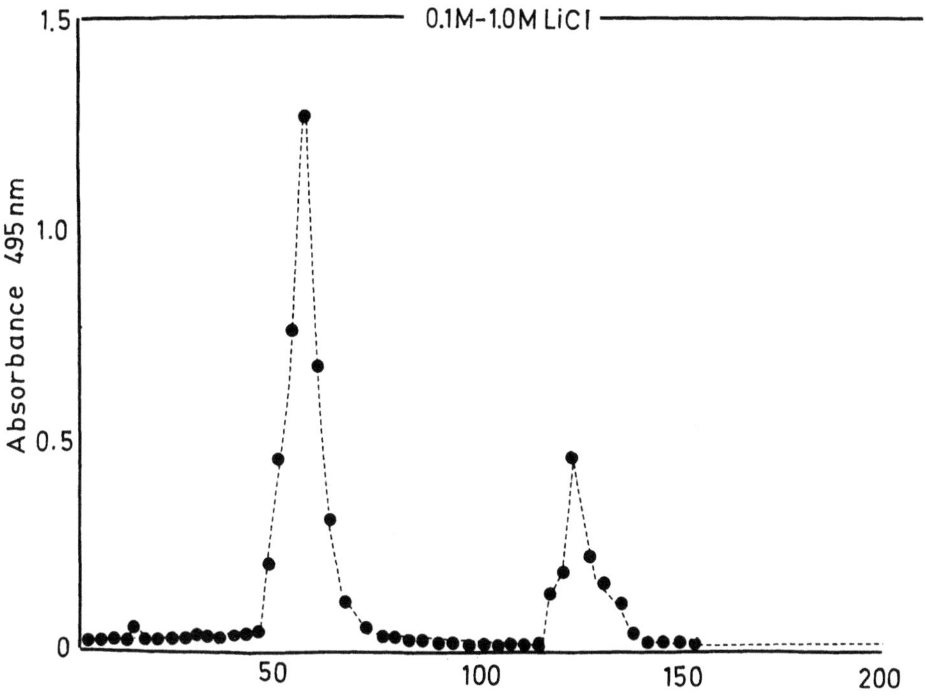

Fig. 1. Fractionation of the Sepharose 2B purified glycoprotein on a column (1.8 x 52 cm) of DEAE-cellulose. The glycoproteins from the column were eluted with a gradient of 0.1M to 1M lithium chloride. Fractions of 2.5 ml were collected and every third fraction was examined for the presence of carbohydrates by the phenol-sulfuric acid procedure (33).

agents i.e., dithiothreitol and tributyl phosphine and iodoacetamide and 1,3-propanesulthone respectively. In both the conditions utilized the alkylation was incomplete. In the event of an external protein constituting a bridge between the glycoprotein fragments [14] the extensive proteolytic enzyme treatment and subsequent alkylation would remove cysteine. In these circumstances it will be relevant to consider that there exists no link protein and cysteine form a component part of the glycoprotein. It is known that the presence of excessive amounts of neuraminic acid and sulfate groups in the glycoprotein will limit the approach of degrading enzymes as well as some reducing and alkylating reagents. This phenomenon may explain the incomplete reduction and alkylation, as has been observed for the luteal phase glycoprotein too [15].

TABLE III

CARBOHYDRATE COMPOSITION OF PURIFIED OLIGOSACCHARIDE FRACTIONS

Carbohydrates	Fractions											
	F-1		F-2		F-3		F-4		F-5		F-6	
	%	M/R[b]	%	M/R[b]	%	M/R[b]	%	M/R[b]	%	M/R[b]	%	M/R[a]
Fucose					18	0.9			12	0.9		
Galactose	26	1.1	24	1.0	22	1.0	50	2.9	30	2.1		
Mannose											41	1.0
N-Acetyl-glucosamine	34	1.1			28	1.1	22	1.1	16	0.9	43	0.9
N-Acetyl-galactosamine			35	1.1					19	1.1		
N-Ace tylgalac-tosaminitol	32	1.0	31	1.0	26	1.0	21	1.0	18	1.0		

[a]Molar ratio relative to mannose

[b]Molar ratio relative to N-acetylgalactosaminitol

Reductive β-elimination and preparation of oligosaccharides

The oligosaccharides from DEAE-cellulose column fraction one, the main purified glycoprotein, were isolated as previously described [10] with minor modification. The glycoprotein was incubated at 48°C for 16 hours in a solution prepared by dissolving $NaBH_4$-$NaB[^3H_4]$ (50 mCi) in 0.05M NaOH to give a final concentration of 2M $NaBH_4$. An aliquot from the reaction mixture was withdrawn and after, reduction, followed by acid hydrolysis showed an appropriate increase in alanine and appearance of α-aminobutyric acid (Table II). The reaction mixture was then cooled to 4°C, adjusted to pH 5.4 with 4M acetic acid. The mixture was applied to a column of AG 50-X8 (H^+) ion-exchange resin. Reduced oligosaccharides were eluted from the column with water and then with 0.1M acetic acid. The water and acid eluates were freed from borate ions separately by repeated evaporations with methanol. The β-eliminated material, obtained by water elution from the ion-exchange column was separated into neutral and acidic oligosaccharides on a column of AG 1X2 ion-exchange resin. The water washings were combined and lyophilized. The residue in pyridine-acetic acid buffer, pH 5.4, was fractionated on a column of Bio-Gel P-6 equilibrated with 0.1M pyridine-acetic acid. The elution profile of the oligosaccharides is shown in Fig.2. The carbohydrate composition of the oligosaccharide fractions, determined by gas liquid chromatography (glc,15), is given in Table III.

Characterization of oligosaccharide Fractions

Oligosaccharide Fraction-1: This fraction was further purified by ion-exchange chromatography on DEAE-Sephadex A-25 and by high performance liquid chromatography (HPLC) on Lichosorb-NH_2 (Merck). The Lichosorb-NH_2 column was washed with a linear gradient of acetonitrile-water. Oligosaccharide F-1, a trisaccharide consisting of a galactose, an N-acetylglucosamine and an N-acetylgalactosaminitol residue was one of the major oligosaccharides. Upon periodate oxidation-borohydride reduction all of the galactose was destroyed, whereas N-acetylglucosamine was recovered unchanged. The N-galactosaminitol was quantitatively converted to N-acetylthreosaminitol.

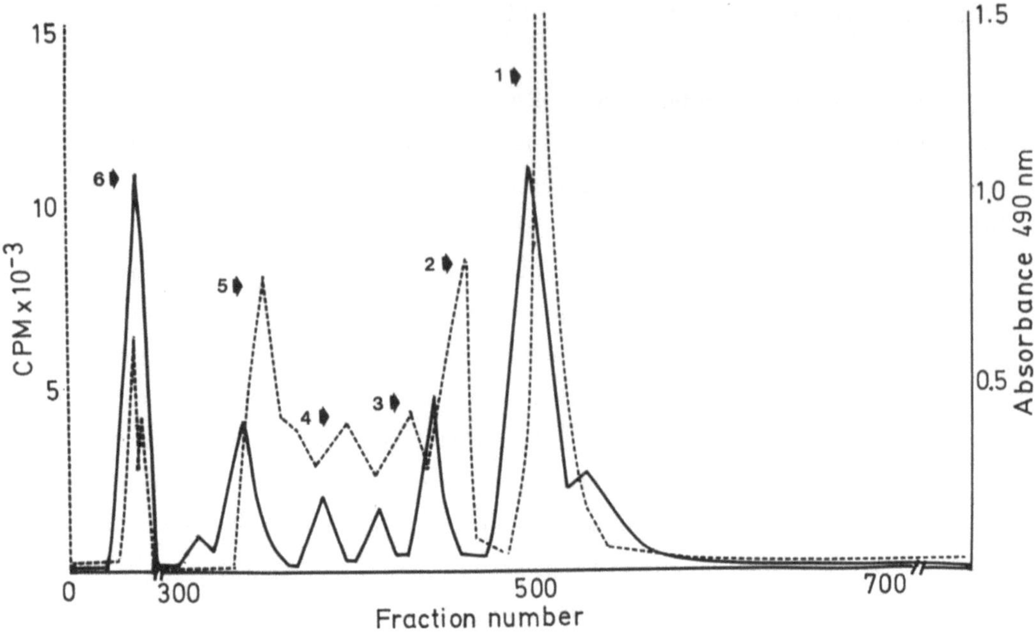

Fig. 2. Gel filtration of the neutral oligosaccharides on a column (2.4 x 190 cm) of Bio-Gel P-6. The oligosaccharides were eluted from the column with 0.1M pyridine-acetic acid (pH 5.4). Fractions of 1.8 ml were collected and every fourth fraction was examined for the presence of carbohydrates (●---●, pehnol-sulfuric acid) and tritium (●——●).

Methylation of the oligosaccharide F-1 using Hakomori procedure [16], and subsequent glc-mass spectrometry showed the presence of 2,3,4,6-tetra-O-methyl-galactose, 2-deoxy-3,6-di-O-methyl-2-(N-methylacetamido)-glucose and 2-deoxy-1,4,5,6-tetra-O-methyl-2-(N-methylacetamido)-galactitol. Extensive treatment of the oligosaccharide with β-galactosidase (Aspergillus niger, Calbiochem) removed approximately 60% of the galactose, whereas treatment with α-galactosidase (A.niger) had no effect on the oligosaccharide. These combined results suggest the presence of structure F-1 proposed in Fig.3.

Oligosaccharide Fraction-2: Also a trisaccharide, consisting of a galactose, an N-acetylgalactosamine and an N-acetylgalactosaminitol residue, was obtained from Bio-Gel P-6 column in very close proximity to oligosaccharide F-1. This oligosaccharide was further purified on DEAE-Sephadex, Bio-Gel P-4, followed by HPLC on Lichosorb-NH_2. The purified oligosaccharide (for composition see Table III) showed a single component on thin layer chromatography (tlc) on silica gel plates in n-butanol-acetic acid-water (3:3:2). Periodate oxidation-borohydride reduction completely eliminated N-acetylgalactosamine and converted N-acetylgalactosaminitol to 2-acetamido-2-deoxy threitol. Galactose was recovered unaltered. Treatment of the oligosaccharide with α-N-acetylgalactosaminidase (*Charonia lampas*) for 64 hours partially (65%) removed N-acetylgalactosamine. Methylation analysis of the oligosaccharide showed the presence of a terminal N-acetylgalactosamine, 3-linked galactose and N-acetylgalactosamine residues. α- and β-Galactosidase from *A.niger* subsequent to α-N-acetylgalactosaminidase treatment did not remove the galactose residue from the oligosaccharide with these results, combined with the observation that a galactose residue next to N-acetylgalactosamine in β-position is resistant to β-galactosaminidase, structure F-2. Fig.3, is proposed for this oligosaccharide.

Oligosaccharide Fraction-3: The crude oligosaccharide fraction was purified on DEAE-Sephadex followed by paper chromatography in ethyl acetate-pyridine-acetic acid-water (5:5:3:1). The isolated oligosaccharide was purified by HPLC as described for oligosaccharide Fraction-1. The carbohydrate composition of the oligosaccharide (Table III) suggested the presence of a tetrasaccharide. Permethylation of the oligosaccharide and characterization of the hydrolytic products as alditol acetates by glc-mass spectrometry showed the presence of a terminal fucose, a 2-linked galactose residue, a 4-linked-N-acetylglucosamine and a 6-linked N-acetylgalactosaminitol. Treatment of the oligosaccharide with α-L-fucosidase (beef epididymis, Sigma) removed a focuse residue. Subsequent treatment of the fucosidase-treated oligosaccharide first with β-galactosidase (*Escherichia coli*, Boehringer) and then a second treatment with purified β-galactosidase (*A.niger*, Calbiochem) removed a substantial amount (75%) of the galactose, leaving behind mainly N-acetylglucosamine and N-acetylgalactosaminitol. Periodate oxidation-borohydride reduction destroyed the fucose and galactose residues, converted N-acetylgalactosaminitol to 2-acetamido-2-deoxyglycerol, and N-acetylglucosamine was recovered unchanged. Methylation analysis and enzyme degradations confirm the results of periodate oxidation, suggesting the N-acetylglucosamine linkage to 0-6 of N-acetylgalactosaminitol. These data support the structure, F-3, for the oligosaccharide proposed in Fig. 3.

treatment of the oligosaccharide with β-galactosidase from C.lampas and from E.coli removed two residues of galactose, and subsequent methylation analysis of the remaining oligosaccharide provided a terminal galactose, terminal N-acetylglucosamine and an N-acetylgalactosaminitol residue linked in O-3 and O-6. This confirmed the sequence of the sugar residues as well as the branching at N-acetylgalactosaminitol, in addition to the anomery of terminal residues. The β -anomeric configuration of the linkage of the galactose residue linked to N-acetylgalactosaminitol has been assigned due to the fact that only a β-linkage for this galactose residue is known [17-21] and also due to the fact that among the acidic oligosaccharides of this glycoprotein this anomeric linkage has been identified by 500-MHz ^{1}H-NMR studies [22]. From the combination of these results structure for oligosaccharide F-4 is proposed in Fig.3.

Oligosaccharide Fraction-5: This oligosaccharide fraction was purified on DEAE-Sephadex A-25 and then on Bio-Gel P-6. The partially purified oligosaccharide was chromatographed on HPLC, and tlc of the eluted oligosaccharide on silica gel plates developed in chloroform-methanol-water (3:2:1) showed the presence of a homogeneous oligosaccharide. The carbohydrate composition of this oligosaccharide showed the presence of a fucose residue, two galactose residues, an N-acetylglucosamine, an N-acetylgalactosamine and an N-acetylgalactosaminitol (Table III). Methylation of the oligosaccharide followed by characterization of the derived alditol acetates showed the presence of 2,3,4-tri-O-methylfucose, 2-deoxy-3,4,6-tri-O-methyl-2-(N-methylacetamido)-D-galactose, 2,4,6-and 3,4,6-tri-O-methyl-D-galactose, 2-deoxy-3,6-di-O-methyl-2-(N-methylacetamido)-D-glucose and 2-deoxy-1,4,5-tri-O-methyl-2-(N-methylacetamido)-D-glucose and 2-deoxy-1,4,5-tri-O-methyl-2-(N-methylacetamido)-D-galactitol. Periodate oxidation-borohydride reduction of the oligosaccharide degraded a fucose, an N-acetylgalactosamine and a galactose residue, and converted N-acetylgalactosaminitol to N-acetylthreosaminitol. From the degradation products an oligomer of oligosaccharide F-4, was isolated and characterized as described for F-4. Treatment of the oligosaccharide with α-L-fucosidase (beef kidney, Boehringer) and subsequent methylation of the fucose-free oligosaccharide followed by characterization of the methylated products showed that the fucose was linked to a galactose residue at O-2. The

F-5 oligosaccharide was treated with α-N-acetylgalactosaminidase (Charonia lampas, Miles) and then with β-galactosidase (Aspergillus niger, Calbiochem). The residual oligosaccharide was purified on Bio Gel P-6 and methylated in the usual manner.. Characterization of the methylated alditol acetates suggested the presence of a terminal fucose and a galactose, a 2-linked galactose, a 4-linked N-acetylglucosamine and a 3,6-linked N-acetylgalactosaminitol. The results of periodate oxidation, methylation and enzymic degradation studies lead to structure F-5 (Fig.3) for this oligosaccharide.

Oligosaccharide Fraction-6: This oligosaccharide fragment had limited tritium incorporation, and methanolysis of the oligosaccharide suggested the presence of mannose and N-acetylglucosamine in addition to small quantities of galactose, fucose, N-acetylgalactosamine and few amino acids. The oligosaccharide was purified on a column of Cocanavalin-A-conjugate into at least two fragments. The major tritium-containing component was an oligosaccharide with tritium incorporated in N-acetylglucosaminitol. The other fraction was treated with endo- β-N-acetylglucosaminidase [20], and subsequent carbohydrate analysis showed only a minor increase in N-acetylglucosamine. The two fractions were combined and rechromatographed on Bio-Gel P-6 and then Concanavalin-A-Sephadex-G-25. The partially purified oligosaccharide was further chromatographed on HPLC. The purified oligosaccharide, isolated in very small quantity, showed the presence of mannose and N-acetylglucosamine in almost equimolar quantity (Table III). Methylation of the oligosaccharide followed by characterization of the derived alditol acetates showed therein the presence of terminal mannose, 3,6-linked mannose, terminal N-acetylglucosamine, 4-linked N-acetylglucosamine and 4-linked N-acetylglucosaminitol. Periodate oxidation-borohydride reduction and subsequent analysis of the carbohydrates of the degraded oligosaccharide suggested it to be a trisaccharide, consisting of a mannose, an N-acetylglucosamine and an N-acetylglucosaminitol residue. Extensive treatment of the oligosaccharide with purified α-mannosidase (Jack bean, Sigma) removed two mannose residues. The residual oligosaccharide was again treated with α-mannosidase with no appreciable effect. Methylation of the mannosidase-degraded oligosaccharide showed the presence of a terminal mannose, a 4-linked N-acetylglucosamine and oligosaccharide was extensively treated with β-galactosidase. The carbohydrate analysis of the enzyme-treated oligosaccharide suggested a loss of a galactose residue, indicating a terminal, β-linked residue. Consecutive treatment of the oligosaccharide with α-*N*-acetylgalactosaminidase and then with α-*L*-fucosidase removed from the oligosaccharide an *N*-acetylgalactosamine and a fucose residue. Methylation analysis of this oligosaccharide fragment, enzyme degraded product, cleared the location of *N*-acetylgalactosamine and fucose as a galactose that was linked in O-2 and O-3. Treatment of the residual oligosaccharide with α -galactosidase suggested the removal of two galactose residues, and subsequent methylation of the remaining oligosaccharide showed the presence of terminal *N*-acetylglucosamine residues and 3-linked galactose residues in addition to a 3,6-linked *N*-acetylgalactosamine and a 3-linked *N*-acetylglucosamine. However, the molar ratios of the methylated sugars indicated two residues each of terminal *N*-acetylglucosamine and 3-linked galactose for one residue of 3,6-*N*-acetylgalactosaminitol. Based on the results of enzymic degradation and on the molar ratios of the methylated sugars [23] the structure given below for the oligosaccharide fragment, free of fucose and *N*-acetylgalactosamine, is proposed.

Gal 1 $\xrightarrow{\beta}$ 4 GlcNAc 1 $\xrightarrow{\beta}$ 3 Gal 1 $\xrightarrow{\beta}$ 3

> GalNAc-ol

Gal 1 $\xrightarrow{\beta}$ 4 GlcNAc 1 $\rightarrow$ 3 Gal 1 $\rightarrow$ 4 GlcNAc 1 $\rightarrow$ 6

However, this structure does not conclusively identify the position of a terminal galactose and 4-linked *N*-acetylglucosamine residue in the original oligosaccharide due to the limited value of the response factor [23] used to obtain the molar ratio of the methylated sugars in glc-mass spectrometry.

Periodate oxidation-borohydride reduction of the enzyme-treated oligosaccharide, a hexasaccharide, and characterization of the degraded fragments by compositional analysis and methylation studies suggested the presence of two fragments:(a) galactose linked N-acetylthreosaminitol and (b) N-acetylglucosamine linked to galactose. These findings do support the partial structure shown above; however, complete study of this oligosaccharide does not clearly indicate the linkage of a galactose and an N-acetygalactosamine. Therefore, a partial structure for this oligosaccharide, F-1, is proposed in Fig.5.

Oligosaccharide Fraction-2: This oligosaccharide was purified by chromatography on HPLC, and carbohydrate composition (Table IV) suggested it to be a pentasaccharide consisting of a galactose, a fucose, an N-acetylglucosamine, an N-acetylgalactosamine and an N-acetylgalactosaminitol. Periodate oxidation and characterization of the degraded products, by analysis of the carbohydrate composition and methylation, established the presence of a fragment (i) and N-acetylglucosamine.

(i) Gal 1 → 3 ThrNAc-ol

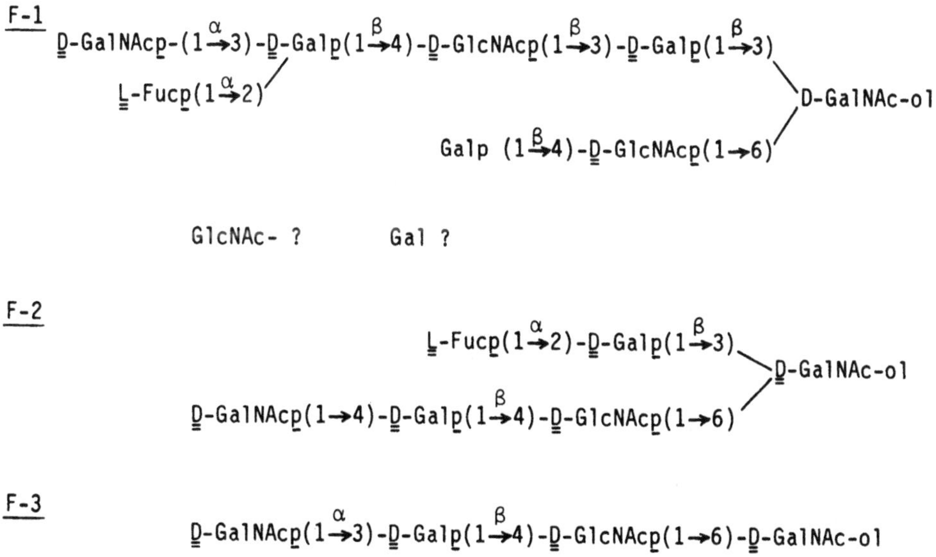

Fig. 5. Proposed partial structures for bovine gall bladder oligosaccharides.

that is responsible for the parallel alignment of the mucus, which allows channels of least resistance for sperm migration. The involvement of the glycoprotein in the maintenance of channels has been demonstrated by cross reaction of the mucus with anti-glycoprotein antibody (11). It is probable that carbohydrate chains that have N-acetylneuraminic acid in close proximity to the protein moiety cause parallel alignment of the glycoprotein, whereas those with N-acetylneuraminic acid remote from the protein core due to continous movement caused by ionic repulsion directs sperm migration.

Extensive treatment of partially purified cervical glycoprotein with proteolytic enzymes caused only a minor change in the molecular size of the major glycoprotein (10). The change in molecular size indicates that the enzymes degraded the original glycoprotein to a similar extent but to a limited degree. This suggests that only some portions of the glycoprotein are relatively easily accessible to enzymes, whereas the remaining are well protected by the carbohydrate chains, particularly those chains that contain sialic acid (10). It has been reported that several glycoproteins, serum as well as secretory type, lose their resistance towards proteinases after treatment with neuraminidase (27-29). It appears that N-acetylneuraminic acid, whether in vicinity or remote from the protein moiety, provides a necessary chemical structure for an important physiological function.

The glycoproteins with neutral oligosaccharide carbohydrate chains or the neutral oligosaccharides within a glycoprotein with hybrid chains (acidic as well as neutral) may contribute to two functions. The neutral carbohydrate chains of the glycoprotein may have been involved in retaining definite protein conformation that imparts to the glycoprotein molecule a physiological activity as well as to give rigidity to the protein. Similar functions have been attributed to the carbohydrate chains of vesicular stomatitis virus G protein (1) and to erythrocytes (2). This assumption need be further verified and possibly a procedure similar to that applied to the virus protein, i.e. inhibition of glycosylation, will be the method of choice.

The other function for the glycoprotein with the neutral oligosaccharide chains which has been proposed is to protect the cervical tissue from invasive action by acting as an acceptor for N-acetylneuraminic acid. It had been demonstrated that spermatozoa will penetrate squamous cells, and that these cells take on characteristics of cancer cells (30). If sperm have the ability to penetrate diploid cells and induce transformation, then some mechanism in the female reproductive tract must function to prevent this phenomenon. It has been proposed that the layer of mucin in immediate contact with mucosal cells of the female genital tract is freed from N-acetylneuraminic acid by mucosal membrane-bound N-acetylneuraminidase (31), thus exposing mucin acceptor molecules for sperm surface N-acetylneuraminic acid, a molecule that endows the sperm with the ability to penetrate cells (32). The fact that cervical carcinoma is more frequently observed at the squamo-columnar cells junction, a place where mucus is scant with excess of bacteria producing a variety of enzymes, including N-acetylneuraminidase, indicates that the neutral oligosaccharide-containing glycoproteins may provide a natural protection to the exposed tissue.

The natural or biological phenomenon involves a cascade of events. The begining of the understanding of these events focusses on the molecular structure, and establishment of chemical structure may initiate the unravelling of the phenomenon.

Acknowledgements

The authors wish to thank Dr. Vernon N. Reinhold, Harvard Medical School, Boston, MA., for the mass spectrum. Part of this investigation (gall bladder) was spported by a research grant from the Pakistan Science Foundation (Grant B-BU/MED 89).

References

1. Gibson, R, Kornfeld, S and Schlesinger, SJ 1981, **J.Biol.Chem.**, 256, 456-462.

2. Bretscher, MS 1973, **Science** 181, 622-629.

3. Bashir-Farahmand, J, Taft, PD and McArthur, JW 1976, **Acta Cytol** 20, 167-173.

4. Hafez, ESE 1977, **Adv Exp Med Biol.** 89, 19-38.

5. Nasir-ud-Din, Walker-Nasir, E, McArthur, JW, Nash, TE, Shalev, M and Jeanloz, RW 1982, **Fertil Steril.** 37, 431-435.

6. Moghissi, KS 1972, **Fertil Steril.** 23, 295-306.

7. Iacobelli, S. Garcia, N and Angeloni, C 1971, **Fertil Steril.** 22, 727-734.

8. Hatcher, VB, Schwarzmann, GOH, Jeanloz, RW and McArthur, JW 1977, **Fertil Steril.** 28, 682-688.

9. Nasir-ud-Din, Jeanloz, RW, Vercellotti, JR and McArthur, JW 1979, **Carbohydr Res.** 75, 349-356.

10. Nasir-ud-Din, Jeanloz, RW, Vercellotti, JR and McArthur, JW 1981, **Biochim Biophys Acta** 678, 483-496.

11. Nasir-ud-Din, Nash, TE, Jeanloz, RW. McArthur, JW and Gminski, D 1979, **Fertil Steril.** 32, 230-232

12. Ruegg, UT and Ruddinger, J 1977, **Meth Enzymol.** 47 E, 111-116.

13. Ruegg. UT and Ruddinger, J 1977 **Meth Enzymol.** 47 E, 116-122.

14. Marshall, T and Allen, A 1977, **Biochem Soc Trans.** 5, 436-439.

15. Reinhold, VN 1972, **Meth Enzymol.** 25, 244-249.

16. Hakomori, S 1964. **J.Biol Chem.,** (Tokyo) 55, 205-207.

17. Bahl, OP 1969, **J. Biol. Chem.** 244, 575-583.

18. Thomas, DB and Winzler, RJ 1969, **J. Biol.Chem.** 244, 5943-5946.

19. Vliegenthart, JFG, Dorland, L and van Halbeek, H 1983, **Adv Carbohydr Chem Biochem.,** 41, 209-374.

20. Kobata, A 1979, **Analyt Biochem.** 100, 1-14.

21. Van den Eijnden, DH, Evans, NA, Codington, JF, Reinhold, VN, Silber, C and Jeanloz, RW 1979, **J.Biol Chem.,** 254, 12153-12159.

22. Nasir-ud-Din, Jeanloz, RW, Lamblin, G, Lhermitte, M, Roussel, P. van Halbeek, H and Vliegenthart, JFG, Manuscript in preparation.

23. Tai, T, Yamashita, K and Kobata, A 1975, **J.Biochem.** (Tokyo), 78.

24. Moghissi, K 1966, **Fertil Steril.** 17, 663-675.

25. Gibbons, RA 1959, **Biochem J.** 73, 209-217.

26. Mason, PL in "Cervical Mucus in Human Reproduction", (M.Elstein, K.S. Moghissi and R.Broth, Eds.) 1973, Scriptor, Copenhagen, pp 82-92.

27. Gottschalk, A and Fazekas de St. Goth, S 1960, **Biochim Biophys Acta** 43, 513-519.

28. Codington, JFC, Sanford, BH and Jeanloz, RW 1970, **J.Natl.Canc.Inst.** 45, 637-647.

29. Montreuil, J 1975, **Pure Appl Chem.** 42, 431-447

30. Bendich, A. Borenfreund, E.Witkins, SS, Beju, D and Higgins, PJ 1976, **Nucleic Acid Res Med Biol.** 17, 43-75.

31. Pricer, WE., Jr and Ashwell, G 1971, **J Biol Chem.** 246, 4825-4833.

32. Daunter, B and Counsilman 1980, **Europ J Obstet Gynec reprod Biol.** 10, 141-161.

33. Dobois, M, Gilles, KA, Hamilton, JK, Rebers, PA and Smith, F 1956, **Anal Chem.** 28, 350-356.

Cyclization Approaches to the Synthesis of Natural Products

Kathlyn A. Parker
Department of Chemistry, Brown University, Providence, RI 02912, USA

The synthesis of complex molecules and the development of synthetic methodology are complementary processes; indeed the syntheses which prove most efficient and therefore attain a touch of elegance are those in which there is a particularly adept match of target structure and the key step or steps.

This paper will describe first a project in which the target structure stimulated the investigation of a synthetic method and then a project where the method, discovered with the aid of serendipity, inspired application in total synthesis.

Part I

This work described here was initiated as an approach to the synthesis of vitamin D metabolites, e.g., 1α,25-dihydroxy vitamin D_3 (1).

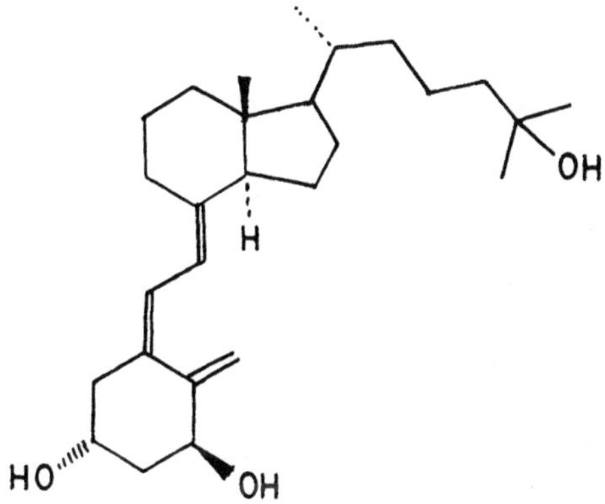

1

Natural Product Chemistry
Ed. by Atta-ur-Rahman

When we began thinking about the synthesis of these compounds in 1980, there was really nothing in the literature which could be considered a practical approach. In fact, almost the entire history of total synthesis approaches to vitamins D_2 and D_3 could be attributed to Lythgoe. Lythgoe had reported many pioneering efforts but he had yet to integrate these into a coherent, stereo- and regiospecific synthesis.[1] A practical total synthesis has to be short and so we considered only approaches which we thought might be developed into extremely efficient routes.

The obvious places to divide the target into two building blocks for a convergent synthesis are between C-7 and C-8 and between C-8 and C-9. (In fact the C-8 ketone derived from degradation of vitamin D had been reelaborated to vitamin D and related compounds.) These dissections isolate the two construction problems which must be faced.

The first of these is the synthesis of the hydroxylated A-ring. We haven't yet had an opportunity to deal with this but I'd just like to point out that approaches to the synthesis of the A-ring of $1\alpha,25(OH)_2D_3$ are limited because many likely precursors have the same oxidation state as a substituted benzene. Two clever solutions to this problem have appeared recently.[1a,b]

The second construction problem is the generation of the trans hydrindane ring system, a long-standing challenge in organic synthesis. The hydrindane system of vitamin D derivatives is a particularly complex case because it contains four contiguous chiral centers.

The key step in our approach to this problem is shown below. It seemed at least possible that a 1,3,8-nona-triene **2** would undergo cycloaddition to afford the desired hydrindene **3**.

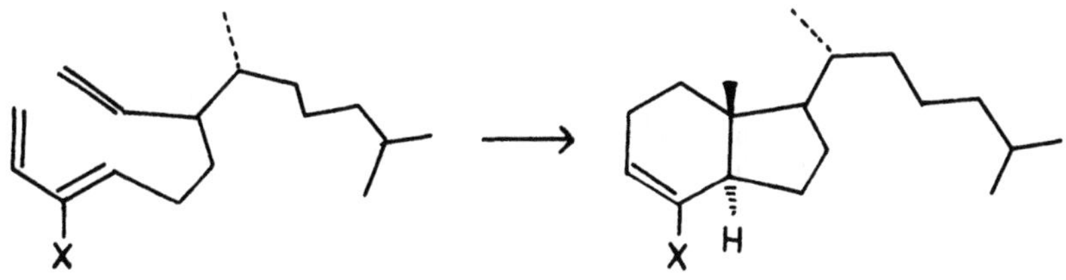

2 3

When we began this work there was little known about the stereochemistry of intramolecular Diels Alder reaction of nonatrienes so this was a high risk venture;[2] after all the triene in question has the potential of closing to one or more of four stereoisomers.

A question which we have not discussed is that of setting up relative stereochemistry at C-17 and C-20. The solution to this problem is based on the enolate Claisen rearrangement and is illustrated in the synthesis of our first substrates (alcohol 5 and ester 4).

Cyclization of alcohol 5 gave a 23% yield of a 1:1 mixture of two products, shown by independent synthesis to be 6a and 7a. Cyclization of ester 4 under similar conditions (6 hr., sealed tube, benzene, 200°) gave a quantitative yield of a mixture of bicyclic esters 6b and 7b.

The Diels Alder reaction of trienes 5 and 4 therefore gave only the two products in which the β-configuration at C-17 (steroid numbering) had been produced. Inspection of models held in position to resemble the possible transition states suggest that interactions between a C-17α alkyl group and the connecting chain are responsible for the absence of the C-17α substituted products.

If we could now destabilize TS2 (relative to TS1), for example by introduction of a substituent, we should see an improved conversion to the desired <u>trans</u> hydrindene. Consideration of the interactions in these two

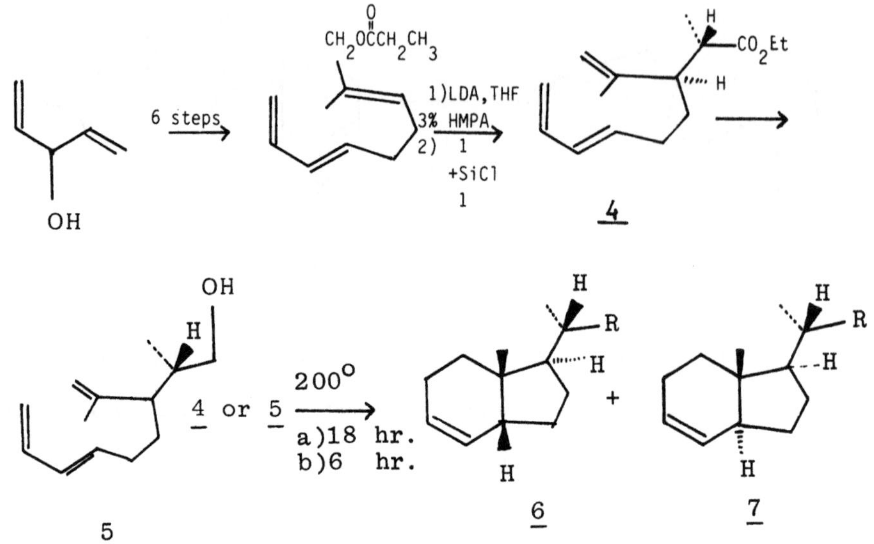

a) R=CO_2Et 100% (1:1)
b) R=CH_2OH 23% (1:1)

Scheme 1

from K.A.Parker and T.Iqbal, J.Org.Chem., 47, 337 (1982)

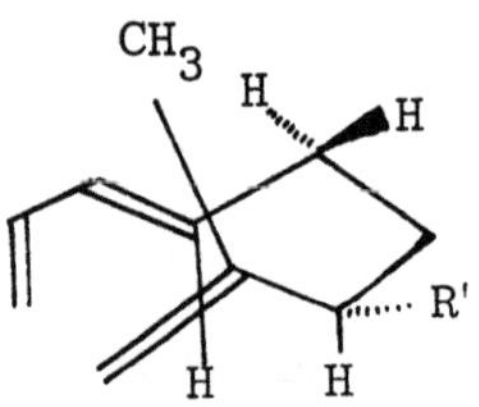

TS 1

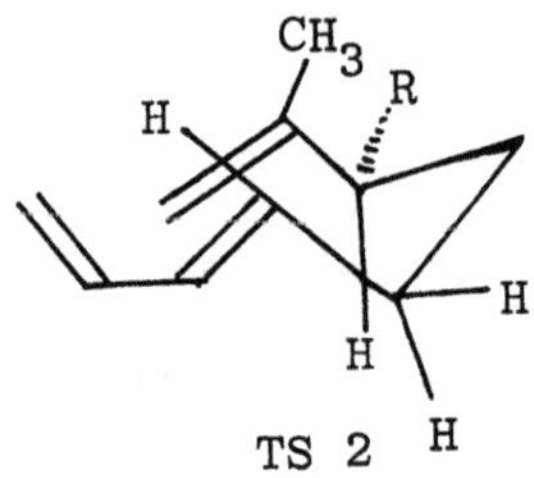

TS 2

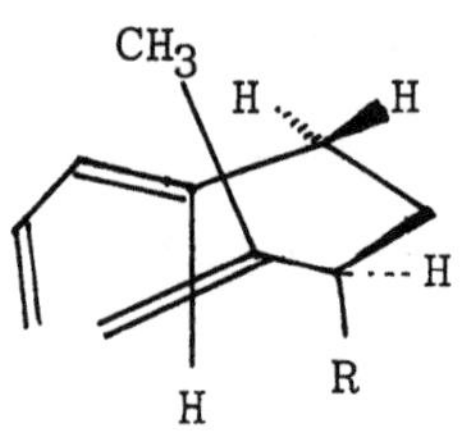

TS 3

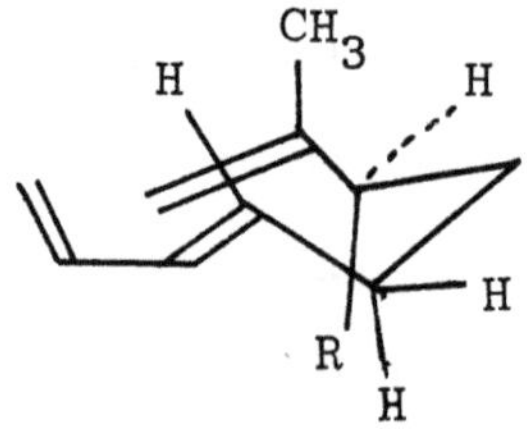

TS 4

transition states led us to believe that a substituent at C-3 of the diene system would disfavor cyclization to the <u>cis</u> product.[3]

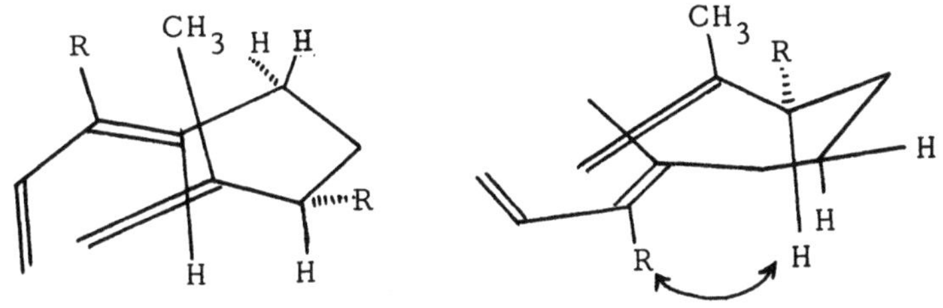

TS1 TS2

In order to have maximum flexibility in testing these 3-substituted trienes, we prepared aldehyde 8 (Scheme 2), a generally useful precursor.

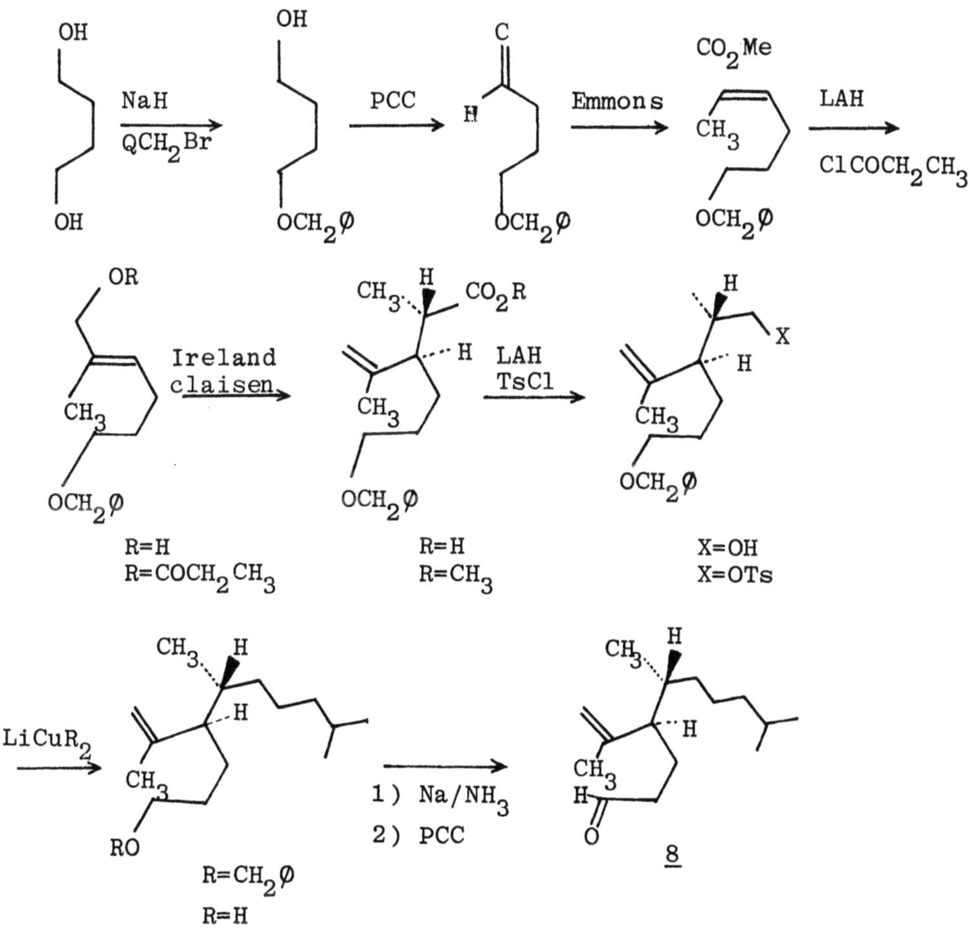

Scheme 2

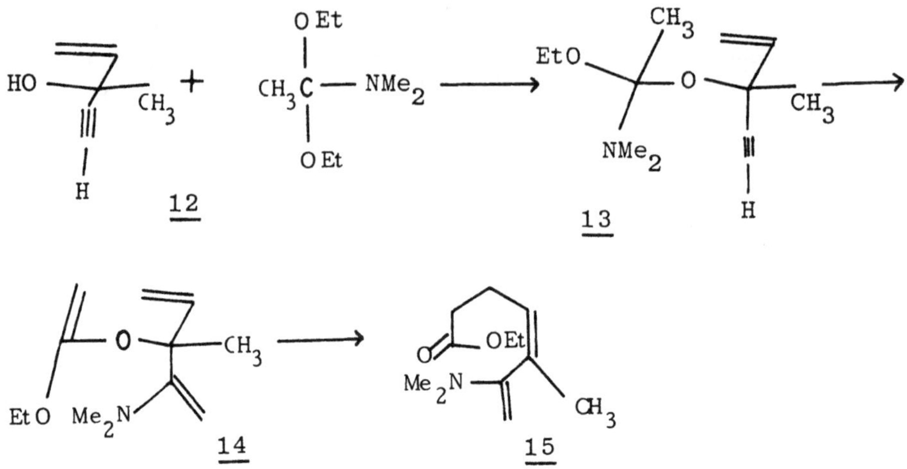

Scheme 4

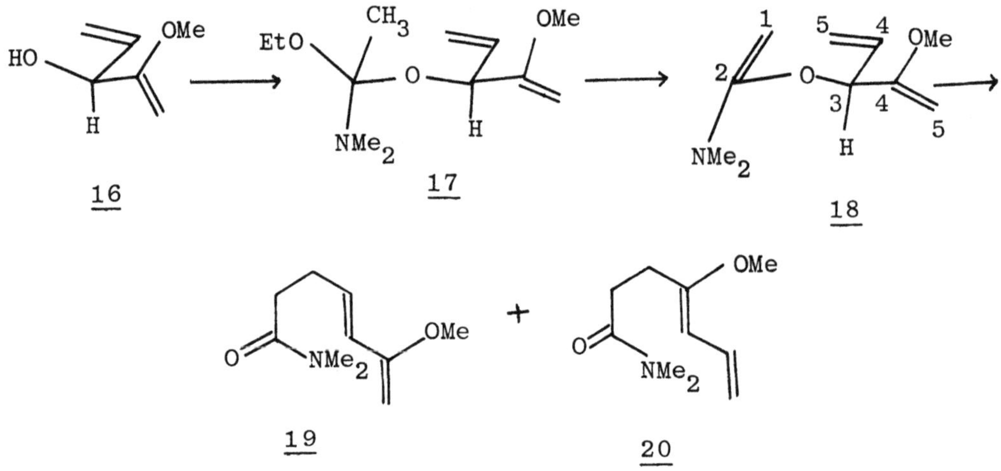

(97.3)

Scheme 5

at C-4 of the Claisen skeleton slows rearrangement significantly; rearrangement of dienols becomes regioselective. An alkyl group at C-4 of the Claisen skeleton has a small retarding effect on the rearrangement <u>unless</u> the substrate also bears an (<u>E</u>)-substituent at C-1; then apparently the steric cost becomes significant and reaction is slowed significantly. An alkyl group at C-5 has a much smaller effect. Thus, rearrangement of ester enol <u>22</u> affords <u>23</u> almost exclusively.

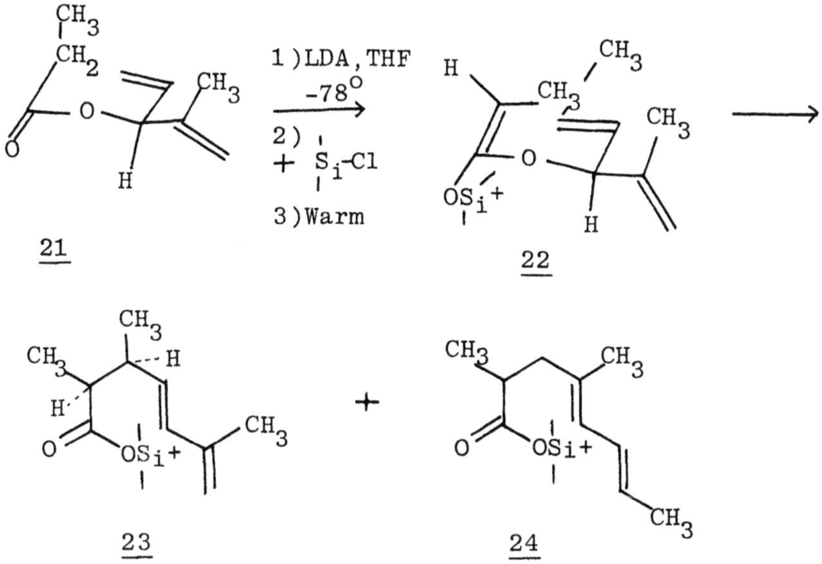

Scheme 6

We are now engaged in a synthesis of biflora-4,20(19),15-triene (25), a component of the defense secretion of the termite Cubitermes umbrates.

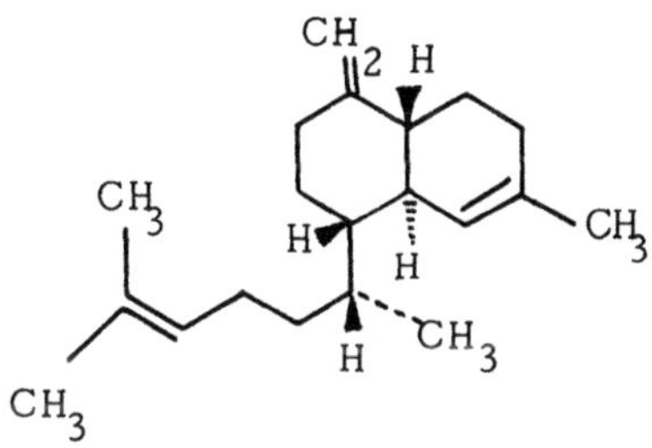

25

In this scheme as in the vitamin D chemistry discussed above, we link the Ireland-type Claisen rearrangement (now regio- as well as stereospecific), with the intramolecular Diels Alder reaction as a strategy for generating 3-4 contiguous chiral centers. This synthesis has progresed as far as the key rearranged compound 26. We believe that only well-precedented chemistry is required to complete the synthesis.

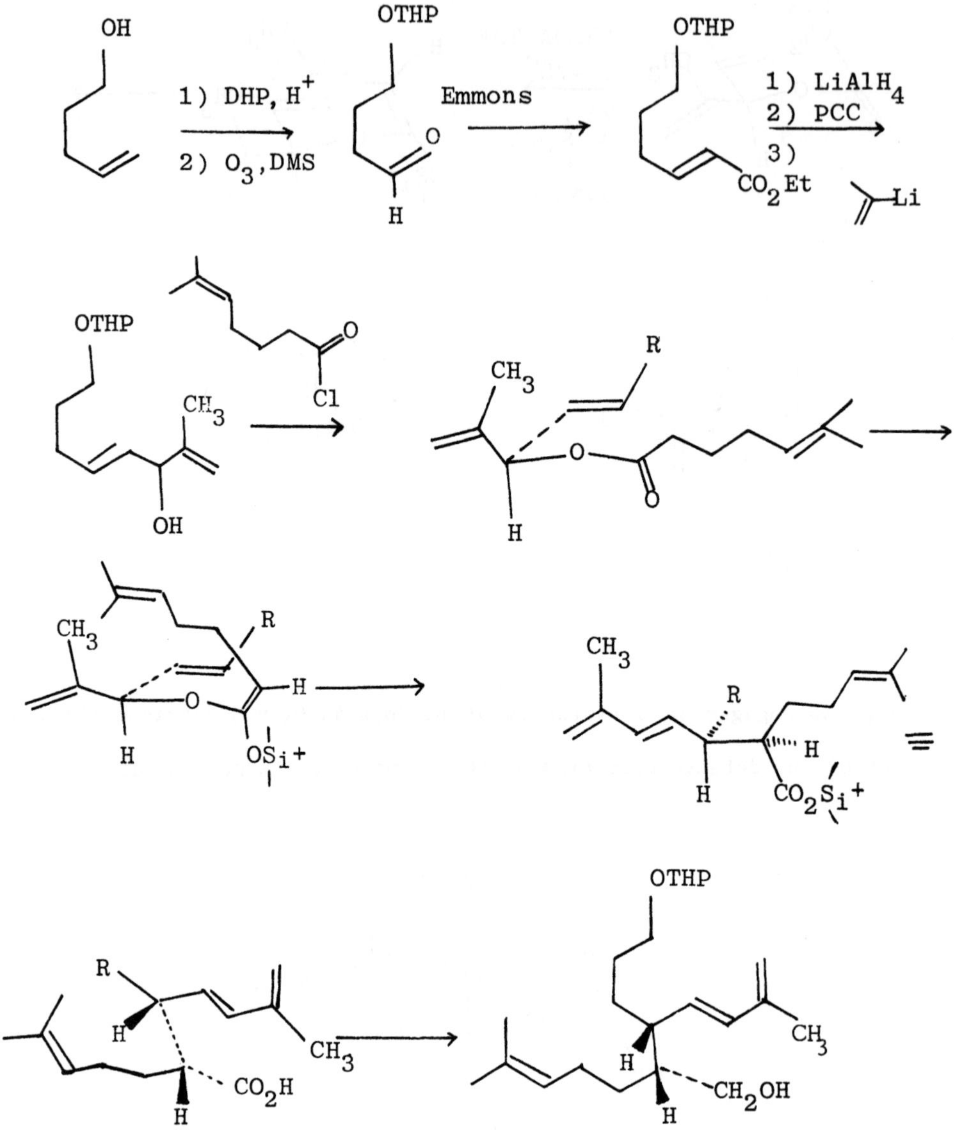

OH
1) DHP, H+
2) O3,DMS
OTHP
O
H
Emmons
OTHP
CO2Et
1) LiAlH4
2) PCC
3)
Li
OTHP
CH3
OH
O
Cl
CH3
R
O
O
H
CH3
R
O
H
H
OSi+
CH3
R
H
H
CO2Si+
≡
R
H
CH3
CO2H
H
OTHP
CH3
H
CH2OH
H

Scheme 7 26

References

1. Since that time, two notable total syntheses have been achieved:

(a) Kocienski, P. J., Lythgoe, B., J. Chem. Soc., Perkin I, 1980, 1400; Lythgoe, B., Waterhouse, P., J. Chem. Soc., Perkin I, 1980, 1405; Harrison, R. G., Lythgoe, B., Wright, P. W., Tetrahedron Lett., 1973, 3649.

(b) Baggiolini, E. G., Iacobelli, J. A., Hennessy, B. M., and Uskokovic, M. P., J. Am. Chem. Soc., 1982, 104, 2945.

2. A thorough product study by Roush on a similar system had been reported: W. Roush, J. Org. Chem., 1979, 44, 4008. Roush's data have proven extremely useful in the analysis of our own system.

3. The same tactic has been employed by Wilson in a very similar approach to the vitamins D. Wilson, S. R., Hague, M. S., J. Org. Chem., 1982, 47, 5413.

Chemical Defense and Self-Defense in Termites

Glenn D. Prestwich

Department of Chemistry, State University of New York, Stony Brook, NY 11794, USA

Summary

Termite soldiers are unique among insects in the variety of chemical weapons and morphological adaptations they use for colony defense. The three main types of chemical defenses are: (1) biting, with the injection of an oily or toxic material into the wound; (2) brushing, with the topical application of an irritant or contact poison onto the cuticle of an attacker; and (3) glue-squirting, in which a viscous, terpenoid secretion is ejected at an aggressor.

The oily secretions of the first group have been identified as normal alkanes and alkenes unusual new irregular diterpenes, and very large-ring (up to C_{38}) macrocyclic lactones. The contact poisons of the second group are electrophilic lipids, such as nitroalkenes, vinyl ketones, and β-ketoaldehydes. A substrate-specific reductase allows autodetoxication of these poisons by conspecific workers. The third group, the glue-squirting nasutes, produces an array of bi-, tri- and tetracyclic diterpenes based on intramolecular cyclizations of cembrene-A. The identification of these chemicals and their importance in termite biology and evolution will be discussed.

Introduction

A chemical arms race has been under way in the insect world for over 100 million years. Predators attempt to subdue their quarry with toxic venoms or to lure them to destruction with seductive perfumes. Prey species counter with the ejection of irritating, sticky, hot, or poisonous chemicals or by being inedible. Unique among the chemically-defended insects are the termite soldiers[1], a caste of warriors whose heads and bodies are so thoroughly modified into weapons that they can neither sustain themselves nor reproduce. No other insect order, not even the

Natural Product Chemistry
Ed. by Atta-ur-Rahman

army ants, have such a totally specialized and dependent full-time army with such a diverse arsenal of chemical weapons[1] and delivery systems[3] as can be found in the Isoptera. Termite soldiers truly are, as Maurice Maeterlinck wrote 60 years ago in The Life of the White Ant, "a caste of nightmarish monsters which recall the most fantastic revelries of Hieronymous Bosch". The alates fly away to form new colonies, and the soldiers defend. Communication among individuals of the colony is based on the exchange of chemical signals, both by smell (olfaction using the antennae) or taste (contact chemoreception). Building, food-finding, nestmate recognition, trail-following, growth, alarm, and defense all involve specific chemical cues.[4]

Termites, often incorrectly called "white ants", evolved quite separately from the social ants, bees, and wasps in the Hymenoptera. Descended from cockroach-like ancestors over 150 million years ago, termite social organization evolved around the need for food sharing to exchange the bacterial and protozoan symbionts necessary to digest cellulose. Over 2000 termite species have been named worldwide (95% are tropical), and most serve a vital role in the recycling of dead cellulosic (woody) materials. Their ecological function in the soils of the semi-arid tropics has been likened to that of earthworms in temperate regions.

The variety of bizarre defensive mechanisms exhibited by termite soldiers (Fig. 1) arises from the expendability of the sterile individuals whose self-sacrifice is essential to the survival of the reproductives. Termite soldiers are thus walking weapons and can execute their kamikazi defensive roles by tactics which include biting, snapping, hole-plugging, squirting, oozing, daubing, defecating, and exploding. I will discuss a number of these in detail, highlighting how chemical structures combine with morphological and biochemical adaptations to produce consummate defensive machines.

Mechanical Defenses

Some termite soldiers have only mechanical weaponry. Primitive dry-wood termites like Incisitermes and Neotermes from Florida use heavy, serated jaws (mandibles) to inflict gashlike wounds

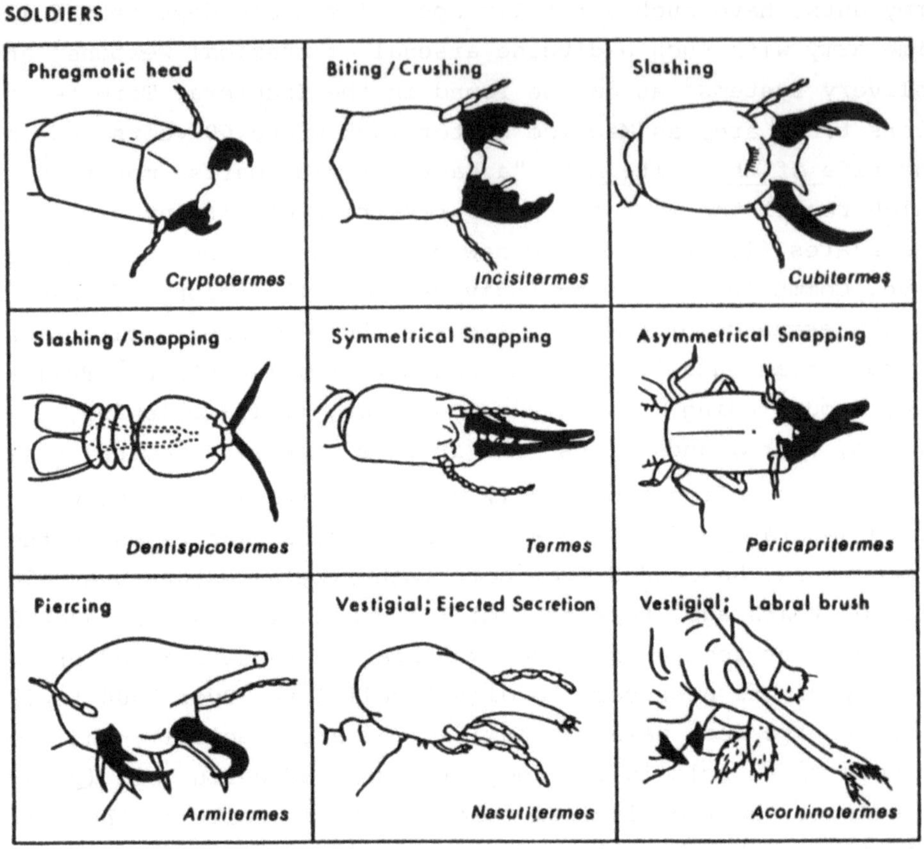

Figure 1. Reproduced with permission from the Annual Review of Ecology & Systematics, Vol. 14, © 1983 by Annual Reviews Inc.

on attackers. Highly-evolved humivorous termites in South America (Neocapritermes, Termes) have huge rectangular heads with mandibles longer than their bodies; when disturbed, they snap their jaws violently with a loud click, and a nearby attacker is battered by the impact. The cosmopolitan tropical drywood termite Cryptotermes has soldiers with phragmotic heads and sharp mandibles; soldiers use their heads like drain plugs, guarding the narrow entrance into the colony's labyrinthine passageways with a dangerous trap door.

A dramatic example of explosive self-sacrifice occurs in the soldierless termites (Anoplotermes), whose abdomens are ringed with constricting muscles. When attacked the abdominal contents are squeezed up towards the thorax, the cuticle ruptures suddenly, and the attacker is bathed with a malodorous and foul-tasting mixture of feces and gut contents.

Chemical Defenses

The mechanical defenses seem primitive when compared with the chemical weaponry of many higher termites. Quennedey and Deligne[5] have studied the morphology and glandular structures of the "frontal weapons" of termite soldiers in detail, and they recognized three primary methods of chemical defense: (1) biting, with the simultaneous introduction of an oily or toxic material into the wound; (2) daubing, with the use of an enlarged paint brush-like upper lip (labrum) to apply a contact poison onto the cuticle of an attacker; and (3) glue-squirting, in which an irritating, viscous entangling agent is sprayed onto an aggressor.

"Biting and Injecting" Soldiers

The first chemical defense method, biting with the addition of a toxic or irritating oil to the wound, has evolved independently numerous times in the higher termites (family Termitidae) and in the subterranean termites (family Rhinotermitidae). Frontal (defense) gland contents are released into a fresh wound, and the would-be assailants "bleed" to death. Soldiers of three genera will serve to illustrate the chemical and morphological variety evolved to this purpose: Macrotermes and its paraffin-olefin wax mixture; Cubitermes and its structurally unique diterpene hydrocarbons; and Armitermes, with its cyclic esters of unprecedented ring size.

Termites of the genus Macrotermes build 2-4 meter high, mounds which dominate the landscapes of Africa and southern parts of Asia. These fungus-gardening termites rely on a symbiotic basidiomycete to help break down cellulose and lignins of grasses and to provide nutrient-enriched food for the colony. Colony defense in this species is the job of a sterile female caste. Small soldiers escort workers (sterile males) during food-gathering and mound repair, while large soldiers repel attacks on the inner hive containing the young larvae and the royal pair. These soldiers have evolved hard, shield-like heads and sharp, scissoring mandibles for defense. The chemical secretions of these soldiers are made and stored in a cephalic (head) defensive gland called frontal gland, and are exuded through a pore called the fontanelle. Both large and small soldiers have the same chemicals, but large soldiers have 500 times more, comprising almost 10% of the soldier's dry weight.

The oily chemical mixture of this species consists of long-chain normal alkanes and (Z)-9 alkenes possessing 21 to 35 carbon atoms, and the exact composition varies with species and location. This material is nontoxic to ants and humans alike, either by ingestion or by topical application. In combat with ants, blind soldiers bite frantically, trying to snip off legs with their powerful jaws. Heat generated by the muscular action melts the waxy secretion into a free-flowing oil which then oozes from the frontanelle down the rostrum (forehead) and out onto the labrum (upper lip) which overlays the mandibles. Every bite, therefore, includes a generous coating of the oily hydrocarbon mixture onto the wound. These hydrocarbons soften the ant's cuticle (which actually includes smaller quantities of the same hydrocarbons as waterproofing) and interferes physically with the initiation of coagulation and resclerotization required for normal wound healing.

Another higher termite which uses the biting-antihealant technique is Cubitermes, a genus of African humivores in the subfamily Termitiniae. These soldiers have squarish orange heads, forked labra, and saber-like mandibles. Their secretion, which oozes vertically down the rostrum, consists of hydrocarbons with twenty-carbon molecular skeletons called diterpenes. Of the five major diterpenes (identified from over 15 components) found in seven Cubitermes species, four are unique in nature to this group of termites (Figure 2).[6]

The genus Armitermes is in the subfamily Nasutitermitinae which contains the glue-squirters (to be discussed later). This primitive genus has both a nozzle-like protrusion of the rostrum and sharp-pointed ice tong-like mandibles. Instead of squirting, it uses its long nose to apply droplets of an oily

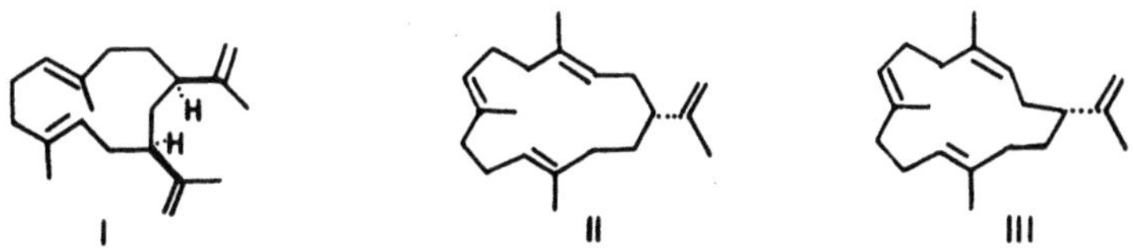

I II III

IV V

Figure 2

secretion to the puncture wounds made by the mandible. The act of biting is often suicidal for these soldiers, since they become inextricably "stapled" to the attacker.

The Armitermes secretion contains macrocyclic lactones with 22 to 38 carbon atoms in a single ring (Figure 3). Solitary bees have 18 to 24 carbon lactones which are used in cell linings. The larger carbon rings are unique to termites as far as we know. The interspecific variation and structure determinations of these macrolides was recently reviewed.[4,7]

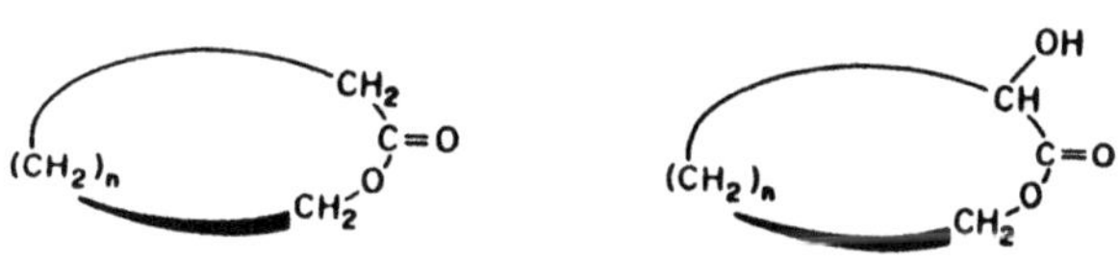

1
n = 19, 21, 23, 25, 27, 29, 31, 33, (35)

2 a, n = 21
b, n = 23

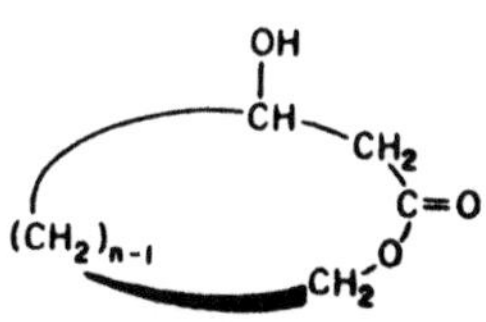

3 a, n = 21
b, n = 23

Figure 3

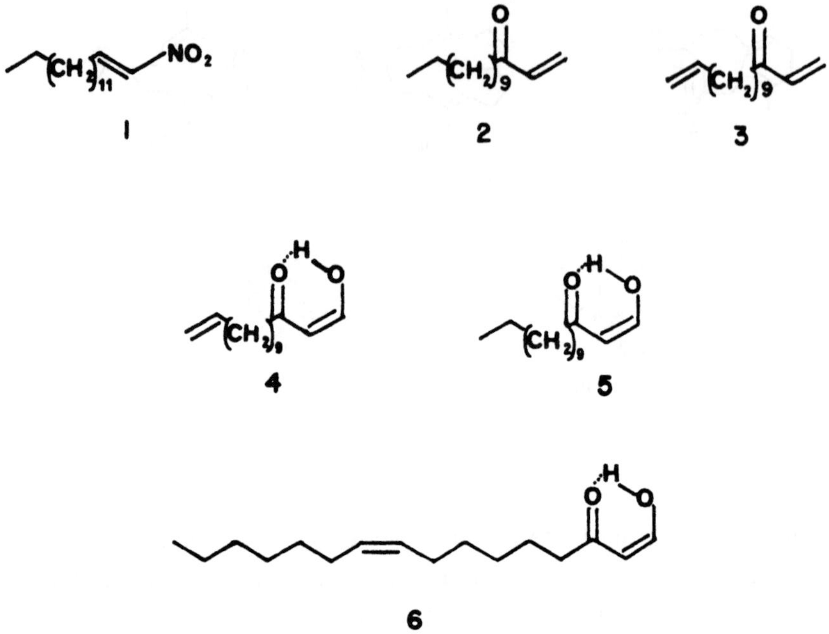

Figure 4

Poison-brushing Soldiers

Soldier termites in the subfamily Rhinotermitinae have enlarged daubing-brush upper lips with which they apply pungent-smelling lipophilic (oil-soluble) contact insecticides to the surface of an attacking arthropod. The toxic secretion is stored in voluminous cephalic and abdominal reservoirs, which in some soldiers can constitute over 50% of their dry weight. The chemistry of these poisons has been recently reviewed (Figure 4).[8]

The monomorphic soldiers of *Prorhinotermes simplex* (subfamily Prorhinotermitinae) produce a cephalic defense secretion which consists of >90% of a single nitroalkene, (1) The remaining 10% is a sesquiterpene hydrocarbon recently identified as (Z)-α-farnesene. Major and minor soldiers of the termites *Schedorhinotermes putorius* and *S. lamianianus* produce up to 35% of their dry weight as a mixture of 3-alkanones, 1-alken-3-ones (2) and α,ω-alkadien-3-ones (3) which act as contact poisons to attacking ants. Recently, we found that the most advanced genera of this subfamily, *Rhinotermes* and *Acorhinotermes* produce β-ketoaldehydes (4-6). These contact poisons – the nitroalkenes, the vinylketones, and the β-ketoaldehydes – are all highly reactive agents and are capable of reaction with biologically important nucleophilic sites containing hydroxyl, sulfhydryl, or amino groups.

The C_{14} β-ketoaldehydes 4 and 5 were isolated as >80% of the crude hexane extract of the "nasutoid" minor soldiers of the Neotropical species Rhinotermes hispidus and R. marginalis. The corresponding C_{13} ketones 2-tridecanone and 12-tridecen-2-one were also isolated, suggesting that the β-ketoaldehydes arise from reductive cleavage of β-keto fattyacyl-coenzyme A derivatives. In the major soldiers, however, these ketoaldehydes and ketones were completely absent. This contrasts with the more primitive Schedorhinotermes species, in which both major and minor soldiers have copious amounts of chemically identical secretions.

The Guyanese termite Acorhinotermes subfusciceps has undergone a secondary loss of the major soldier caste, and thus has only nasutoid minor soldiers. Hexane extracts of these soldiers gave the sixteen-carbon β-ketoaldehyde 6 as the major product and ca. 5% of (Z)-8-pentacecen-2-one as the minor product. These compounds can be formally derived from reductive cleavage of the β-ketopalmitoleoyl-CoA in a highly parsimonious fashion. That is, defense substance production may result from the harnessing of the fatty acid degradation enzymes to provide lipid-like contact poisons.

The morphological trend in this subfamily is the regression of the mandibles with the prolongation of the labrum into a daubing brush. The importance of the mandibulate major soldiers decreases until the caste is eventually lost in Acorhinotermes. The chemical instability and thus the inherent reactivity of the defense chemicals which the nasutoid minor soldiers deploy increases with evolutionary advancement. Finally, it appears that selection favors the minimum amount of additional enzymic machinery to carry out the preparation of the defensive chemicals. As a result, the more advanced termites employ chemicals which are structurally more similar to normal intermediates in fat metabolism.

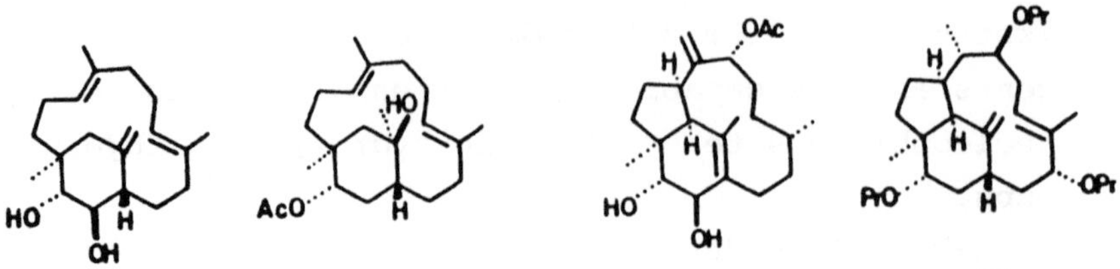

VI VII VIII IX

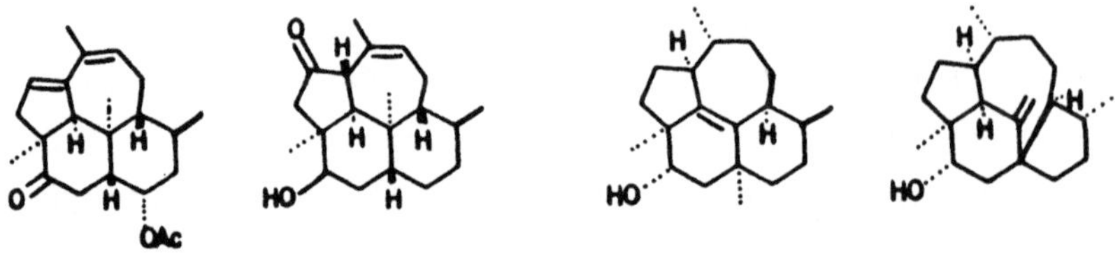

X XI XII XIII

Figure 5

The ability of rhinotermitid termite workers of these genera to survive in the presence of toxins secreted by their respective soldiers requires the existence of enzymes capable of detoxifying these reactive compounds. We found that termite worker of Prorhinotermes simplex and Schedorhinotermes lamanianus possess substrate-specific alkene reductases which, in the presence of reduced nucleotide cofactors, catalyze the reduction of the electron-deficient double bonds of the unsaturated electrophilic groups. The saturated compounds are subsequently recycled in vivo via catabolism to acetate.

Glue-squirting Soldiers

The Nasutitermitinae is the most abundant and diverse subfamily of termites in the world; the soldiers of this subfamily have provided a chemically intriguing and diverse set of new diterpenoid, and more recently, macrolide defensive chemicals.

The soldiers of the advanced genera of this subfamily posses an elongate rostrum, called the nasus, which is used to eject a viscous and sticky terpenoid secretion. These secretions are sticky because they contain high concentrations of hydrogen-bond, dome-shaped diterpenes (Fig. 5) in a monoterpene solvent. The structures of the tricyclic trinervitanes (e.g., VIII,IX)

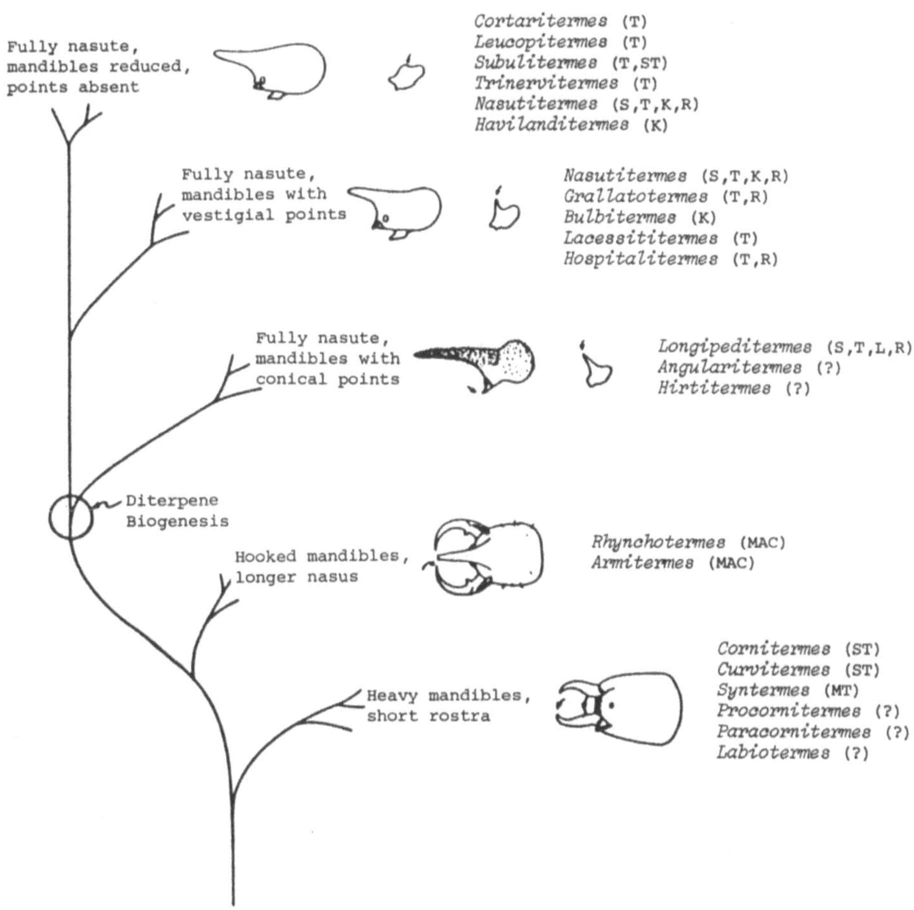

Figure 6. Reproduced with permission from the Annual Reviews of Ecology & Systematics, Vol. 14, © 1983 by Annual Reviews Inc.

and tetracyclic kempanes (e.g., X,XI) inspired further investigations into their biosynthetic origins and a search for biosynthetic key intermediates. The recent discoveries of bicyclic secotrinervitanes (e.g., VI,VII) in Nasutitermes and Longipeditermes and the spirotetracyclic longipane (XIII9 from Longipeditermes provide further evidence for a stepwise progression from the monocyclic 14-membered ring cembrene-A to bicyclic, tricyclic, and tetracyclic carbon skeletons. Furthermore, alternative ring-forming pathways may occur in the tricyclic to tetracyclic conversion as shown by the existence of the 1,2-methyl shifted skeleton of the rippertanes (e.g., XII). Unique functionalization of the tetracyclic kempane skeleton (e.g., XI) occurs in Nasutitermes octopilis from Guyana and this may be responsible for the unpalatability (to ants) of this litter-dwelling species relative to sympatric arboreal Nasutitermes species.

Intraspecific variation of defense chemicals in isolated populations of Trinervitermes gratiosus and T. bettonianus gives rise to genetically distinct chemical "races". In the latter species, the chemical composition of the soldier secretion was shown to be invariant under conditions in which field-collected alates were raised to incipient colony status in the laboratory and fed grass from the region of a different T. bettonianus chemotype. This was also taken as further evidence of the de novo biosynthetic origin of these secretions in the frontal glands.

Biosynthesis was unambiguously demonstrated by injection of ^{14}C-labeled acetate and mevalonate derivatives into the abdomens of N. octopilis soldiers using micropipettes. Purified mono- and diterpenes showed 0.05-0.3% incorporation of label in 12 - 24 hr, indicating relatively rapid and efficient incorporation of the precursors.

The subfamily Nasutitermitinae is alleged to be a classical example of parallel diphyletic evolution in which regression of the soldier mandibles and elongation of the rostrum to form the nasus occurred independently along two branches of this subfamily. However, we found identical diterpenes occur in advanced nasutes at the ends of both alleged phyletic lines after over 70 million years of evolution from mandibulate ancestors. In view of the absence of diterpenes in the primitive mandibulate nasutes Syntermes, Cornitermes, Armitermes, and Rhynchotermes, we had suggested that the diphyletic hypothesis for the origins of the Nasutitermitinae was improbable. We proposed an alternative monophyletic route (Fig. 6) in which all glue-squirting non-mandibulate nasutes diverged from a common diterpene-producing ancestor which evolved prior to the separation of West Godwana into proto-Africa and proto-South American in the Cretaceous.

There is growing excitement in fields which integrate chemistry and biology. In the study of termite behavior and evolution, chemists are now helping to solve long-standing biological mysteries. In return, termites have provided a rich harvest of new natural products as well as many unique structural and biochemical modifications for defense. For every answer, we find two new questions, and the study of the chemistry and ecology of these primitive social insects will continue to provide fascination and surprises in the years to come.

Acknowledgments. I thank the NSF (DEB-7823257, CHE-7925081 and CHE 8304012) for grants in support of this research, the Alfred P. Sloan Foundation for a Research Fellowship (1981-85), and the Camille and Henry Dreyfus Foundation for a Teacher-Scholar Award (1981-86). I am also grateful for the hospitality of our Pakistani hosts, the organizational expertise of Profs. P. Le Quesne and Atta-ur-Rahman, and the support of NSF International Programs for this symposium-workshop.

References

1. Prestwich, G.D., Scientific American 249, 78 (1983).
2. Prestwich, G.D., J. Chem. Ecol. 5, 459 (1979).
3. Prestwich, G.D., Ann. Rev. Entomol. 29, 201 (1984).
4. Prestwich, G.D., Ann. Rev. Ecol. System. 14, 287 (1983).
5. Deligne, J., Quennedey, A., and Blum, M.S. In H.R. Herman, ed., Social Insects, Vol. 2, 1-67. New York: Academic Press, 1982.
6. Prestwich, G.D., J. Chem. Ecol. 1984, in press
7. Prestwich, G.D., Tetrahedron 38, 1911 (1982).
8. Spanton, S.G. and Prestwich, G.D., Tetrahedron 38, 1920 (1982).

Some Recent Isolation Structural and Synthetic Studies on the Constituents of Indigenous Medicinal Plants

Atta-ur-Rahman
H. E. J. Research Institute of Chemistry, University of Karachi, Karachi-32, Pakistan

Abstract: Isolation and structural studies carried out on medicinal plants of Pakistan are described. The synthesis of a number of biologically interesting natural products and their derivatives is also presented.

The bulk of the populations of the Afro-Asian countries, particularly those living in villages, rely on the indigenous medical system to provide relief from disease. Systematic scientific investigations, particularly during the current century, have resulted in the identification of a growing number of active constituents many of which are now routinely used in modern medicine. These include reserpine for the treatment of cardiac arrhythmias, vincamine as a vasodilator, vinblastine and vincristine as anti-tumour agents, etc. Isolation, structural and synthetic studies have, accordingly, been directed in many laboratories around the world, including ours, to isolate new natural products which would prove to be valuable chemotherapeutic agents. Some of the recent studies carried out by us are briefly presented here.

Isolation and Structural Studies on Berberis aristata

Berberis aristata DC(Berberidaceae) is a shrub found in the northern mountainous regions of Pakistan and India as well as in the Nilgiri Hills of southern India. The extracts made from the root bark are known as "rasaut" and are used in the traditional system of medicine for the

Natural Product Chemistry
Ed. by Atta-ur-Rahman

treatment of jaundice and skin diseases. As a result of careful isolation studies, two new alkaloids, "karachine" [1] and "taxilamine" [2] have recently been isolated (37,38). Karachine is the first naturally occurring berbinoid of this skeletal system and is the most complex of more than 50 protoberberine alkaloids presently known. Its structure [1] has been elucidated largely on the basis of its high resolution mass and 360 MHz (FT) nmr spectra, and the positioning of groups confirmed by Nuclear Overhauser Effect studies.

The uv spectrum of karachine, λ_{max} EtOH 226 and 285 nm(log ε 3.90 and 3.62), was suggestive of a tetrahydroprotoberberine. The mass spectrum shows the molecular ion at m/z 433, and the base peak at m/z 336. The latter peak fits exactly for the molecular ion of berberine or epiberberine and is formed by loss of 97 mass units from the molecular ion via cleavage alpha to the nitrogen atom (C-14 to C- bond), followed by a retro-Diels-Alder process. The m/z 97 fragment corresponds to C_6H_9O, or, more specifically, to 2 moles of acetone minus the elements of water. A sharp absorption band at 1710 cm^{-1} in the ir spectrum ($CHCl_3$) denoted the presence of a non-conjugated carbonyl. The 360 MHz (FT) nmr spectrum in $CDCl_3$ presented a complex pattern, but allowed for the tentative assignment of expression [1] to karachine.

In order to settle conclusively the nature of the substitution pattern in aromatic-rings A and D, n.o.e. study was carried out. Irradiation of the C-10 methoxyl singlet at δ 3.77 resulted in an overall 11.6% increase in the area of the δ 6.52 and δ 6.55 ring D aromatic doublet of doublets. Alternatively, irradiation of the H-1 singlet at δ 6.73 gave a 2.8% increase of δ 2.70 and 2.72 doublet of doublets assigned to the C-ε protons, as well as to a 5.6% increase of the signal at δ 3.07 due to H-13. Significantly, irradiation of either the H-1 or H-4 singlets at δ 6.73 and δ 6.17 respectively led to no observable n.o.e. for the methoxyl absorptions. Further support for the structure of karachine has come from its borohydride reduction, and analysis of the mass and nmr spectra of the corresponding alcohol. The structure assigned by us to karachine has recently been confirmed by synthesis (66).

Karachine must arise by the condensation of berberine **[3]** with 2 moles of acetone and accompanying loss of water, as suggested in the Scheme 1. It is the first naturally occurring berbinoid incorporating acetone units (42). It is a true alkaloid and not an artefact of isolation since (a) optically active, as well as inactive, naturally occurring adducts of the related benzophenanthridine alkaloids with acetone are known (62), (b) no acetone was used during the isolation process, and (c) various attempts on our part to obtain karachine by condensation of berberine with acetone at varying pH's were to no avail.

Recently another bridged protoberberine alkaloid, valachine has been isolated from Berberis validiviana to which structure **[58]** has been assigned (41). Valachine probably arises in the plant by condensation of berberine with acetaldehyde anion followed by a second condensation with acetone anion as indicated in Scheme 2. Bridged protoberberines such as karachine and valachine represent a new class of natural products.

Taxilamine **[2]** was isolated by chromatography of the alkaloidal fraction (4) using neutral alumina. Besides a consistent u v spectrum, the 360 MHz (FT, $CDCl_3$) ^{1}H-nmr spectrum of taxilamine shows H-5 and H-8 as singlets at δ 7.15 and δ 7.40, respectively; H-3 and H-4 as a doublet of doublets at δ 8.46 and δ 7.66 (J_{vic} = 5.5 Hz); and H-5' and H-6' as another doublet of doublets at δ 6.44 and δ 7.28 (J_{vic} = 9.1 Hz). The 4 methoxyl signals appear as singlets at δ 3.92, δ 3.96, δ 3.97 and δ 4.06. This spectrum bears a distinct resemblance to that reported for rugosinone (71). The mass spectrum of taxilamine confirmed the molecular formulation of $C_{20}H_{19}O_6N$ and the structure **[2]** assigned.

Taxilamine **[2]** is the fourth member of its class of pseudobenzyl isoquinoline alkaloids and must probably have been formed in nature through oxidative rearrangement of palmatine to supply initially polycarpine. Hydrolytic N-deformylation followed by further oxidation would then afford taxilamine (54).

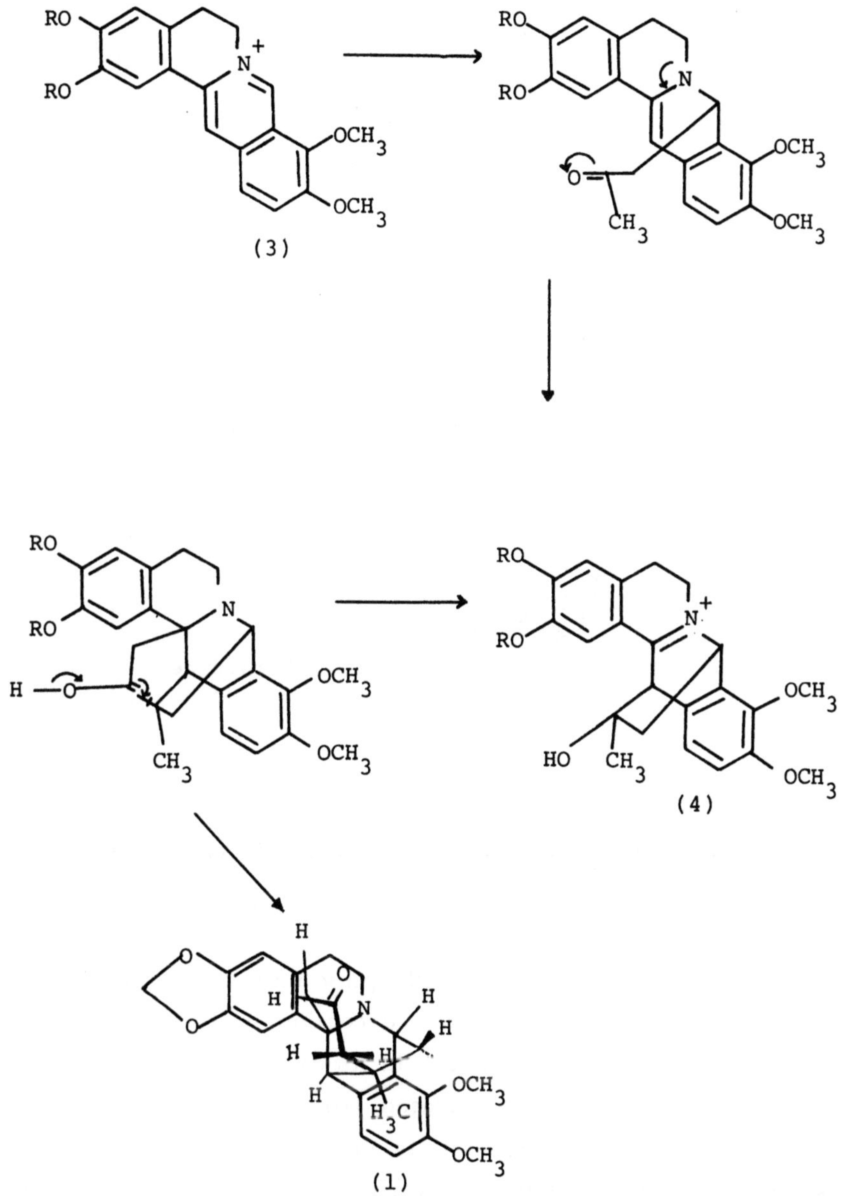

RO
RO
N+
OCH3
OCH3
(3)
RO
RO
N
OCH3
O
CH3
OCH3
RO
RO
N
H
O
OCH3
CH3
OCH3
RO
RO
N+
OCH3
HO
CH3
OCH3
(4)
H
O
O
H
O
N
H
H
H
H
H
H3C
OCH3
OCH3
(1)

Scheme 1

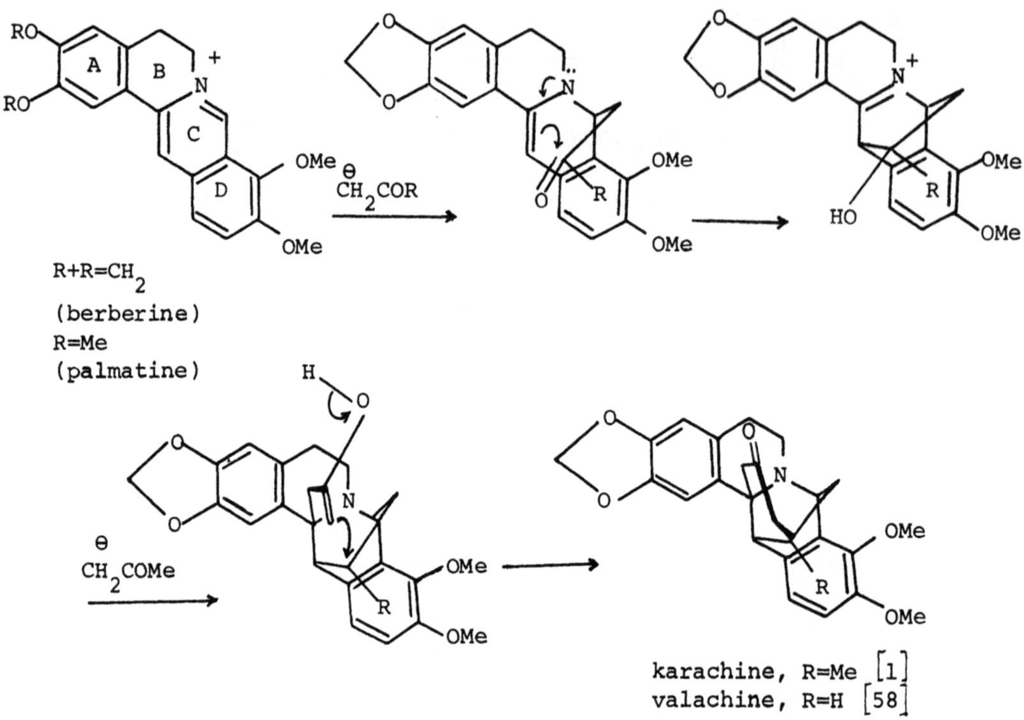

Scheme 2

Isolation and Structural Studies on the Chemical Constituents of Fagonia indica

Fagonia indica Linn. is a small spiny undershrub which is widely distributed in Pakistan. An aqueous decoction of the leaves and young twigs is a popular remedy for cancer in its early stages. A new sapogenin "nahagenin" [5] (3) has been isolated from the hydrolysed extracts of the aerial parts of the plant, and its structure has been elucidated on the basis of a 400 MHz ^{1}H-nmr spectrum, a 100 MHz CMR spectrum and high resolution mass spectrum.

The substance analyzed for $C_{30}H_{48}O_4$ (confirmed by high resolution mass spectrometery, m/z = 472.3740 mass, 472.3552 for $C_{30}H_{48}O_4$). Major peaks in the mass spectrum occurred at m/z 454, 436, 424, 409, 395 and 261. The ir spectrum ($CHCl_3$) showed peaks at 1740 cm^{-1} and 3460 cm^{-1} suggesting a α-lactone and hydroxy groups. The substance readily afforded a diacetate (m/z = 556), but was found to be remarkably inert to attempted hydrolysis of the lactone. The CMR showed no olefinic protons. The CMR recorded on a 100 MHz instrument confirmed the presence of 30 carbons. The carbon atoms in the A and B rings were readily recognised by comparison with corresponding signals of known pentacyclic triterpenoids (45). Eight quaternary centres and 6 methyl groups were also identififed. The CMR displayed a resonance at δ 177.29 for the carbonyl carbon, and 3 resonances at δ 84.72 (s), δ 76.54 (dd) and δ 71.92 (d) for the oxygen-bearing carbons C(20), C(3) and C(23) respectively. On the basis of these spectral data, structure **[5]** was assigned to nahagenin which has been confirmed by an unambiguous structure determination by a single crystal X-ray diffraction analysis carried out by Prof. Clardy and co-workers.

Isolation and Structural Studies on Buxus papilosa

Buxus papilosa (Buxaceae) is a shrub which occurs abundantly in the northern regions of Pakistan. Extracts of Buxus species have been used since ancient times for the treatment of a wide variety of diseases including malaria and venereal diseases. Buxus papilosa has found use in the indigenous system of medicine as a febrifuge, for relief of rheumatism and for the treatment of a number of other ailments. Seven new alkaloids, papilamine **[6]** (4), papilicine **[7]** (5), moenjodaramine **[8]** (6), harappamine **[9]** (7), karachicine **[10]** (8), papilinine **[11]** (9), Buxaminol-G **[59]** (72) and a known alkaloid irehine **[60]** (73) has recently been isolated by us from the leaves of this plant and their structures elucidated on the basis of their spectral data as well as their derivatives.

The uv spectrum of moenjodaramine showed absorption maxima at 207, 237, 245 and 254 nm, characteristic of the presence of a 9(10 - 19) abeo-diene system (46). An identical uv spectrum is encountered in buxamine E, buxaminol E and papilamine (4). The ^{1}H-nmr spectrum ($CDCl_3$) showed 3 singlets, corresponding to the 3 tertiary methyl groups at δ 0.71, δ 0.75 and δ 1.03. The secondary (C-21) methyl group resonated as a doublet at δ 0.88 (J = 6 Hz). A 3-proton singlet resonating at δ 2.1

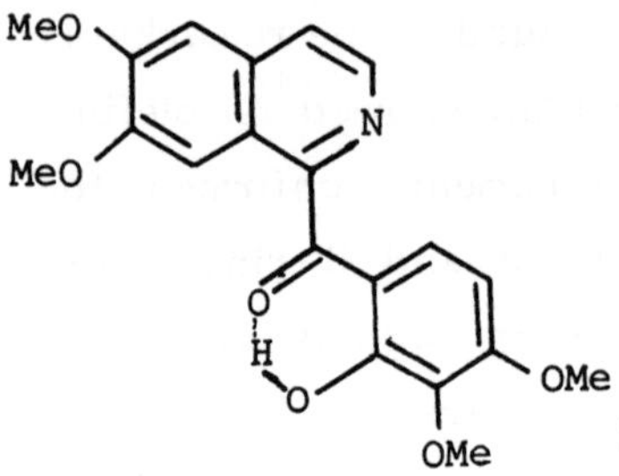

(2)

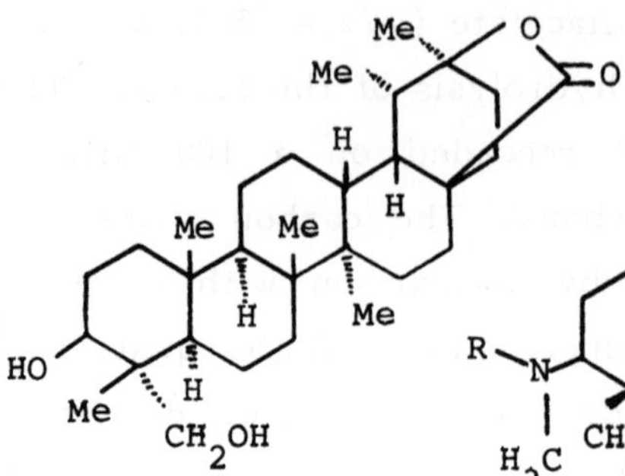

(5)

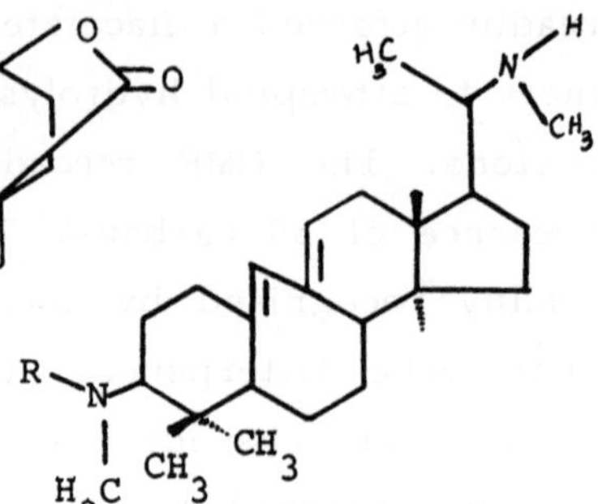

Papilamine (6) R=H
Papilicine (7) R=CH_3

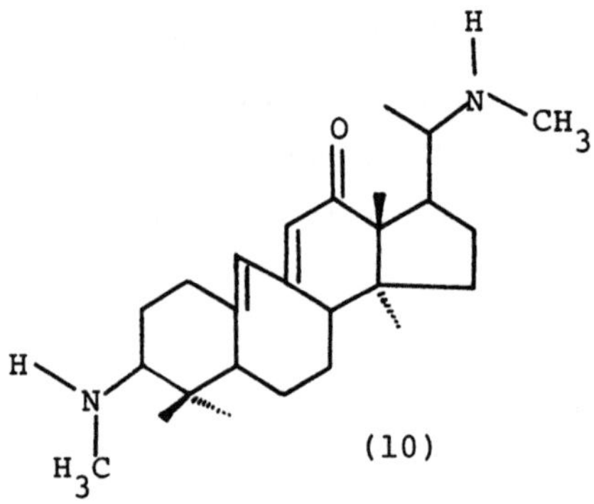

(10)

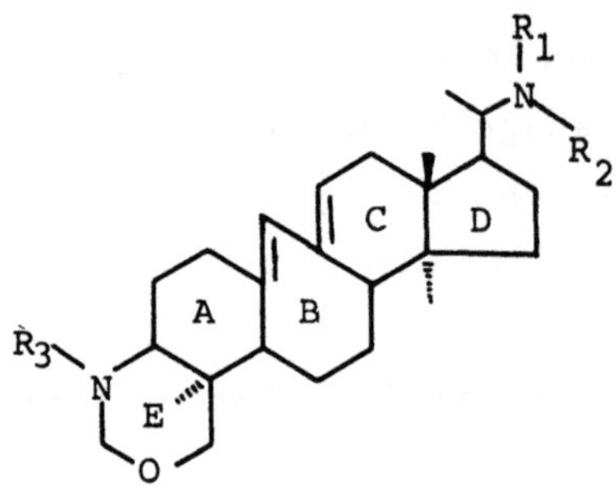

Moenjodaramine (8) $R_1=R_2=R_3=CH_3$

I.R: ν_{max} 1360, 2940 and 1595 cm^{-1}

U.V: λ_{max} 207, 237, 254 nm

Mass: m/e 426.3609 (M^+), 58.0650 (100%)

NMR: δ 0.71 (s,3H), 0.75 (s,3H), 1.03 (s,3H), 0.88 (d,3H)

Harappamine (9) $R_1=H$, $R_2=R_3=CH_3$

I.R: ν_{max} 3400, 2540, 1650 cm^{-1}

U.V: λ_{max} 238, 246

Mass: m/e 412.3454 (M^+), 58.0660 (100%)

NMR: δ 0.68 (s,3H), 0.76 (s,3H), 1.03 (s,3H), 0.72 (d,3H), 5.98 (s,1H), 5.56 (m,1H)

Papilinine (11) $R_1=CH_3$, $R_2=R_3=H$

was assigned to the $-NCH_3$ group, while another peak resonating at δ 2.2 and integrating for 6 protons was assigned to the $-N(CH_3)_2$ group attached to C-20. A set of AB doublets resonating at δ 3.24 and δ 3.82 was assigned to C-29 methylene protons α - to the C-3 nitrogen. A singlet at δ5.55 was ascribed to the isolated olefinic proton at C-19 while a multiplet centred at δ 5.55 was assigned to the C-11 olefinic proton.

The mass spectrum of the compound afforded the molecular ion at m/z = 426.3609 which corresponded to the formula $C_{28}H_{46}N_2O$ (calcd. 426.3609). The substance showed a base peak at m/z 58.0650 corresponding to the composition $C_3H_8N^+$ which suggested the loss of $CH_2N^+(CH_3)_2$ characteristically encountered in alkaloids bearing a $-N(CH_3)_2$ grouping on ring A, and which may be formed in moenjodaramine by intramolecular proton transfer and cleavage. Another peak at m/z 57.0625 corresponded to the fragment $CH_2{=}^+N(CH_2)CH_3$. A peak at m/z 85.0883 was in accordance with the composition $C_5H_{11}N$ (calcd.85.089) which was attributed to $(CH_2)_2CH{=}N^+(CH_3)_2$ formed by the cleavage of ring A along with the side chain. A peak at m/z 72.0810 having the composition $C_4H_{10}N^+$ corresponded to the loss of $CH_3CH{=}N^+(CH_3)_2$ commonly encountered in alkaloids bearing a $-CH(CH_3)-N(CH_3)_2$ grouping on ring D (68). Another peak at m/z 71.0734 having formula $C_4H_9N^+$ was assigned to the fragment $CH_2-CH{=}N^+(CH_3)_2$ formed by cleavage of ring A along with the side chain.

In the light of the above studies, structure **[8]** has been assigned to moenjodaramine. This substance has previously been reported as a synthetic product prepared from desoxy-16-buxidienine C (47) but it has not been isolated. A second alkaloid, harappamine, was similarly established to have structure **[9]**.

Moenjodaramine **[8]** and harappamine **[9]** are the first representative members of a new class of pentacyclic natural products bearing both a tetrahydrooxazine ring and a 9(10→19) abeo-diene system.

(a) Isolation and Structural Studies on the Chemical Constituents of Catharanthus roseus

Studies on the alkaloids of Catharanthus roseus have resulted in the isolation of an alkaloid to which structure **[12]** was originally assigned. The substance afforded a uv spectrum which was typical of a dihydroindole system, showing absorption maxima at 212, 246 and 303 nm and minima at 276, 266 nm. The ir spectrum showed the presence of an ester carbonyl absorption at 1730 cm^{-1} . The mass spectrum was very similar to that reported for vindolinine (39) and 19-epi-vindolinine (55). A high resolution mass measurement on the molecular ion afforded the exact mass to be m/z 336.1837 in agreement with the formula $C_{21}H_{24}N_2O_2$. The CMR spectrum of the alkaloid **[12]** (broad-band and off-resonance) showed interesting similarities to the CMR spectra reported for 19R-vindoline (2), 19S-vindoline and 16-epi-19R-vindolinine. The ester carbonyl carbon resonated at δ 173.47, whereas the methyl of the ester group resonated at δ 52.6 (q). The substance afforded 4 doublets for the tertiary aromatic carbons, and 2 singlets for the 2 quaternary aromatic carbon atoms. A characteristic singlet appeared at δ 81.36 corresponding to the quaternary carbon atom α-to the indoline nitrogen (10).

The ^{1}H-nmr spectrum **[12]** recorded on a 200 MHz instrument showed the presence of a doublet at δ 0.62 (J = 7.4 Hz) which is assigned to the C-18 methyl protons. The proton adjacent to the carbomethoxyl function resonated as a double doublet at δ 3.18 (J_1 = 12.2 Hz, J_2=5.8 Hz). A double-doublet at δ 6.41 was assigned to the olefinic proton at C-15, showing coupling with the vicinal olefinic proton and an allylic coupling with C-3 proton (J_1= 10 Hz, J_2 = 2.8 Hz). The other olefinic proton at C-14 resonated as a doublet of double doublets at δ 5.84 (J_1 = 10 Hz, J_2 = 5.2 Hz, J_3 = 1.8 Hz). The chemical shift of δ 0.62 for the methyl group is consistent with a 19S-configuration as the methyl group of 19S-vindolinine resonates at δ 0.57 while the methyl group in 19R-vindolinine resonates at δ 0.95.

A series of NOEDS measurements established that the alkaloid **[12]** was 19S-vindolinine and the the ^{13}C values reported by other workers were incorrect. Our studies have moreover revealed a remarkable isomerisation of 19S- to 19R-vindolinine on standing in chloroform. A plausible mechanism for this conversion is presented in Scheme 3.

19S-vindolinine, when refluxed in benzene for 3 h in the presence of equimolar amount of lead tetraacetate, was found to be smoothly transformed to 2 faster running products. The major product formed in 70% yield afforded a normal indolic uv spectrum. The ir spectrum (KBr) showed bands at 1655 cm^{-1} and 1730 cm^{-1}, which were assigned to N_b-CHO and $-CO_2CH_3$ groups respectively. The mass spectrum showed M^+ at 352.1783 (calcd. for $C_{21}H_{24}N_2O_3$, 352.1786) and other major peaks at 320, 293, 214, 169 and 154. The ^1H-nmr spectrum ($CDCl_3$) showed resonances at δ 1.23, (3H, d, J=5.6 Hz, C=CH-C$\underline{H}_3$), δ 3.67 (3H, s, OCH_3), δ 5.46 (1H,q, J=5.6 Hz, C=C$\underline{H}$-CH_3), δ 5.7 - 6.1 (2H, m, HC = CH), δ 7.6-6.9 (4H, m, aromatic), δ 8.00 (1H, s, N_b-CHO) δ 8.35 (1H, s, NH). Irradiation at δ 5.46 resulted in the collapse of the methyl group at δ 1.23 to a singlet.

The above spectroscopic data were identical with those for **[16]**, a product previously reported to be formed from 19-iodo-tabersonine on heating with sodium acetate in DMF (40). In order to confirm the structure of the oxidation product, 19S-vindolinine **[12]** was oxidized with iodine under conditions previously described for the oxidation of its diastereoisomer (59). This afforded the corresponding 19-iodo-tabersonine in quantitative yields. Treatment of the latter with sodium acetate in hot DMF afforded [16]. A direct spectroscopic and chromatographic comparison of the product formed by lead tetraacetate oxidation with that prepared from 19-iodotabersonine unambiguously established its structure. A plausible mechanism for the formation of **[16]** is presented in Scheme 4.

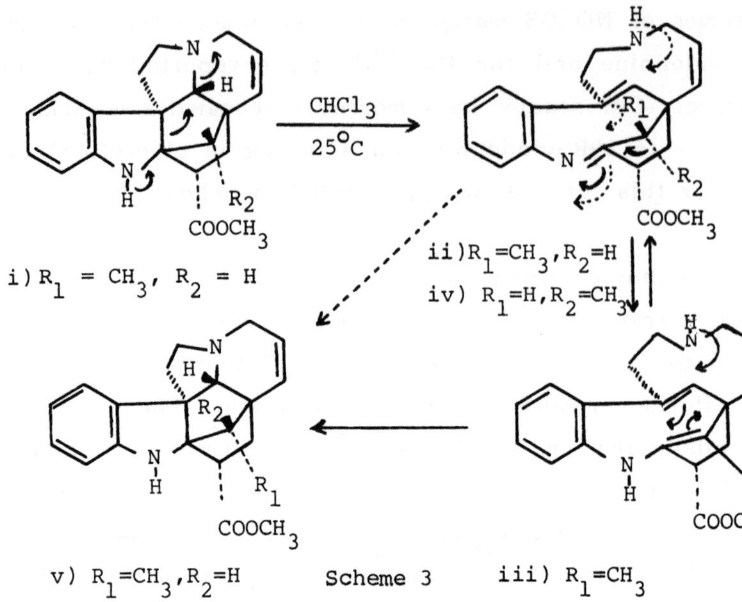

Scheme 3

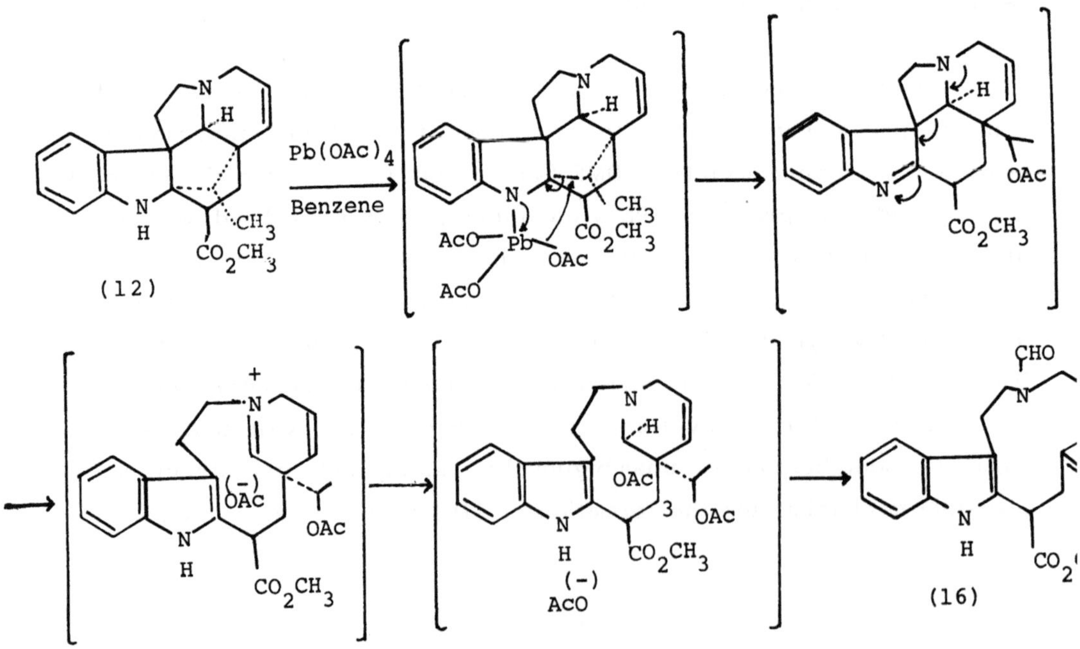

Scheme 4

The facile formation of **[16]** from **[12]** is biogenetically interesting particularly in view of the occurrence of the binary indole alkaloids such as catharine **[18]** (60) in which one of the moieties bears a distinct resemblance to **[16]** & raises the interesting possibility that the indole moiety of catharine may arise by a parallel process occurring in a binary precursor alkaloid such as **[17]** (scheme 5). More recently we have reported a new pseudoindoxyl alkaloid, rosamine **[19]** and its structure has been elucidated on the basis of its spectral data (11). The uv spectrum showed absorption maxima at 228, 255 and 390 nm and minima at 246, 280 and 340 nm, characteristic of the pseudoindoxyl system. The ir spectrum indicated presence of ester carbonyl at 1720 cm^{-1} and conjugated carbonyl group at 1690 cm^{-1}. The mass spectrum showed the molecular ion peak at m/z 352 and the base peak was observed at m/z 135. The prominent peaks in the ^{1}H-nmr spectrum occurred at δ 1.0 (t J= 7.3 Hz), δ 3.19 (s) δ 4.55 (d, J = 4.4 Hz).

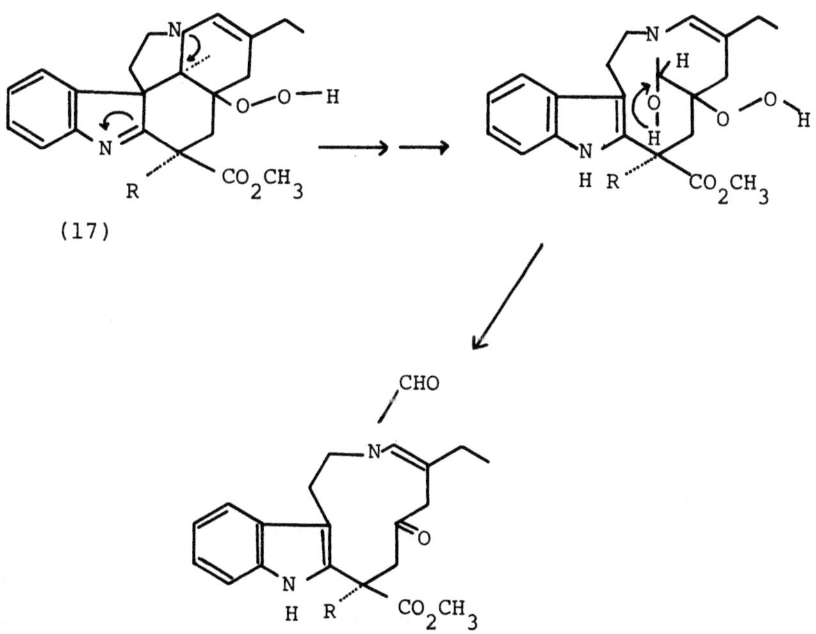

(18) R=10-vindolinyl

Scheme 5

(b) A Rapid Procedure for the Isolation of Catharanthine, Vindoline and Vinblastine

The binary indole alkaloids vinblastine **[20]** and vincristine **[21]** are among the most potent chemotherapeutic agents known to man for the treatment of several different types of cancer including Hodgkins disease, acute leukaemia in children and choriocarcinoma. As both alkaloids occur only in minute traces in the leaves of Catharanthus roseus, they cost several thousand dollars per gram. This has attracted the attention of a number of groups towards developing facile procedures for their isolation and synthesis.

Our earlier efforts in this field (12) have led to 2 different syntheses of vinblastine based on functionalisation of the olefinic bond of catharanthine before (13) (Scheme 6), or after (14) (Scheme 7) (15,16) coupling with vindoline. These syntheses employed a novel modification of the polonovski reaction developed by Potier and co-workers (52). A similar approach to vinblastine has subsequently been reported by the French group (56).

As the above synthetic routes were all based on catharanthine and vindoline as starting materials, it was important to develop an isolation procedure which could afford these alkaloids in bulk without having to resort to extensive chromatography. We have now developed a rapid isolation procedure for the isolation of catharanthine, vindoline and vinblastine. This procedure has been tested and found to be extremely satisfactory both at laboratory and pilot-plant levels. This should produce catharanthine, vindoline and vinblastine much more readily, as well as stimulate further research on the development of new anti-tumour derivatives of these oncolytic alkaloids. The procedure that we have developed (17,18), involves extraction of the alkaloids with pH 3 phosphate buffer, selective precipitations by use of appropriate solvents and selective extractions. For vinblastine, a direct and simple isolation procedure has been developed which does not involve any chromatography, but affords pure vinblastine sulphate in yields of 40 gm per ton of leaves. This is 20 times higher

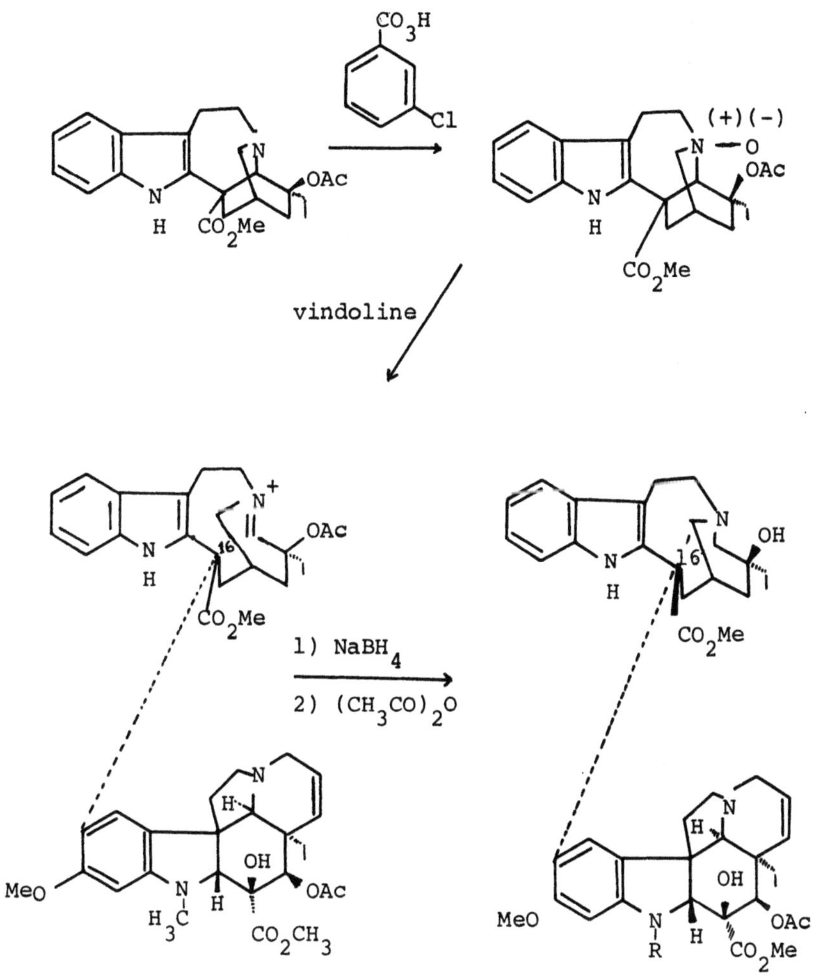

vinblastine (20) R=CH_3
vincristine (21) R=CHO

Scheme 6

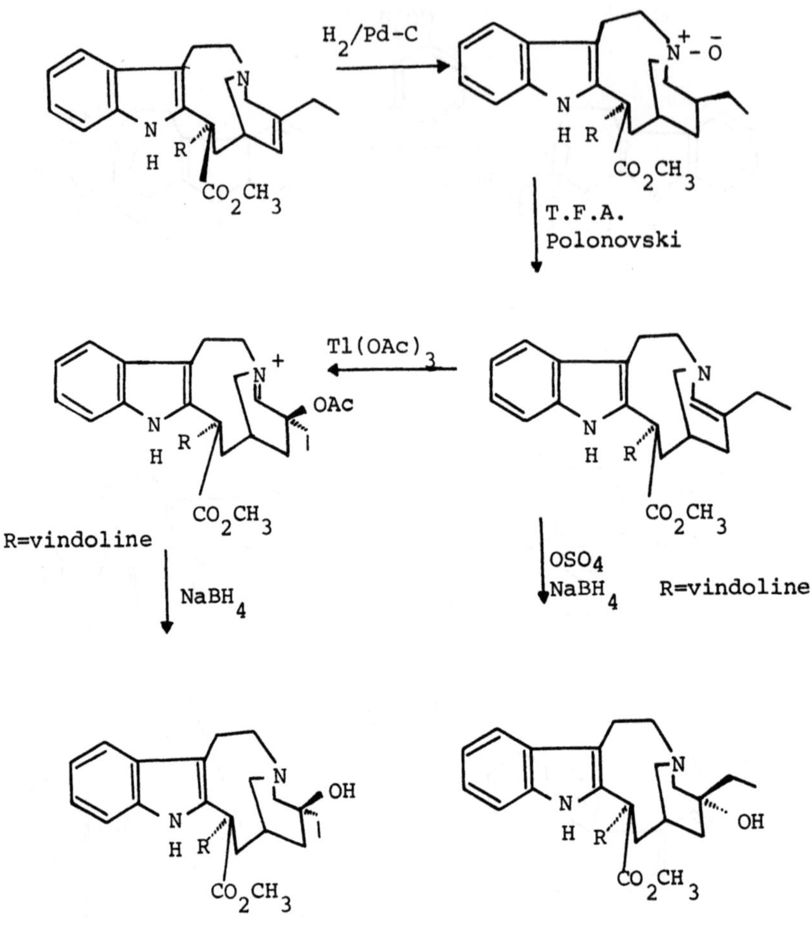

(20) R=vindoline

R=vindoline

Scheme 7

than has been internationally attainable previously (previous yields 2 grams per ton , UNCTAD/GATT booklet on Medicinal Plants, 1982). Pilot plant investigations are being carried out.

Isolation and Structural Studies on the Chemical Constituents of Betula utilis

Betula utilis is a tree commonly found at high altitudes in the temperate Himalayas extending from Chitral eastwards to Azad Kashmir, and in Sikkim and Bhutan. The infusion of its bark has found wide use in indigenous medicine as an antiseptic, carminative, and for hysteria. Our interest in the systematic investigation of the chemical constituents of Pakistani medicinal plants has led us to a chemical investigation of the bark of Betula utilis. This has resulted in the isolation of a new triterpenoid "Karachic acid" **[22]** the structure of which has been solved on the basis of chemical and spectroscopic studies (19).

Isolation and Structural Studies on the Chemical Constituents of Cucumis prophetarum

The isolation of a number of cucurbitacins with cytotoxic properties prompted us to investigate the active principles present in the fruits of Cucumis prophetarum (Cucurbitaceae), a plant locally known as "Choti indrayan" or "Khar indrayan". It is a perennial trailing herb with ellipsoidal echinate fruits. The plant grows wild in various regions of Pakistan, India, Saudi Arabia and tropical Africa. The fruits are used in indigenous medicine as an emetic and purgative. It is known to contain cucurbitacins B and D and traces of cucurbitacins G and H.

As a result of isolation studies carried out on the fruits of this plant, we have isolated a new cucurbitacin, cucurbitacin Q-1, **[23]** which closely resembles cucurbitacins O and P in its structure (20). The cytotoxicity of these cucurbitacins against Eagles KB strain of human carcinoma of the nasopharynx has been demonstrated and it has been shown that the side-chain double bond and tertiary acetate are essential for cytotoxic activity.

Isolation and Structural Studies on Rhazya stricta

Rhazya stricta Decaisne (Apocynaceae) is a small glabrous erect shrub which grows profusely in the north-western region of the Indo-Pakistan subcontinent. It is used by the traditional practitioners as a bitter tonic for sore throat, in fever, in general debility and as a curative for chronic rheumatism. As a result of isolation and structural studies carried out by us, a number of new alkaloids have been isolated and their structures elucidated. These are strictalamine **[24]** (21), rhazinol **[26]** (1), rhazimol **[27]**, strictamine-N-oxide **[28]** (22), rhazicine **[29]** (23), rhazimal **[61]** (74) and strictimine **[30]** (24). While working on the legumes of this plant we have recently reported two indole alkaloids, isovallesiachotamine **[31]** (25) and tetrahydroalstonine **[32]** (26). Their structures have been elucidated on the basis of their spectral data. Space does not permit a detailed discussion.

Another new alkaloid rhazimine **[33]** recently isolated by us has been found to possess a novel skeletal system in which the bond between C-2 and C-3 is broken (22). The alkaloid probably arises in the plant by an intramolecular Polonovski reaction from an ajmaline-type precursor (Scheme 11).

Studies on the Chemical Constituents of Loranthus grewinkii

Siddiqui and co-workers had previously reported a new triterpenoid, "loranthol" from the berries of Loranthus grewinkii, a parasite found widely distributed on pear, apricot and almond trees. The gum from these berries is widely used in the indigenous system of medicine as a general tonic, relaxant and laxative. This triterpenoid has been re-isolated and its structure **[34]** has been elucidated on the basis of chemical and spectroscopic studies (28).

A Formal Total Synthesis of (+)-Vincamine and (+)-Eburnamonine

In view of the reported pharmacological properties of vincamine **[41]** and eburnamonine (67), we have developed a new route to these alkaloids

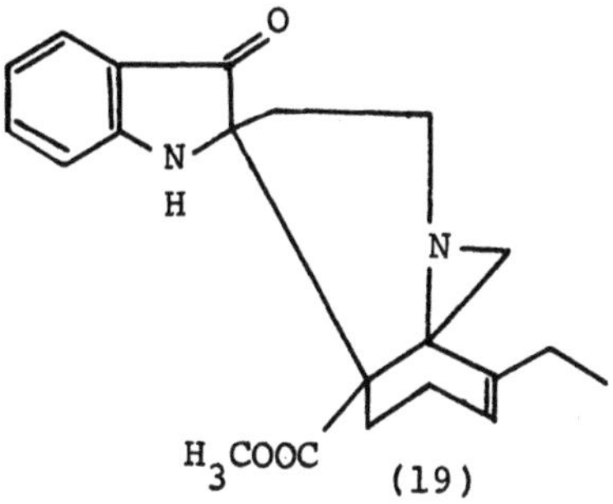

(19)

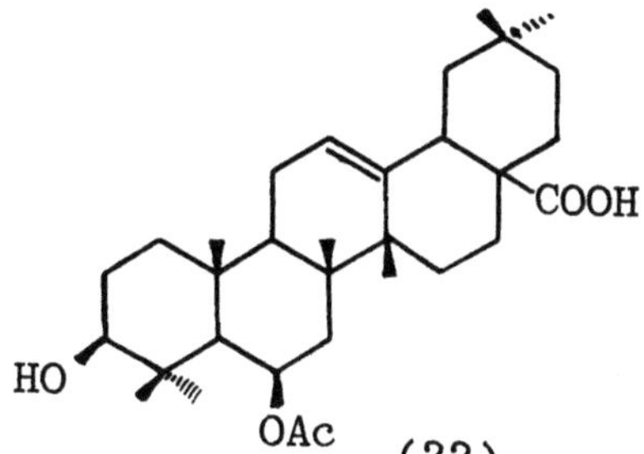

(22)

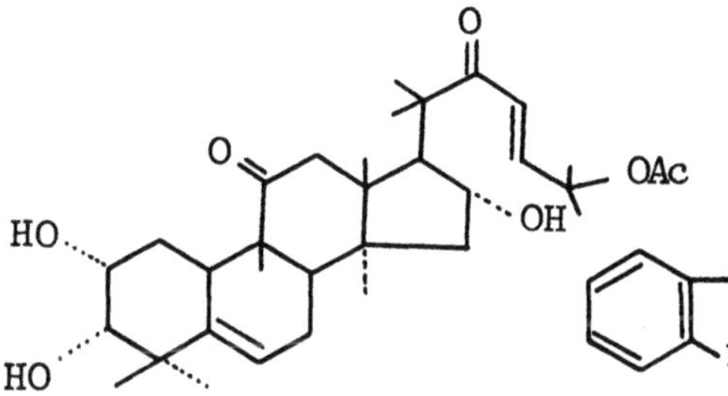

(23)

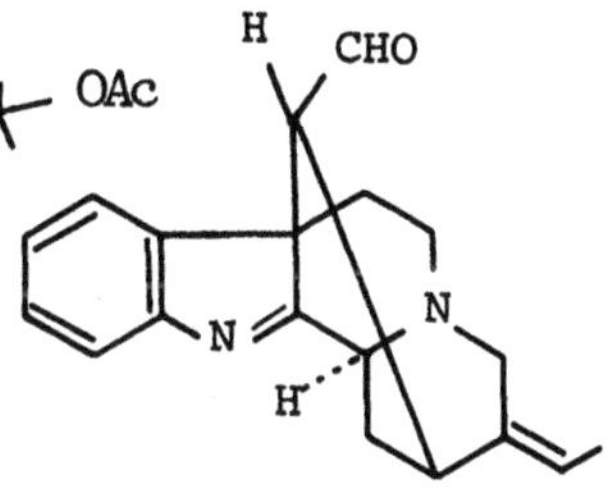

(24)

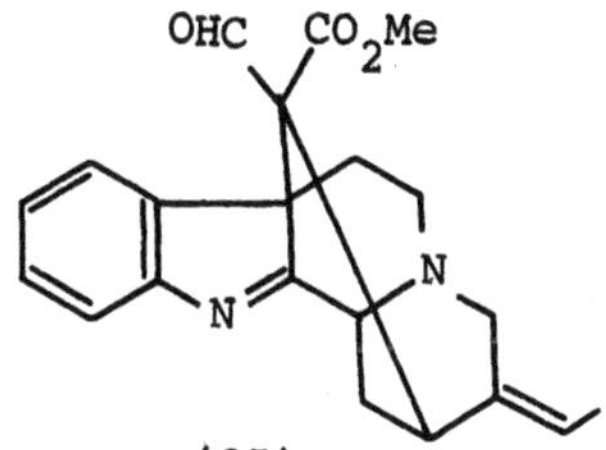

(25)

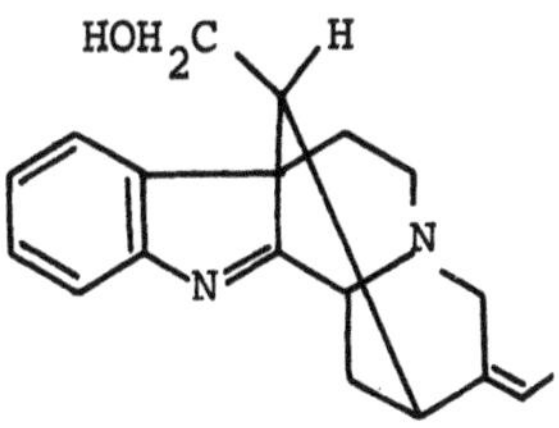

(26)

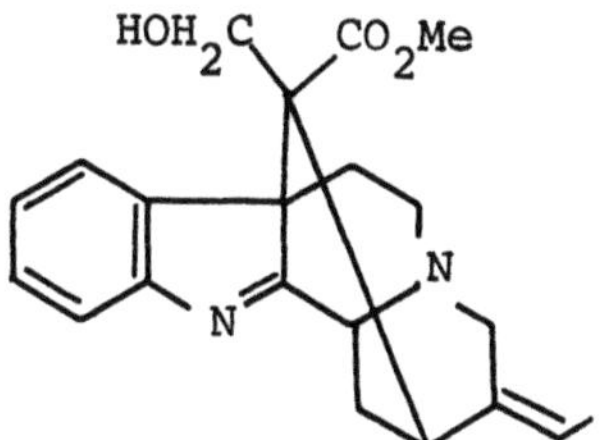

(27)

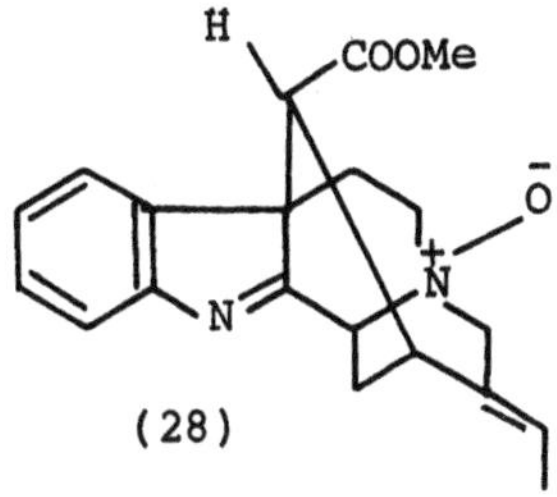

(28)

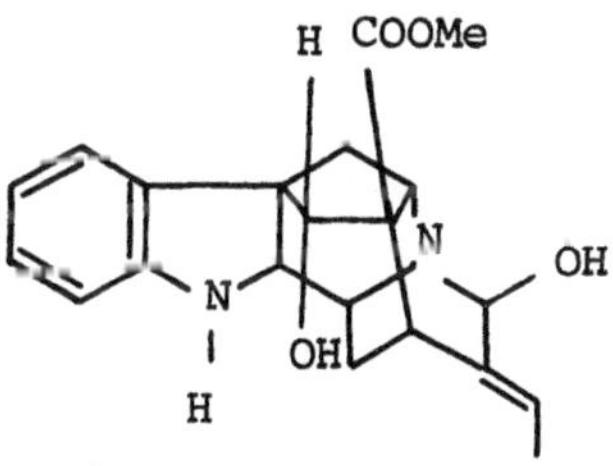

(29)

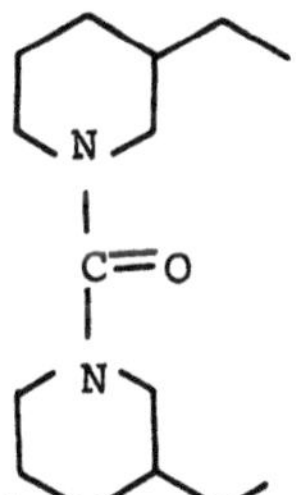

(30)

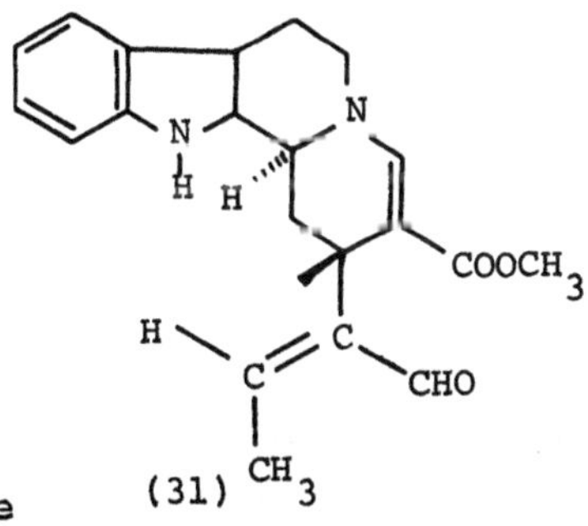

(31)

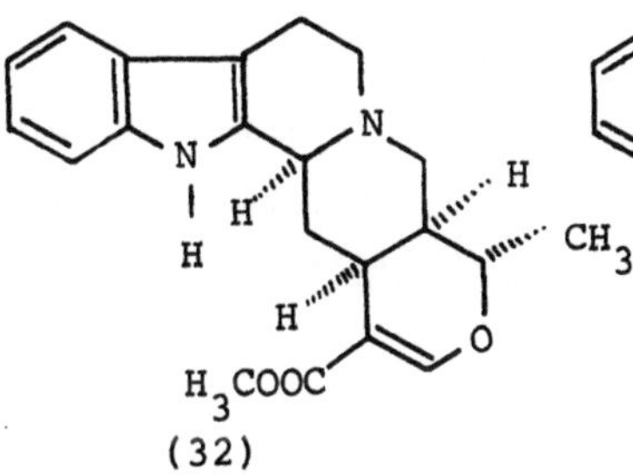

(32)

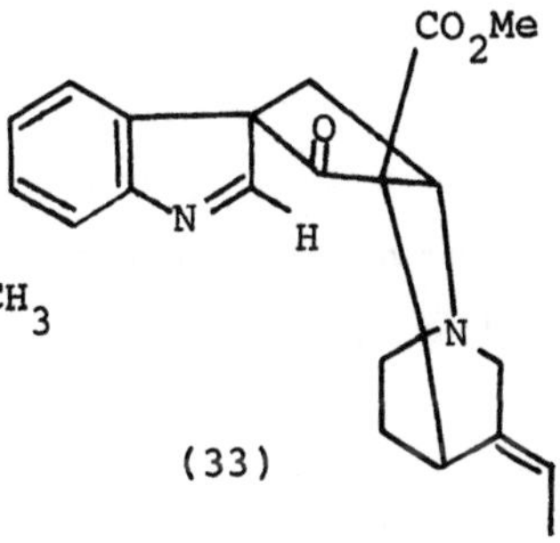

(33)

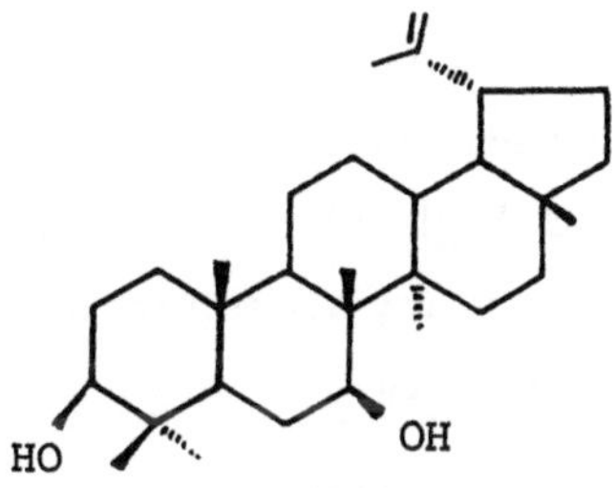

(34)

involving a novel synthesis of the key intermediate **[35]** a compound which contains 4 of the requisite 5 rings and 2 of the 3 chiral centers.

Butyraldehyde and morpholine were condensed by the method of Stork (63) to afford the morpholinoenamine **[36]**. Alkylation of **[36]** with methyl acrylate gave the dialkylated product **[37]** which was hydrolysed with aq. acetic acid to give **[38]** in 45% yield.

Condensation of **[38]** with tryptamine afforded a gum which on purification through a silica column gave a mixture of 2 diastereoisomeric compounds, which could readily be separated by preparative layer chromatography as **[39]** and **[40]** (Scheme 8) each of which crystallised as white needles m.p. 185-186°C and 230°C respectively.

Lithium aluminium hydride reduction of the diastereoisomeric mixture of **[39]** and **[40]** afforded the amine **[35]** in 20% overall yield. **[35]** is convertible to vincamine **[41]** and eburnamonine **[42]** by one of several routes (48,69,43). This thus formally constitutes a total synthesis of these alkaloids (29).

A Total Synthesis of N-Methyl Secodine

The currently accepted biosynthetic route to the indole alkaloids envisages the mediation of 14, 21-dehydrosecodine **[43]** which can undergo intramolecular Diels-Alder reactions in 2 different ways to afford the Iboga alkaloid catharanthine **[44]** or the Aspidosperma alkaloid tabersonine **[45]** (Scheme 9) (70,30). Inspite of intensive efforts by several groups, the synthesis of dehydrosecodine has still not been accomplished because of its high susceptibility to oxidation, dimerization and polymerisation; the synthesis of N-benzyldehydrosecodine has recently been reported (51). Secodine has, however, been synthesised (49,52,61,57) and a number of approaches to the indole alkaloids involving the intermediacy of the secodine system have been studied (50,64). We have recently developed a short and high yield synthesis of N-methyl secodine based on a facile Friedel-Crafts acylation reaction at the indole 2-position (31) which is shown in Scheme 10.

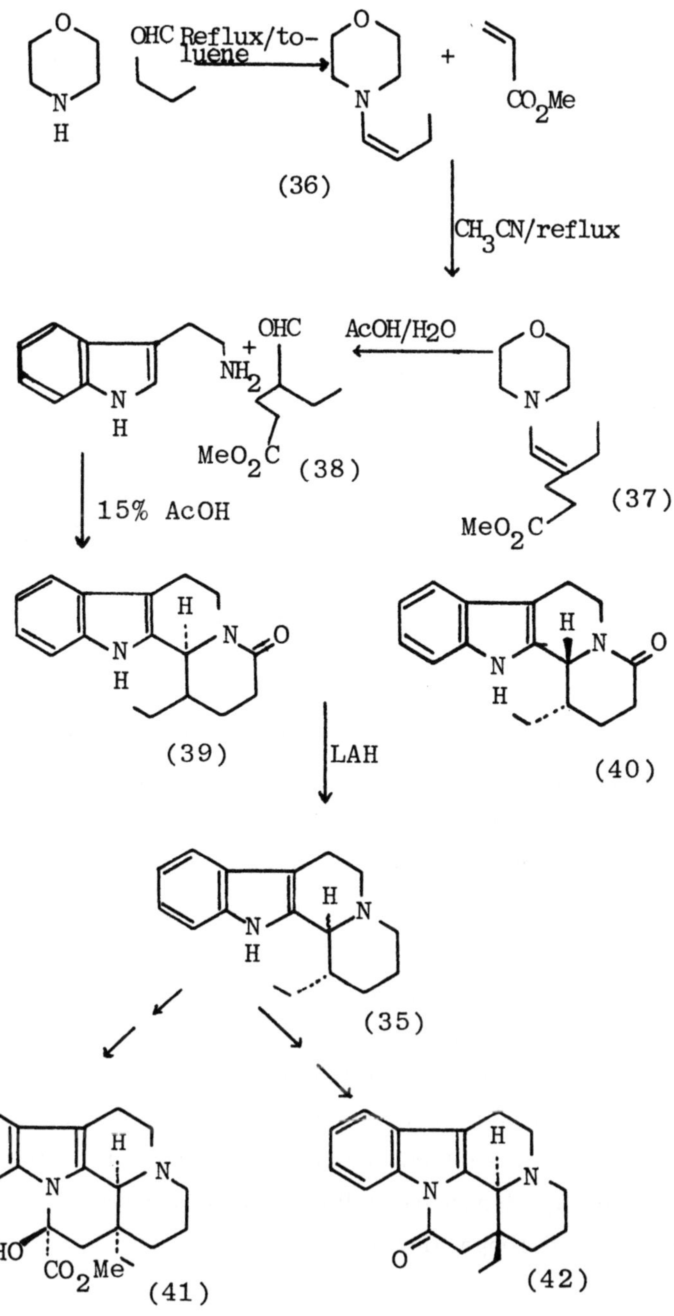
OHC
Reflux/toluene
(36)
CO_2Me
CH_3CN/reflux
AcOH/H2O
OHC
NH_2
MeO_2C
(38)
(37)
15% AcOH
(39)
LAH
(40)
(35)
HO
CO_2Me
(41)
(42)

Scheme 8

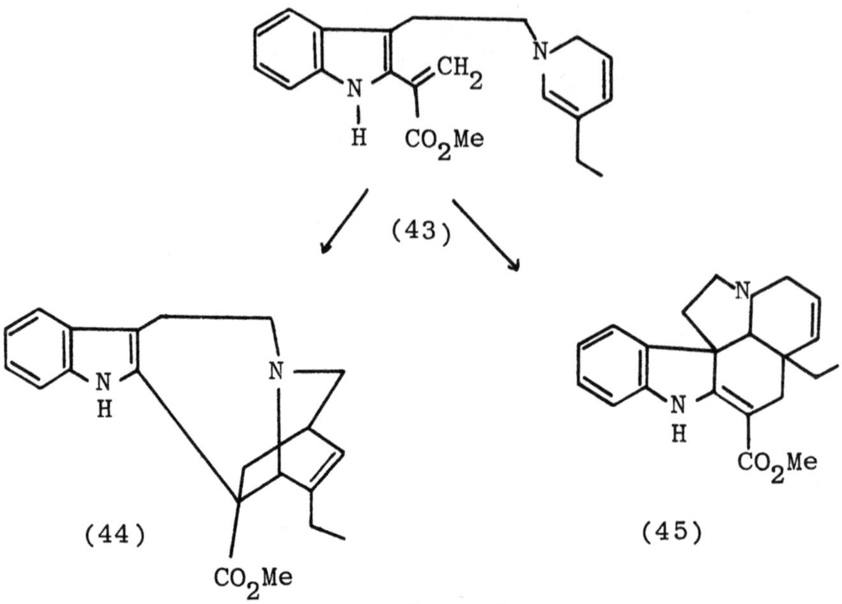

Scheme 9

N-methyl secodine **[49]** when refluxed in acetonitrile for 8 h afforded the 2-hydroxy carbazoles **[51]** as a major product (yield 80%). The facile conversion of the secodine derivative to the carbazole system **[52]** (Scheme 10) suggests the intermediacy of N-methyl dehydrosecodine **[50]** in the reaction which may have been formed through aerial oxidation. The generation of carbazole derivatives has previously been reported (65) and it has been proposed that 2-hydroxy carbazole is formed via dehydro-secodine by an intramolecular rearrangement and hydrogen transfer mechanism. A parallel project aimed at synthesising N-benzylsecodine has also been carried out. This synthesis represents the shortest route reported to date to the secodine system. Attempts are presently underway to generate the corresponding dehydrosecodines for biomimetic transformations to the Aspidosperma and Iboga alkaloids.

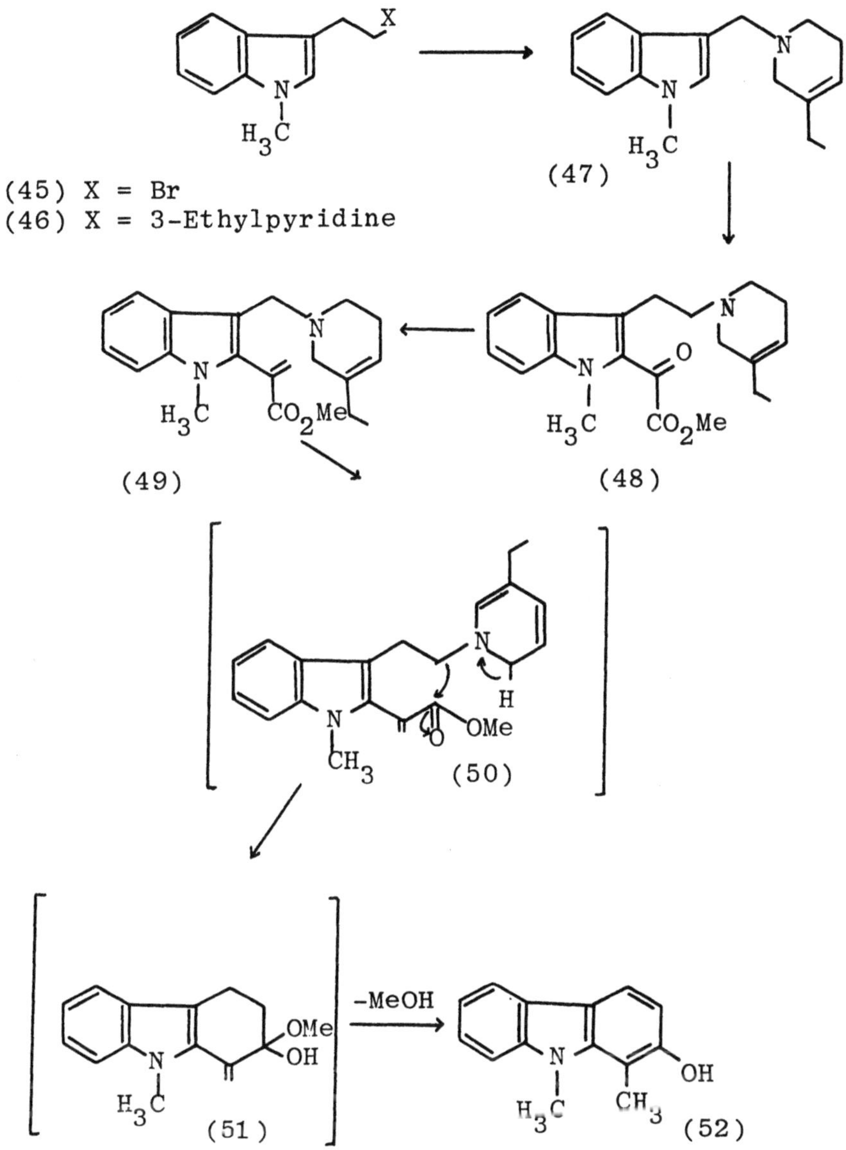
X
N
H3C
(45) X = Br
(46) X = 3-Ethylpyridine
(47)
(48)
O
CO2Me
(49)
OMe
H
CH3
(50)
(51)
OH
-MeOH
(52)

Scheme 10

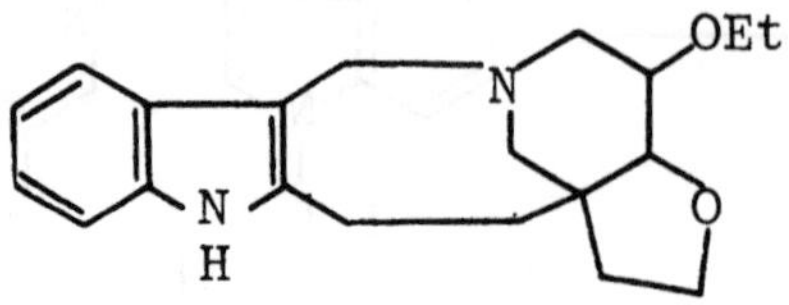

(53)

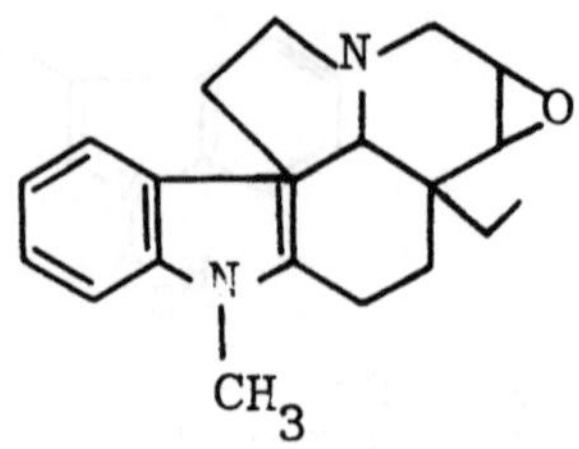

(54)

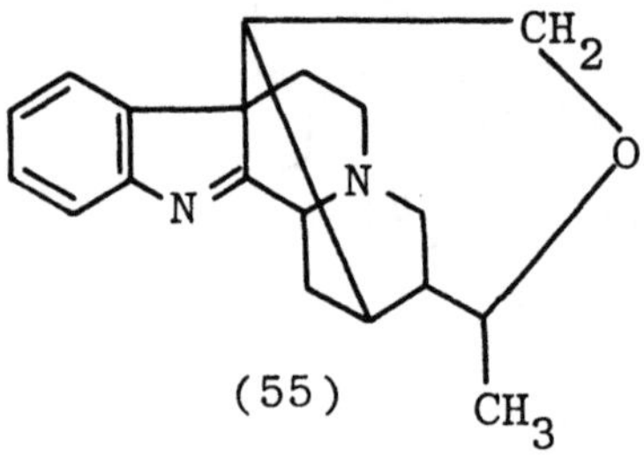

(55)

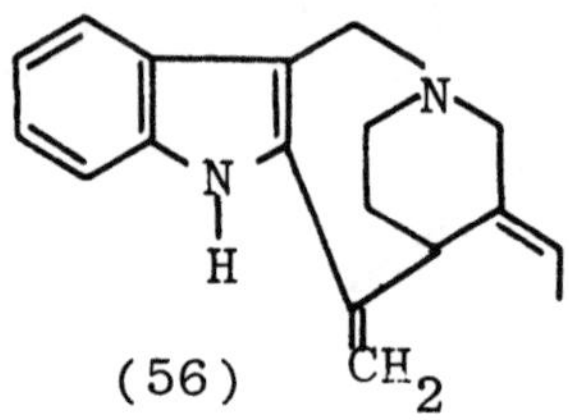

(56)

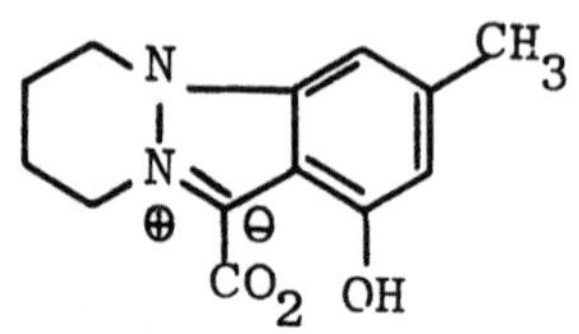

(57)

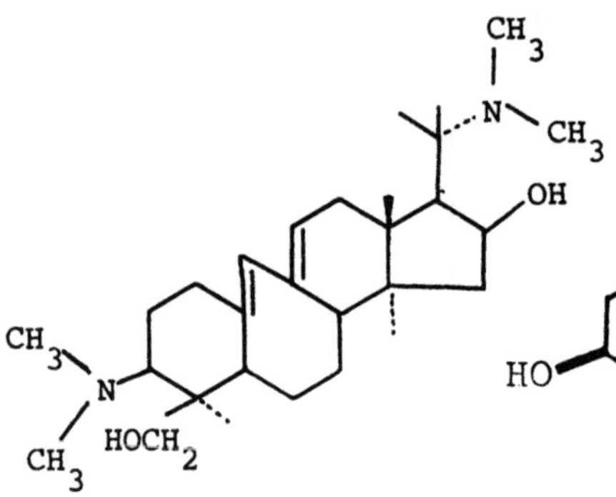

(59)

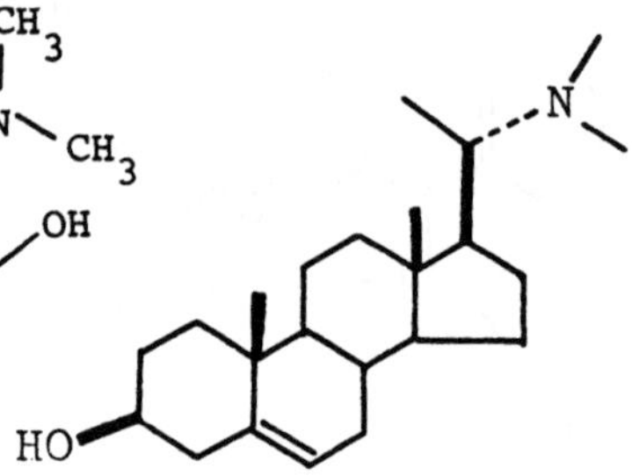

(60)

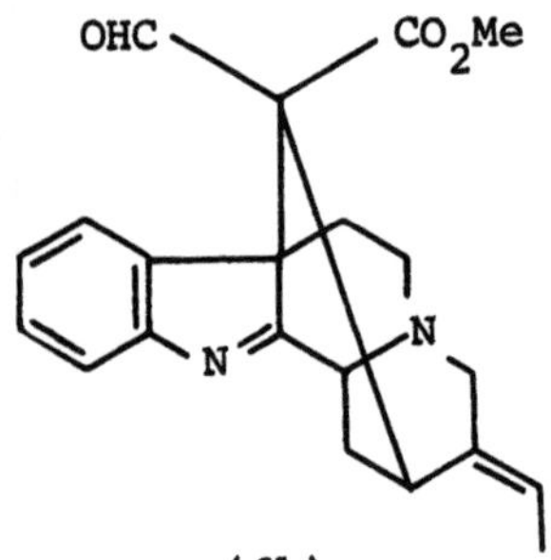

(61)

Isolation and Structural Studies on the Chemical Constituents of Ervatamia coronaria

Ervatamia coronaria (Apocynaceae) is a glabrous evergreen tree commonly grown in the gardens of Pakistan. This indigenous medicinal plant is used for the treatment of ophthalmia, for application on wounds and inflammed parts of the body, as anthelmintic etc. We have recently isolated and identified on the basis of spectral data four alkaloids from this plant namely hyderabadine **[53]** (32), meharanine **[54]** (33) lahoricine **[55]**, (34) and apparicine **[56]** (35).

Isolation and Structural Studies on the Chemical Constituents of Nigella sativa

Nigella sativa Linn. (Ranunculaceae) locally known as kalonji is an indigenous herbaceous plant. It grows in mediterranean countries and is also cultivated in Pakistan. It is used as a spice as well as for the treatment of various diseases.

We have isolated a new alkaloid nigellicine **[57]** (36) from this plant and its structure has been elucidated on the basis of uv,ir, mass, ^{1}H-nmr 2D- ^{1}H-nmr, CMR and by X-ray crystallography.

References

1. Y.Ahmad, K.Fatima, P.W.Le Quesne and Atta-ur-Rahman, **J.Chem.Soc.Pak.,** 1 69 (1979)
2. A.Ahond, M.M. Janot N.Langlois, G.Lukacs, P.Potier, P.Rasoanaino, M. Sangare, N.Neuss, M.Plat, J.Le Men, E.W. Hagaman and E.Wenkert, **J.Am.Chem.Soc.,** 96, 633 (1974)
3. Atta-ur-Rahman, A.A.Ansari and J.Clardy, **Heterocycles,** 19, 217 (1982)
4. Atta-ur-Rahman, S.Farhi, G.A. Miana, M.Nisa and W.Voelter, **Z.Naturforsch.**, (in press)
5. Atta-ur-Rahman, M.Nisa and T.Zamir, **Z.Naturforsch.**, 39b (1),127 (1984)
6. Atta-ur-Rahman and M.Nisa, **Planta Medica** 49(2),126 (1983)
7. Atta-ur-Rahman and M.Nisa, **Heterocycles,** 20(1), 69 (1983)
8. Atta-ur-Rahman and M.Nisa, **Z.Naturforsch.**, 39b, 839 (1984)
9. Atta-ur-Rahman and M.Nisa, **Z.Naturforsch.**, (in press)
10. Atta-ur-Rahman, M.Bashir, S.Kaleem and T.Fatima, **Phytochemistry,** 22, 1021 (1983)
11. Atta-ur-Rahman, M.Bashir, I.Ali and M.I.Choudhary, **Z.Naturforsch.**, 39(9) 1292 (1984)
12. Atta-ur-Rahman and J.H. Mason, **Tetrahedron,** 36, 1057 (1980)
13. Atta-ur-Rahman, A.Basha and M.Ghazala, **Tetrahedron Letters.**, 2351 (1976)

14. Atta-ur-Rahman, **Pakistan Patent No. 126852**, dated 14-2-1978
15. Atta-ur-Rahman, **12th Int.Symp.on Chem.Nat.Prod.(IUPAC)**, C44, 245, (1980)
16. Atta-ur-Rahman, **Proc.Asian Symp.Med.Plants and Spices**, 1, 222; (1981); **Chem.Abstr.**, 95, 98083 p.
17. Atta-ur-Rahman, M.Bashir, K.Fatima and A.H.Mistry, **Pakistan Patent**, application No. 1
18. Atta-ur-Rahman, M.Bashir, M.Hafeez, N.Perveen, J.Fatima and A.N.Mistry, **Planta Medica, 47**, 246 (1983)
19. Atta-ur-Rahman and M.A.Khan, **Phytochemistry**, 14, 789 (1975)
20. Atta-ur-Rahman, V.U.Ahmad, M.A. Khan and F.Zehra, **Phytochemistry**, 12, 2741 (1973)
21. Atta-ur-Rahman, Y.Ahmad, K.Fatima, J.L. Occolowitz, B.A. Solheim, J.Clardy R.L. Garnick, and P.W.Le Quesne, **J.Am.Chem.Soc.**, 99, 1943 (1977)
22. Atta-ur-Rahman and S.Khanum, **Phytochemistry**, 23(3) 709 (1984)
23. Atta-ur-Rahman and S.Khanum, **Heterocycles**, 22(10),2183 (1984)
24. Atta-ur-Rahman and K.Zaman, **Heterocycles**, 22(9),2023 (1984)
25. Atta-ur-Rahman and S.Malik, **J.Nat.Prod. (Lloydia)**, 47, 388 (1984)
26. Atta-ur-Rahman and S.Malik, **Planta Medica.**, 50,283 (1984)
27. Atta-ur-Rahman and S.Khanum, **Tetrahedron Letters** 25(35),3913 (1984)
28. Atta-ur-Rahman, M.A.Khan and N.H.Khan, **Phytochemistry**, 12, 3004 (1973)
29. Atta-ur-Rahman and M.Sultana, **Z.Naturforsch.**, 37b, 793 (1982)
30. Atta-ur-Rahman and A.Bash, **"Biosynthesis of Indole Alkaloids", International Series of Monographs in Chemistry"**, Oxford University Press, England (1983)
31. Atta-ur-Rahman, M.Sultana and I. Hasan, **Tetrahedron Letters**, 24, 1845 (1983)
32. Atta-ur-Rahman and N.Daulatabadi, **Z.Naturforsch.**, 38b, 1310 (1983)
33. Atta-ur-Rahman, A.Muzaffar, N.Doulatabadi, **Z.Naturforsch.**, 38b, 1700 (1983)
34. Atta-ur-Rahman, N.Doulatabadi and A.Muzaffar, **Z.Naturforsch.**, **Z.Naturforsch.**, 39(9), 1289 (1984)

35. Atta-ur-Rahman, N.Doulatabadi and A.Muzaffar, **J.Nat.Prod.**, (accepted for publication)

36. Atta-ur-Rahman, S.Malik, H.Cun-heng and J.Clardy (submitted for publication)

37. G.Blasko, N. Murugesan, A.J. Freyer, M.Shamma, A.A.Ansari and Atta-ur-Rahman, **J.Am.Chem.Soc.**, 104, 2039 (1982)

38. G.Blasko, M.Shamma, A.A.Ansari and Atta-ur-Rahman,**Heterocycles**, 19, 257 (1982)

39. C.Djerassi, S.E. Flores, H. Budzikiewicz, J.M. Wilson, L.J. Durham, J.Le. Men, M.M. Janot, M. Plat, M. Gorman and N.Neuss, **Proc.Nat. Acad.Sci.**, 84, 113 (1962)

40. L.Diatta, R.Z. Andriamialisoa, N. Langlois and P.Potier, **Tetrahedron**, 32, 2839 (1976)

41. S.Firdous, A.J. Freyer, M.Shamma, Atta-ur-Rahman and A. Urzua, **J.Chem.Soc. Chem.Comm.**, (20) 137, (1984)

42. T.R. Govindachari, B.R. Pai, S.Rajerwari, S.Natarajan, S. Chandrasekaran, M.S. Premila, R. Charubala, K. Venkatesan, M.M. Bhadbahade, K. Nagarajan and W.J. Richten, **Heterocycles**, 15, 1463 (1980)

43. J.L. Hermann, R.J. Gregge, J.E. Richman, G.R. Kieczykrwski, S.N. Normandin, M.L. Quesada, C.L. Semmelhack, A.J. Poss and R.H. Schlessinger, **J.Am.Chem.Soc.**, 101, 1540 (1979)

44. M.M.Janot and R.Goutarel, **Bull.Soc.Chim.Fr.**, 2234 (1962)

45. S.A.Knight,**Organic Magnetic Resonance**, 6, 603 (1974)

46. H.Khoung, D.H. Genlier, M.M.Q.K.Huu, E.Stanislas and R.Goutarel, **Tetrahedron**, 22, 3321 (1966)

47. H.F. Khoung, R.Paris, R.Razafindrambao, A.Cava and R.Goutarel, **C.R.Acad.Sci.Pari.**, 273, 558 (1971)

48. M.E. Kuehne, **J.Nat.Prod.(Lloydia)**, 27, 435 (1964)

49. M.E. Kuehne, D.M. Roland, and R.J. Hafter, **J.Org.Chem.**, 43, 3705 (1978)

50. M.E. Kuehne, T.H. Matsko, J.C. Bohnert and C.H. Kirkemo, **J.Org.Chem.**, 44, 1063 (1979)

51. J.P.Kutney, Y. Kartoon, N.Kawammra and B.R. Worth, **Can.J.Chem.**, 60, 1269 (1982)

52. J.P.Kutney, R.Badger, J.F. Beck, H. Basshardt, F.S. Mantough, V.E. Ridaura-Sanz, Y.H. So, R.S. Sood and B.R. Worth, **Can.J.Chem.**, 57, 289 (1979)

53. N.Langlois, F.Gueritte, Y. Langlois and P.Potier, **J.Am.Chem.Soc.**, 98, 7017 (1976)

54. N.Murugesan and M.Shamma, **Tetrahedron Lett.**, 4521 (1979)

55. H.Mehri, M.Koch, M.Plat and P.Potier, **Ann.Pharm.Fr.**, 30, 643 (1972)

56. P.Manganey, R.Z. Andriamialisoa, N.Langlois, Y. Langlois and P.Potier, **J.Am.Chem.Soc.**,101, 2243 (1979)

57. C.Marazano, J.L. Forrey and B.C. Das, **J.C.S.Chem.Comm.**, 21, 742 (1977)

58. P.Rasoanaivo, N.Langlois and P.Potier, **Tetrahedron Letters**, 3369 (1974)

59. P.Rasoanaivo, A.Ahond, J.P. Cosson, N.Langlois, P.Potier, J.Guilhem, A.Duerind, C.Riche and C.Pascard, **C.R.Acad.Sci.**, 79, 279C (1974)

60. P.Rasoanaivo, N. Langlois and P.Potier, **Tetrahedron Letters.**, 42, 3669 (1974)

61. S.Raucher, S.E. MacDonald and R.F. Lawrence, **J.Am.Chem.Soc.**, 103, 2419 (1981)

62. M.Shamma and J.L. Moniot, **"Isoquinoline Alkaloids Research"** pp. 287-288, Plenum Press, New York (1978)

63. G.Stork, A.Brizzolara, H.Landesmann, J.Szmuskovics and R.Terrel, **J.Am.Chem.Soc.**, 85, 207 (1963)

64. A.I.Scoot,P.C.Cherry and C.C.Wei, **Tetrahedron**, 30, 3013 (1974)

65. A.I.Scott and P.C.Cherry, **J.Am.Chem.Soc.**, 91, 5872 (1969)

66. R.V.Stevens and J.R. Pruitt, **J.Am.Chem.Soc.**, 91, 5872 (1969)

67. W.I.Taylor and N.R.Farnsworth, **"The Vinca alkaloids"**, Dekker, New York (1979)

68. G.R.Waller and O.C. Dermer, **"Biochemical Applications of Mass Spectrometry"**, pp. 83, John Wiley & Sons. New York (1980)

69. E.Wenkert and B.Wickberg, **J.Am.Chem.Soc.**, 87, 1580 (1965)

70. E.Wenkert, **J.Am.Chem.Soc.**, 84, 98 (1962)

71. W.N.Wu., J.H. Beal and R.W.Doskotch, **J.Nat.Prod. (Lloydia)**, 43, 143 (1980)

72. Atta-ur-Rahman, Mehrun Nisa and Kishwar Jahan, **Phytochemistry**; (accepted for publication)

73. Atta-ur-Rahman, Mehrun Nisa and J.Clardy, **J.Nat.Prod.** (Lloydia) (submitted for publication)

74. Atta-ur-Rahman, Y.Ahmad, K.Fatima, J.L.Occolowitz, B.A. Solheim, J.Clardy, R.L.Garnick, P.W. Le Quesne, . **J.Am.Chem.Soc.**, 99, 1943 (1977)

A Chemical Power of Movement in Higher Plants

Hermann Schildknecht

Organisch-Chemisches Institut der Universität Heidelberg, Im Neuenheimer Feld 270, D-6900 Heidelberg, FRG

1. Phytodynamic

In 1880 appeared the "Power of Movement in Plants" as one of the later volumes of the collected works of Charles Darwin. Five years before his death in 1882, Darwin started a series of painstakingly planned experiments with about a dozen genera from completely different plant families on the circumnutation. His own discovery of the principle of evolution compelled him to do so, and his main conclusion was therefore:

> "It is hardly possible to doubt that plants must derive some great advantage from this remarkable powers of movements"

which may be favourable to the survival of the species.

1.1 Plant Mimicry

Sensitive plants were much discussed by philosophers, like mimosa even in the pre-Christian era. However, only recently we were able to see a clear defense reaction in the fact that at the slightest touch the pairs of pinnules folded together, first the pinnae and then the whole leaf pressing close to the stem and the whole plant pressing itself to the ground. We agree with Hassenstein that this behavior represents a beautiful example of plant mimicry.

Natural Product Chemistry
Ed. by Atta-ur-Rahman

1.2 Plant Sleep

Androsthenes, the admiral of Alexander the Great, described very early the plant sleep in the case of Tamarindus indica L.

In his "Plant Geography" the Greek philosopher Theophrastus already compared the phenomenon of "plant sleep" with the phenomenon of movement of the sensitive plant, and thus unknowingly presented mimosa as the index plant for all following research works - also on the chemical power of movements in plants.

1.3 The Physiological Clock

Analysis of leaf movements in the early eighteenth century provided the first clue that organisms have an "internal clock" (De Mairan 1729 - Zinn 1759) and subsequent studies from Pfeffer (1873) and Darwin (1881) are providing useful information of periodic nyctitropic movements of leguminoseae leaflets.

2. Turgoractive Factors

In volume II of Pfeffer's Plant Physiology one can read that "it is presumed precise chemical stimuli that plays a prominent role in the autonomous movement". Here reference is made to chemistry, the material basis of all stimulation processes.

2.1 Bioassay of the Active Principle of Seismonasty and Nyctinasty

When a leaf is separated from the mimosa plant, and, after restoration, is placed in a solution of the supposed active principle, the sap stream transports the phytohormone to the active site, i.e. to the pulvini, where the biomembranes respond so strongly that each pair of pinnules rapidly folds up through visible turgor changes until the whole leaf has assumed its typical irritated posture.

The primary as well as the tertiary pulvini are irritated by the active substances. The response of the primary pulvinus is manifested by a drooping of the whole leaf stalk.

We have also observed that the pulvini usually react one after the other corresponding to their position and the acropetal progress of the chemonastic stimulus.

The reaction behavior induced by the chemonastic stimulus of a mimosa crude extract can be demonstrated as a function of its concentration in a number of tests. Despite individual variation, which can be always observed in the bioassay, a clear decrease in the reaction time with declining concentrations can be seen.

When calculating the average values of the time of the first reaction movement and plotting them as a function of each concentration a hyperbola becomes evident.

2.2 The First Irritants from Mimosa pudica L.

In order to elucidate the complex mode of the stimulus response in the Mimosaceae, we started in 1970 with the study of free amino acids by using different chromatographic methods.

As some of these amino acids are active according to Fitting, we determined the amino acids quantitatively in fresh plants. Proving Fittings's hypothesis, we measured the stimulating effect of some amino acids and some of their mixtures.

It is remarkable that the extracts free of amino acids cause a disciplined behavior of the pinnules, which close exactly one after the other.

Therefore we had to look for further leaf movement factors not belonging to the amino acids.

2.3 Periodic Leaf Movement Factors (PLMF)

When a plant with pinnate leaves exhibits the above-mentioned "sleep movement", and its pinnules are joined to the stalk by pulvini, the leaf extract will most certainly react chemonastically.

Acacia karroo is not sensitive, however, it also folds up its pinnae at night, when it looks like a stimulated mimosa.

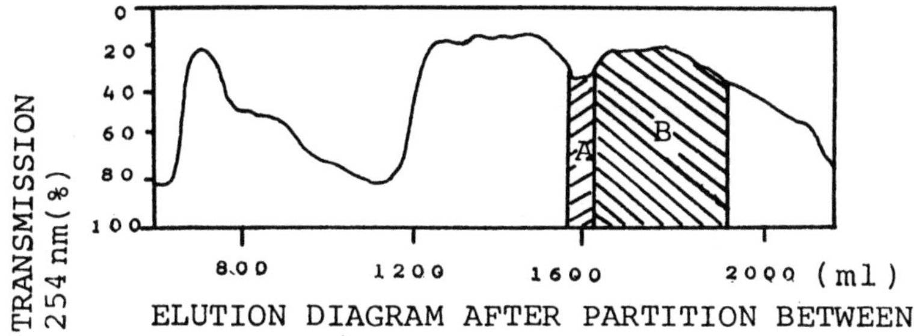

ELUTION DIAGRAM AFTER PARTITION BETWEEN n-BUTANOL/WATER ON BIOGEL P-2
ELUENT: H_2O

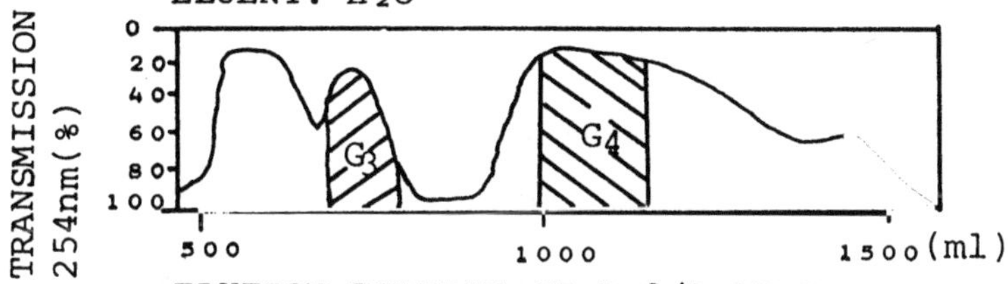

ELUTION DIAGRAM OF P-2/B ON SEPHADEX G-10
ELUENT: $H_2O(G_3,G_4)$

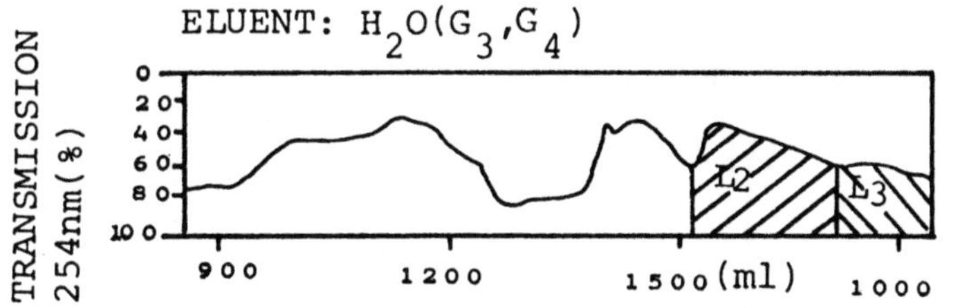

ELUTION DIAGRAM OF G_4 ON SEPHADEX LH-20
ELUENT: MEOH/H_2O 96:4 (L_2,L_3)

ELUTION DIAGRAM OF THE ACTIVE PEAK L_3 ON SEPHADEX LH-20 ELUENT: ETOH/H_2O 90:10

Fig. 1. Isolation of PLMF 1

For the isolation and purification (Fig. 1) of the hitherto unknown PLMF a dried water extract of <u>Acacia karroo</u> was dissolved in water, a methanol precipitation was carried out and the filtrate was subjected to liquid extraction with n-butanol and water. The concentrated water extract was chromatographed on polyacrylamid after it has been filtrated through the same material.

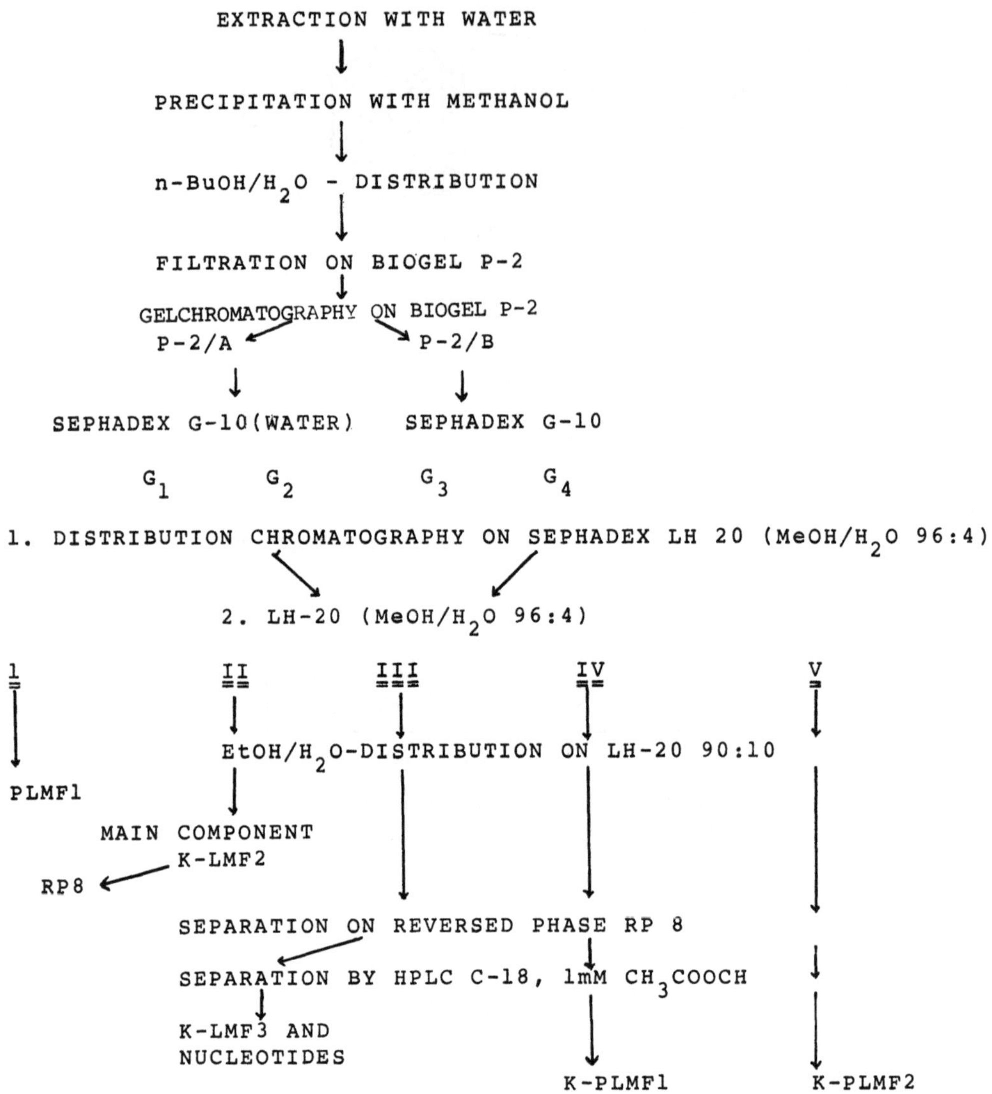

Scheme 1. Separation scheme of PLMFs

Separation from amino acids was achieved by filtration with Sephadex. The final separation was accomplished by partition chromatography on Sephadex LH 20. The active material was eluted with the last fraction, where we found 2'.3' cyclic GMP and AMP. The cyclic nucleotides were characterized by UV, IR, ^{1}H-NMR and ^{13}C-NMR spectra.

Two further nucleotides with nearly the same chromatographic behavior were identified as 3' AMP and 3' GMP. These open nucleotides are most probably no artefacts.

NUCLEOTIDES	AUTHENTIC SUBSTANCE t_r [MN]	M.PUDICA	A.KARROO
2',3'-cAMP	10,3	10,4	10,3
3'-AMP	8,6	8,7	8,7
2',3'-cGMP	4,25	4,3	4,3
3'-GMP	3,6	3,65	3,6

25 cm NUCLEOSIL C-18, 5µ,8mm, 1mM HOAc, 2400 PSI, 5ml/min, UV: 254 nm.

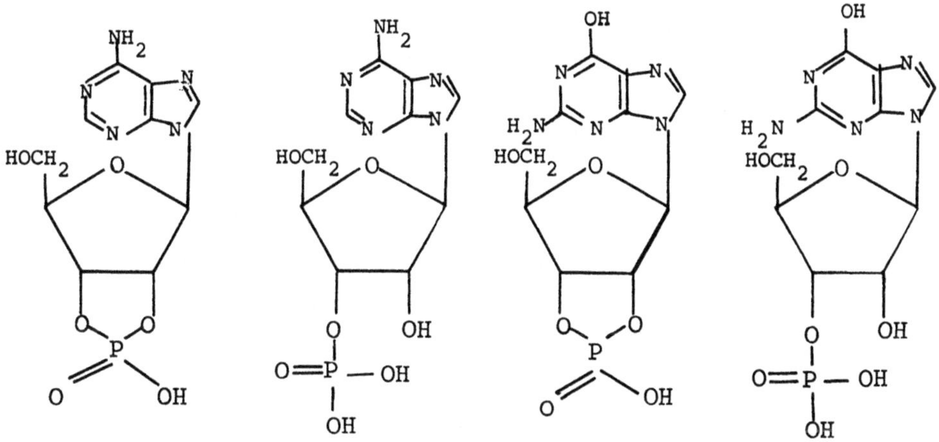

Fig. 2. Comparison of retention time (HPLC) of nucleotides from active fractions of Mimosa pudica and Acacia karroo

From the last highly active fractions of chromatography we isolated by HPLC on reversed phase two leaf movement factors PLMF 1 and 2.

2.3.1 PLMF 1

In the preliminary spectroscopic tests the UV spectrum displayed three absorption maxima at 217, 245 and 285 nm, which indicate a derivative of gallic acid. The mass spectrum with ions at m/z 170, 153, 126, 107 and 79 confirmed this suggestion.

The appearance of m/z 63,9609, which gave evidence for a SO_2 group, was especially important. We also found in the IR spectrum a S=O stretching vibration at 1248 cm^{-1} and an absorption at 818 cm^{-1}, which is typical for sulfates of primary alcohols.

We observed in the mass spectrum of the trimethyl-silylated active compound from Acacia karroo, which was carried out with accurate mass measurement, that the covalently bonded sulfate was displaced by the silylation reagent.

The covalently bonded sulfate was definitely proved by the negative ion fast atomic bombardment (FAB) mass spectrum. We used a beam of highly accelerated Xenon atoms as agent for sputtering ionization and embedded the compound in a matrix of glycerol. The spectrum exhibited the key fragments m/z 411, 241, 169 and 97, which correspond to the ions of gallic acid glucoside sulfate minus one, to gallic acid glucoside sulfate minus gallic acid, to gallic acid minus one and to HSO_4^-.

In the ^{1}H-NMR spectrum of the PLMF 1 of Acacia karroo the signal at 7.12 ppm corresponds to the resonance of the two ring protons H-2 and H-6 of gallic acid. The relation existing between the intensity of the aromatic and the aliphatic protons suggests a hexose, which was identified as glucose by GC-MS analysis of the trimethylsilylated products of the hydrolysis.

Thus, the signal at 5.05 ppm with a coupling constant of 7,5 Hz can be identified as the resonance of the anomeric proton of a D-glucopyranose with ß-glucoside linkage. Since we found only one singulet at 7.12 ppm, we had to assume a gallic acid with symmetric substitution. An ester linkage was improbable, because the two ring protons H-2 and H-6 of gallic acid showed a chemical shift of 7.3 ppm. Therefore linkage via the OH group at C-4 was likely.

Since we do not find a signal at 4.32 ppm in the proton resonance spectrum of the PLMF 1 with an intensity of two protons, we have to assume a substitution at the CH_2OH group of the glucose.

The proposed gallic acid glucoside-6'-sulfate was confirmed by the ^{13}C-NMR spectrum. We assigned the signals above 110 ppm to the ^{13}C-atoms of gallic acid. The signal at 106.7 ppm verifies the ß-glucosidic linkage of C-1' of the glucopyranose. In the narrow region between 62 and 64 ppm the signal of the hydroxymethyl group of hexoses appears. A shift of about 6 ppm to lower field is anticipated in case of esterification with sulfate. We therefore expected the corresponding signal of the PLMF 1 between 68 and 70 ppm. According to this assumption, we found in the off-resonance spectrum a triplet at 69.4 ppm.

On the basis of spectroscopic investigations (Table 1) it was possible to assign the structure for the newly found PLMF 1 and a very similar structure for the PLMF 2 (Table 2).

These are single highly active movement compounds, which cause a specific mimosa reaction even in a concentration of 10^{-7} mol/l.

Table 1. Spectroscopic structure elucidation of PLMF 1

METHOD	MEASURED	RESULT
UV MS	λ_{max}: 217, 285 nm m/z: 79.0178, 107.0131, 125.0242, 126.0337, 153.0199, 170.0197	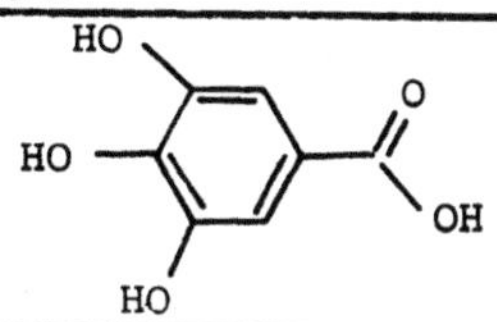
MS MS (TMS) IR	m/z: 63.9609 (SO_2^+) 227 $\nu_{S=O}$ 1248 cm^{-1} γ_{COS}: 818 cm^{-1}	$-CH_2-OSO_3^{\ominus}$
MS NEGATIVE ION FAB	m/z: 411, 241, 169, 97	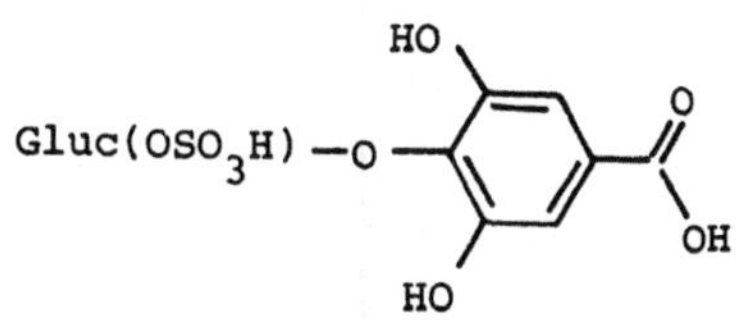
75.46 MHz- ^{13}C-NMR	δ: 69,4	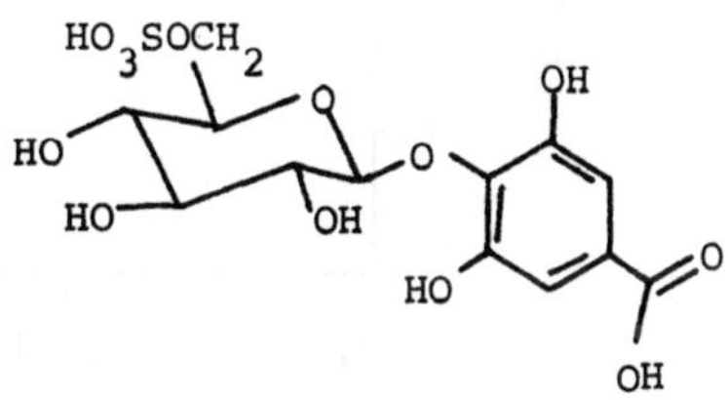

The proposed structure of PLMF 1 was definitely proved by synthesis. The 3.5-diacetyl gallic acid methylester was converted by a "classical" Koenigs-Knorr reaction to the peracetylated methyl glucosylgallate. After two steps of saponification we got the sulfate by reaction with a pyridine-sulfate-trioxide complex.

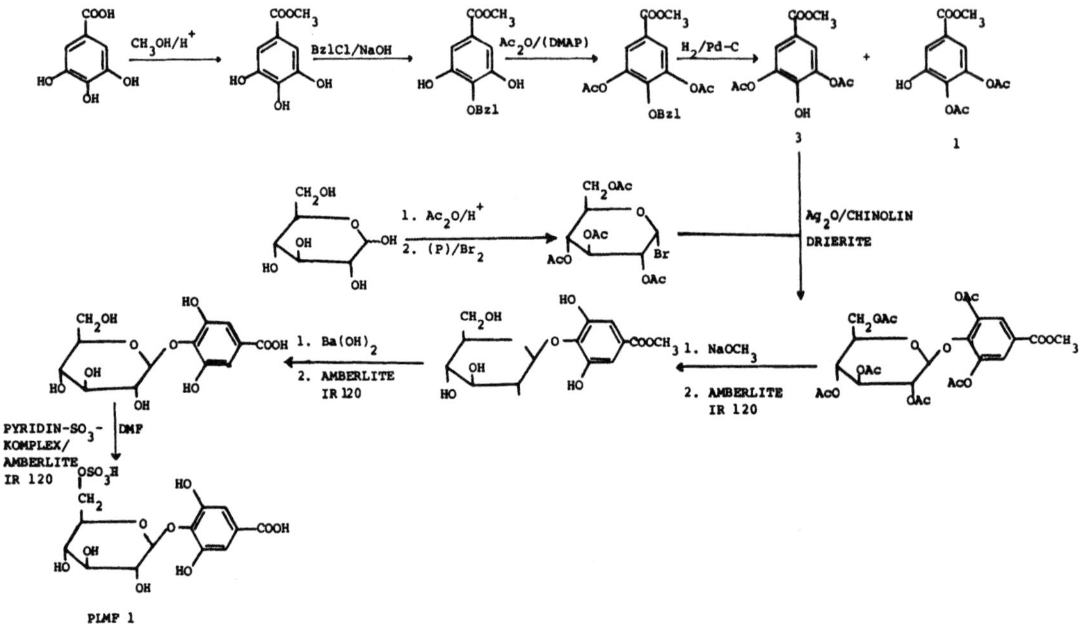

The ^{1}H-NMR spectra of the natural and synthetic compounds are identical, and the dose-response curve (Fig. 3) of a synthetic sample is similar to the crude mimosa extract.

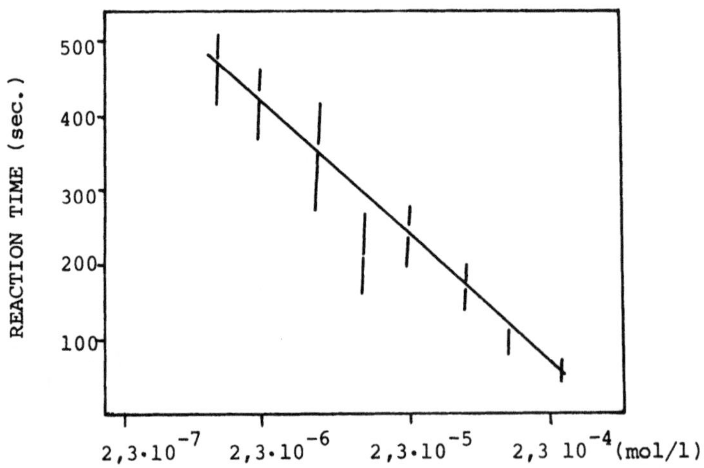

Fig. 3. Semi-logarhythmic dose-response curve of synthetic PLMF 1

2.3.2 LMFs from M. pudica L.

Despite all the intensive work on the non-sensitive acacia species, the sensitive plant M. pudica still remains an important object for our research work on the movement phenomena of plants.

By an unavoidably lengthy, yet rational separation process (Scheme 2) we were able to isolate from 23 kg of fresh mimosa from our plantation at Hainsbachweg in Heidelberg 0.8 mg of a compound which also exhibited a considerable nastic activity.

EXTRACTION OF THE FRESH PLANT WITH DOUBLY DISTILLED WATER

↓

EXTRACTION OF THE DRY EXTRACT WITH WATER

↓

nBuOH/H_2O-PARTITION

↓

MeOH-PRECIPITATION

↓

TWOFOLD GEL CHROMATOGRAPHY OVER
ULTROGEL AcA 202 (H_2O)

↓

GEL CHROMATOGRAPHY ON SEPHADEX G-10 (H_2O)

↓

PARTITITION CHROMATOGRAPHY ON LH-20 (MeOH/H_2O 96:4)

LB-1 LB-2 LB-3 LB-4 LB-5

↓ (LB-3, LB-4)

MEDIUM PRESSURE SEPARATION ON RP8 (CH_3COON/MeOH/H_2O)

↓

PREPARATIVE HPLC ON C-18 (CH_3COOH/$MeOH_2O$)

↓

M-LMF 5 M-LMF 4

Scheme 2. Isolation of the LMF from Mimosa pudica L. Separation by medium pressure reversed phase chromatography on Merck Lithoprep RP8, HPLC separation on Polygosil C-18,5 µm

Chromatographic and spectroscopic studies indicated that the active compound of Mimosa pudica was the PLMF 1 isolated from Acacia karroo.

However, the extract of M. pudica L. contains, aside from the PLMF 1, a further leaf movement factor called M-LMF 5. Preliminary UV and ^{1}H-NMR spectroscopic investigations on small amounts of sample immediately showed a high degree of similarity in the arene and sugar moities of the two mimosa hormones. However, subsequent comparative TLC and HPLC studies revealed distinct differences in the polarity and chemical behavior of the two compounds. It followed from these investigations that the sugar moiety of M-LMF 5 has to contain an acid group whose pK_a value compares with that of acetic acid.

The IR spectrum of M-LMF 5 shows a pronounced band between 1660 and 1730 cm^{-1}, which is caused by superposition of two carbonyl-absorptions: that of the gallic acid CO group and that of an aliphatic CO group.

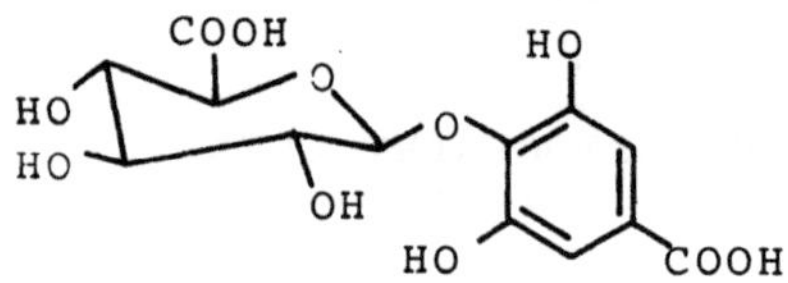

M-LMF 5

UV (H_2O)	:218,248,288
NMR (O_2O)	:7,14s,5,09d, 3,97d,3,69m
MS(hexasilylated)	:825(M^+-CH_3),745,655,465,458
	443,399,385,375,370,285

Fig. 4. Structure of L-LMF 5 with analytical data

The ^{13}C-resonances at -175,7 ppm (C-6) and -172,3 ppm (C-7) confirm the presence of the acid groups in the molecule. Both groups must be free since large downfield shifts occured in the potassium salt. The singulet of H-2,6 at 7.07 ppm in the ^{1}H-NMR spectrum points to symmetrical substitution of the aglycone moiety. The ^{13}C-NMR signal at -106,8 ppm is characteristic for C-1' of a β-D-O-glucosidically linked pyranose.

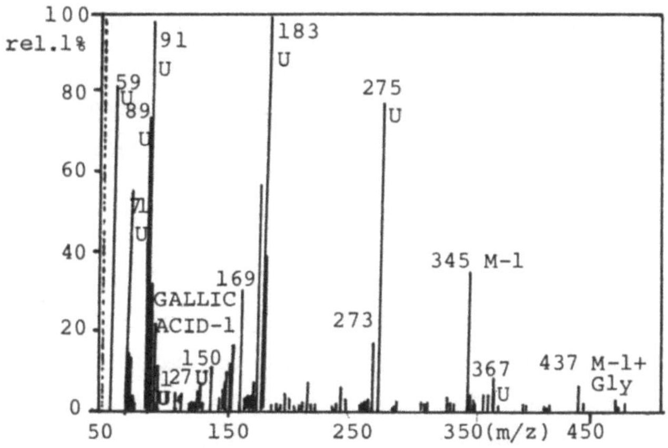

Fig. 5. Negative Ion FAB mass spectrum of M-LMF 5

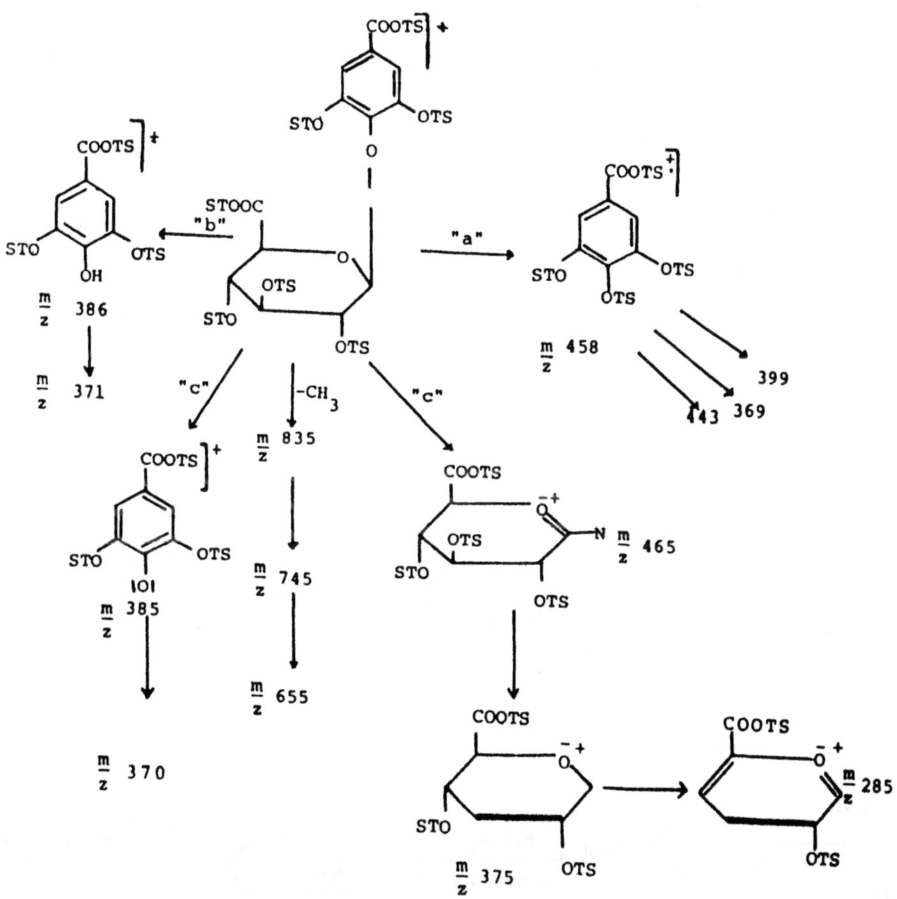

Fig. 6. Fragmentation of the persilylated M-LMF 5

Separated from the multiplet of three protons (δ = 3.69), in the ^{1}H-NMR spectrum a doublet from H-5' at 3.97 ppm ($J_{4,5}$ = 8,5 Hz), appeared typical for a hexauronic acid. Since also this signal is split by diaxial coupling, the ß-D-glucopyranosiduronic acid in M-LMF 5 must be 1'-O-glucosidically linked with the 4-OH group of the gallic acid.

The structure of the sugar and aglycone of the phytohormone could be confirmed by GC and GC-MS analysis after acid hydrolysis and silylation. The data obtained on mass spectrometric investigation of the unhydrolyzed molecule were also consistent with the proposed structure. The low-resolution FAB (fast atomic bombardment) measurement shows m/z 345 for M-1 in the spectrum of the negative ions. In the high resolution MS of persilylated M-LMF 5 we obtained $M-CH_3$ as highest mass (m/z 835, 3065, $C_{33}H_{67}O_{11}Si_7$, error 0,3 mmu).

Starting from the D-glucurono-3,6-lactone we synthesized the 4-O-(ß-D-glucopyranosiduronyl)gallic acid (M-LMF 5), from which we first prepared the methyl bromin-2,3,4-tri-O-acetyl-α-D-glucopyranuronyl acid.

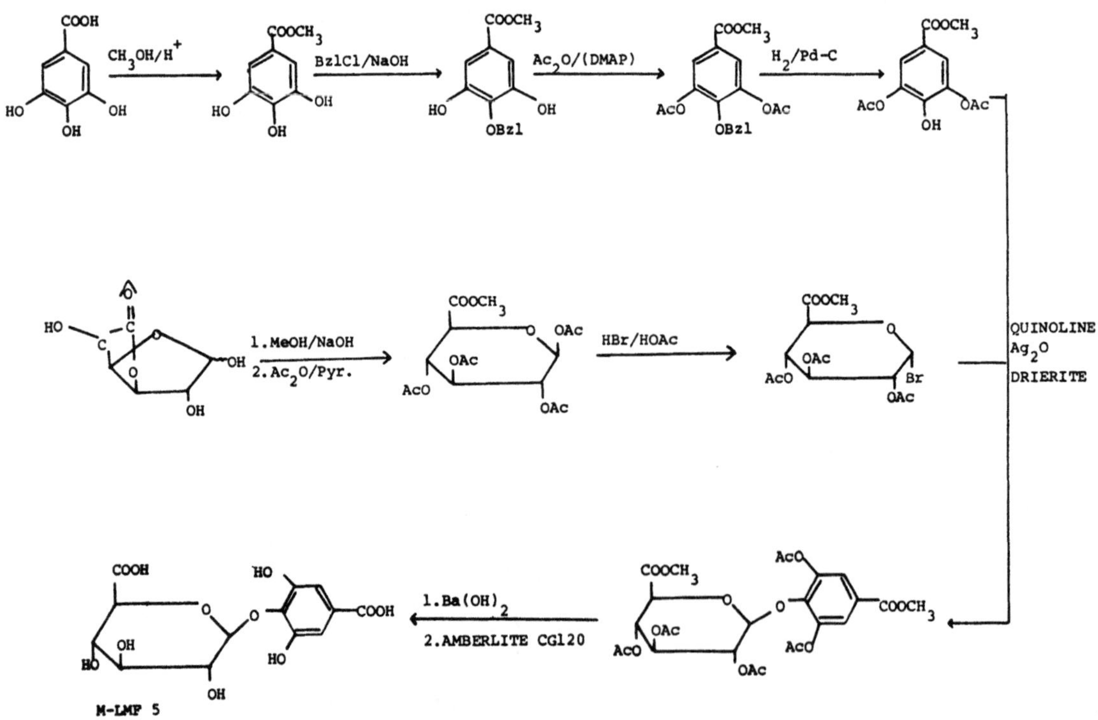

The protected aglycone of M-LMF 5 was prepared as described by us. By glucosidation in quinolin with silver oxide we obtained the peracetylated dimethylester of M-LMF 5. The removal of the protecting groups by $Ba(OH)_2$, followed by ion-exchange and medium pressure chromatography, provided the desired product in the H-form. The potassium salt was prepared by another cation exchange.

This new leaf movement factor of the mimosa is only apparently similar to the already known factors such as the PLMF 1, since its chemical and biological behavior is largely determined by the carboxy group in the sugar moiety, which is much less acidic than the sulfuric ester group.

3. Turgorins - A New Class of Phytohormones

The aforementioned leaf movement factors, the turgorins, are - like other phytohormones - also active even in a concentration of 10^{-5} to 10^{-7} of a molar solution. In addition, they show a clear structure-activity relationship. The activity decreases when the free phenolic OH groups are absent (Table 2).

Table 2. Structure-activity relationship of the turgorins

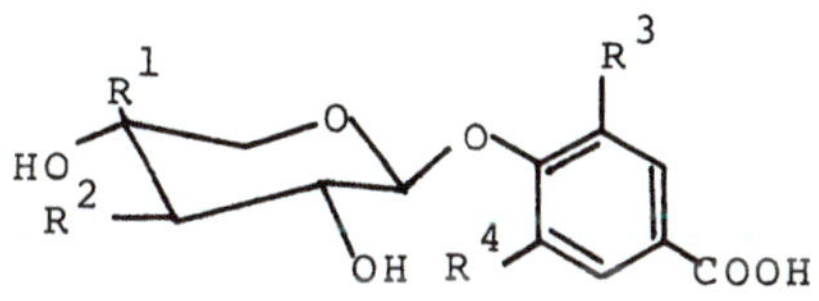

TURGORIN	STRUCTURE R^1	R^2	R^3	R^4	MINIMUM CONC.[mol/l]
PLMF 1	CH_2OSO_3H	OH	OH	OH	$2,33\text{-}10^{-7}$
PLMF 2	CH_2OSO_3H	OSO_3H	OH	OH	$1,96\ 10^{-7}$
S-PLMF 2	CH_2OSO_3H	OH	OH	H	$2,42\ 10^{-6}$
M-LMF 5	COOH	OH	OH	OH	$2,75\ 10^{-6}$
PLMF-synth	CH_2OSO_3H	OH	H	H	$2,51\ 10^{-5}$
LMF-synth.	COOH	OH	H	H	$1,57\ 10^{-3}$

Plant hormones, or more technically, phytohormones, are chemical messengers for plant behavior. In a stricter sense, phytohormones are defined as naturally occuring molecules which affect the physiological process at a concentration of 10^{-5} M or less and exhibit activity in a spectrum of plant species (Table 3).

Table 3. Occurence of PLMF 1 in Fabaceae, Oxalidaceae, and Malvaceae. t = reaction time of 30 mg of solid crude extract in 1 ml of water

PLANT	t(s)	PLMF 1	LOCATION
MIMOSA PUDICA	25	0.04mg/kg	HEIDELBERG
ACACIA KARROO	30	005mg/kg	SOUTH AFRICA
ACACIA DEALBATA	40	" "	SOUTHERN FRANCE
ALBIZIA JULIBRISSIM	50	" "	HEIDELBERG
GLEDITSIA TRIACANTHOS	60	001mg/kg	HEIDELBERG
SAMANEA SAMAN	40	" "	INDIA
OXALIS STRICTA	40	" "	HEIDELBERG
OXALIS DEPPEI	40	" "	HEIDELBERG
ABUTILON GRANDIFLORUM	25	2mg/kg	HEIDELBERG

t=REACTION TIME OF MIMOSA LEAVES WITH 30 mg SOLID CRUDE EXTRACT IN 1ml WATER

3.1 Turgorins from Robinia pseudacacia (Fabaceae)

The leaf movements of Robinia pseudacacia are caused by different stimuli. It is remarkable that it raises and droops its pinnules according to the type of stimulation. Upon intensive sunshine, the leaves move upwards and thus diminish the surface subjected to the light. In the extreme case, the leaves stand vertically (photonasty) in order to protect the chlorophyll. Pinnules fixed to their normal posture begin to suffer within two days when the plant was exposed to intensive sunshine.

We were able to confirm that the raising of the pinnules is not an effect caused by temperature. When the light is directed only to two pinnules with a UV lamp (366 nm), they raise within two minutes. The neighbouring leaves which are not illuminated remain in their normal horizontal posture, although they are exposed to the same temperature. Thus, a glowing immersion-heater placed near the leaves causes no movement, the leaves wilt when the temperature becomes too high.

The sleeping posture (nyctinasty) looks different: the leaves move downwards at night so that the leaves' lower sides face each other. At the same time the leaf stalks raise by 3 to 4°. Nyctinasty probably aims at protecting the leaf's surface against complete cooling caused by radiation towards the dark sky or by cold winds, and in the case of dryness against evaporation. The direct approach of the leaves provides mutual protection. Leaves that were fixed and thus could not assume their sleeping posture showed more damages done by frost than others.

In addition, the plant also shows thermonasty and a slight seismonasty. These stimulants lead to the drooping of the leaves. When a small tree kept at room temperature is exposed to a flood of cold air, it droops its leaves within two minutes. The same occurs when the plant is vigorously beaten or shaken. Younger trees show an especially pronounced nyctinastic behavior. Older trees are mainly photonastically sensitive.

The leaves and stems of Robinia pseudacacia were tested for substances, which in the case of Mimosa pudica showed a specifically chemonastic activity. We extracted 90 kg of plant material in a pilot plant with water/ethanol 80 : 20. The extract was pre-purified by methanol precipitation and liquid-liquid extraction with n-butanol.

The active water phase of the n-butanol-water distribution was freeze-dried. We tried to dissolve half of this extract in methanol and separated the soluble compounds by column chromatography. Pre-filtration, gel filtration twice repeated on Sephadex G 10 and partition chromatography on Sephadex LH 20 led to a slightly active zone, to a blue-flourescing fraction and to a highly active region.

After several failures we could isolate from it about 50 µg of PLMF 1 by means of medium pressure and repeated preparative high-pressure liquid chromatography. The quantity of the second active substance of Robinia pseudacacia was not sufficient for a purification. The retention times on the HPLC with different eluents of this substance were identical with those of PLMF 2.

3.2 Turgorins of the Oxalidaceae

It has been known since Plinius the Elder's time that oxalis-type plants also behave nyctinastically, and under certain circumstances even react thigmonastically or seismonastically. This is observed particularly nicely in the case of Oxalis deppei, the four-leaf clover, which folds its leaves together like an umbrella, not only at night but also in full sunshine.

The first attempts made in the search for LMF in this family of plants, which are quite different from the Fabaceae, have revealed that here also the chemical principles underlying nyctinastic movement are compounds at least related to the already known factors. Thus, an oxalis leaf folds its pinnules when it is placed in a solution of PLMF 1.

It took O. deppei two to four hours to assume its sleeping posture. During the first hour, we could only observe a very slight drooping, which then quickened until the leaf was closed by 3/4 and slowed down again. It was remarkable that the leaves first showed the irritated posture, which later changed to the sleeping posture when the leaf had moved downwards by 3/4. The irritated posture is characterized by the fact that the leaves move downwards only at the pulvinus whereas sleeping posture means that the leaves do not only move downwards but fold up also at the leaf axis. The nasties are well shown in Fig. 7., a picture of Oxalis acetosella.

Even O. purpurea, which usually does not show any movement, exhibited weak reactions when it was placed in a PLMF 1 solution. The maximum reaction occured within three to four hours.

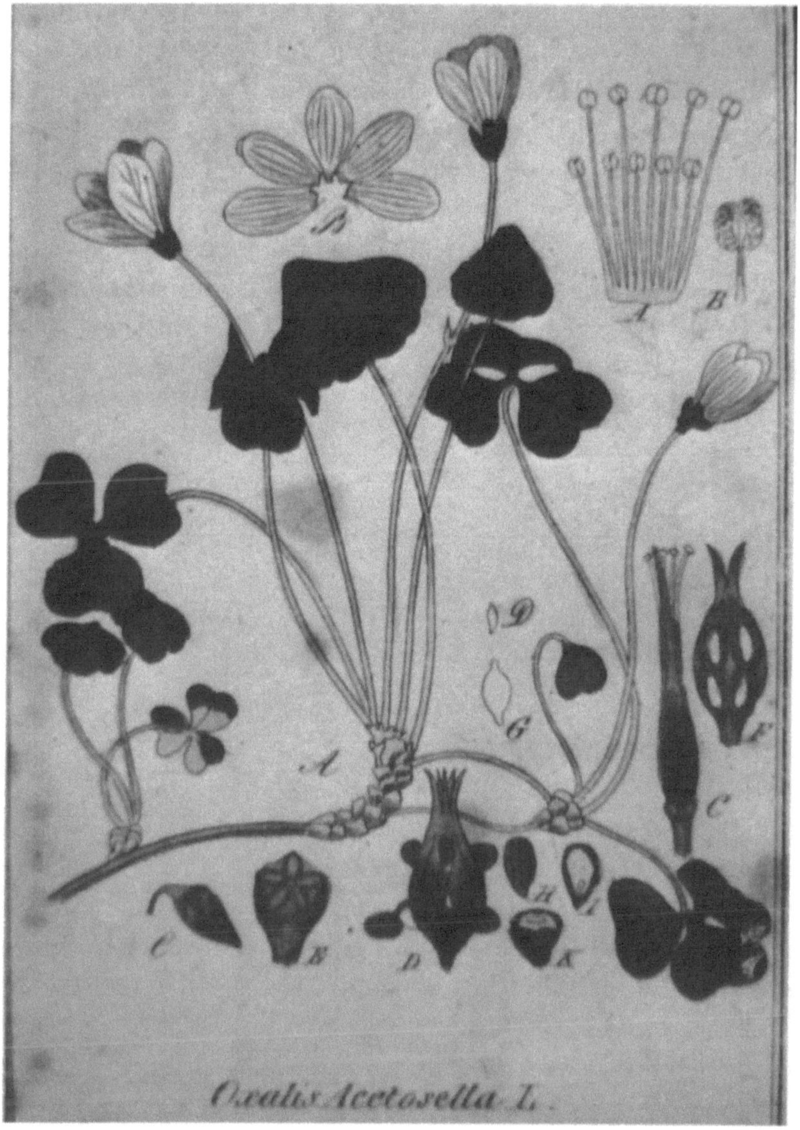

Fig. 7. Oxalis acetosella showing the three movement postures

3.2.1 Turgorins of Oxalis stricta

In the light of our observations on the nasties of Oxalidaceae, it is not surprising that we have isolated a leaf movement factor S-PLMF 1 from Oxalis stricta, the sorrel, which is identical with the PLMF 1 (Table 4). A second factor S-PLMF 2, which could be isolated by the usual methods of enrichment, exhibited in a bioassay the well-known leaf movement reaction, and, following preliminary spectroscopic studies, it could be concluded that it was likewise chemically related to the previously described LMFs.

Table 4. Comparison of the retention times (HPLC) of diverse PLMFs. Column: 25 X 0,8 cm, Nucleosil C-18, 5 µm; detection: λ = 210 nm; System I: 20% CH_3OH/0,05% CH_3COOH; 3 ml/m; System II: 1 mM CH_3COOH; 5 ml/min

	tr[min] system I	tr[min] system II
S-PLMF 1	15,1	5,8
PLMF 1	15,3	5,9
S-PLMF 2	11,9	4,3
PLMF 2	42,2	3,1

By means of GC-MS investigations on the silylated hydrolyzate we found that the aglycone of S-PLMF 2 was not gallic acid but protocatechuic acid. Glucose was also found to be a structural unit of S-PLMF 2. As in the case of the other LMFs, the total structure of S-PLMF 2 has been determined nmr spectroscopically.

Hence, here once again the synthesis served as a means of proof of the structure; the end product, 4-O-(6-O-sulfo-ß-D-glucopyranosyl)protocatechuic acid, proved to be identical with the natural substance, both spectroscopically and in the bioassay.

This structural variant of PLMF 1 exhibited specific activity at a concentration of 0.001 to 0.005 mg/l in the biological test system; the absence of a hydroxy group in the aglycone obviously leads to a reduction of the activity by a factor of more than 10.

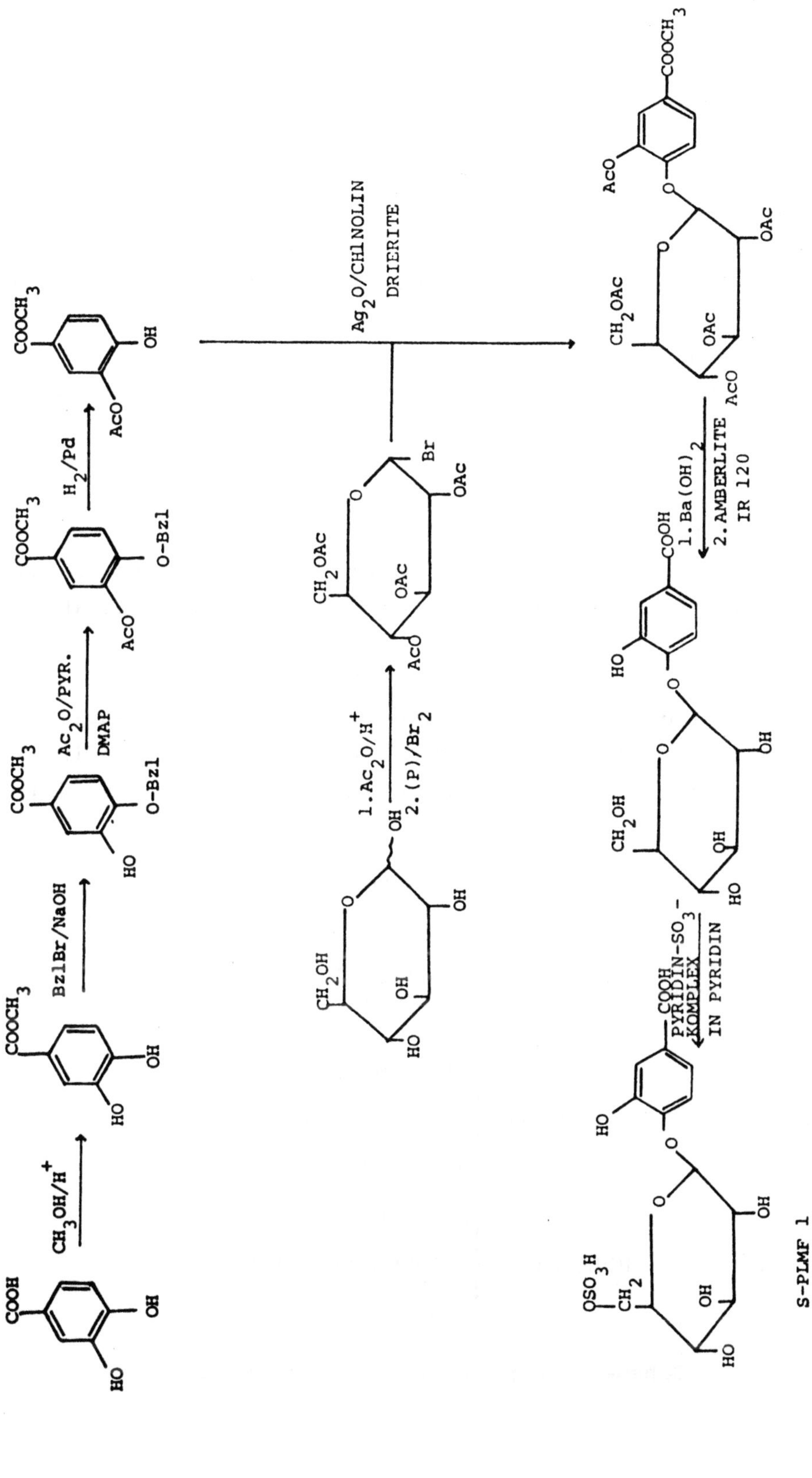
COOH
HO
OH
CH_3OH/H^+
$COOCH_3$
BzlBr/NaOH
O-Bzl
Ac_2O/PYR.
DMAP
AcO
H_2/Pd
CH_2OH
CH_2OAc
OAc
Br
1. Ac_2O/H^+
2. (P)/Br_2
Ag_2O/CHINOLIN
DRIERITE
1. $Ba(OH)_2$
2. AMBERLITE
IR 120
PYRIDIN-SO_3-
KOMPLEX
IN PYRIDIN
OSO_3H
CH_2
S-PLMF 1

3.3 A new Turgorin from Glycine max. (Fabaceae)

The soybean Glycine max. is a well-known short-day plant and its leaflets show pronounced movement between day and night.

However, a sufficient number of observations of this plant's leaf movement by Holdsworth had been collected to reveal that they have no diurnal rhythm at all. The leaf angle shows striking and immediate nastic responses to light and darkness, and even to changes of light intensity; but kept in uniform light the leaves never shut and in darkness they never open.

I think this is the reason why we could not find the leaf movement factors in the extracts of Glycine max., which are responsible for the endogenous rhythms of the plants mentioned before.

Already the first extractions of the suggested leaf movement factors revealed that a highly polar compound must be present, which was further enriched by a corresponding gel chromatography (Scheme 3).

The UV and ^{1}H-NMR spectra recorded from all fractions pointed to the fact that these active compounds must be substituted derivatives of cinnamic acid (Fig. 8). This is confirmed by the finding that the chromatographically pure active frac-

H_2O-Extraction
↓
Extraction with MeOH
↓
Twofold Gel chromatography over
Sephadex G10
↓
Partition chromatography on
LH 20 (MeOH/H_2O)
↓
Partition chromatography on
↓
LH20 (EtOH/$H_2$0)
↓
HPLC on C-18 (H_2O)
↓
HPLC on C-18 (H_2O/MeOH/HOAc)
↓
G-PLMF1

Scheme 3. Separation Scheme of G-PLMF 1

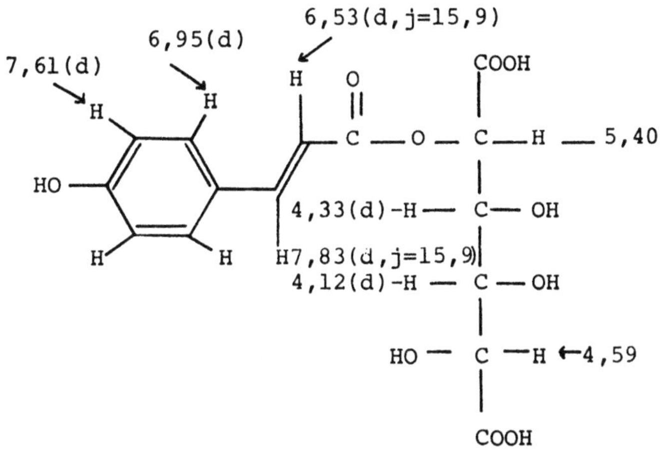

G-PLMF 1 (Glycine max.)

ms (hexasilylated)	: 773 (M^+-CH_3) $C_{32}H_{61}O_{10}Si_6$ high resoln.
ir (KBr)	: 2900,2490,1730,1690,1600,1510,1280, 980,820
UV(MeOH	: 228,295s,311
(MeOH+NaOMe)	: -, 305,358
nmr in D_2O	

Fig. 8. Structure of G-PLMF 1 with analytical data

tion is photounstable. However, the active limiting concentration of G-PLMF 1 was comparable to that of a 2,5 X 10^{-6} molar solution.

When elucidating the structure, the instability of the active compound made it impossible to record a ^{13}C-NMR spectrum. The protone resonance spectrum, however, revealed quite clearly the signals for a trans-cinnamic acidic system and in the upfield region for a galactaric acid. By means of a computer simulation we could prove that the sugar acid is substituted at C-2, whereby the p-coumaric acid functions as an acidic component so that a free phenolic OH group must be present. The IR spectrum could be correspondingly interpreted. PLMF 1 was especially well characterized by mass spectrometry, as such and also because of its hydrolysis products. According to that, G-PLMF 1 is the 2-O-(4-hydroxy-trans-cinnamoyl)galactaric acid.

It could be possible that G-PLMF 1 was only formed at nightfall and thereby causes the nyctinastic movement. However, daylight causes the formation of a cis/trans equilibrium which is no longer active.

4. Final Remarks

All the above-described results show that we have found the chemical code for nyctinasty, photonasty and seismonasty as well, and perhaps for the "internal clock" of plants.

Here the activity of the turgorins is structurally linked with para-hydroxylized aromatic carbonic acids.

Furthermore it is a rule that the substituent which is not linked with glucose must not have another neighbouring substituent. This is again a new hint to a specific hormone receptor.

Turgoractive substances will not be described completely until other plants from very different families will have been tested, too (Table 5). We have already been able to show that - by means of the aforementioned bioassay - turgorins can be detected in plants of Portulacaceae and - what is especially interesting - in plants of Vitaceae showing tendril curling.

Table 5. Plants containing turgorins

PLANT-SPECIMEN	– FAMILY	t[s]
WISTERIA SINENSIS	FABACEAE	120
PORTULACA OLERACEAE	PORTULACACEAE	40
VITIS GONGYLODES	VITACEAE	90

t=REACTION TIME OF MIMOSA LEAVES WITH 30mg SOLID CRUDE EXTRACT IN 1ml WATER

TURGORINS

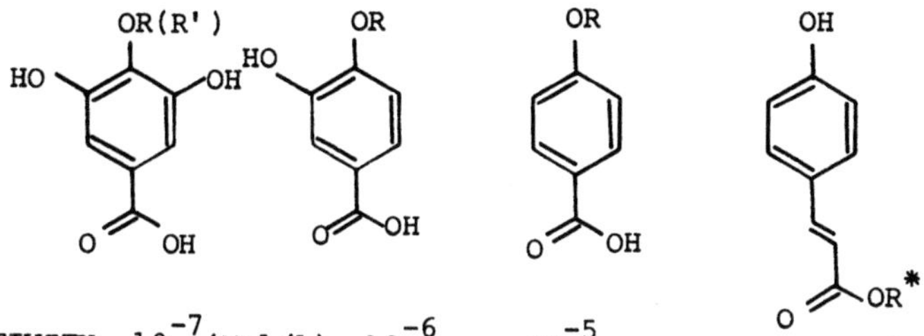

ACTIVITY 10^{-7}(Mol/l) 10^{-6} 10^{-5} 10^{-6}

R=Gluc (OSO_3H)

10^{-6}

R=Gluc (COOH)

R*=Galactaric acid

INACTIVE NATURAL SUBSTANCES

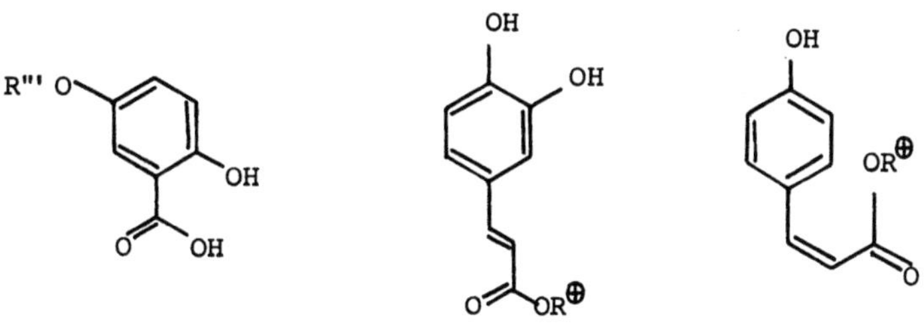

R'''-Glucoapioside

Fig. 9. Structure-activity relationship between turgorins and related substances

Acknowledgements

I would like to express my appreciation to my fellow investigators on turgorins, who, in discussions and by criticism, have helped me so much and whose experimental results serve to illustrate an hitherto quite unknown field of natural products. I also want to thank expecially the "Deutsche Forschungsgemeinschaft" and the "Fonds der Chemischen Industrie" for their support of my research.

References

Ch. Darwin, The Power of Movement in Plants, 1 st. Edit., J. Murray, London (1880).

H. Schildknecht: Über die Chemie der Sinnpflanze *Mimosa pudica L.*, Springer, Berlin 1978.

H. Schildknecht, Angew. Chem. Int. Ed. Engl. 20 (1981) 164-184

H. Schildknecht, Angew. Chem. Int. Ed. Engl. 22 (1983) 695-710

Natural Product Synthesis Via Functionally Substituted Vinyl Carbanions

Richard R. Schmidt

Fakultät Chemie, Universität Konstanz, D-7750 Konstanz, FRG

SUMMARY

Direct C-lithiation of functionally substituted acrylates yields short routes for the synthesis of butenolides, tetronates, α-hydroxybutenolides, and α-hydroxytetronates. Approaches to the synthesis of aspertetronins, gregatins, ascorbic acids, chlorothricolide, and 3-deoxy-D-*manno*-2-octulosonate (KDO) are discussed

INTRODUCTION

Functionally substituted vinyllithium compounds are of great interest as reactive intermediates especially if other unprotected groups are also present [2-5]. Our investigations with ß-substituted acrylic acids and derivatives have demonstrated that these compounds are accessible to direct lithiation of the α- and/or ß-position (Scheme 1) [2-4]. This means that the generation of stable vinyllithium derivatives prevails over the expected 1,2- or 1,4-addition of the lithiating agent to the acrylate system. The accessibility to a twofold lithiation leads to a preparative versatility, which for structural reasons is not available to aryl- and hetaryl systems.

Natural Product Chemistry
Ed. by Atta-ur-Rahman

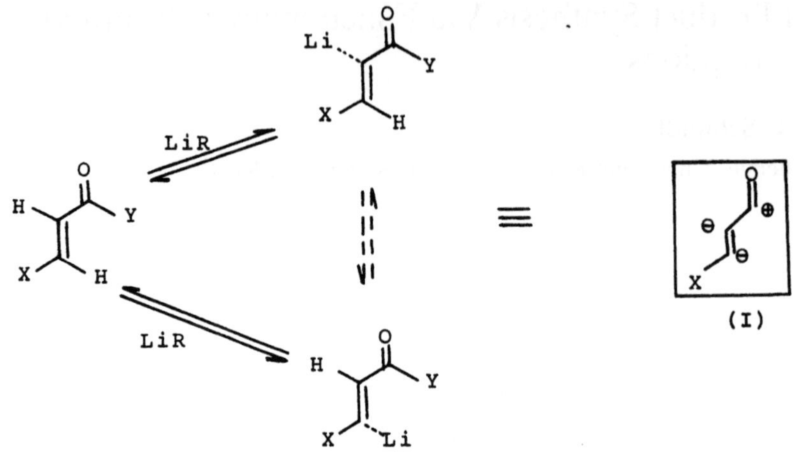

X = OR, NR_2, SR, Y = OR, OH, NR_2, NHR

Scheme 1

The regioselectivity is mainly determined by the ß-alkoxy, ß-dialkylamino, ß-N-acyl-N-alkylamino, and ß-alkylmercapto groups on one side and the carboxylic functionality on the other side [6,7]. In addition, some acrylic acid derivatives could be lithiated selectively in either the α- or ß-positions by use of kinetic or thermodynamic control [6-8].

Reaction with electrophiles and especially with electrophilic - nucleophilic compounds (aldehydes, ketones, azomethines, esters, α,ß-unsaturated carbonyl compounds, cumulenes, etc.) afforded short syntheses for butenolides, tetronates, α-hydroxybutenolides, α-hydroxytetronates, pyrrolenines, cyclopentenones, oxalacetates, and maleic acid derivatives (Scheme 2) [3,4]. Thus from the acrylate systems in stepwise reaction the valuable synthon I (Scheme 1) is generated, which is a structural unit of many natural products.

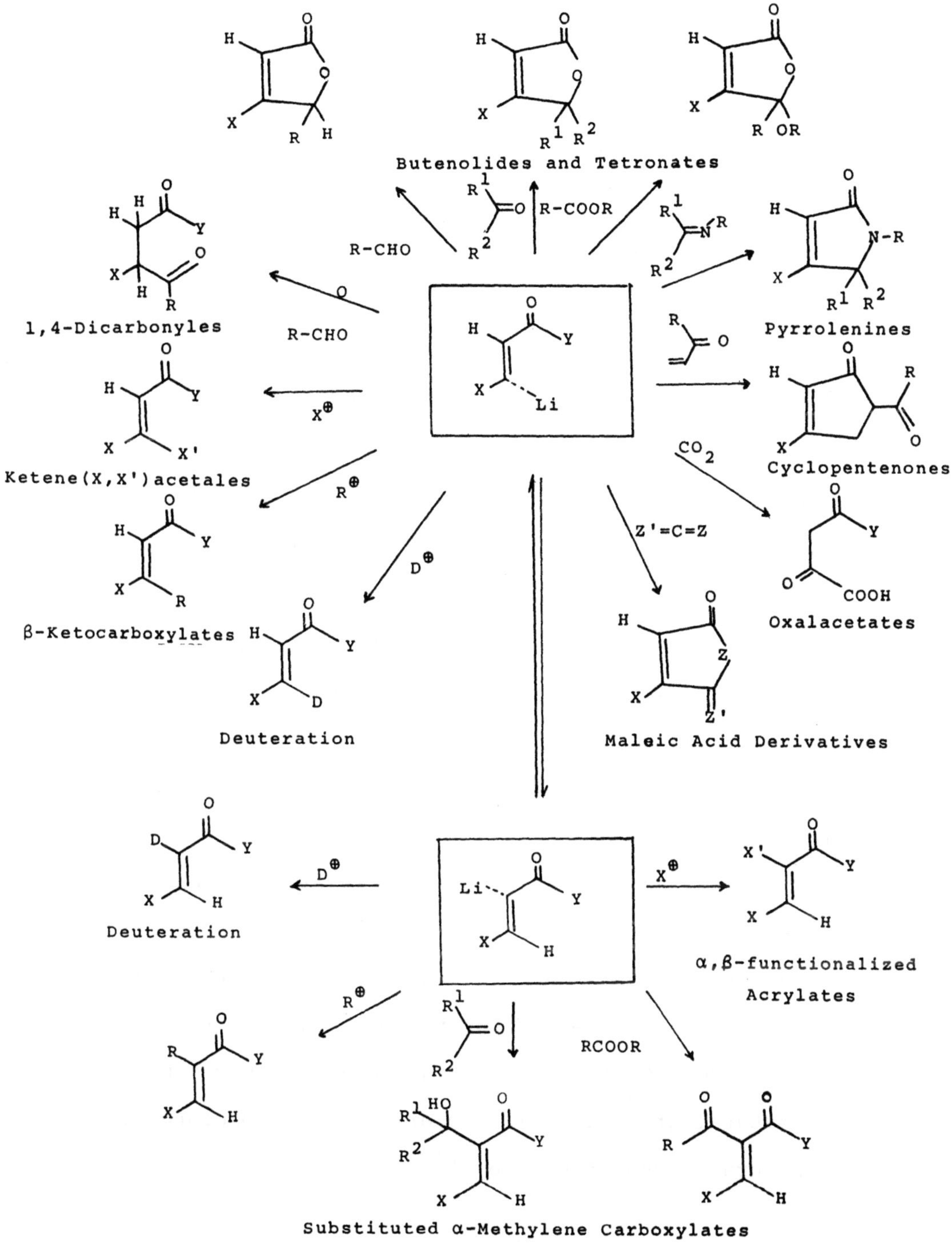

X,X' = OR, NR_2, SR; Y = OR, NR_2; Z,Z' = O,S,NR; R = Alkyl, Aryl

Scheme 2

A wide variety of substituents are comparible with this lithiation: alkyl, aryl, electron-donating and electron-withdrawing substituents in α- or ß-position conducting to interesting regio- and stereoselectivities [3,4]. The compounds obtained are portions of many natural products. Approaches to the synthesis of natural products with butenolide structures like aspertetronins, gregatins, ascorbic acids, and the macrolide antibiotic chlorothricolide and syntheses for the carbohydrate 3-deoxy-D-*manno*-2-octulosonate (KDO) will be outlined.

BUTENOLIDE AND TETRONATE SYNTHESES: ASPERTETRONINS AND GREGATINS

The ease of butenolide and tetronate formation from ß-functionally substituted acrylates and carbonyl compounds was used quite successfully [3,4]. However, the observation that these compounds are accessible to a subsequent α-lithiation was especially rewarding [9]. This methodology seemed to be of use for a short synthesis of the antibiotically active aspertetronins and gregatins, for which tetronate (furan-2-one) structures were proposed (Scheme 3). Reaction of the lithiated acrylate species with (E,E)-3.5-octadien-2-one gave the α-unsubstituted tetronate. α-Lithiation with LDA and reaction with acetic anhydride led to the expected α-substituted tetronate in an overall two-step reaction [9]. However, by these investigations it was demonstrated, that the natural gregatin B is isomeric to the compound obtained and has a furan-3-one structure [9,10].

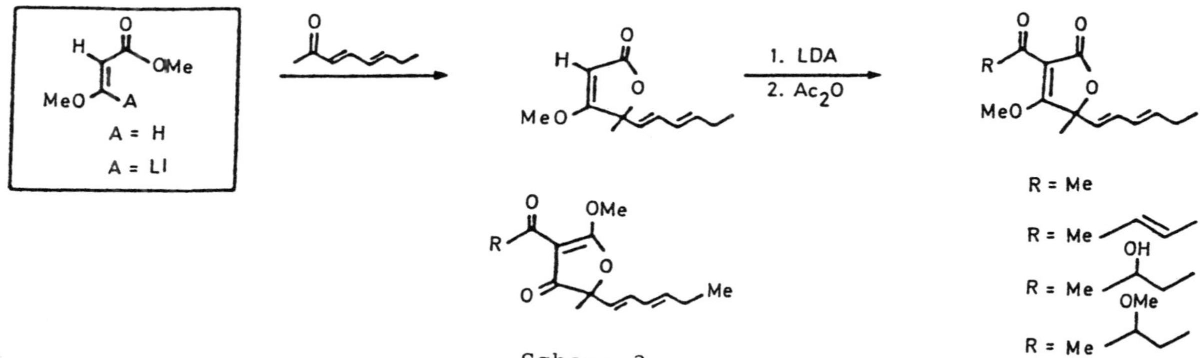

Scheme 3

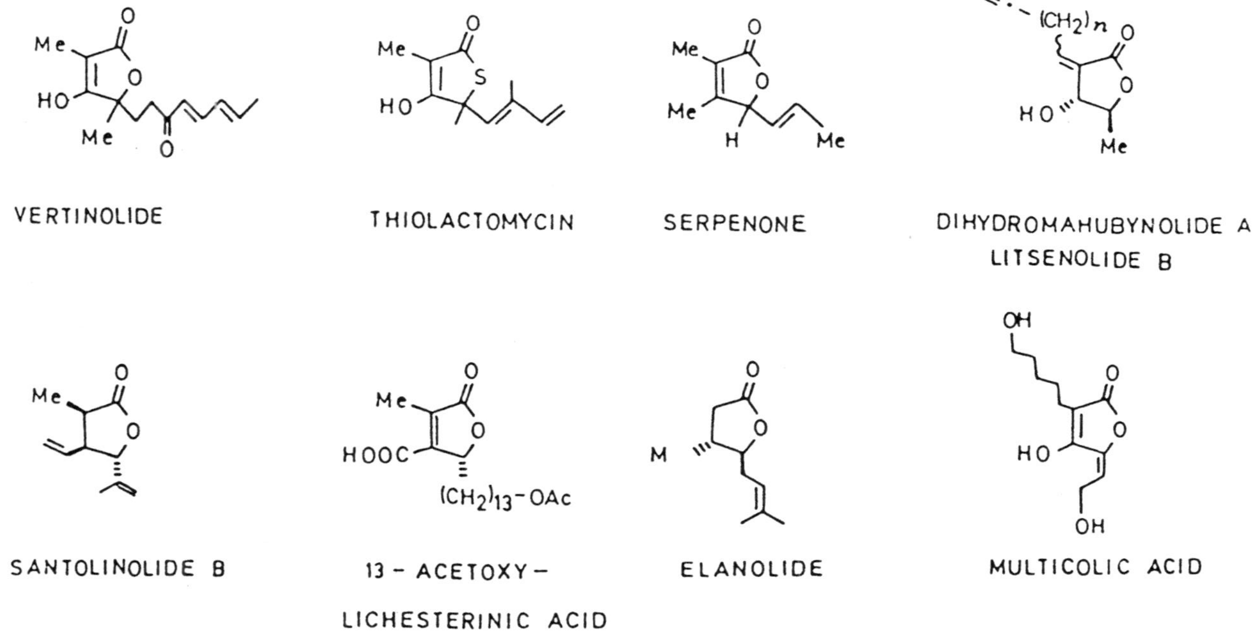

Scheme 4

This efficient method for the preparation of substituted butenolide and tetronate derivatives should be applicable to the synthesis of other naturally occurring examples of this class of compounds. A variety of such compounds is given in Scheme 4, for which short syntheses using this methodology can be devised. However, our interest was devoted to more complex natural products containing the α-hydroxy-butenolide and α-hydroxy-tetronate unit as an important building block.

INVESTIGATIONS TOWARDS THE SYNTHESIS OF CHLOROTHRICOLIDE

The macrolide antibiotic chlorothricin (Scheme 5) was isolated from *Streptomyces antibioticus* [11]. This antibiotic is active against gram-positive bacteria because of inhibition of pyruvate carboxylase and maleate decarboxylase. The aglycon chlorothricolide has some of the original activity of chlorothricin. Several investigations towards the synthesis of partial structures of this complicated molecule have been reported [11,12]. The retrosynthetic analysis of the aglycon chlorothricolide (Scheme 5) leads to two substructures. One of these contains an α-hydroxy-tetronate moiety, which should be directly obtainable from a cyclohexenone precursor and a ß-lithiated acrylate species.

The complex macrolide antibiotics kijanimicin [13] and tetrocarcin [13] have aglycon structures quite similar to chlorothricin. Therefore the outlined strategy to the synthesis of chlorothricolide should be especially rewarding because the corresponding aglycon structures kijanolide and tetronolide could be obtained via a similar route (Scheme 6).

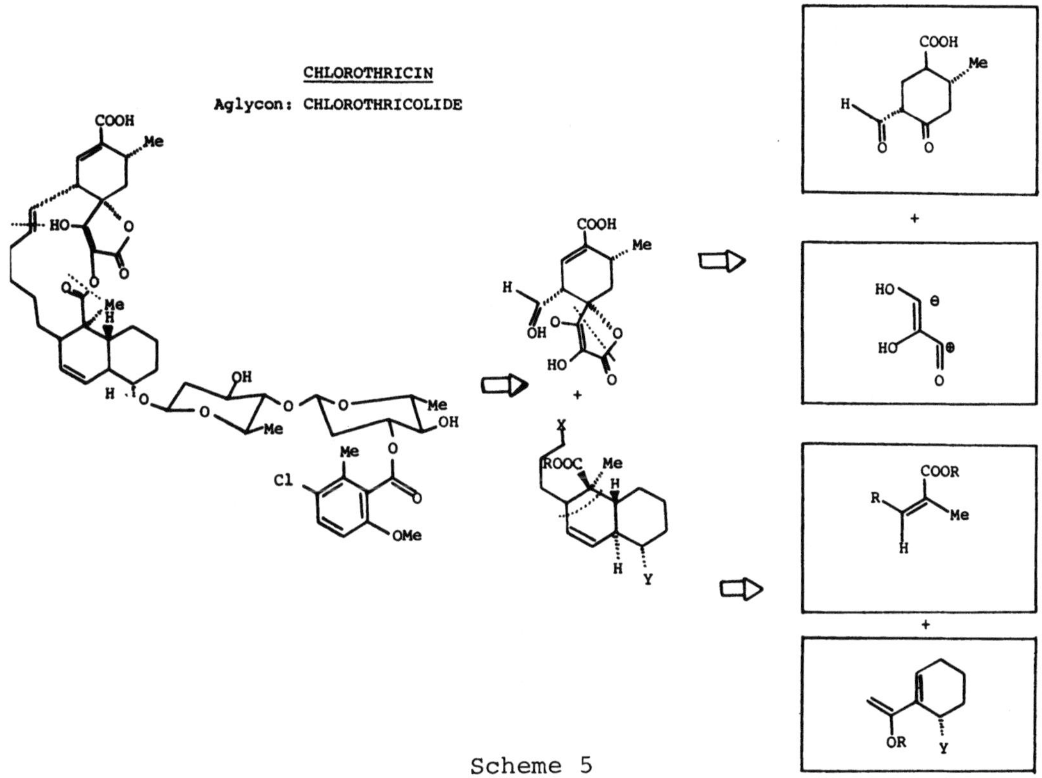

Scheme 5

To prove the feasibility of this synthetic approach the cyclohexenone derivative shown in Scheme 7 was required. For a regioselective reaction of the ketone carbonyl group of this polyfunctional molecule the aldehyde functionality had to be protected, the carboxylic group should be esterified, and the CC-double bond had to be introduced at a later stage otherwise the cis-arrangement of the substituents would be lost. A cycloaddition route starting from pent-3-en-2-one and acrylate gave the required cyclohexanone derivative [14]. Investigations towards the introduction of the missing CC-double bond were carried out successfully. The methodology used should be effective at a later stage of the total synthesis of chlorothricolide.

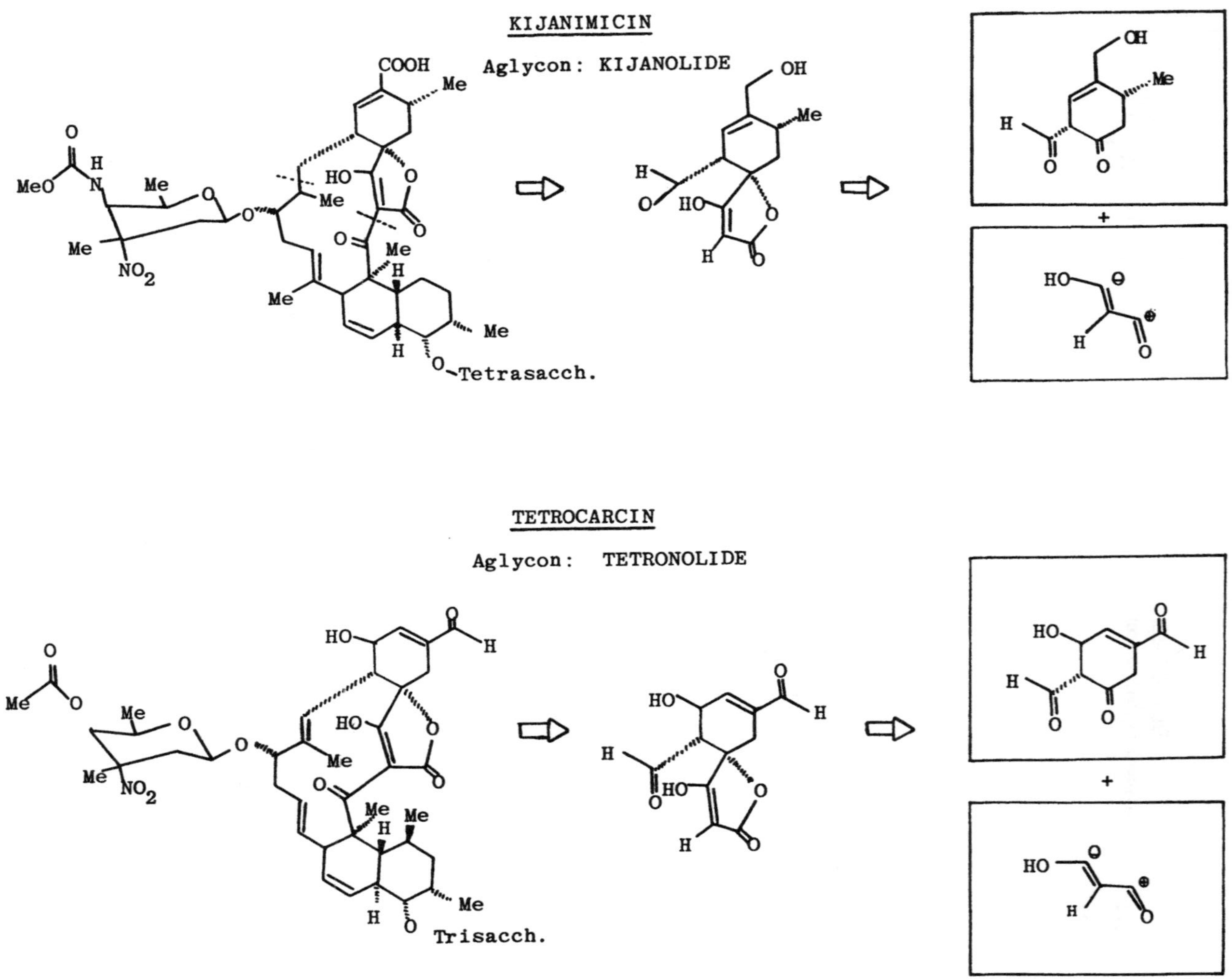
KIJANIMICIN
Aglycon: KIJANOLIDE
COOH
Me
OH
MeO
O
H
N
Me
O
HO
Me
NO$_2$
O-Tetrasacch.
TETROCARCIN
Aglycon: TETRONOLIDE
HO
H
Me
Trisacch.

Scheme 6

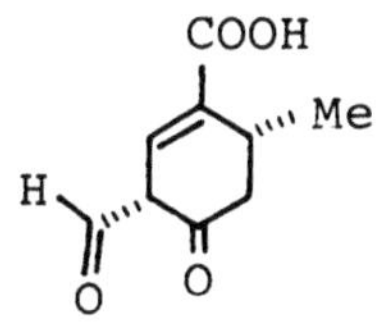

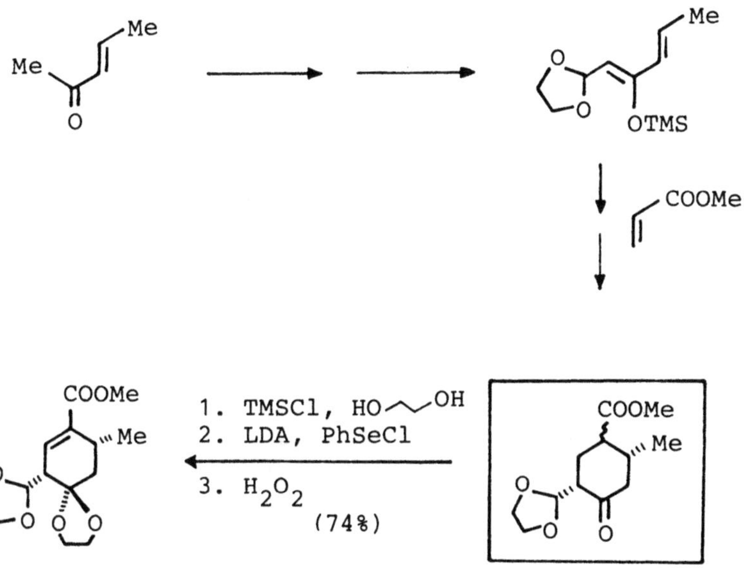

Scheme 7

The low reactivity of cyclohexanone and derivatives in reactions with ß-lithiated α,ß-functionally substituted acrylates [15] was reason to use the more reactive ß-lithiated ß-functionally substituted acrylates for the formation of the tetronate moiety. The missing α-hydroxylic group could be introduced successfully via subsequent α-lithiation and hydroxylation as shown in Scheme 8 [14]. In addition, diastereospecific attack of the vinyllithium species at the cyclohexanone derivative was observed leading according to X-ray analysis [16] exclusively to the desired diastereoisomer.

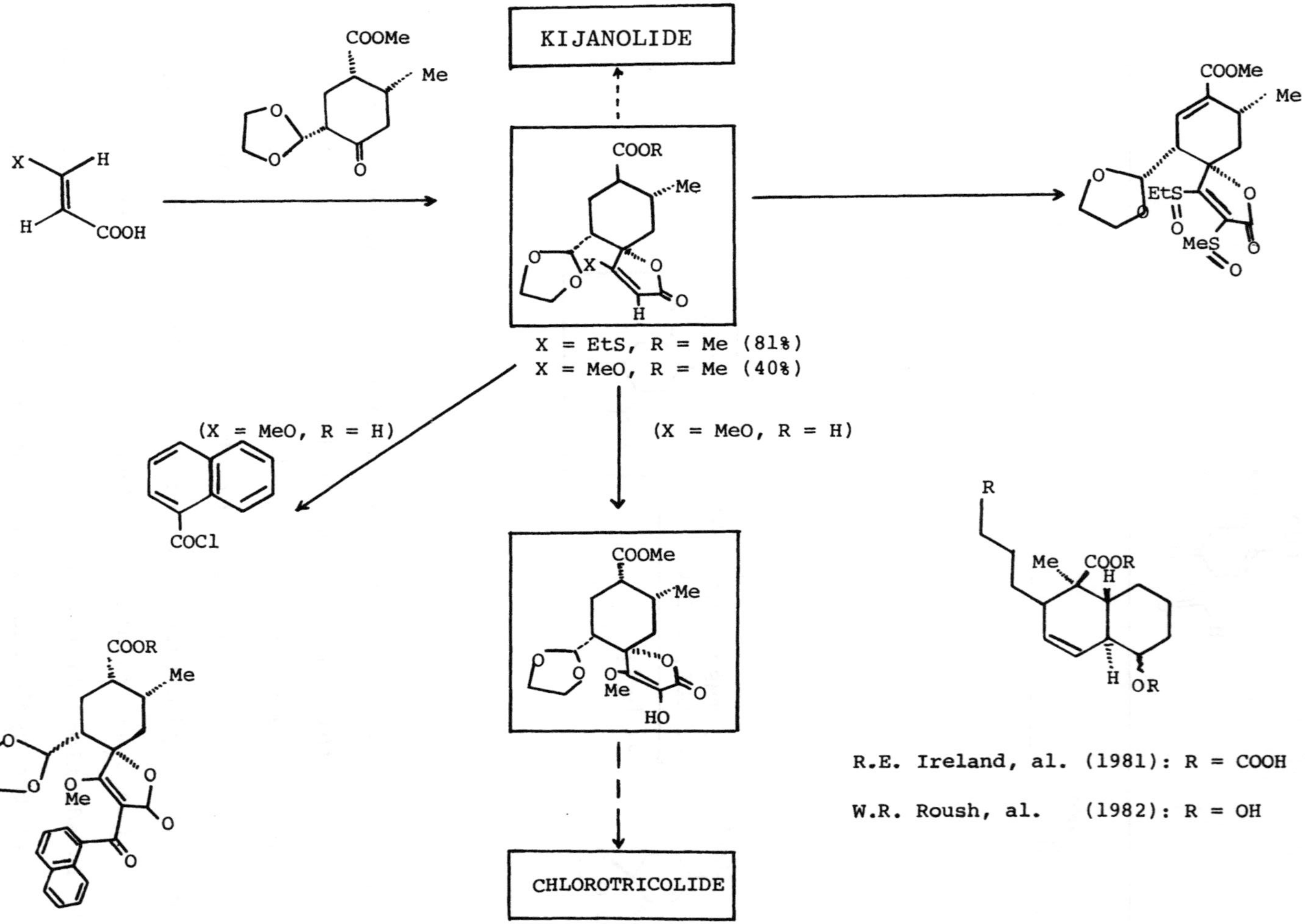
KIJANOLIDE
COOMe
Me
COOH
COOR
X = EtS, R = Me (81%)
X = MeO, R = Me (40%)
(X = MeO, R = H)
(X = MeO, R = H)
COCl
EtS
MeS
HO
CHLOROTRICOLIDE
R
OR
R.E. Ireland, al. (1981): R = COOH
W.R. Roush, al. (1982): R = OH

Scheme 8

Syntheses of analogs of the missing dehydrodecaline moiety have been already published by the groups of Ireland and Roush [11,12]. Therefore the construction of the macrolide antibiotic chlorothricolide should be possible. The synthetic strategy outlined is applicable to the synthesis of kijanolide and tetronolide as well. The successful α-acylation of the spirotetronate species is a crucial step for this aim.

SYNTHESIS OF 3-DEOXY-D-*MANNO*-2-OCTULOSONATE (KDO)

A straight forward retrosynthetic scheme for the formation of L-ascorbic acid is shown in Scheme 9. Application to 2.3-dimethoxy acrylic acid and 2.3-O-isopropylidene-D-glyceraldehyde led to a convenient ad diastereoselective formation of 2.3-di-O-methyl-D-araboascorbate and -D-ascorbate in a two step procedure (54 %, 2:1) [15]. The structural assignment was possible by comparison of physical properties with literature values. Demethylation was carried out with boron tribromide. This model reaction for a new synthesis of KDO gave preferentially the desired *erythro*-diastereoisomer.

KDO is an integral part of the lipopolysaccharides of gram-negative bacteria [17]. Syntheses of 3-deoxy-D-*manno*-2-octulosonate (KDO) from D-mannose derivatives by Wittig type chain elongation required additional steps for the introduction or liberation of the 2-keto group [18,19]. Syntheses from D-arabinose and derivatives led to disadvantageous diastereoisomer formation and separation [20]. The interest in the chemistry of KDO was reason to develop an efficient synthesis with lithiated functionally substituted acrylates as pyruvate anion equivalents.

L-ASCORBIC ACID: RETROSYNTHETIC SCHEME

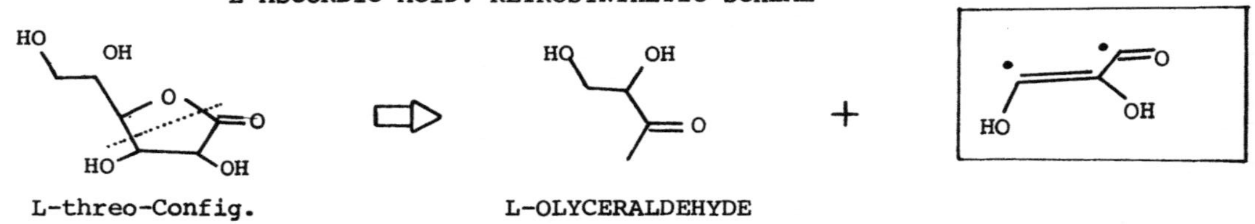

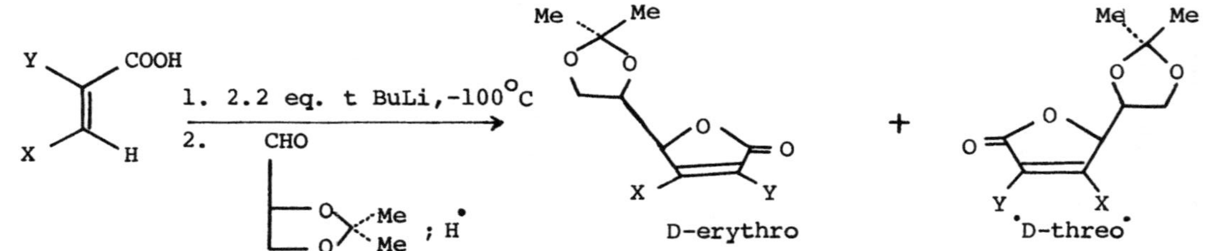

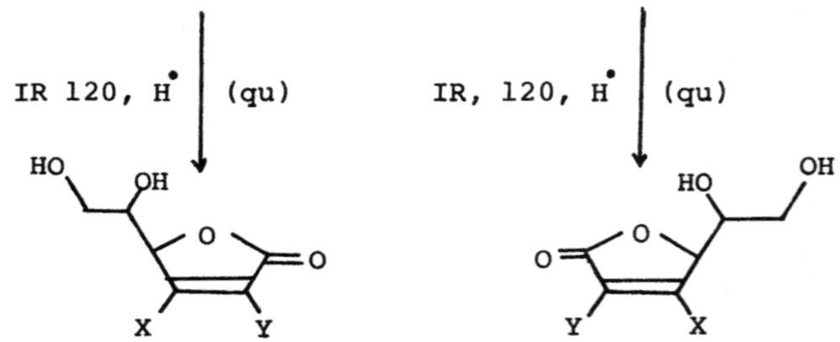

	X	Y	%	erythreo/threo
(a)	EtS	H	60	Z : 1
(b)	PhS	He.	66	Z : 1
(c)	EtS	MeO	60	Z : 1
(d)	MeO	MeO	54	Z : 1

X, Y = MeO: $|\alpha|^{20}_{589}$ - 38.2 (MeOH) $|\alpha|^{20}_{580}$ - 30.9 (MeOH)

Lit.: $|\alpha|^{20}_{589}$ - 37 (MeOH)

Scheme 9

The retrosynthetic scheme 10 outlines a short and efficient KDO-synthesis via an α-hydroxybutenolide intermediate. However, a prerequisite is the generation of β-lithiated species from α-alkoxyacrylates or equivalents, which react with O-protected D-arabinose diastereoselectively to the D-*manno*-isomer.

The results shown in scheme 11 [15] demonstrate, that α-methoxy acrylate is directly lithiated with tert.-butyllithium at the vinylic β-C-atom. However, with protected D-arabinose as electrophile modest yields and low diastereoselectivities were obtained. Deisopropylidenation of the D-*manno*-isomer with acidic ion exchange resin was nearly quantitative, demethylation, however, was frustrated with byproduct formation, an observation made already with structurally similar compounds. Therefore this short route to KDO was not efficient. Structural proof for the intermediates was obtained by transformation of the deisopropylidenated intermediate into a penta-O-acetyl derivative synthesized by Unger and his group [17].

More convenient and comparable with the successful Collins KDO synthesis [18] is the route shown in scheme 12 [15]. Starting from β-alkylthio acrylates and protected D-arabinose (reactions (d)-(f)) a 2.3-dideoxy-D-*manno*-octonate derivative is obtained, which is identical to a compound reached after five steps in the Collins approach [18]. Additional three steps are required for the introduction of the 2-keto groups and for the deprotection and transformation to KDO. The results with various lithiated acrylates clearly show that the stereoselectivities obtained depend on the functional groups at the acrylate system. The amount of

KDO: RETROSYNTHETIC SCHEME

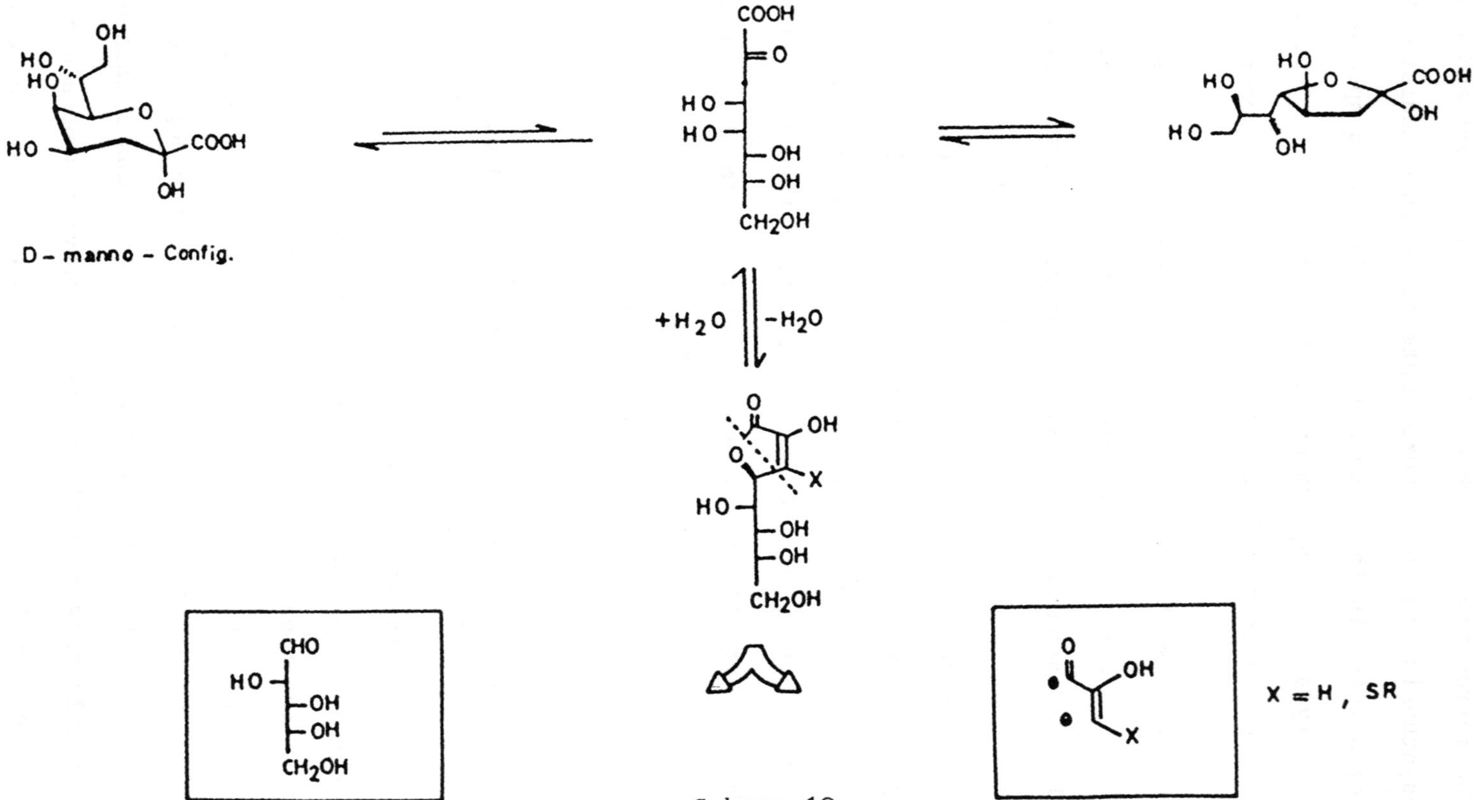

Scheme 10

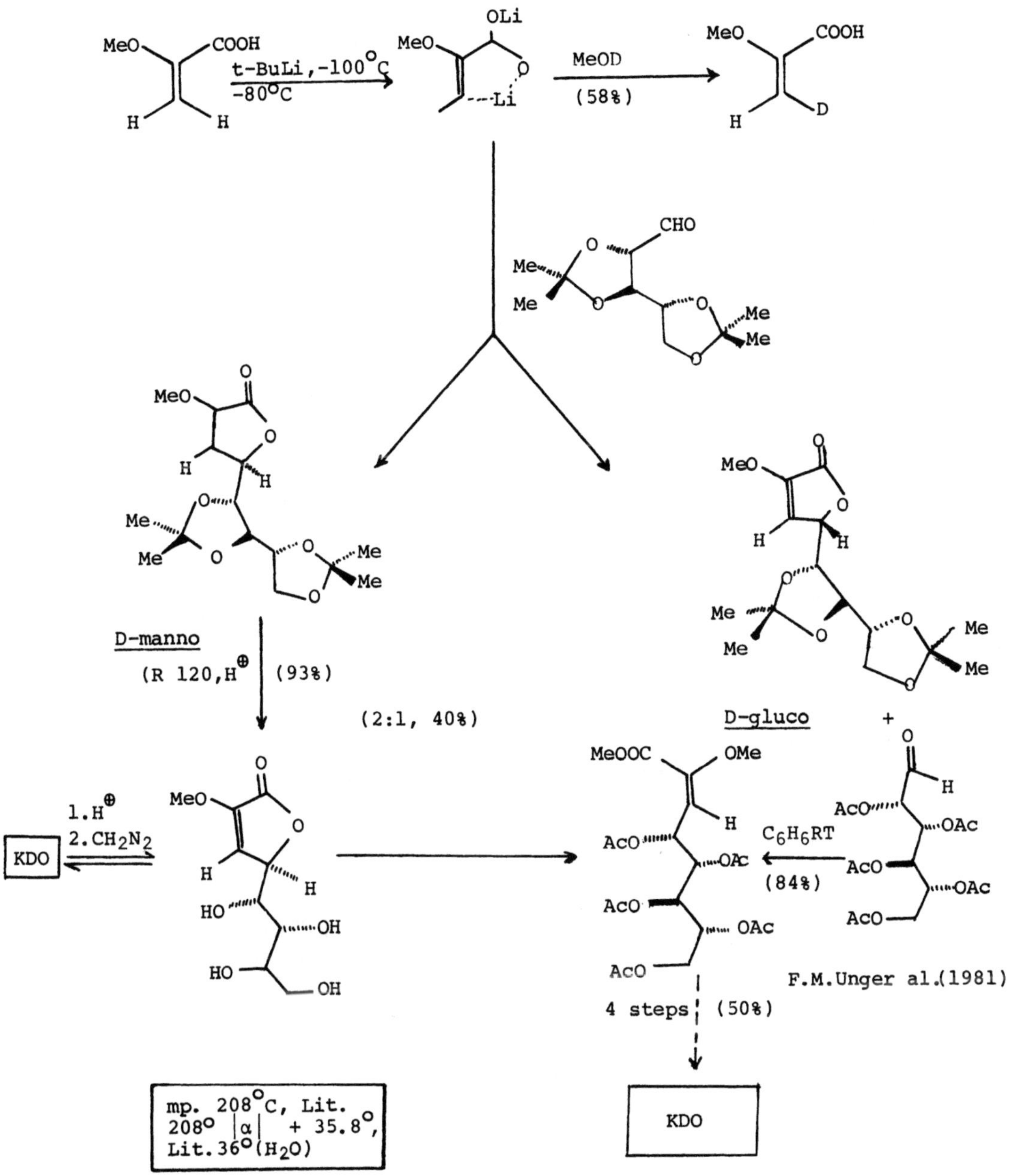
MeO
COOH
t-BuLi,-100°C
-80°C
OLi
MeOD
(58%)
D
CHO
Me
D-manno
(R 120,H⊕
(93%)
(2:1, 40%)
D-gluco
MeOOC
OMe
1.H⊕
2.CH2N2
KDO
AcO
OAc
C6H6RT
(84%)
HO
OH
F.M.Unger al.(1981)
4 steps
(50%)
mp. 208°C, Lit.
208° |α| + 35.8°,
Lit.36°(H2O)
KDO

Scheme 11

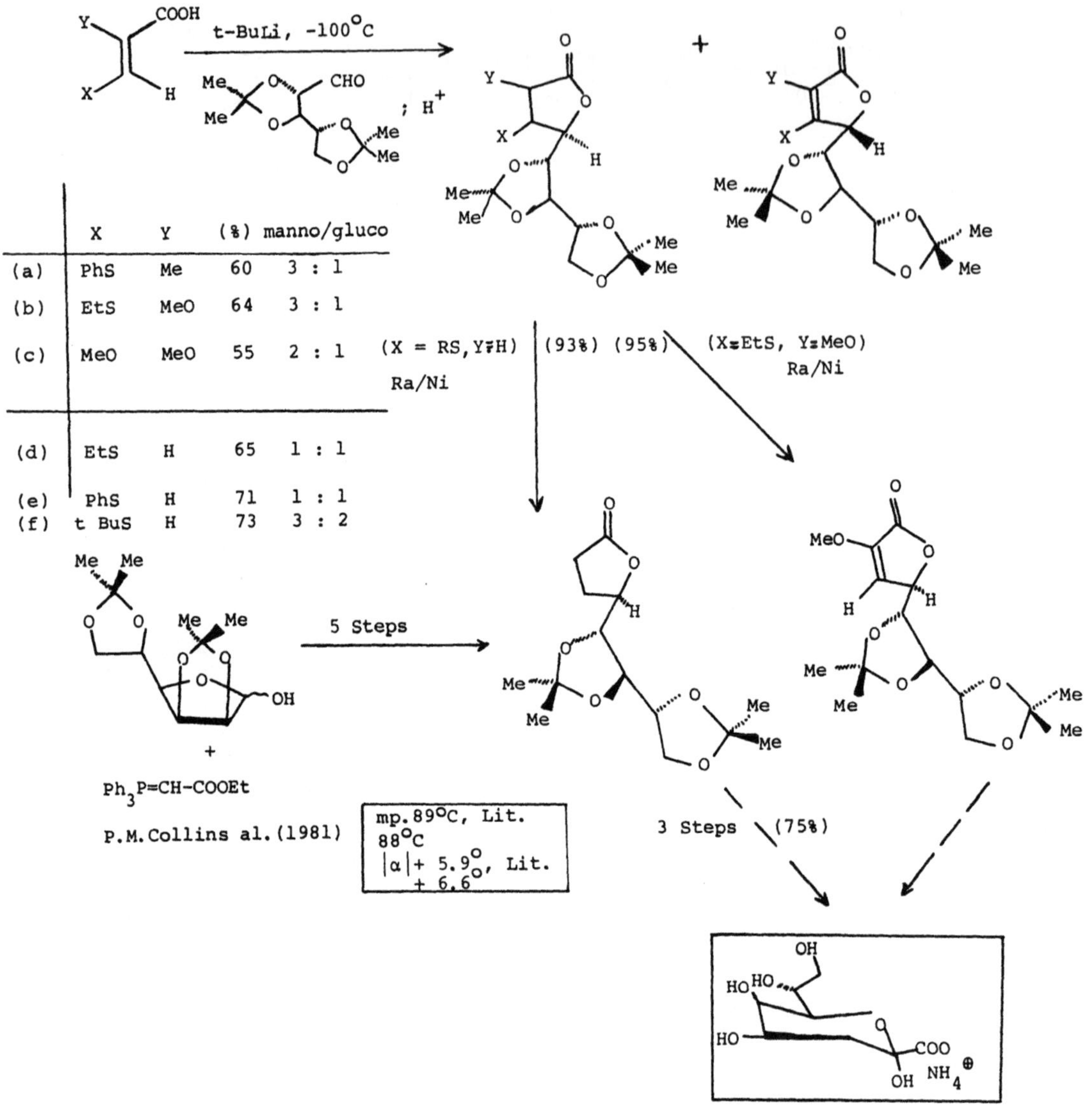

	X	Y	(%)	manno/gluco
(a)	PhS	Me	60	3 : 1
(b)	EtS	MeO	64	3 : 1
(c)	MeO	MeO	55	2 : 1
(d)	EtS	H	65	1 : 1
(e)	PhS	H	71	1 : 1
(f)	t BuS	H	73	3 : 2

Scheme 12

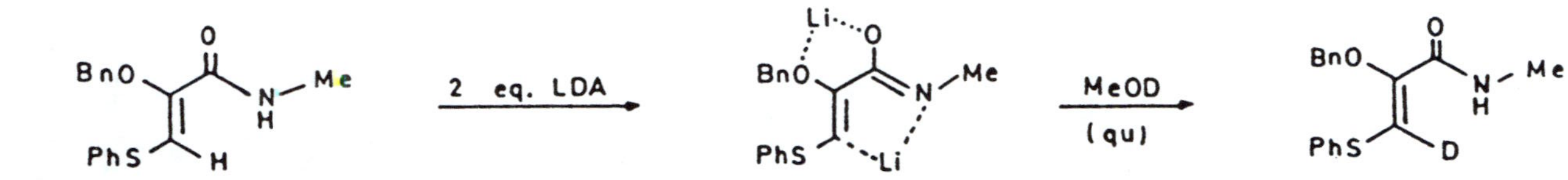

Scheme 13

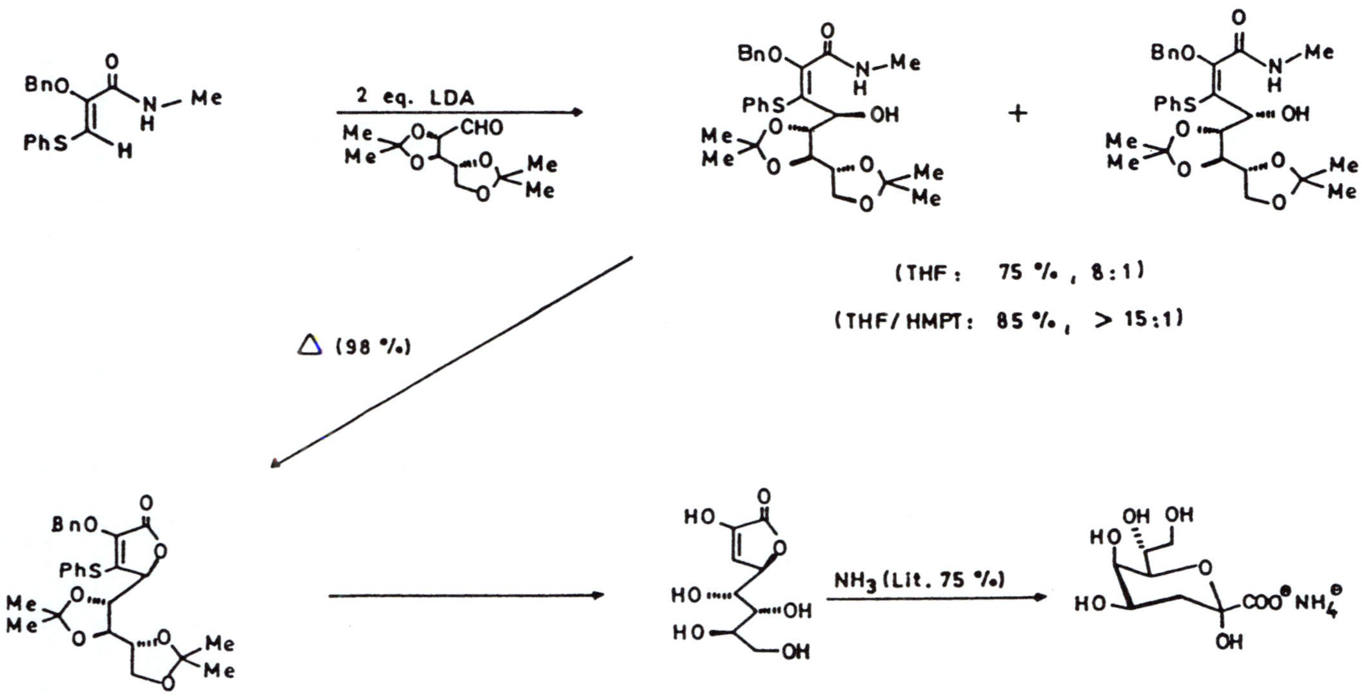

Scheme 14

the *manno*-isomer formed is positively influenced by bulkier groups at the α-C-atom. In addition, the reactivity of α-functionally substituted ß-lithiated acrylates is increased by the presence of ß-alkylthio or ß-arylthio substituents.

This experience was used in the specifically designed acrylate system shown in scheme 13 [15]. The α-benzyloxy group was chosen for the introduction and ease of liberation of the 2-keto-functionality. The ß-phenylthio group was selected to increase the reactivity of the ß-lithiated species and the N-methyl amide group was suggested to increase the diastereoselectivity in diastereoface selections due to the rigidity of the overall system.

Deuteration experiments already demonstrated the ease of formation of the dilithiated species in scheme 13 from the easily accessible acrylamide derivative [15]. Reaction of this dilithiated species with 2.3:5.6-di-O-isopropylidene-D-arabinose gave the desired D-*manno*-octulosonate derivative depending on the solvent system used in high yield and in excellent diastereoselectivity (Scheme 14) [21]. Clean crystallisation of this compound avoided any chromatographic separation of the diastereoisomers. Steric and stereoelectronic effects favor *Re*-attack of the vinyllithium derivative in the (M)-conformer over *Si*-attack. For the (P)-conformer both reactions seem to be less favored [22]. This way the preferential *manno*-isomer formation can be rationalized. The transformation of the intermediate obtained into KDO was carried out in five convenient steps [21].

ACKNOWLEDGEMENTS

Financial support from the Deutsche Forschungsgemeinschaft and the Fonds der Chemischen Industrie is gratefully acknowledged. - I am particularly indebted to my coworkers Rainer Betz, Rolf Hirsenkorn and Okiko Miyata.

REFERENCES

1. Vinyl Carbanions, Part 20. - For part 19, see ref. 21.
2. R.R. Schmidt, Lect.Heterocycl.Chem. 4, 97 (1978); Suppl.J. Heterocycl.Chem. 15 (1978), and references cited therein.
3. R.R. Schmidt and R. Hirsenkorn, Tetrahedron 39, 2043 (1983), and references cited therein.
4. R.R. Schmidt, Bull.Soc.Chim.Belg. 92, 825 (1983), and references cited therein.
5. H.W. Gschwend and H.R. Rodriguez, Org.Reactions 26, 1 (1979).
6. R.R. Schmidt, J. Talbiersky, and P. Russegger, Tetrahedron Lett. 1979, 4273.
7. R.R. Schmidt and H. Speer, Tetrahedron Lett. 1981, 4259.
8. R.R. Schmidt and J. Talbiersky, Angew.Chem. 89, 891 (1977); Angew.Chem.Int.Ed.Engl. 16, 851 (1977).
9. O. Miyata and R.R. Schmidt, Tetrahedron Lett. 23, 1793 (1982).
10. N.G. Clemo and G. Pattenden, Tetrahedron Lett. 23, 589 (1982).

11. S.E. Hall and W.R. Roush, J.Org.Chem. 47, 461 (1982), and referendes cited therein.

12. R.E. Ireland, W.J. Thompson, G.H. Sronji, and R. Effer, J.Org.Chem. 46, 4863 (1981), and references cited therein.

13. A.K. Mallams, M.S. Puar, and R.R. Mossmann, J.Am.Chem.Soc. 103, 3943 (1981); T. Tatsuya, K. Masaji, S. Kuniikatsu, and T. Fusao, J.Antibiot. 35, 979 (1982), and references cited therein.

14. R. Hirsenkorn, unpublished results.

15. R. Betz, unpublished results.

16. We are grateful to O. Scheidsteger and G. Huttner for carrying out an X-ray analysis.

17. F.M. Unger, Adv.Carbohydr.Chem.Biochem. 38, 323 (1981), and references cited therein.

18. P.M. Collins, W.G. Overend, and T. Shing, J.C.S.Chem.Commun. 1981, 1139.

19. D. Sharon, R.S. Sarfati, D.R. Strobach, and L. Szabo, Eur.J. Biochem. 11, 364 (1964); M.B. Perry and D.T. Williams in Methods in Carbohydrate Chemistry, R.L. Whistler and J. N. BeMiller, eds., Academic Press, New York 1967, Vol. 7, p. 44.

20. M.A. Ghalambor, E.M. Levine, and E.C. Heath, J.Biol.Chem. 241, 3207 (1966); C. Hershberger, M. Davis and S.B. Binkley, ibid. 243, 1585 (1968); C. Hershberger and S.B. Binkley, ibid. 243, 1578 (1968); N.K. Kochetkov, B.A. Dimitriev, and L.V. Bachinovs ky, Carbohydr.Res. 11, 193 (1969).

21. R.R. Schmidt and R. Betz, Angew.Chem., submitted for publication.

22. Comparable results were obtained in aldol reactions of α-alkoxy aldehydes: D.A. Evans, J.V. Nelson, and T.R. Taber in Topics in Stereochemistry, N.L. Allinger, E.L. Eliel, and S.H. Wilen, eds., John Wiley and Sons, New York 1982, Vol. 13, p. 1, and references cited therein.

The Alkaloids of *Zizyphus sativa*

A. H. Shah[1], G. A. Miana[1], and R. Tschesche[2]

[1]Department of Chemistry, Gomal University, D. I. Khan, Pakistan
[2]Institut für Organische Chemie und Biochemie, Universität Bonn, D-5300 Bonn, FRG

Abstract: From the bark of Zizyphus sativa Gaertn seven previously undescribed cyclopeptide alkaloids have been isolated. The structures of these new compounds were elucidated by spectroscopic methods applied to the alkaloids and the some of their transformation products, and by chemical degradation reactions.

Introduction

The cyclopeptide alkaloids are an expanding group of polyamide plant bases which incorporate an alkoxystyrylamine unit as an integral part of 13-, 14-, or 15-membered heterocyclic rings. The size of the rings has been adopted as the basis for the classification of this group of natural products. The larger group containing 14 members is again separated via the contained β-hydroxyamino acid; namely, those with β-hydroxyleucine (frangulanine type), with β-hydroxyphenylalanine (integerrine type), and with trans-3-hydroxyproline (amphibine-B type). Cyclopeptide alkaloids are commonly found in the plants of family Rhamnaceae [1].

Results and Discussion

In continuation of our work on cyclopeptide alkaloids [2] from the plants of family Rhamnaceae we now report the alkaloids of Zizyphus sativa Gaertn. The plant under our investigation was collected from

Natural Product Chemistry
Ed. by Atta-ur-Rahman

Hazara district of Pakistan. Extensive chromatography of the crude bases furnished seven previously unknown cyclopeptide alkaloids. The three known compounds isolated were: frangufoline, nummularin-B and mucronine-D. The new compounds are named as: sativanine-A, -B, -C, -D, -E, -F and -G. The column chromatography fraction no.12 contains further new compounds which have been detected by the mass spectrometry of this fraction. Work is in progress on this fraction. Here we wish to give details about the new compounds isolated from this plant.

Sativanine-A (1)

Sativanine-A was isolated from the polar fraction by thin layer chromatography on silica gel plates as colourless substance. It melted at 80° and was easily soluble in methanol, chloroform and dichloromethane. The IR spectrum of (1) displayed characteristic secondary amide bands, conjugated C=C double bond, an aryl ether and N-methyl groups. The UV spectrum showed end absorption with shoulders at 250 and 280 nm which revealed the presence of 14-membered ring system [3].

The elementary composition of (1) was determined by high resolution mass spectrometry as $C_{30}H_{40}N_4O_4$. The mass spectrum of (1) Fig.1 follows the typical fragmentation pattern for the 14-membered cyclopeptide alkaloids [4].

The ions a (m/z 114), i (m/z 135), m (m/z 131 and the amine fragment $[H_2N^+{=}CH{-}CH(CH_3)_2]$ show the presence of N,N-dimethylleucine or isoleucine, p-hydroxystyrylamine, β-phenylserine and valine. The fragment c (m/z 229) indicates that p-hydroxystyrylamine unit is linked with β-phenylserine. On the other hand ions f (m/z 224) and h (m/z 260) demonstrate that p-hydroxystyrylamine unit is bound on one side through oxygen with β-phenylserine and on the other side with valine through nitrogen. The ions k (m/z 230) and l (m/z 202) represent the direct linkage between β-phenylserine and valine; thus proving the structure of the 14-membered ring system.

Table A. High resolution mass spectrometry of Sativanine-A (1)

Ion	Formula	Div.	Found	Intensity (%)
M^+	$C_{30}H_{40}O_4$	-0.5	520.3050	0.06
a	$C_7H_{16}N$	-0.7	114.1275	100
b	$C_{26}H_{31}N_4O_4$	0.8	463.2353	1.45
c	$C_{13}H_{13}N_2O_2$	2.6	229.1003	0.15
d	$C_{12}H_{13}N_2O$	-0.3	201.1025	0.32
e	$C_{20}H_{23}N_2O_2$	-0.2	323.1758	0.56
f	$C_{15}H_{14}NO$	0.5	224.1081	0.98
g	$C_{23}H_{25}N_3O_4$	-	407	0.5
h	$C_{14}H_{16}N_2O_3$	0.4	260.1165	1.53
i	C_8H_8NO	0.3	135.0687	13.59
j	$C_{22}H_{24}N_2O_3$	0.1	364.1788	0.68
k	$C_{14}H_{16}NO_2$	-1.6	230.1165	0.34
l	$C_{13}H_{16}NO$	-2.2	202.1210	0.68
m	C_9H_7O	-0.3	131.0494	7.66
n	$C_4H_{10}N$	2.4	72.0837	3.54

The fragment ions a (m/z 114) and b (m/z 463) are of high intensity. They are formed due to the breakage of C-C bond adjacent to the N,N-dimethylamino group (α-cleavage). The ion a analysed as $C_7H_{16}N$ forms the base peak of the spectrum. This gave a daughter ion at m/z 85 by the loss of ethyl radical with a strong meta stable ion for that process at m* = 63.37.

This fragmentation pattern is encountered for N,N-dimethylisoleucine but not for leucine. The principal fragmentations observed are listed in Table A and assignments are depicted in Scheme I. The elementary composi-

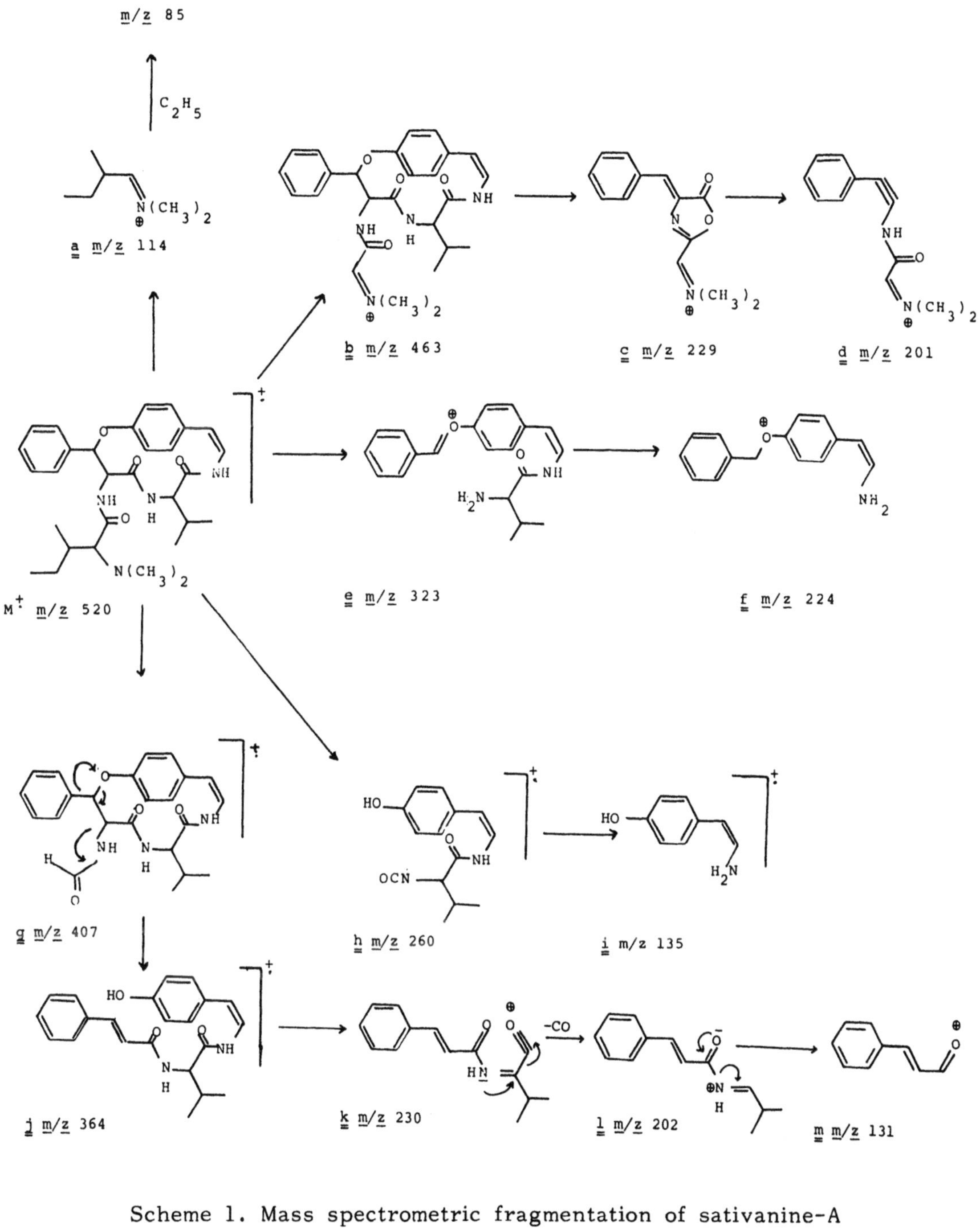

Scheme 1. Mass spectrometric fragmentation of sativanine-A

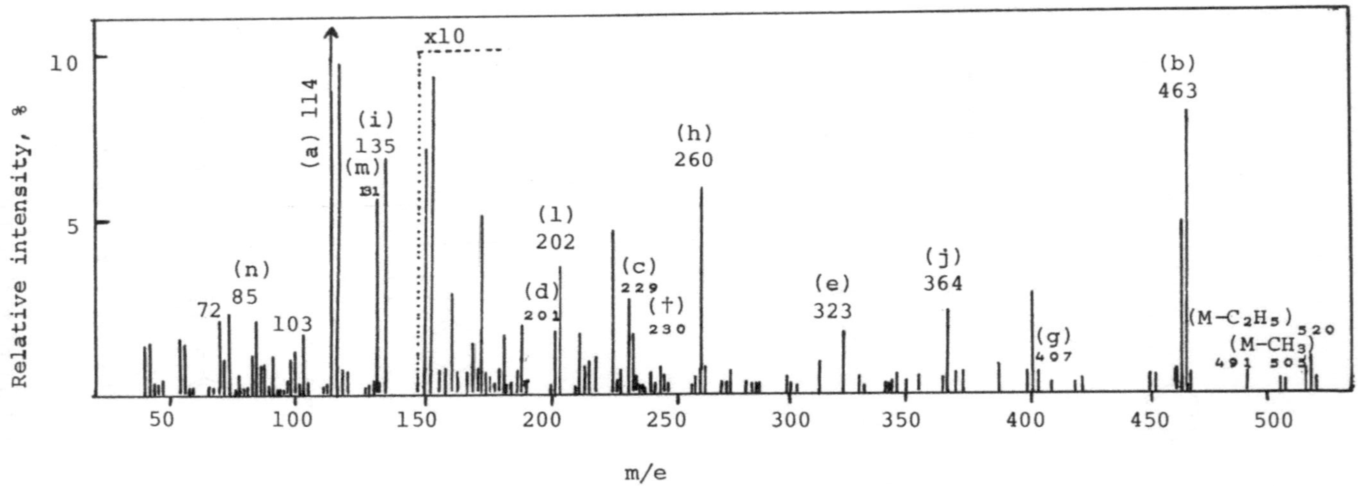

Fig.1. Mass spectrum of sativanine-A(1) at 70eV

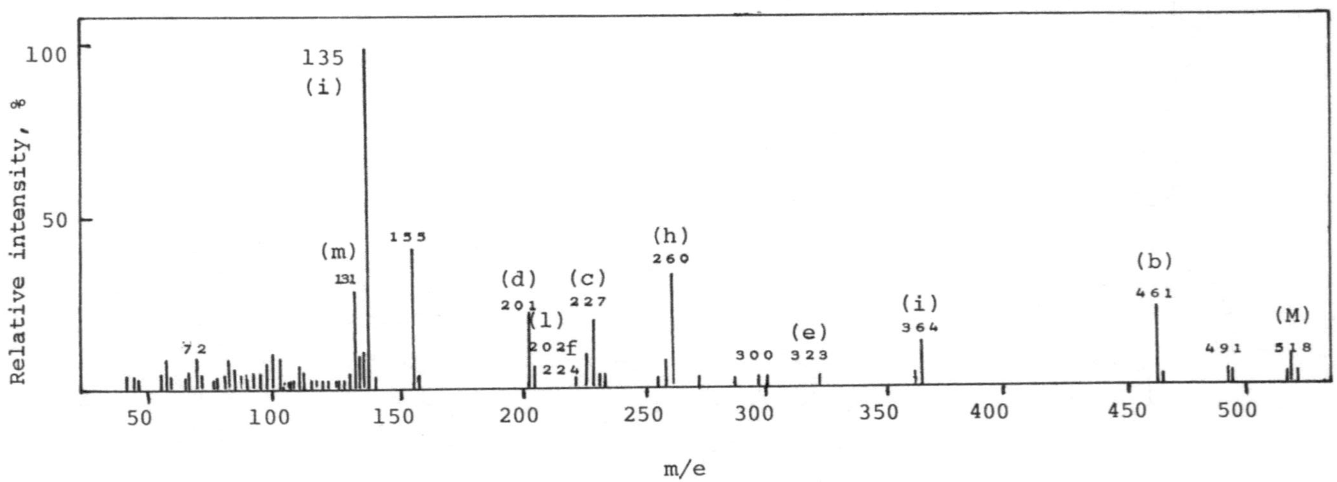

Fig.2. Mass spectrum of sativanine-B(2) at 70eV

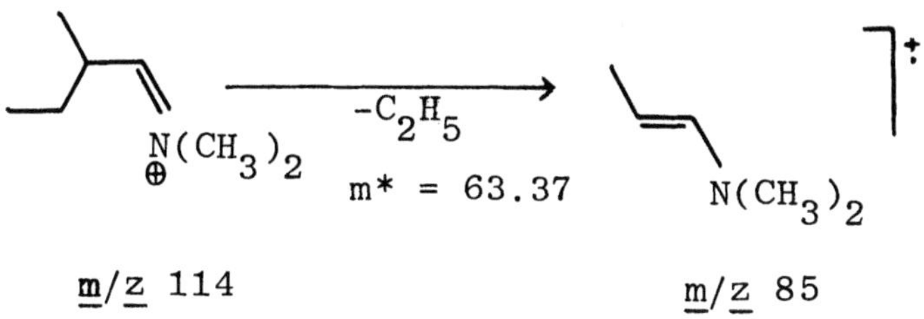

tion of all fragments was substantiated by high resolution mass measurements. The structure (1) is confirmed for sativanine-A.

Sativanine-B (2)

Sativanine-B (2) is a colourless amorphous substance which is easily soluble in methanol and chloroform. Its IR spectrum gave the usual bands for cyclopeptide alkaloids. The UV spectrum of (2) revealed the presence of 14-membered ring cyclopeptide alkaloid [4].

By high resolution MS the elementary composition of (2) was determined as $C_{30}H_{38}N_4O_4$ (M^+518). It shows 14 double bond equivalents which indicates the presence of an additional ring in the side chain. The principal fragments observed and assignments are exhibited in Table-II and Scheme-II. The mass spectrum of (2) differs greatly from those of 14-membered cyclopeptide alkaloids but resembles closely that of nummularine-G [5] which carries an additional ring in the side chain.

Through the breakage of butyl group on the terminal amino acid(α-cleavage) the primary fragment ion m/z 461 is formed. Because of the very small amount of the material it was not possible to confirm whether this amino acid was the derivative of leucine or isoleucine. On the breakage of the cyclic side chain, the ions m/z 155 ($C_8H_{15}N_2O$) and j (m/z 364) are formed. The ion m (m/z 131), i (m/z 135) and m/z 72 indicate the presence of phenylserine, p-hydroxystyrylamine and valine. The fragment i forms the base peak of the spectrum. The ions f (m/z 224) and h (m/z 260) show that p-hydroxystyrylamine unit is bound on one

Table B. High resolution mass spectrometry of Sativanine-B (2)

Ion	Formula	Div.	Found	Intensity (%)
M^+	$C_{30}H_{38}N_4O_4$	-0.5	518.2888	2.84
b	$C_{26}H_{29}N_4O_4$	1.2	461.2201	11.84
c	$C_{13}H_{11}N_2O_2$	1.3	227.0834	32.03
d	$C_{12}H_{13}N_2O$	-0.3	201.1024	12.90
e	$C_{20}H_{23}N_2O_2$	-	323	0.5
f	$C_{15}H_{14}NO$	0.2	224.1078	1.29
h	$C_{14}H_{16}N_2O_3$	-0.2	260.1159	6.27
i	C_8H_9NO	-0.6	135.0678	100
j	$C_{22}H_{24}N_2O_3$	0.5	364.1788	7.7
l	$C_{12}H_{15}NO$	-1.1	202.1176	4.50
m	C_9H_7O	-3.7	132.0493	1.81
n	$C_4H_{10}N$	-0.2	72.0811	7.32

side with phenylserine and on the other side with valine. The linkage of the side chain with phenylserine is confirmed by fragments c (m/z 227) and d (m/z 201). The identity of each fragment was proved by high resolution mass spectrometry.

Sativanine-C(3)

Sativanine-C crystallized from $CHCl_3$/Ether, is a colourless substance with m.p. 113-114°. It is easily soluble in chloroform, methanol and dichloromethane but insoluble in Pet.ether. It gave orange colour with Dragendorff reagent.

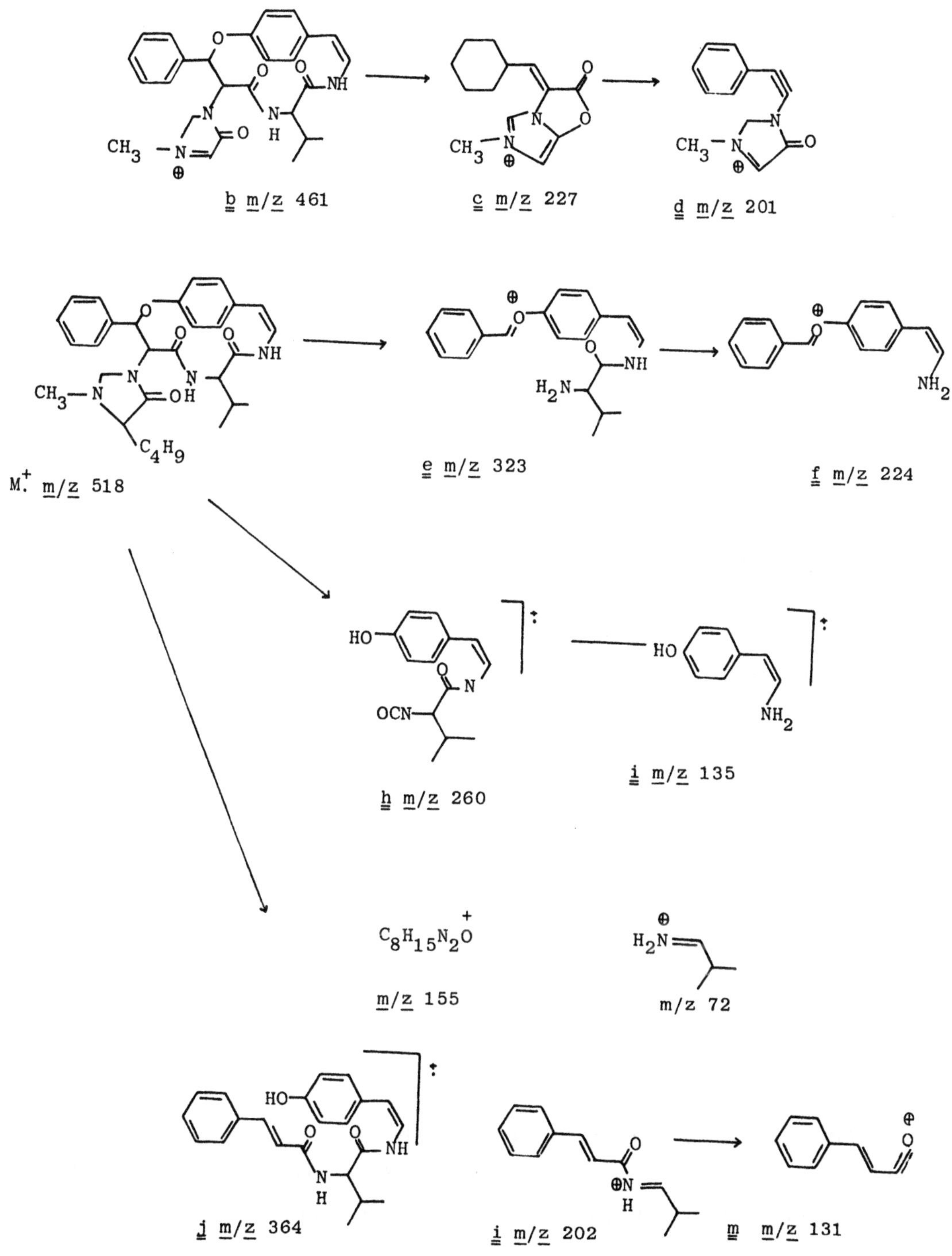

Scheme 2. (Mass spectrometric fragmentation of sativanine-B)

Its UV spectrum (measured in MeOH) gave absorption maxima at 320 and 258 nm, chracteristic of 13-membered ring containing cyclopeptide alkaloids [6].

The IR spectrum displayed characteristic secondary amide bands and absorption attributed to a conjugated (C=C) double bond, an aryl ether and a N-methyl group.

The molecular formula was determined by high resolution mass spectrometery as $C_{29}H_{43}N_5O_6$. From the mass spectrum of this compound Fig.3 valuable and important informations about the basic structural units can be obtained. Specially the lower portion of the mass spectrum is more interesting. The mass spectrum of (3) resembles that of 13-membered cyclopeptide alkaloid; mucronine-D [6]. The base peak at m/z 58 represents the amine fragment a (CH_3-NH=CH-CH_3) which indicates that the end amino acid could be N,N-dimethylglycine or N-monomethylalanine. The amine fragment ($C_4H_9CH=N^+H_2$), representing leucine or isoleucune is found at m/z 86. The ions at m/e 96 and m/z 165 show the presence of proline and substituted hydroxystyrylamine unit.

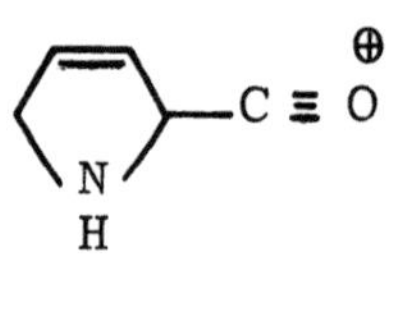

m/z 96

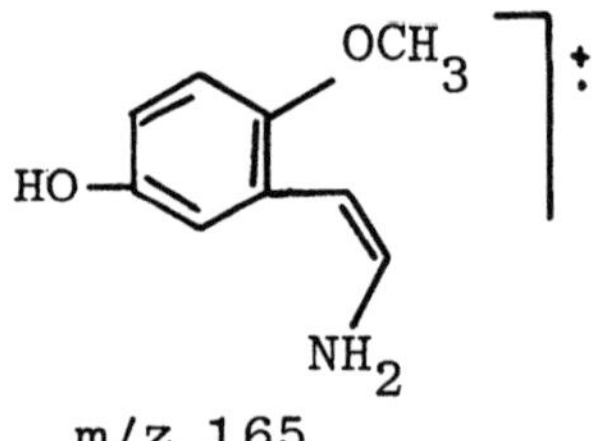

m/z 165

The fragments formed by the stepwise breakage of the side chain and the ring system are very important from structural determination point of view. Specially the fragments which carry two units are more informative.

Sativanine-C(3)

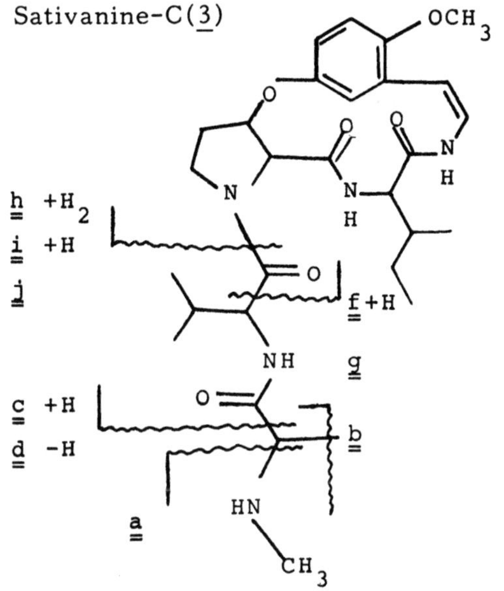

Ion	m/z
a	58
b	542
c	500
d	598
e	457
f	401
g	400
h	374
i	373
j	372

In this way the ions l (m/z 185) and m (m/z 157) shows that the end amino acid (N,N-dimethylglycine or N-monomethylalanine) is bound with the intermediate amino acid valine; while n demonstrates the linkage of the valine with hydroxyproline and so with the 13-membered ring system.

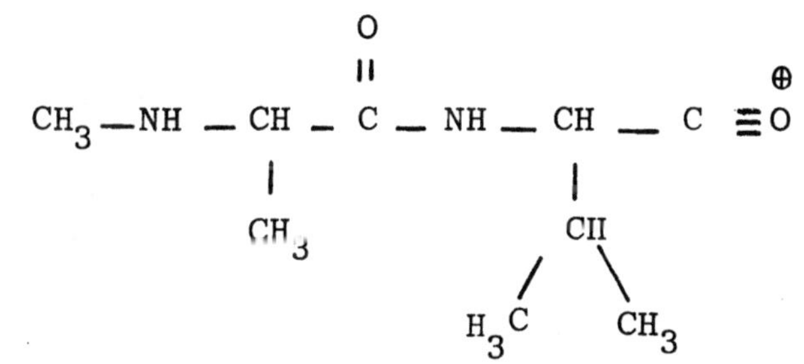

l (m/z 185)

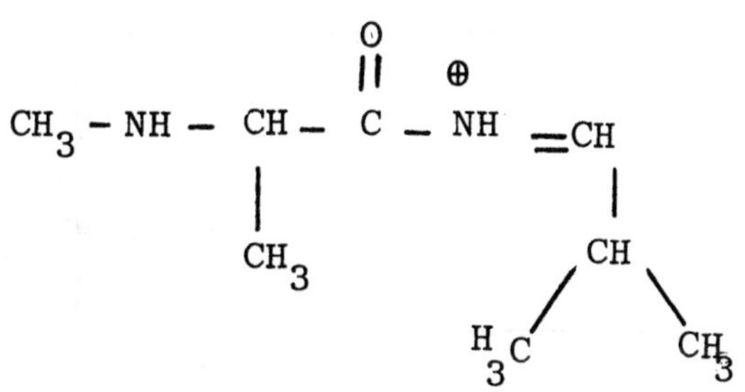

m (m/z 157)

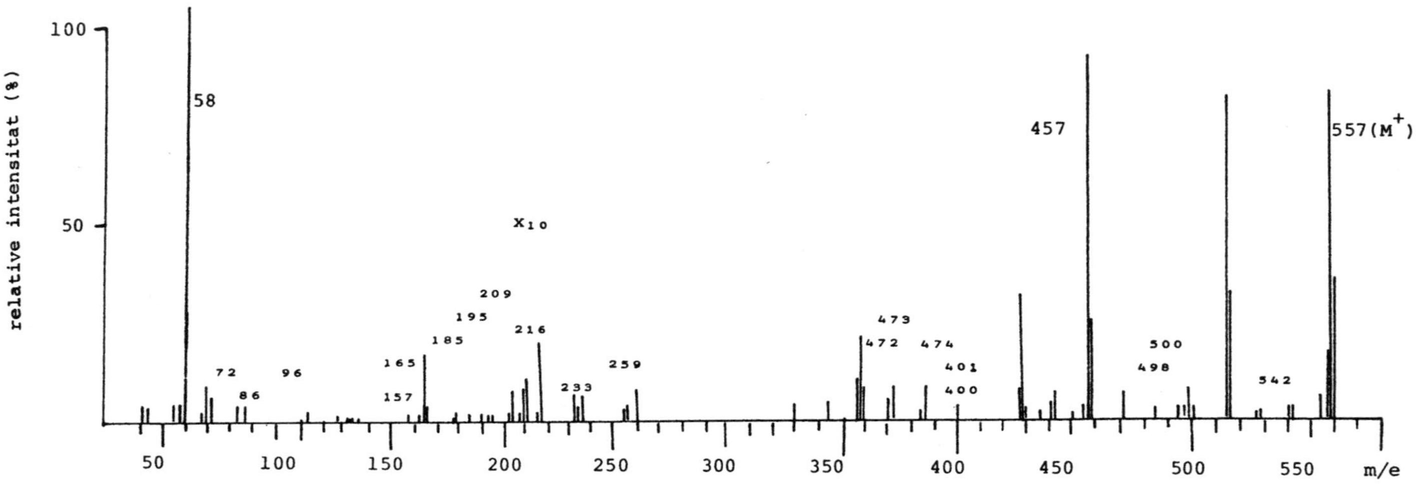

Fig.3. Mass spectrum of sativanine-C

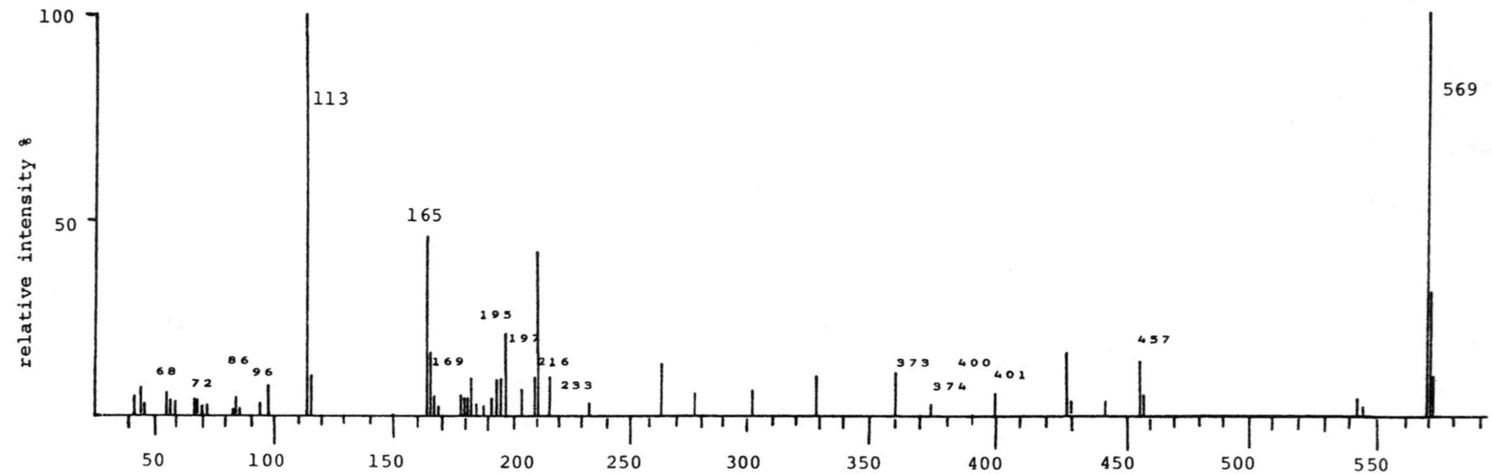

Fig.4. Mass spectrum of sativanine-D

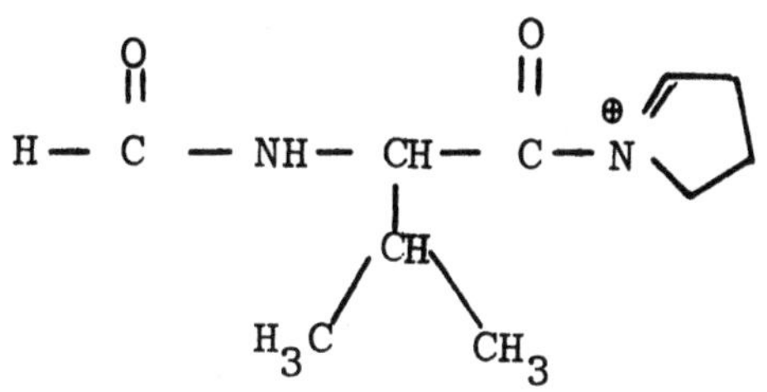

n (m/z 195)

From the fragment ion p it could be seen that hydroxyproline (from the carboxyl group side) is linked with C-6 amino acid (leucine or isoleucine) of the 13-membered ring.

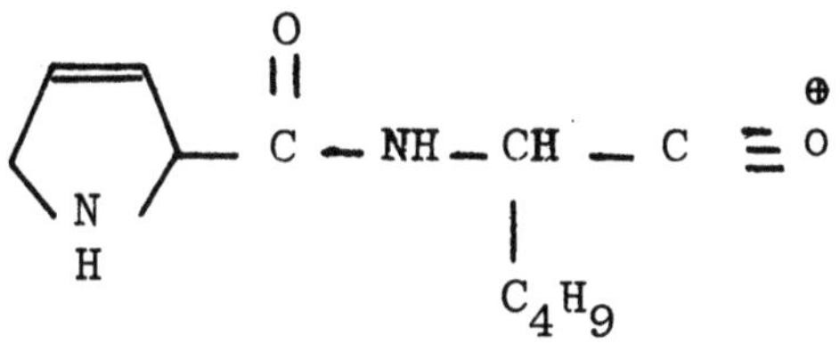

p (m/z 209)

The fragment ions r, s and t demonstrate the linkage of hydroxyproline with styrylamine unit through an ether linkage.

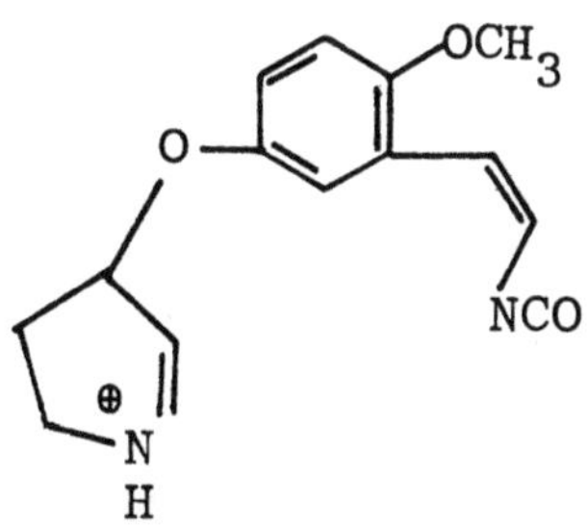

r (m/z 259)

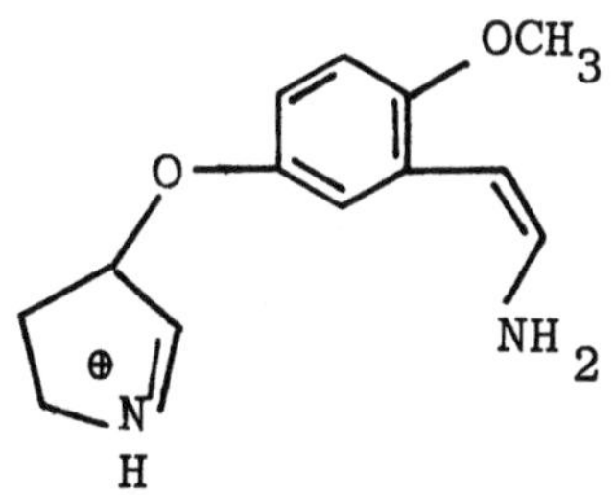

s (m/z 233)

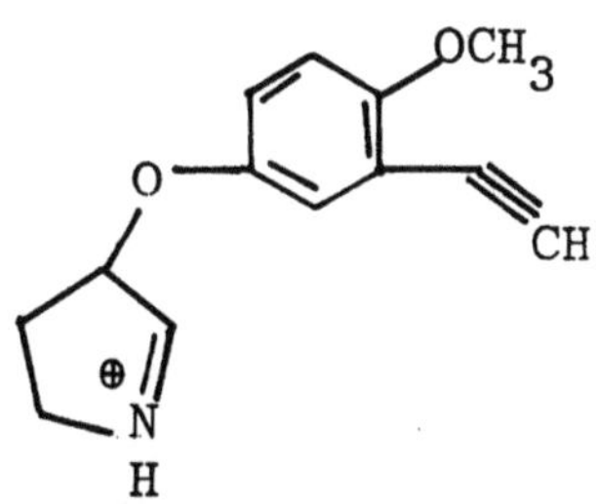

t (m/z 216)

Through the interpretation of mass spectrum still two questions remained open i.e. whether the terminal amino acid is N,N-dimethylglycine or N-monomethylalanine; and secondly C-6 amino acid of the 13-membered ring is leucine or isoleucine.

To confirm the end amino acid (3) was teated with a mixture of formic acid/acetic anhydride. A new compound m.p. 130-131° was formed. This derivative gave very faint colour with Dragendorff reagent and was found less polar on TLC plates than (3). With the help of mass spectrometry its molecular formula was determined as M^{+} 585, $C_{30}H_{43}N_5O_7$. It was confirmed as N-formylsativanine-C (4).

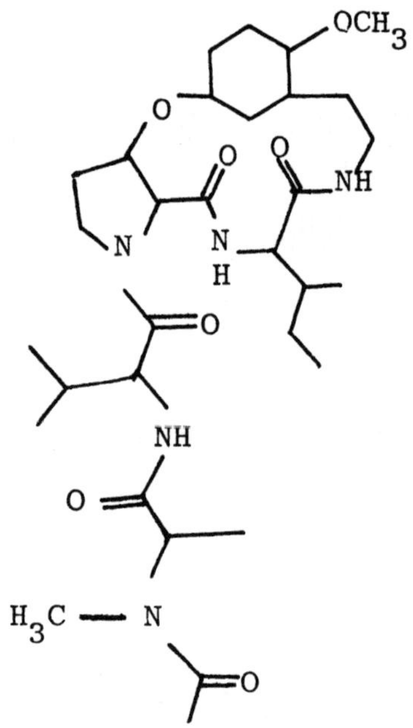

N-Formylsativanine-C (4).

The structure of (4) was confirmed by detailed mass spectrometric studies. The N-formylalanine carrying fragments are: l and m found at m/z 213 and m/z 185 respectively.

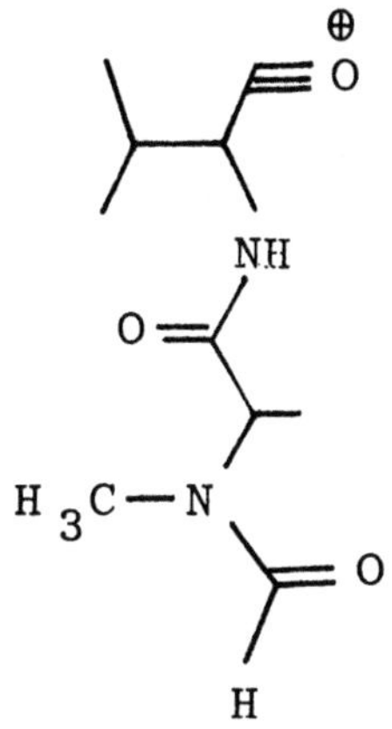

l (m/z 213)

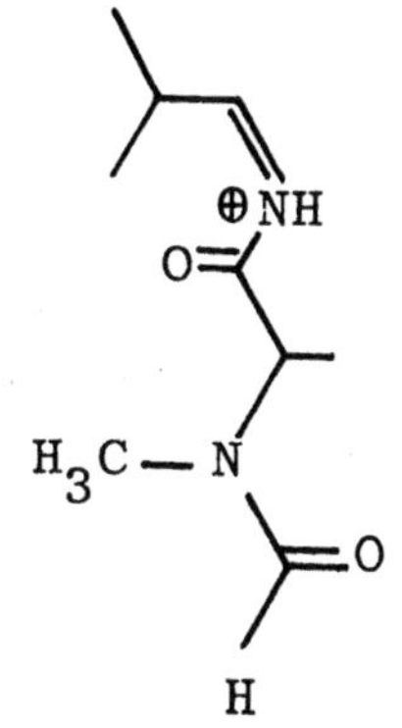

m (m/z 185)

Through the breakage of the end N-formyl methylalanine, the base peak m/z 114 of the mass spectrum is found. On its fragmentation by two times elimination of CO, the ions of higher intensity m/z 86 and m/z 58 are formed. It proofs that the end amino acid in (4) was N-monomethylalanine.

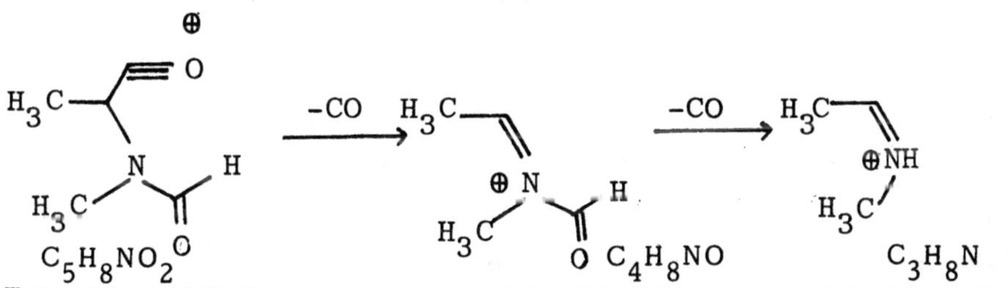

m/z 114 (100%) m/z 86 (50.4%) m/z 58 (37.3%)

Finally the structure of sativanine-C was confirmed as (3) by its acid hydrolysis. In the hydrolysate N-monomethylalanine, isoleucine and valine were confirmed by paper chromatography and direct comparison with the authentic samples.

Sativanine-D (5)

By high resolution mass spectrometry the elementary composition of (5) was determined as $C_{30}H_{43}N_5O_6$. The IR spectrum showed the usual secondary amide and conjugated (C=C) double bond absorption, plus bands attributed to phenolic ether and N-methyl group. The UV spectrum revealed the presence of a 13-membered ring system.

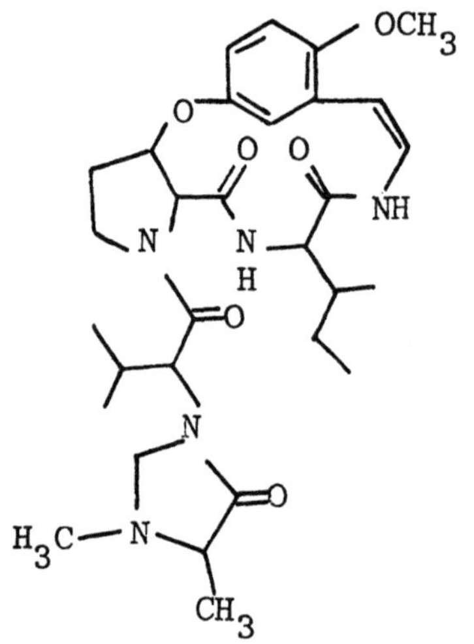

Sativanine-D (5)

The mass spectrum of (5) Fig.4 differ greatly from those of other 13-membered cyclopetide alkaloids but it resembles the spectra of 14-membered cyclopeptide alkaloids nummularine-G and sativanine-B which contain an additional ring in the side chain.

$M^{+\cdot}$(m/z 569) whose elementary composition by high resolution mass spectrometry was given as $C_{30}H_{43}N_5O_6$, appears as a peak of high intensity. The fragment ions a, b, c and d which appear due to the breakage of the side chain and end amino acid in the usual cyclopeptide alkaloids, do not appear in the spectrum of (5). Instead of this the base peak of the spectrum is formed due to the resonance stabilized fragment m/z 113 with ($C_5H_9N_2O$) elementary composition.

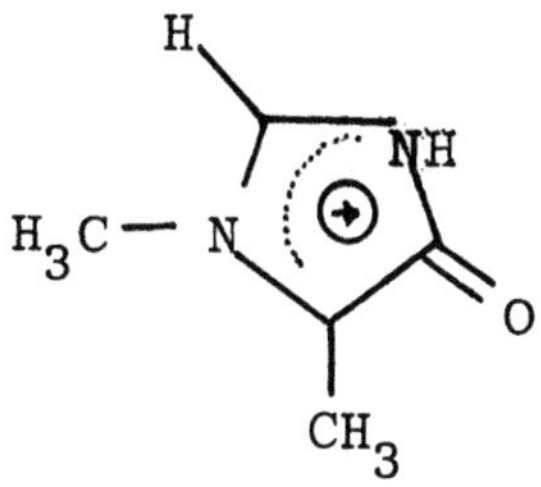

m/z 113

Other fragments containing this additional ring are the ions l (m/z 197) and m (m/z 169).

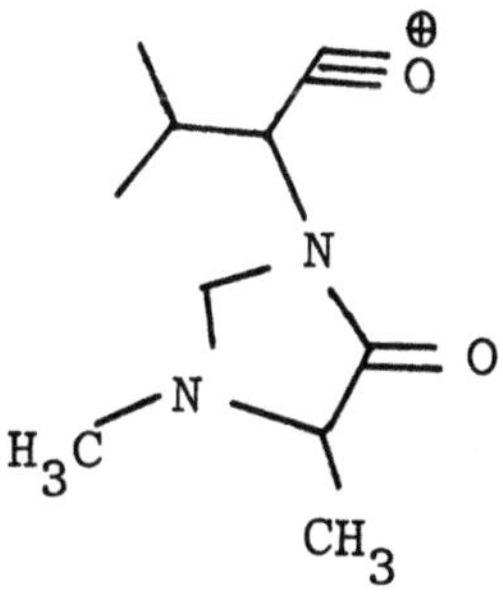

l (m/z 197)

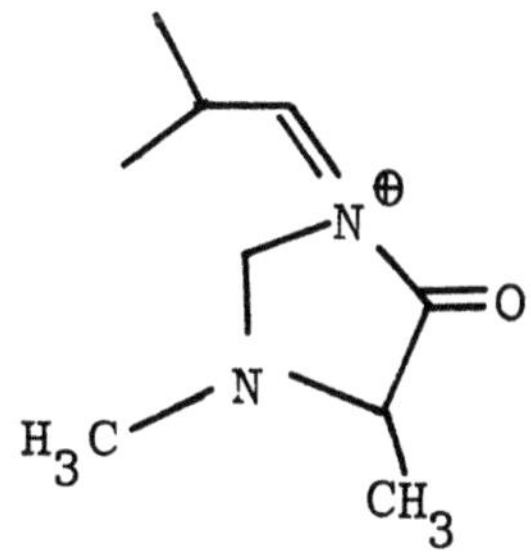

m (m/z 169)

By the breakage of above fragments the following ions are formed which carry the full 13-membered ring system of the alkaloid (5).

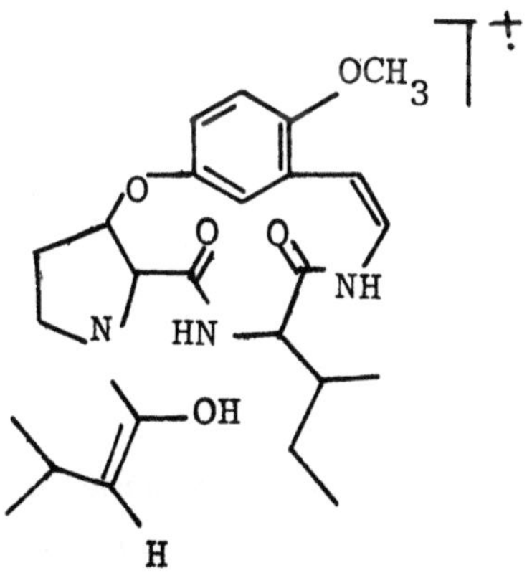

e (m/z 457)

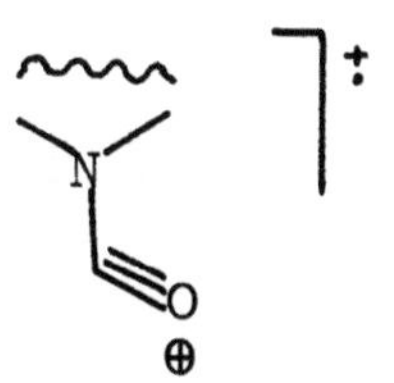

g (m/z 400)

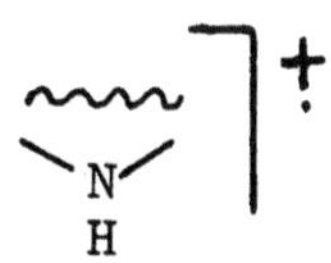

h (m/z 373)

The fragment n (m/z 195) formed through the breakage of ether and carbamide bonds of the hydroxyproline confirms that the side chain amino acid valine is linked with hydroxyproline.

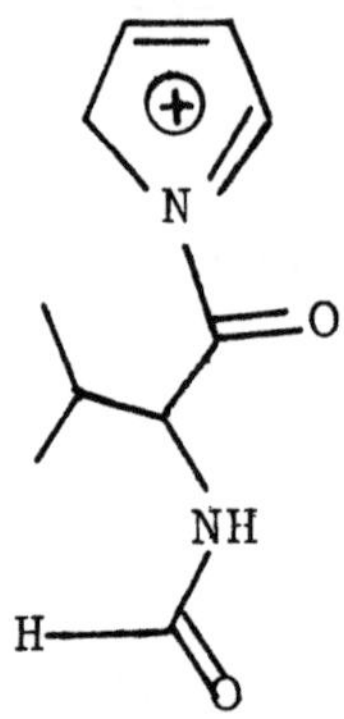

n (m/z 195)

The linkage of hydroxyproline on one side with C-6 amino acid (leucine or isoleucine) and on the other side with styrylamine unit is demonstrated by theions p (m/z 209), s (m/z 233) and t (m/z 216).

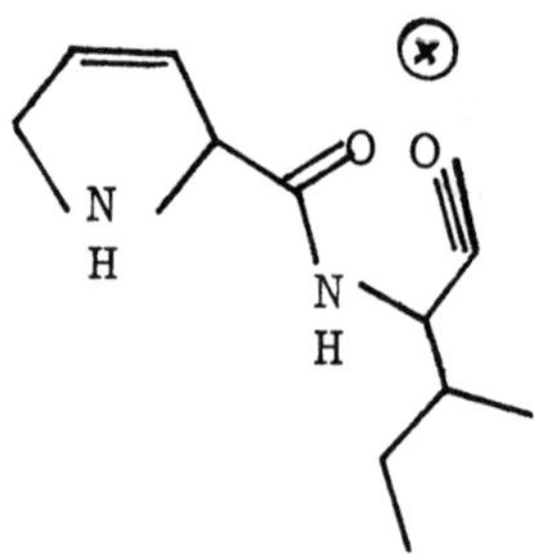

p (m/z 209)

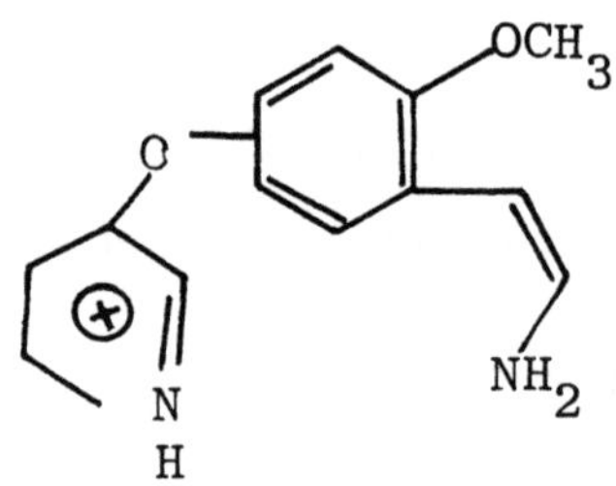

s (m/z 233)

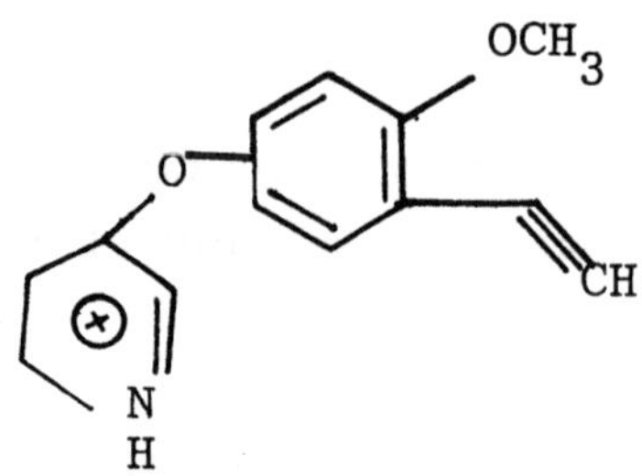

t (m/z 216)

The linkage of the C-6 amino acid (leucine or isoleucine) with styrylamine unit and so the whole ring system is given by the ion u (m/z 304).

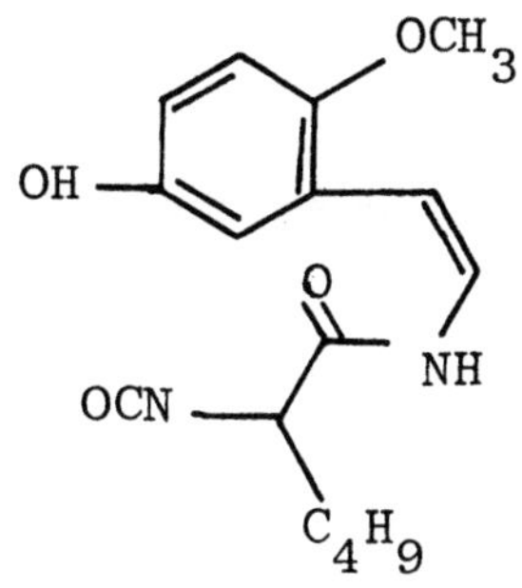

u (m/z 304)

The C-6 amino acid was confirmed to be isoleucine by the hydrolysis of (5) with 6N HCl. In the hydrolysate isoleucine, valine and N-methylalanine were confirmed on paper chromatography and comparision with the authentic samples. Thus proving structure (5) for sativanine-D.

(5) is the first representative of its type in the growing list of 13-membered cyclopeptide alkaloid a derivative namely; nummularine-B-Cycl (6) was prepared from nummularine-B, which gave the analogous fragmentation on electron impact.

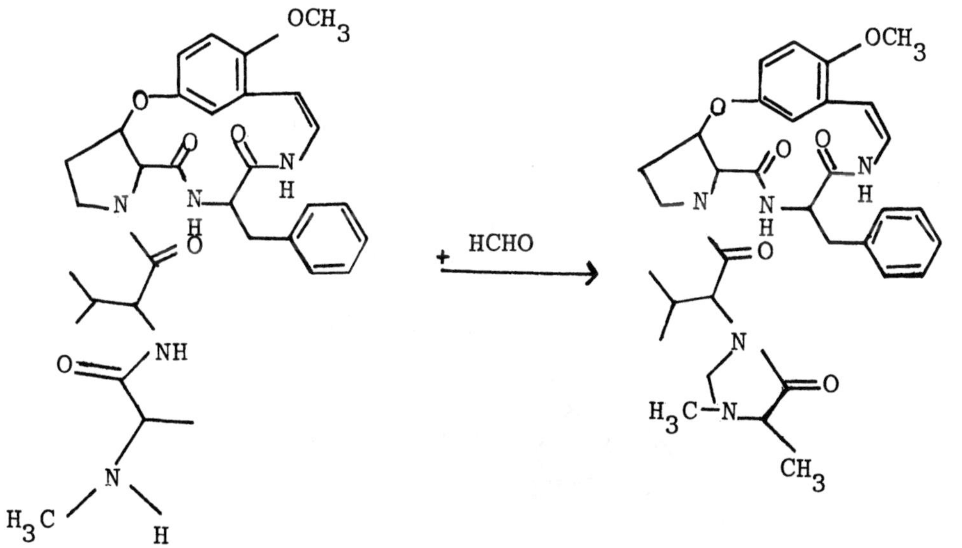

Nummularine-B

Nummularine-B-Cycl (6).

Sativanine-E (7)

The IR spectrum (7) showed the typical bands for cyclopeptide alkaloids. The UV spectrum gave absorption maxima characteristic for 13-membered cyclopeptide alkaloids, at 321 and 268 nm. Shoulders at 289, 282 and 272 nm are typical for a tryptophan unit [7].

The PMR spectrum showed at δ 2.48 and 3.79 ppm singulets for the N,N-dimethylamino and methoxy groups respectively. The doublet at 5.92 (*J* 9Hz) is assigned to the olefinic proton at the carbon atom next to the aromatic ring, the other olefinic proton of the styrylamine function appears as quatret at 6.94 (J_1=9 Hz, J_2=12 Hz). The rest of the signal complex between 6.5 and 8.5 ppm represents 11 protons. The sextet at 5.32 (J_1=6.4 Hz, J_2=3.2 Hz) is assigned to proton at C-3, the doublet at 4.42 (J=3.2 Hz) to the C-2 proton of the hydroxyproline and the triplet at 4.22 (*J* 4.4 Hz) to the C_a-proton of leucine.

The presence of leucine was confirmed by paper chromatography of the acid hydrolysate of (7) and comparison with the authentic sample.

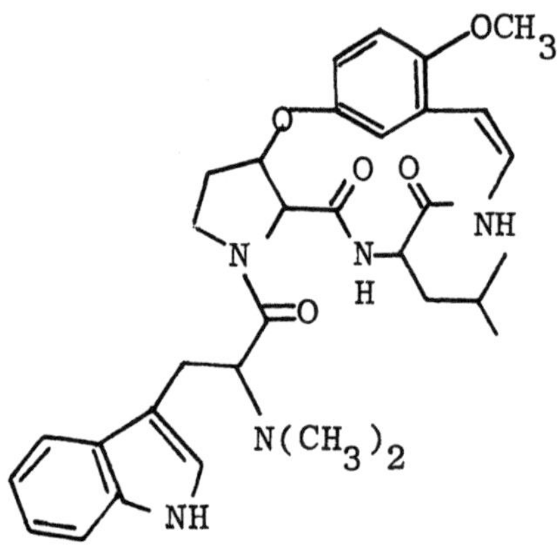

Sativanine-(7)

The molecular formual of (7) was determined by high resolution mass spectrometry to be $C_{33}H_{41}N_5O_N$. The mass spectrum of (7) Fig.5 resembles that of nummularine-C which carries a short side chain [8]. -Cleavage yields the two most intense ions a m/z 187 ($C_{12}H_{15}N_2$) and b m/z 457 ($C_{24}H_{33}N_4O_5$). The typical fragments for hydroxyproline m/z 68 (C_4H_6N), leucine m/z 86 ($C_5H_{12}N$) and the methoxy styrylamine group m/z 165 ($C_9H_{11}NO_2$) reveals the identity of the units that form the 13-membered ring, f m/z 401 ($C_{21}H_{27}N_3O_5$), g m/z 400 ($C_{21}H_{26}N_3O_5$), h m/z 374 ($C_{20}H_{28}N_3O_4$), i m/z 373 ($C_{20}H_{27}N_3O_4$), p m/z 209 ($C_{11}H_{17}N_2O$), g m/z 181 ($C_{10}H_{17}N_2O$), s m/z 233 ($C_{13}H_{17}N_2O$) and u m/z 304 ($C_{10}H_{20}N_2O_4$) show their linkage. The α-cleavage product a fragments further into m/z 144 and m/z 130, typical for tryptophan derivatives.

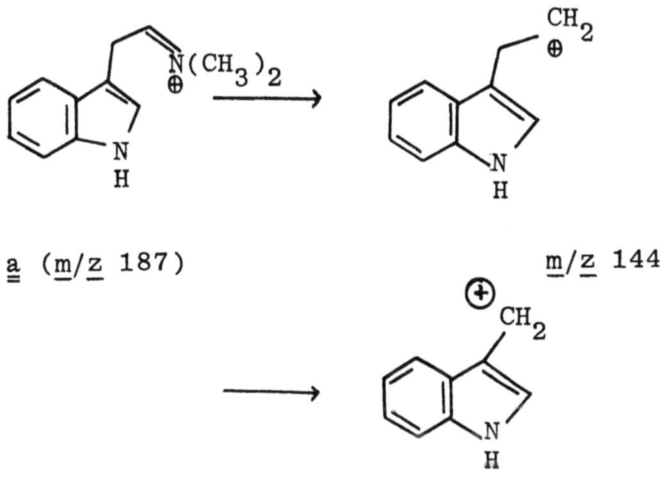

The identity of each fragment was proved by extensive high resolution mass spectrometry.

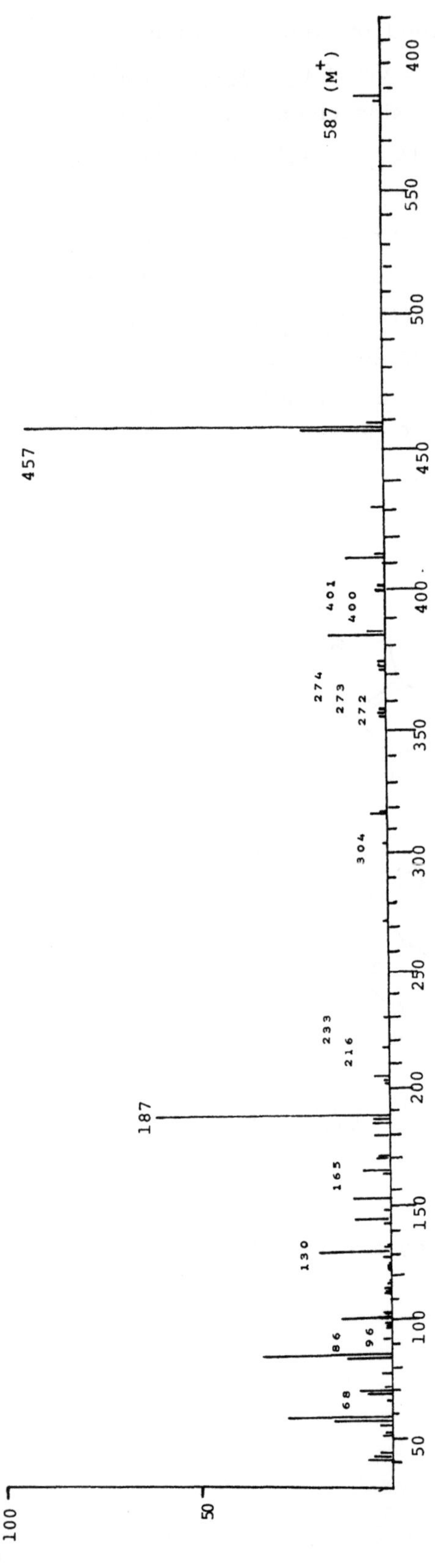

Fig.5. Mass spectrum of sativanine-E

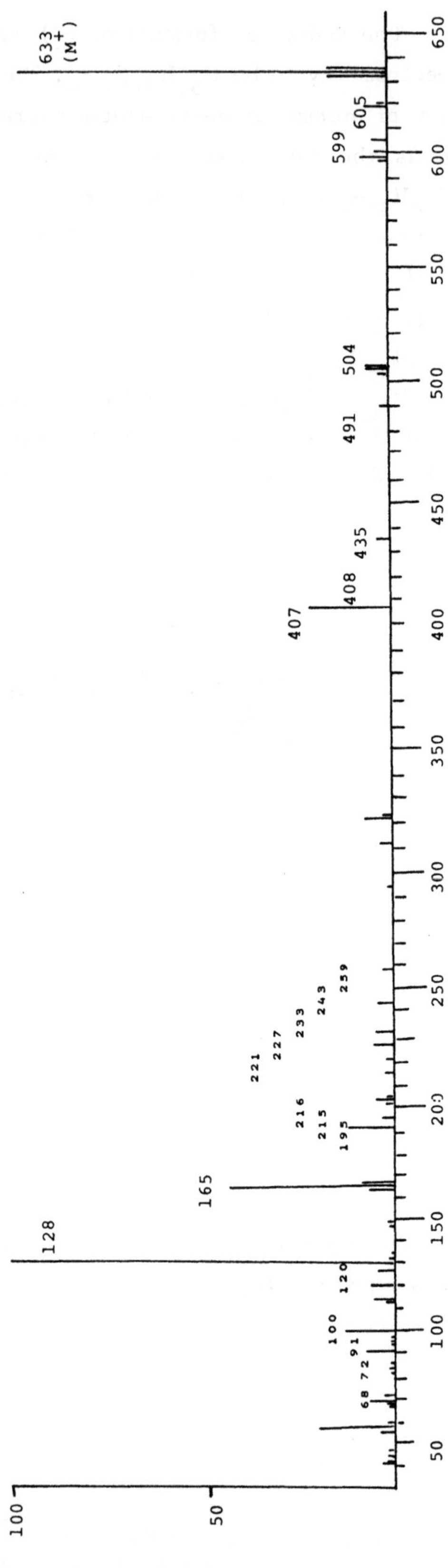

Fig.6. Mass spectrum of sativanine-F

Sativanine-F (8)

Isolated by consective TLC and HPLC, this alkaloid gave very faint colour with Dragendorff's reagent. The IR spectrum showed the typical absorptions for cyclopeptide alkaloids. The UV spectrum gave the characteristic absorption maxima at 320 and 268 nm for the 13-membered ring system.

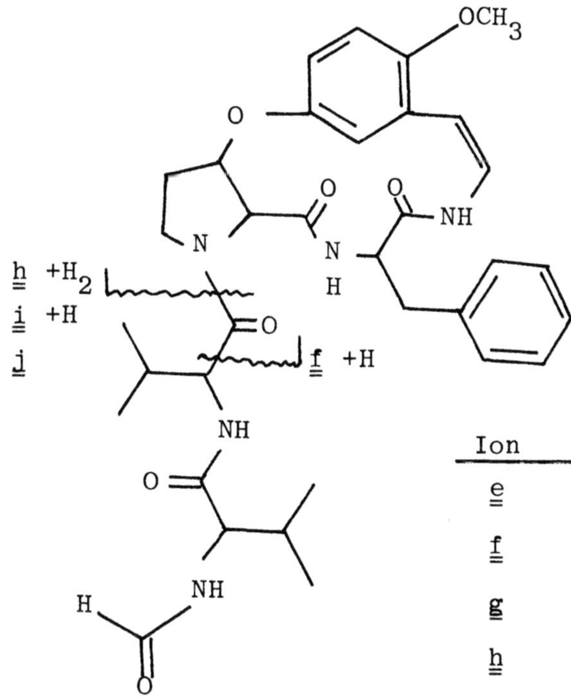

Sativanine-F (8)

Ion	m/z
e	491
f	435
g	-
h	408
i	407
j	406

The mass spectrum of this new cyclopeptide alkaloid Fig.6 which carries a N-formyl rest on the terminal amino acid differs fundamentally from the spectra of all so for known cyclopeptide alkaloids. Since (8) contains no basic amino group the ionized molecule is fairly stable, thus giving an intense molecular ion, and the usual α-cleavage product a and b are absent. Instead of this the base peak is produced by the cleavage of the peptide bond between N-formylvaline and valine yielding the ion m/z 128 whcih thenm eliminates twice CO to give m/z 100 and m/z 72.

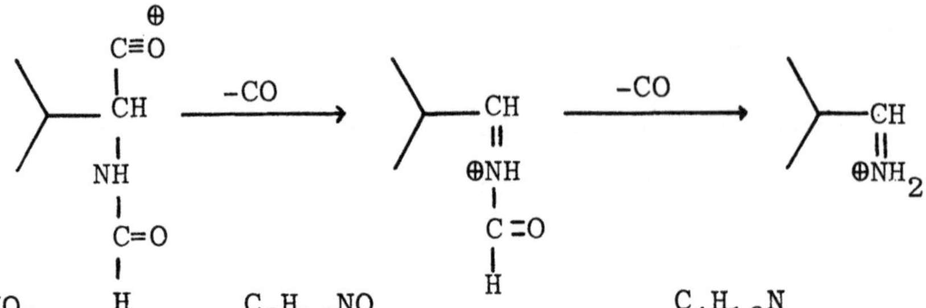

$C_6H_{10}NO_2$ — m/z 128 (100%)

$C_5H_{10}NO$ — m/z 100 (18%)

$C_4H_{10}N$ — m/z 72 (3.6%)

The same cleavage with a hydrogen rearrangement gives the counterpart m/z 506. The peaks at m/z 68,72,120 and 165 indicate the presence of hydroxyproline, valine, phenylalanine and substituted styrylamine unit. The composition of the side chain can be deduced from the fragment l, while n and o prove the attachment of the side chain with hydroxyproline.

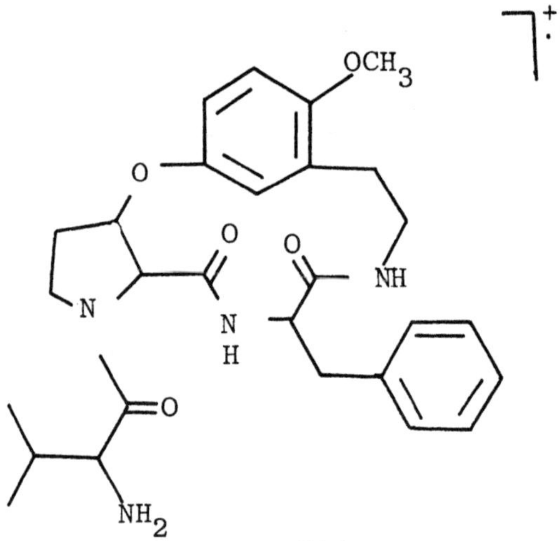

m/z 506

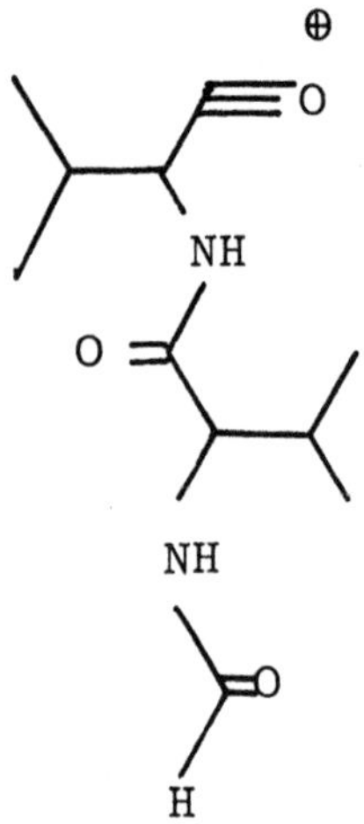

$C_{11}H_{19}N_2O_3$

l (m/z 227)

Hydroxyproline is joined to o-methoxy-p-hydroxystyrylamine by an ether linkage, as can be seen from the peaks r, s, and t, and also to phenylalanine by an amide bonding as can be deduced from the fragments p and q. The whole ring system is represented by the ions e, f, h and i.

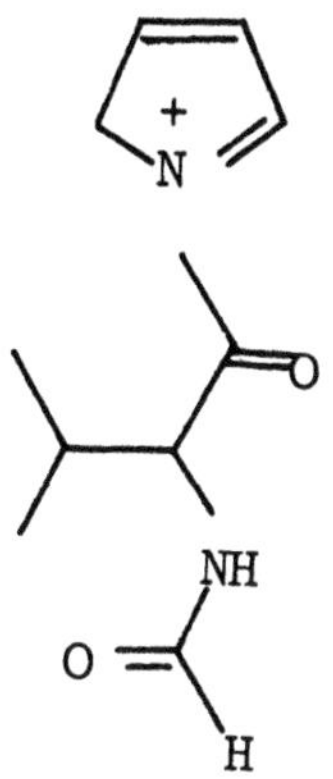

$C_{10}H_{15}N_2O_2$

n (m/z 195)

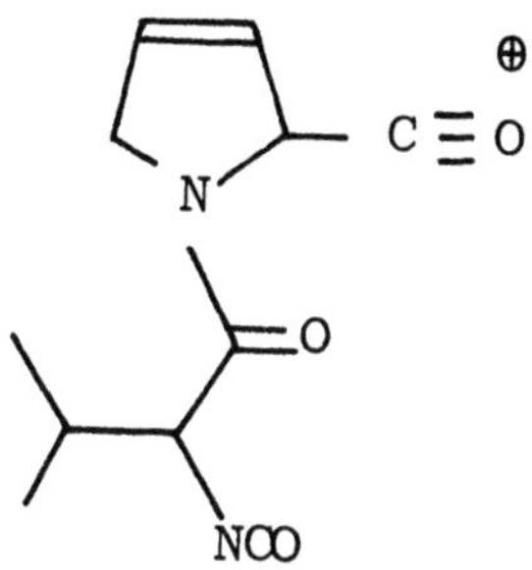

$C_{11}H_{13}N_2O_3$

o (m/z 221)

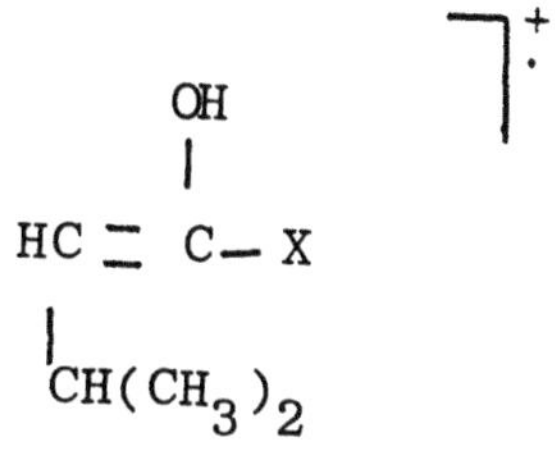

$C_{28}H_{33}N_3O_5$

e (m/z 491)

In order to assure that (8) is not an artefact produced during extraction with methanol; an other portion of the barks was extracted with benzene-ammonia-ethanol (100:1:1). (8) was isolated again which proves that it is nautrally occuring compound.

(8) is the first naturally occuring N-formyl cyclopeptide alkaloid. In order to confirm the mass spectral fragmentation of N-formyl cyclopeptides N-formylsativanine-C, N-formylnummularine-B and N-formylmauritine-C were prepared and studied. They gave the same characteristic fragmentation [8].

Sativanine-G (9)

Sativanine-G, m.p. 92°, $C_{28}H_{42}N_4O_5$ (M^+514.3168) was recognized to be a 13-membered cyclopeptide alkaloid from its UV spectrum. The IR spectrum exhibited bands for -NH, secondary amide, $-OCH_3$, NCH_3, C=C and aryl ether. On hydrolysis it gave N,N-dimethylisoleucine and isoleucine. Mass spectral peaks of (9) corresponds with those of nummularine-C [9], with the only exception that the base peak of the former is 34 mu lower than that of the later. These data reveal that N,N-dimethylisoleucine instead of N,N-dimethylphenylalanine as the end amino acid and isoleucine instead of leucine as the amino acid residue bound to the styrylamine moiety, (9) is a new addition to the growing list of 13-membered cyclopeptide alkaloids which belongs to nummularine-C type.

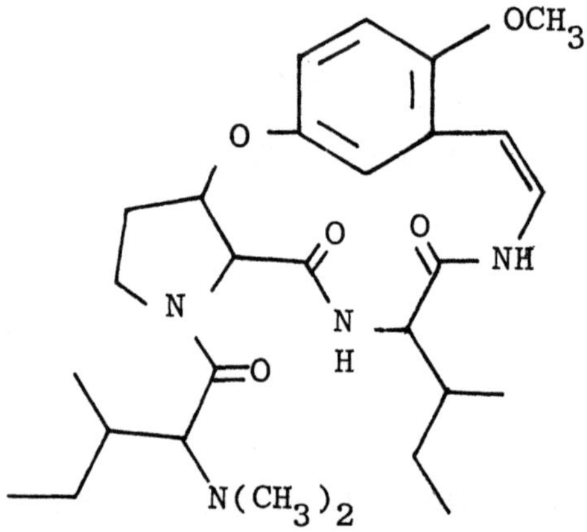

Sativanine-G (9)

Experimental

The plant material was collected from the Hazara district of Pakistan. Carry-14 (UV), Perkin-Elmer 221 (IR), Bruker WP 90 (NMR), Kratos MS-30 and MS-50 instruments were used. Extraction and chromatographic separation and analysis of the different fractions were carried out as usual [8].

References

1. R.Tschesche, E.U.Kauffmann. The Alkaloids, R.H.F. Manske, Academic Press, New York, Vol. XV,(1975)

2. For Part-36, see reference: A.H.Shah, G.A.Miana, V.B.Pandey, R.Wagner, R.Tschesche, **J.Chem.Soc.**, 1984 (in press)

3. R.Tschesche, J.Rheingans, H.W.Fehlhaber, G.Legler, **Chem.Ber.**, 100, 3924 (1967)

4. H.W.Fehlhaber, **Z.Analyt.Chem.**, 235, 91 (1968)

5. R.Tschesche, M.Elgamal, G.Eckhardt. **Chem.Ber.**, 110, 2649 (1977)

6. R.Tschesche, S.T.David, J.Uhlendorff, H.W.Fehlhaber. **Chem.Ber.**, 105, 3106 (1972).

7. R.Tschesche, Ch.Spilles, G.Eckhardt. **Chem.Ber.**, 107, 686 (1974).

8. A.H.Shah, Dissertation Bonn University (1982).

Some Contributions to Researches on Medicinal Plants in a Historical Context

Salimuzzaman Siddiqui
H. E. J. Research Institute of Chemistry, University of Karachi, Karachi-32, Pakistan

At the inaugural session of the Symposium I had briefly referred to the increasing interest of the medical profession in medicinal plants of the subcontinent, starting with the isolation of Conessine from Holarrhena antidysenterica by Dr. Hains in 1858, at the Medical College in Bombay. I also traced the various stages of progress in this direction, which ultimately led to the establishment of the Research Institute at the A&U Tibbi College Delhi under my direction, for undertaking studies in herbal drugs employed in the Ayurvedic and Unani traditional systems of medicine. In consultation with the physicians of these two systems, a number of medicinal plants were selected for studies in their chemical constituents, out of which I would particularly deal here with the alkaloids of Rauwolfia serpentina and H.antidysentera ,the resinol of Bhilawan (Semecarpus anacardium) (marking nuts), and the oil, leaves and fruits of Melia azadirachta (Neem). This will be appropriate, because my interest in the constituents of these plants has continued with me in Pakistan to this day. In more recent years I have been also involved in the alkaloidal constituents of Peganum harmala which grows wild in the non-arable areas of our region. In the context of these studies, an extension of von Braun (BrCN) reaction on alkaloids as well as simpler bases has led, through their respective cyanamides, to a whole new series of derivatives with pharmacophoric groupings, namely, urea, diamines, carbinol amines, and, in steroidal bases, also a guanidine function.

Natural Product Chemistry
Ed. by Atta-ur-Rahman

Rauwolfia Alkaloids [1-12]

In consideration of the fact that the roots of Rauwolfia serpentina had been used over a decade and half by Masihulmulk Hakim Ajmal Khan in the treatment of mental ailments, and according to Dymock they had a fairly rich alkaloidal content, I selected it for systematic studies in its constituents. After a great deal of negative work carried out on stored roots available from the market, I made it a point to obtain freshly harvested and shade dried roots from the Bihar Province. It was from the alcoholic extracts of this material that it was possible to isolate and characterise nine new alkaloids namely Ajmaline, Ajmalinine, Ajmalicine, Isoajmaline, Neoajmaline, Serpentine, Serpentinine and two weaker bases m.p. 220° and 340°. Some of these alkaloids have become world famous in the treatment of cardiovascular diseases and mental ailments. The main alkaloid was given the name ajmaline in memory of late Hakim Ajmal Khan. Initial work on the structure of Ajmaline was later extended by Robinson, whose comprehensive studies led to its formulation as I, which was subsequently modified by Woodward as II.

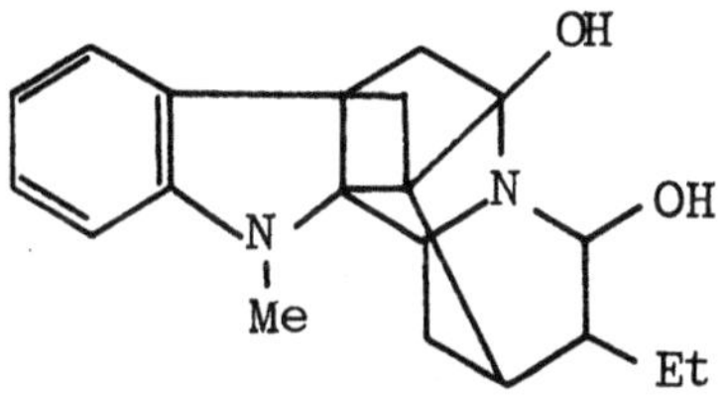

I

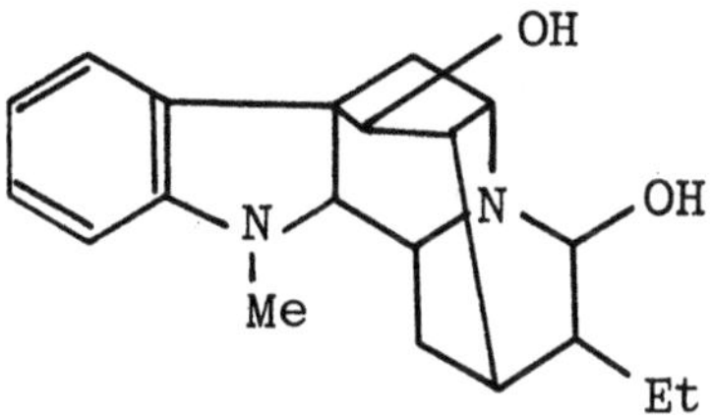

II

More recently, in view of the increasing importance of ajmaline as a drug of choice in the treatment of cardiac arrhythmias, studies in the correlation of its structure and activity led to the hypothesis, that any chemical intrusion that stabilizes either the carbinolamine or the aldehyde-imine structure of ajmaline, greatly reduces its cardiac activity. Thus diacetyl ajmaline in which the carbinol-amine ring is stabilized as well as N-methyl ajmaline in which the aldehyde group is fixed, show very meagre anti-arrhythmic activity, indicating that the carbinol-amine aldehyde-imine tautomeric structure serves as what may be termed as a cardiophoric grouping.

On the basis of an observation in the course of these studies, that mono-nitroajmaline is about twice as active as the mother base, the nitration studies were extended to reserpine. Following highly critical experimental conditions, it was possible to obtain 1 -, 12-and 9-mononitro derivatives, the first two in farily good yields. Pharmacological studies have shown that 1-nitroreserpine has the same order of activity as reserpine, without its undesirable side effects. On the other hand, 12-nitroreserpine has lower hypotensive action, but is much longer lasting when administered to unanaesthetised dogs, over ninety hours as against ca. twenty-four in the case of reserpine.

Holarrhena antidysenterica

During the 1930's, systematic studies in the alkaloids of Holarrhena antidysenterica were undertaken on account of its long reputed importance in the treatment of amoebic dysentery and other intestinal ailments. As a result of these investigations six new crystalline bases were isolated and their mutual relationship with respect to conessine was established on the following pattern.

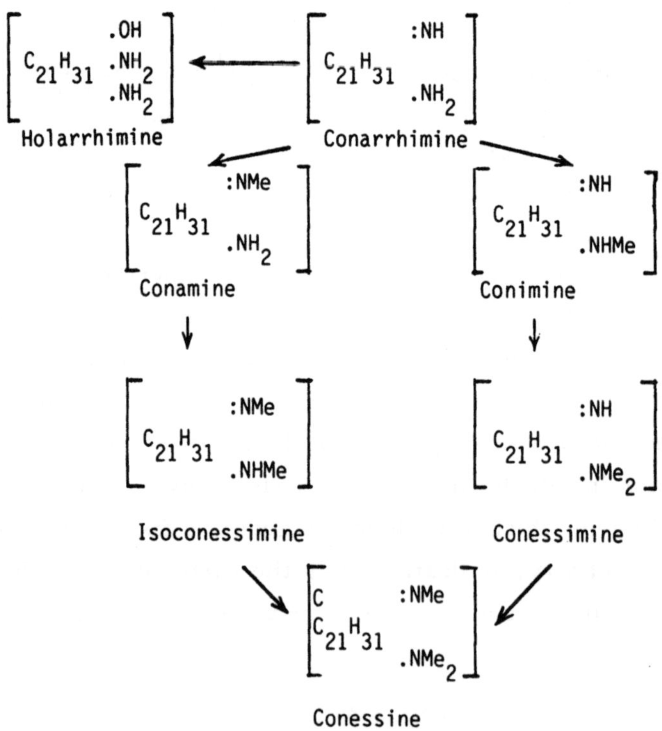

According to this pattern conarrhimine was considered as the mother base, which leads on the one hand to conessine and its various nor bases through methylation, and on the other, to holarrhimine through the hydrolytic fission of its nitrogen ring. These conclusions were mainly based on studies in the action of BrCN on conessine, which led to the mono- and di -nor bases identical with isoconessimine and conimine, both being convertible to conessine on methylation with formaldehyde/formic acid. It may be noted in this context that methylation of the petroleum ether soluble fraction of the total alkaloids raised the yield of conessine to 0.4% as against a maximum of 0.1% reported by earlier workers.

Following the observations of Simonsen and Spaeth relating to Hofmann and Emde degradation of conessine, these studies were extended to isoconessine and it was noted that while conessine gives apoconessine on Hofmann degradation which on further degradation after Emde gives a hydrocarbon, later identified by Haworth as pregna-3,5, 2-triene, isoconessine is quantitatively reconverted on its Hofmann degradation as a ditertiary base with the elimination of two molecules of methyl alcohol. Correlating this contrast between the behaviour of conessine and isoconessine with the observation of von Braun et.al. on the negative influence of β - γ-unsaturation on the N-stability of radicals attached to the nitrogen atom, it was considered probable that the isomerization of conessine was due to a shifting of the double bond.

It was further observed that on heating conessine and isoconessine hydroiodides at 310° and 325° respectively, an identical hydrocarbon $C_{21}H_{30}$ with three double bonds, is obtained which was named as conessen. Comprehensive studies in it subsequently carried out by Haworth led to the establishment of the structure of conessine. On the other hand, the structure of iso- and neo-conessine, obtained on the treatment of conessine with sulphuric acid [13] were established by Janot [14].

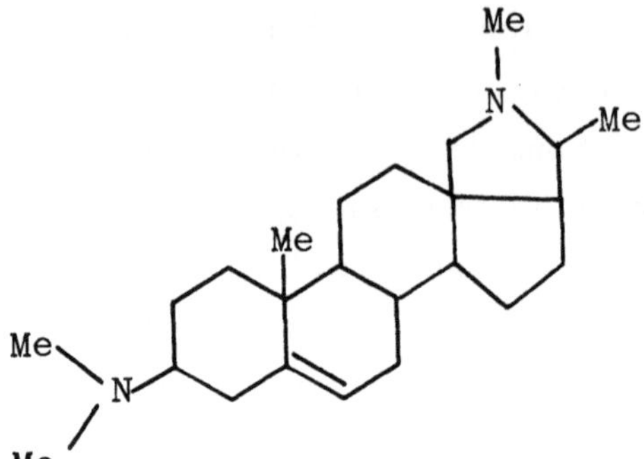

Conessine

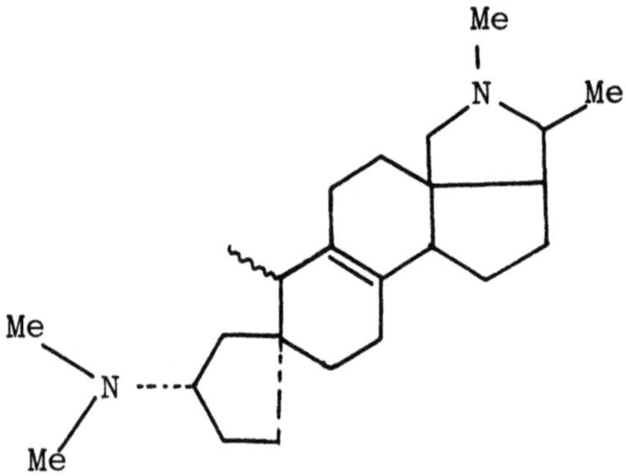

Neoconessine

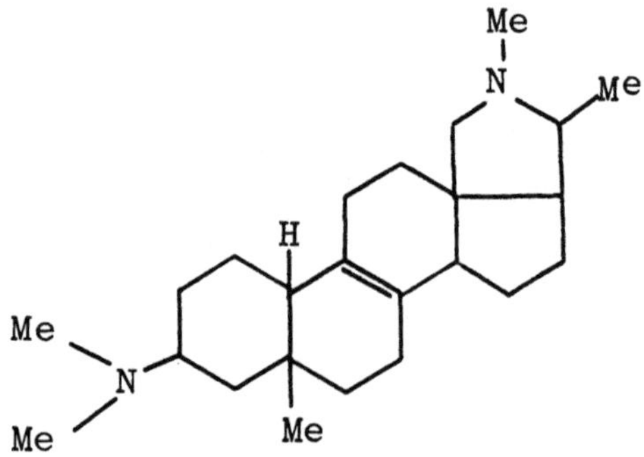

Isoconessine

Reverting to the early work on the reaction of cyanogen bromide with conessine and isoconessine, it has been recently possible through highly critical experimental conditions to convert the monocyanamides of the two bases to their amides on mild acidic hydrolysis, diamines on reduction with zinc and hydrochloric acid, and carbinol amines from the latter on reaction with nitrous acid. On the other hand, they also furnished the guanido derivatives on treatment with concentrated ammonia [15].

Later on, with the object of ascertaining the applicability and limitations of these reactions, these studies were extended to a number of bases including *Harmala*, *Rauwolfia* and *Ephedra* alkaloids, as well as simpler aliphatic and aromatic amines. [16,17]

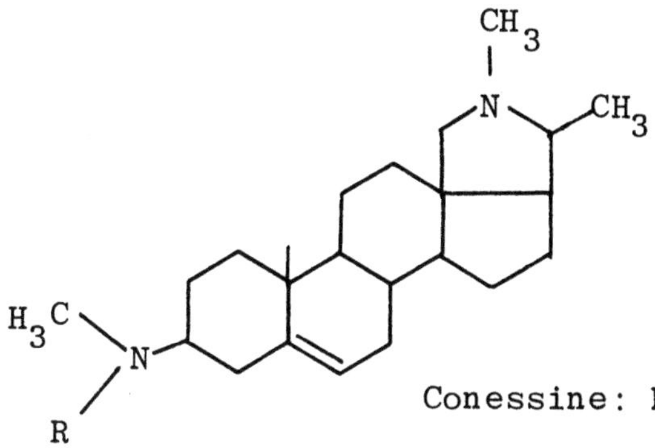

Conessine: R = CH_3

$-CH_2NH_2$	N-Aminomethylisoconessimine	170°C
$-CH_2OH$	N-HYdroxymethylisoconessimine	97°C
$-\overset{\displaystyle N}{\overset{\parallel}{C}}-NH_2$	N-Amidoisoconessimine	155°C
$-\overset{\displaystyle O}{\overset{\parallel}{C}}-NH_2$	N-Guanidoisoconessimine	145°C

In continuation of studies in the correlation of structure and activity relationships, extensive pharmacochemical studies carried out on the main alkaloidal bases of Peganum harmala namely harmine and harmidine/harmaline led to a whole series of new potentially pharmacophoric derivatives of harmine, harmidine, harmol, harmidol, tetrahydroharmine and amino tetrahydroharmine. Further studies in this direction are being persued.

Semecarpus anacardium

Owing to the medicinal importance attributed to Semecarpus anacardium (Bhilawan) systematic investigation of the resinol occurring in the pericarp of its nuts, was undertaken in 1931, which resulted in the isolation of a phenolic substance Bhilawanol through fractional distillation under reduced pressure. Studies in the structure of Bhilawanol revealed it as a catechol derivative with a $C_{15}H_{27}$ unsaturated straight side chain, carrying an average of two double bonds. It was later established that it is closely

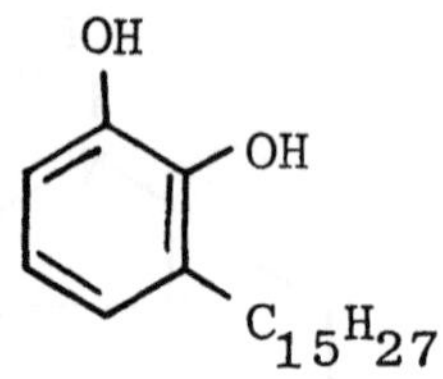

Bhilawanol

related to uroshiol, obtained by Majima [17] from the Japanese lacquer varnish ki-urushi (Rhus vernicifera) and differs from it only in the number and position of the double bonds in the side chain. These findings subsequently led to the industrial development of the resinol for the production of stoving enamels, varnishes, and moulding compositions during the World War II. In a recent reinvestigation of the petroleum ether insoluble, alcohol soluble component of Bhilawan nuts, two dimeric flavonoids have been isolated, the exact structure of which is under investigation.

The utter neglect of this potent drug in modern medical practice has been mainly due to its toxicity and vesicent action on external application or internal administration. Studies in its chemical constitution had to some extent made it possible to control its toxicity. Thus it was found that the free hydroxyl groups and the two double bonds in the side chain of Bhilawanol are together responsible for its vesicent action. When both these active centres are elimintated as for instance in diacetyl or dimethyl-tetrahydro-bhilawanol, a completely non-vesicent product results. Unfortunately, studies in the correlation of structure and activity in respect of its antirheumatic and a wide range of other therapeutic properties claimed for it have not so far been undertaken.

Melia azadirachta

In view of the curative properties attributed in folklore and traditional medicine to Melia azadirachta (Nim), it has been subjected to chemical and therapeutic studies from about the beginning of the current century.

These studies were mainly concerned with the fatty acid components and other mostly amorphous substances isolated from the oil. Thus, it was only as late as 1942 when the isolation of two crystalline triterpenoids, nimbin and nimbinin [18] from the nim oil was reported through solvent partitioning technique, without involving any chemical intervention. Their structures were subsequently established through investigations carried out by several groups of workers [19]. These investigations stimulated great deal of interest among many groups of workers in the constituents of various parts of the Nim tree, but it was first in 1972 when studies in the fresh, undried, fruit pulp were undertaken by us, as a result of which the isolation and structure of three new crystalline triterpenoids namely, nimolicine, nimoline and 17 -hydroxyazadiradione was reported. Further studies in various parts of neem tree (fruits, leaves and twigs) subsequently carried out in view of the increasing interest in its constituents throughout the world, have so far resulted in the isolation and structure elucidation of three additional triterpenoids namely nimolicinol, 6 α-hydroxyazadirone and azadirachtol, along with some flavonoidal constituents quercetin, isorhamnetin (unreported), nimbocetin, and an aromatic ester.

Nimbinine, $C_{28}H_{34}O_6$;

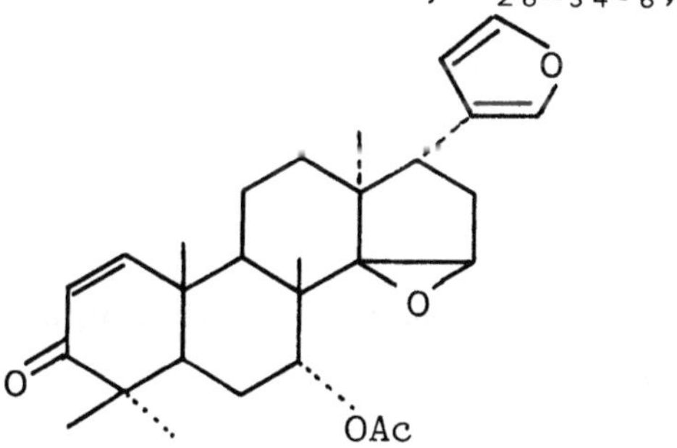

Nimbin, $C_{30}H_{36}O_9$;

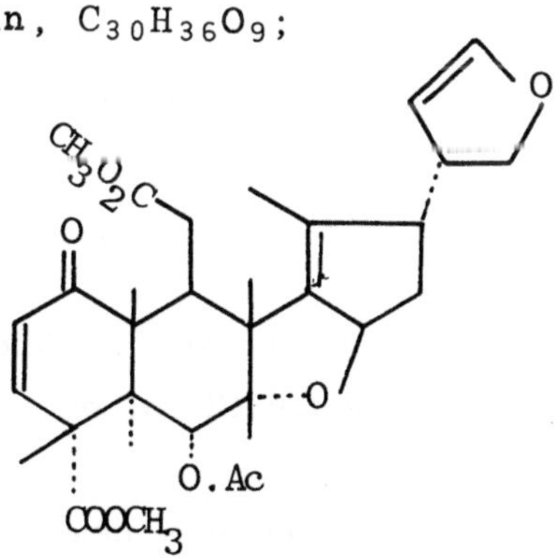

In respect of the experimental procedures employed in the isolation of plant constituents, it may be noted that fresh undried uncrushed plant materials have been employed for studies in their constituents, in order to avoid structural changes which are likely to occur in them through aerial oxidation and enzymatic action in the process of drying and storage. In the isolation of individual constituents also, classical procedures have been employed with the advantage that the new chemical constituents are obtained at the laboratory level in sizeable quantities, and not just in miligrams through column chromatographic techniques. It is noteworthy that this provides the possibility of studies in the correlation of structure and pharmacological activity, which is of vital importance in arriving at new therapeutic agents based on plant materials.

References

1. Chemical examination of the roots of Rauwolfia serpentina Benth, S.Siddiqui and R.H.Siddiqui, **J.Indian.Chem.Soc.,** 8, 667-80 (1931)

2. The alkaloids of Rauwolfia serpentina Benth Part - I Ajmaline series, S.Siddiqui and R.H.Siddiqui, **J.Indian Chem.Soc.,** 9, 539-44 (1932)

3. The alkaloids of Rauwolfia serpentina Benth. Part - II Ajmaline series, S.Siddiqui and R.H.Siddiqui, **J.Indian.Chem.Soc.,** 12, 37-47 (1939)

4. A note on the alkaloids of Rauwolfia serpentina Benth. S.Siddiqiui, **J.Indian.Chem.Soc.,** 16, 421-22 (1939)

5. Isolation of a new hypotensive factor from the roots of Rauwolfia serpentina Benth. S.Siddiqui, **Chemistry and Industry (London),** 1270-71 (1957).

6. Studies in the alkaloids of Rauwolfia serpentina Benth, and the mode of their occurrence Part I: A new hypotensive factor and other alkaloidal complexes from the roots, S.Siddiqui, **Pakistan J.Sci.Ind.Res.,** 1, 3-5 (1958)

7. Studies in the Alkaloids of Rauwolfia serpentina Benth, and the mode of their occurrence Part - II. S.Siddiqui, S.A.Warsi and M.Ala-ud-Din, **Pak.J.Sci.Ind.Res.,** 2, 80-85 (1959).

8. Studies in the Ajmaline series, Salimuzzaman Siddiqui, S.A.Warsi, M.Alauddin and Viqar Uddin Ahmad, **Pak.J.Sci.ind.Res.** 4, 1-3 (1961)

9. Constituents of Rauwolfia vomitoria from East Pakistan, S.Siddiqui and M.Manzur-i-Khuda, **Pak.J.Sci.Ind.Res.,** 4, 1-3 (1961)

10. Studies in the Alkaloids of Rauwolfia Caffra Sonder, Part - I. Isolation of Ajmalicine, Ajmaline, Raucaffrine and three New Alkaloids, Raucaffricine, Raucaffriline and Raucaffridine, S.Siddiqui, N.H.Khan, M.A.Khan, **Pak.J.Sci.Ind.Research.,** 8, 23 (1965)

11. Studies in the alkaloids of Rauwolfia caffra sonder, Part - II, S.Siddiqui, N.H. Khan and M.H. Khan, **Pakistan J.Sci.Ind.Res.,** 9, 210 (1965)

12. Some aspects of studies in the Ajmaline series, S.Siddiqui and Viqar Uddin Ahmad,**Proc.Pak.Acad.Sci.,** 5, 1 (1968)

13. Studies in the conessine series, Part-I, Isomerization of conessine and its nor-bases, S.Siddiqui, **Proc.Indian.Acad.Sci.,**2A, 426-37 (1945); Studies in the conessine series. Part-6. Action of sulphuric acid on conessine and its conversion to a new isomer, neoconessine, S.Siddiqui and S.K.Vasishta, **J.Sci.Ind.Res., India,** 3, 599-63 (1945); Siddiqui, S., Misra, S.C. and Sharma, V.N. **J.Sci. Ind.Res.,** 3(12), 555-558 (1945).

14. Janot, M.M.,Devissaguet, P.,Pais, M., Jarren, F.X., Khuong-Huu, Q and Goutarel, R., **Tetrahedron Letters**, 36, 4375-80 (1966); **Bull.Soc.Chim.**, 4567 (1967); Ibid, 11, 4318-28 (1967)

15. Extension of von Braun (BrCN) reaction to conessine series and other organic bases, Bina S.Siddiqui and Salimuzzaman Siddiqui, **Z.Naturforsch.**, 35(b), 1049 (1980); Extension of von Braun (BrCN) Reaction on Organic Bases, Part-II, Abdul Malik, Nighat Afza, Salimuzzaman Siddiqui, **Z.Naturforsch.**, 37b, 512 (1982); Some Extensions of von Braun (BrCN) Reaction on organic Bases", S.Siddiqui and Bina S.Siddiqui, **Proc.Pak.Acad.Sci.**, 18, 89 (1982)

16. Some Extension of von Braun (BrCN) Reaction on Organic Bases Part-III, S.Siddiqui, Sabira Begum and Bina S.Siddiqui, **Z.Naturforsch.**,37b, 1481 (1982)

17. Majima, et al. **Ber**, 55, 172 (1922)

18. A note on the isolation of three new bitter principles from nim oil (Melia azadirachta), S.Siddiqui, **Current Science**,11, 278-9 (1942); Utilization of nim oil and its bitter constituents (nimbidin series) in the pharmaceutical industry, S.Siddiqui and C.Mitra, **J.Sci.Ind. Research,(India)**, 4, 5-10 (1945)

19. Henderson et al., **Proc.Chem.Soc.**, 269 (1963); Narayanan et al., **Chemistry and Industry**, 1283 (1962) and 322, 324 (1964); M.Harris, et al., **Tetrahedron**, 24, 1517 (1968); Old D.O.C.; C.R.Narayanan, R.V.Pachapurkar and B.M.Sawant, **Tetrahedron Letters.**, 3563 (1967); J.D.Connolly, K.L.Handa and R.McGrindle, **Tetrahedron Letters**, 437 (1968).

Advanced Techniques in Mass Spectrometry

David L. Smith
Department of Medicinal Chemistry, University of Utah, Salt Lake City, UT 84112, USA

Introduction

The first commercial mass spectrometers, developed over 40 years ago for the petroleum industry, were valued for their ability to rapidly characterize mixtures. During the first 25 years, the design of mass spectrometers remained relatively unchanged and most analyses were performed using electron impact ionization and heated inlets. Only during the last 15 years has the enormous potential of mass spectrometry become generally evident. Efforts to make mass spectrometry more informative have resulted in new techniques such as chemical ionization, directly coupled gas chromatography (GC/MS), and high resolution. Sophisticated computer systems are now used to acquire, display and store large volumes of data while advanced ion optics make possible detection of picogram quantities of many substances. These are well developed technologies and are used throughout the world.

Recently, even more advanced techniques such as FAB, MS/MS and LC/MS have been developed to extend the utility of mass spectrometry to higher masses and to further increase its ability to analyze very complex mixtures. These advanced techniques, which may soon be regarded as standard methods, are discussed below.

Fast Atom Bombardment Mass Spectrometry

The major weakness of mass spectrometry has been its inability to analyze substances of low volatility; that is, to be analyzed by mass spectrometry, a substance must have a vapor pressure greater than 10^{-8} torr at a temperature below which it undergoes thermal degradation. Thus, it was the inability to produce gaseous ions representative of molecular structure that re-

Natural Product Chemistry
Ed. by Atta-ur-Rahman

stricted mass spectrometry to the analysis of low mass substances. Instrumentation necessary for analysis of high mass ions was not the limiting factor. Since the upper mass limit of a sector mass spectrometer is proportional to the square of the radius of the magnet, one only needs to use larger magnets to analyze high mass ions. Today there are commercially available mass spectrometers with upper mass limits of 10,000. Once we have the means of producing higher mass ions, mass spectrometers with even larger mass ranges could be produced.

Fast atom bombardment (FAB) mass spectrometry, as described by Barber et al. (1) and Surman and Vickerman (2), has substantially reduced the requirement of sample volatility. For example, conventional, direct probe mass spectrometry can be used to analyze underivatized nucleosides such as adenosine and uridine (3), while FAB mass spectrometry gives molecular ions for oligonucleotides containing as many as 10 nucleotides (4). The magnitude of the long-term effect of FAB on the field of mass spectrometry will probably be comparable to that of coupling gas chromatography to mass spectrometry.

The technique of fast atom bombardment uses high energy particles to desorb charged molecular species from the surface of a liquid in which it is dissolved. Once in the gaseous phase, the ions may be analyzed in the usual manner by either a quadrupole or sector mass spectrometer. FAB mass spectra have many features in common with chemical ionization mass spectra; protonated molecular ions usually dominate the positive ion mass spectrum with a variety of adduct and fragment ions of lower intensity present depending on the nature of the sample. Negative ions may also be produced in abundance with the $M-H^-$ usually predominating in the case of substances which are normally uncharged.

For analyses which are of interest to most analytical chemists, it has been shown that similar results are obtained when either charged or neutral particles are used to effect vaporization of the sample (5-7). Maximum sensitivity is achieved when high intensity (10^3 particles/sec/mm^2) beams of particles having a translational energy between 5 and 10 KeV are used. Suitable beams of ions or neutrals may be produced by thermal ionization or electronic discharges, respectively.

That the majority of the FAB papers in the literature have used a beam of neutral fast atoms, is because such a beam was used in the first work reporting on the observation of high molecular weight, highly polar substances, and does not imply any advantage to using a neutral beam. Devices for producing beams of Cs^+ are somewhat simpler to construct than those required for production of a neutral beam. In addition, the ion beam may be focused directly on the sample and may be conveniently varied over a wide range of intensity. Since the source of a sector mass spectrometer is normally operated at 4 to 8 KV in the positive ion mode, a high voltage must be used to initially accelerate the ions beam to 10 to 15 KeV so that the potential barrier in the ion source may be overcome. Use of a neutral beam circumvents this problem but has the added drawback that the pumping system on the source must be able to handle the relatively high volume of gas (Ca. 1 ml (STP)/min) passing through the discharge source.

An essential requirement for obtaining a good FAB spectrum is complete dissolution of the solute in a liquid matrix. While glycerol has been the most frequently used solvent, any liquid in which the sample is soluble and which has a sufficiently low vapor pressure may be used. The importance of the fluid may be illustrated for the case of gangliosides. No FAB spectrum is obtained when a ganglioside is dissolved in glycerol while excellent spectra are observed if it is dissolved in triethylamine and 1,1,3,3-tetramethylurea (8). Thioglycerol, various pumping fluids, and crown ethers have also been used as solvents (9,10).

Usually from 0.1 μL to 1 μL of solvent in which 1 to 100 μg of solute have been dissolved are placed on the probe tip. Since glycerol is difficult to handle neat, it is often efficacious to first dissolve the sample in an alcohol/glycerol or water/glycerol solution. In this case the volatile solvent is removed in the vacuum chamber leaving a thin film of glycerol on the probe tip. Solubility of the sample may be improved by adding acid, base, or surfactants such as Triton X-100; by sonication or by simply allowing the sample/glycerol mixture to equilibrate for several days.

While the methodology for producing mass spectra of high molecular weight, highly polar substances is well known at this time, a satisfactory theoretical model describing the role of the dispersing matrix is not yet available. However, the processes by which ions are desorbed from the surface is presumably the same as in secondary ion mass spectrometry and may be described as a combination of thermal vaporization and sputtering (11,12).

Thermal vaporization is due to the localized heating caused by the impact of a single high energy particle. According to this model, the energy of the impacting particle is equilibrated among translational degrees of freedom of molecules on the surface. The importance of this model is determined by the binding energy of substances on the surface and by the rate at which heat is dissipated to other parts of the sample. The preponderance of

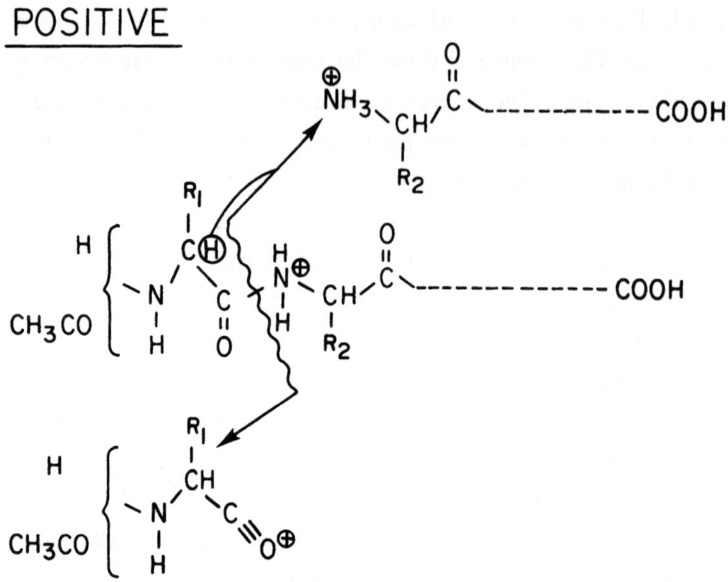

Scheme I

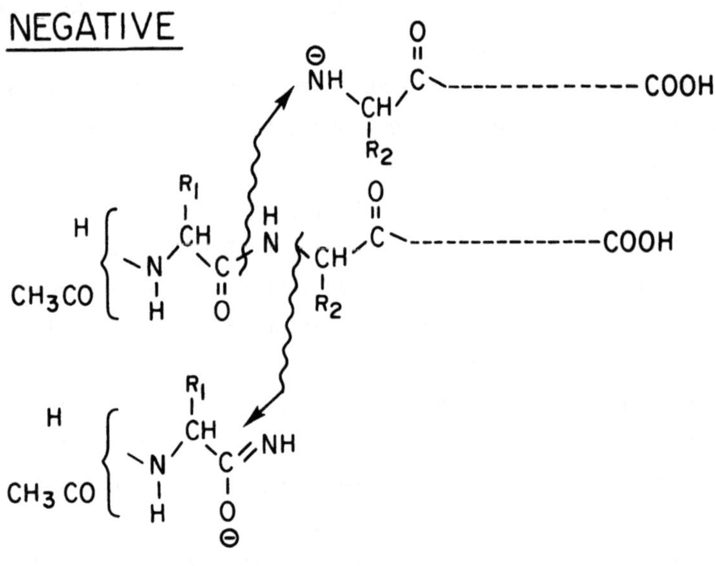

Scheme II

molecular ions in FAB mass spectra suggests that, within the time scale for significant vaporization, the internal degrees of freedom are not substantially excited.

For the impact energies typically used in FAB, a second process, sputtering, probably contributes to ion desorption. According to this model, molecules interact with the impacting particle as hard spheres and desorption occurs during a very short time interval following impact and prior to equilibration of energy. Both thermal vaporization and sputtering are enhanced by the use of heavy projectiles which form shallow energy tracks on the surface in contrast to deep, narrow tracks formed by light particles such as helium. The translational energy of the desorbed particles is different in the two models and may be used to establish their relative importance. In the case of thermal desorption, secondary ions may have energies up to approximately 1 eV while, for sputtering, they may range from 10-50 eV.

Independent of the mechanism responsible for ion desorption, the bombardment of a liquid surface by high energy particles does produce large quantities of high mass ions. FAB mass spectrometry has been especially useful in analyses of small peptides with molecular weights up to 3,000 daltons because, in addition to indicating the molecular weight, the complete amino-acid sequence may also be derived from key fragment ions (13-16). Fragmentation reactions giving ions diagnostic of the sequence are illustrated in Schemes I and II. Protonation at an amide nitrogen gives rise to two types of fragments; formation of acylium ions (N-terminal sequence ions) and ammonium ions (C-terminal sequence ions). In more complex peptides interpretation is greatly facilitated by preparing a 1:1 acetyl: ^{2}H-acetyl derivative from which all N-terminal fragments may be recognized as 1:1 doublets separated by 3 daltons.

Scheme II shows how diagnostic negative ions are formed. These ions may be used to substantiate interpretation of the positive ion spectrum since C-terminus ions will appear 2 daltons lower and N-terminus ions 15 daltons higher than the corresponding ions in the positive ion FAB spectrum. The negative ion FAB mass spectra of peptides usually have intense M-H ions which help to ascertain the molecular weight.

It has recently been shown that FAB mass spectrometry may be used to ascertain the number of readily exchangeable hydrogens in a substance (17). This information provides an important constraint in deducing the structure and is complementary to elemental composition information derived from exact mass measurements. The most reliable results were obtained when the sample was first dried in vacuum and dissolved in a mixture of [hydroxy-$^{2}H_3$]-glycerol: $^{2}H_2O$ (1:1) and analyzed immediately before back-exchange had had an opportunity to occur. The shift in mass of parent and fragment ions from those obtained using normal glycerol gave the number of exchangeable hydrogens. This technique was used to determine the number of exchangeable hydro-

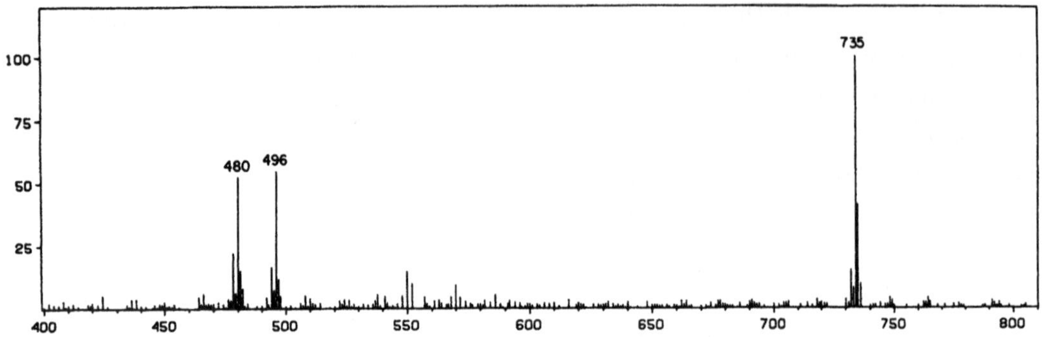

Fig. 1. Positive ion FAB mass spectrum of 1 μg of dipalmitoylphosphatidylcholine dissolved in glycerol

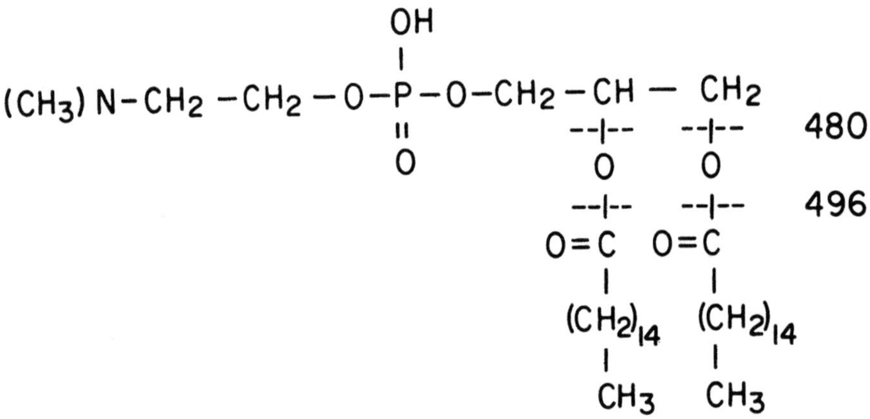

Scheme III

gens in a variety of substances, including nucleotides, polysaccharides and antibiotics, and gave reliable measurements when as many as 29 exchangeable hydrogens were present.

Phospholipids constitute a class of biologically important compounds for which neither conventional nor field desorption mass spectrometry has yielded complete analysis. The positive ion FAB mass spectrum of 1 μg of dipalmitoylphosphatidylcholine is given in Fig. 1. A prominent M+H ion indicates the molecular weight and may be used to deduce the number of carbon atoms and double bonds in the two fatty acid groups. Two structurally important fragment ions which are the result of cleavages illustrated in Scheme III may be used to determine the chain length and degree of saturation in each of the fatty acids. Negative ion FAB mass spectra contain intense ions corresponding to the fatty acid anion, RCO_2^-, and may be used to corroborate

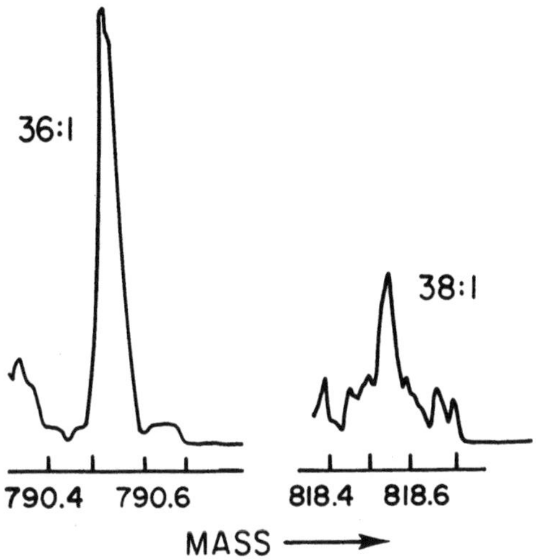

Fig. 2. M+H ions for 10 ng of phosphatidylserine obtained under high resolution conditions (M/ΔM = 10,000). Peaks are the summation of 256 scans

structural information deduced from the positive ion spectrum. Similar results have been obtained for other phospholipids including inositol, serine, ethanolamine and sphingomylin (18,19).

High resolution mass spectrometry may be used in conjunction with FAB to both reduce the background and to determine the elemental composition of peaks of interest. This is illustrated in Fig. 2 which shows peaks for the M+H ions for a two-component mixture of phosphatidylserine obtained using a resolving power of 10,000. Since only 10 ng of material were used, a signal averager and pulse counting electronics were employed to improve the signal-to-noise ratio. Given suitable reference peaks, the accuracy of mass measurements performed in the FAB mode is comparable to that obtained in the electron impact mode.

In addition to its usefulness in overcoming the volatility requirement of conventional mass spectrometry, FAB mass spectrometry will become even more useful as techniques are developed to couple it with high mass analysis capability, collisional activation and liquid chromatography mass spectrometry.

Tandem Mass Spectrometry

Tandem mass spectrometry, often referred to as MS/MS, uses special instrumental operating parameters or configurations to separate and detect fragment ions known to have come from a selected precursor ion (20-22). The most notable successes of tandem mass spectrometry in the area of analytical chemistry fall into two categories: (1) increased specificity for complex mixture analysis and (2) increased structural or sequence information.

An example of increased specificity may be found in the analysis of an enzymatic hydrolysis of an isoaccepting tRNA which has been derivatized via trimethylsilylation. The normal mass spectrum contains a multitude of peaks due to the complex nature of the sample which contains as many as 10 modified nucleosides as well as impurities in the enzyme, tRNA and derivatizing reagents in addition to the four major nucleosides, uridine, cytidine, adenosine, and guanosine. Although unequivocal determination of a minor, modified nucleoside in such a mixture would be impossible using only a low resolution mass spectrum, tandem mass spectrometry, because of its very high specificity, could probably establish the presence of minor constituents.

Tandem mass spectrometry has also been used to obtain structural information when only pseudomolecular ions are produced, as in the case of chemical ionization (22-24), or fast atom bombardment (25,26). This information, when combined with the elemental composition which may be obtained from a separate, exact mass measurement, is often sufficient to describe major structural features of an unidentified substance.

In either case, ions of a mass of interest are normally directed into a gas-filled chamber where collisions occur and result in a vibrationally excited ion which may then undergo unimolecular decomposition to give structurally diagnostic fragments. These fragment ions, along with intact or unreacted parent ions, exit the collision chamber guided by their original momentum, and are mass analyzed by one of several different techniques. One can then proceed to interpret the resulting secondary ion mass spectrum or collisional activation mass spectrum (CAMS) in the usual manner knowing with certainty that all of the ions are derived from the parent ion.

Collision induced dissociation (CID) mass spectra, another term used to describe CAMS, result from the unimolecular decomposition of a vibrationally excited ion and are usually similar to normal electron impact mass spectra (27). Differences between the two types of mass spectra are probably due to non-equilibration of the excitation energy prior to dissociation and may depend on the kinetic energy of the impacting particle. The efficiency with which fragment ions are collected from the collision cell also contributes to the appearance of the secondary ion mass spectrum and depends on several factors, including the mechanism by which the parent ion is initially excited as well as on the degree of excitation of the parent ion.

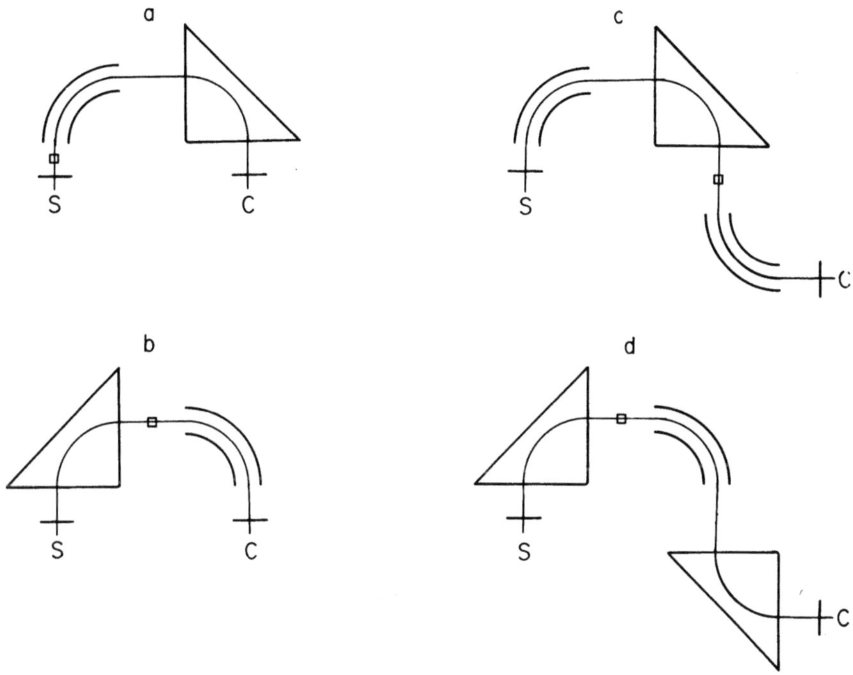

Fig. 3. Schematic representation of some sector mass spectrometers which may be used to produce CID mass spectra. S, C and □ refer to source, collector and collision cell, respectively

A variety of instrumental methods have been shown to be effective in selecting a primary ion beam, inducing the collisional dissociation process, and generating a mass spectrum of the products. Some methods are preferred because they use existing, slightly modified instruments; others, requiring very sophisticated instruments which are just becoming commercially available, clearly give superior results in terms of sensitivity and specificity. Four typical combinations of electric and magnetic sector instruments including the preferred location of the collision cell are illustrated in Fig. 3.

The normal geometry (EB) instrument, illustrated in Fig. 3a, has a collision cell located in front of the electric sector (28-29). A secondary ion mass spectrum is generated by first adjusting the magnet so that ions of a desired mass, M_2, are detected. Holding both the electric sector voltage and magnetic field fixed, the accelerating voltage is scanned upward. Only ions which are accelerated as mass M_1, and then dissociated to give ions of mass M_2 are detected, where the mass of the parent ion, M_1, is given by Eq. (1).

$$M_1 = M_2 \frac{(HV)_2}{(HV)_1} \qquad (1)$$

$(HV)_1$ and $(HV)_2$ are the accelerating voltages required to detect the normal and secondary ions of mass M_2, respectively.

Reverse geometry instruments (29,30), Fig. 3b, were then developed to give similar information but in a more meaningful form. The magnetic field is adjusted so that only ions of a desired mass, M_1, enter the collision cell. The electric sector voltage is then swept downward from its normal value so that fragment ions of decreasing mass (actually energy) are transmitted to the collector. The resulting mass spectrum includes all of the secondary ions, M_2, which were produced by CID of the parent ion, M_1, where the mass of the daughter ion, M_2, is given by Eq. (2).

$$M_2 = M_1 \frac{E_2}{E_1} \qquad (2)$$

E_1 and E_2 are the electric sector voltages required for detecting the parent ion, M_1 and its daughter ion, M_2.

It has subsequently been demonstrated that the same information (i.e., identification of daughter ions of M_1) may be obtained with a normal geometry instrument if the magnetic (B) and electric (E) sectors are scanned simultaneously so that the ratio of E/B is held constant (30-32). The linked scan method attains a higher resolution than either the accelerating voltage or electric sector voltage scanning methods described above and, for this reason, is often used in conjunction with reverse geometry instruments. It should be noted, however, that the linked scan method is relatively insensitive to the kinetic energy of the ions and is, therefore, not suitable for kinetic energy release measurements. In addition, the uncertainty in identification of the parent ion may be as large as several mass units depending on the geometry of the instrument.

A triple sector instrument of the configuration EBE which has a collision cell located between the magnetic and second electric sector has been reported (33) and is illustrated in Fig. 3c. This instrument is normally operated by selecting ions of a desired mass with the first two stages and analyzing CID fragments by scanning the voltage of the second electric sector. The information obtained via this method is similar to that obtained with a reverse geometry instrument described above except that the first two sectors may be used as a conventional high resolution mass spectrometry to select ions of a desired elemental composition for subsequent collisional activation analysis.

Another variation of a triple sector mass spectrometer has the BEB configuration (34) and is illustrated in Fig. 3d. This instrument has the collision cell located between the first magnetic and electric sectors and may therefore use only low resolution to select the parent ion. Since fragment ions are analyzed by two sectors, energy focusing occurs and substantially increases the mass resolution of the secondary ions. Of the instruments illustrated in Fig. 3, this is the only one which can be used to assign the nominal masses of the parent and daughter ions with certainty.

Another mass spectrometer designed specifically for CID/MS uses three quadrupole mass spectrometers which are connected in series (21,35). The quadrupole nearest the ion source selects ions of the desired mass and injects them into the second quadrupole which is operated in the RF-only mode so that ions of all masses are transmitted to the third quadrupole where mass analysis of the CID ions occurs. Since unit mass resolution is attained for both primary and secondary ions, there is no uncertainty in their mass. This instrument is highly versatile and may be scanned in three different ways. If the third quadrupole is set for a mass of interest, the first quadrupole may be scanned to indicate all possible parent ions. Alternatively, the first quadrupole may be set at a mass of interest and the third scanned to indicate all daughter ions. In the third mode of operation, the first and third quadrupoles are scanned simultaneously with a fixed mass difference to indicate all ions which are prone to lose a specified unit of mass.

A new generation of mass spectrometers which combine attractive features of both high resolution sector and quadrupole instruments has recently been announced (36,37). Ions of interest are selected by a double focusing first stage mass spectrometer capable of high resolution. These ions are decelerated to less than 100 eV and injected into an RF-only quadrupole section which functions as a collision cell. Fragment ions, along with unreacted parent ions, are then analyzed by a second quadrupole section. It seems that potentially important analytical applications as well as interesting studies in ion physics using these new instruments will be limited only by their extraordinarily high cost.

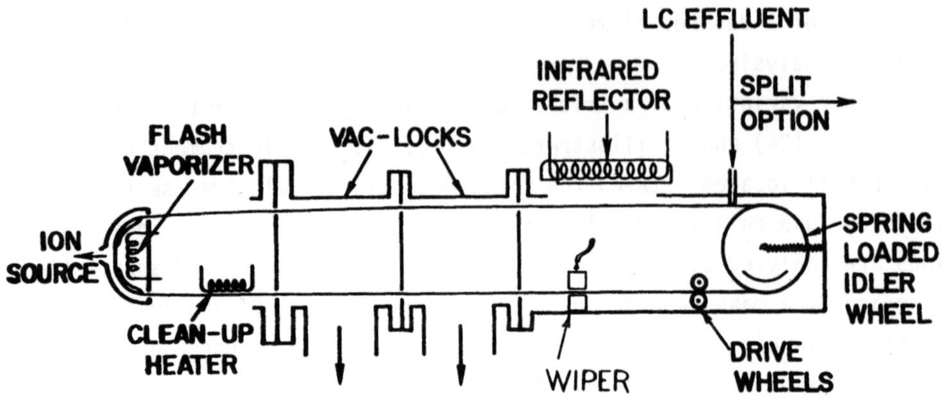

Fig. 4. Schematic representation of moving belt LC/MS interface system (courtesy of Finnigan-MAT)

Liquid Chromatography-Mass Spectrometry

The desirability of coupling the power of mass spectrometry as a detector with the ability of liquid chromatography to separate complex mixtures is obvious. During the last 10 years, considerable effort has gone into the development of combined liquid chromatography-mass spectrometry (LC/MS) with emphasis on retaining full capabilities of both systems. This is not easy because liquid chromatography requires a mass flow of 1-2 orders of magnitude greater than that of gas chromatography. The two primary obstacles to combining liquid chromatography with mass spectrometry are due to the fact that a small amount of solute must be separated from a large amount of solvent and that special means of vaporization and ionization are needed to detect thermally labile/highly polar substances.

In the most widely used LC/MS interface, a moving belt system separates solute and solvent outside the mass spectrometer (38,39) as illustrated in Fig. 4. The effluent from the column is deposited on a continuously moving belt which passes through a solvent evaporation chamber and eventually transports the solute to the ionization chamber where the belt is heated and the sample is vaporized. A combination of vacuum and mild heating are used in the evaporation chamber to remove the solvent and leave behind the solute. The fact that sample ionization can be effected by either electron impact or chemical ionization is a particularly attractive and distinguishing feature of the moving belt system. This method is, however, limited to analysis of substances which can be vaporized intact from a conventional solids probe.

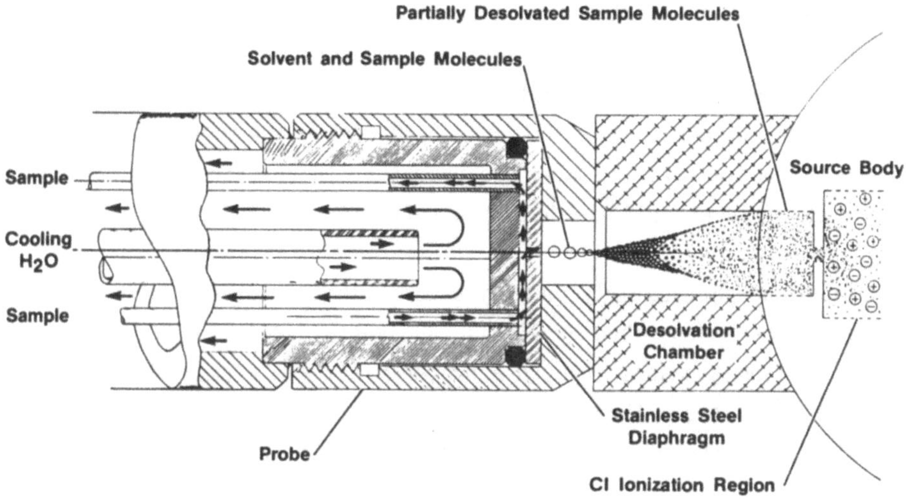

Fig. 5. Schematic representation of a direct inlet LC/MS interface (courtesy of Hewlett Packard)

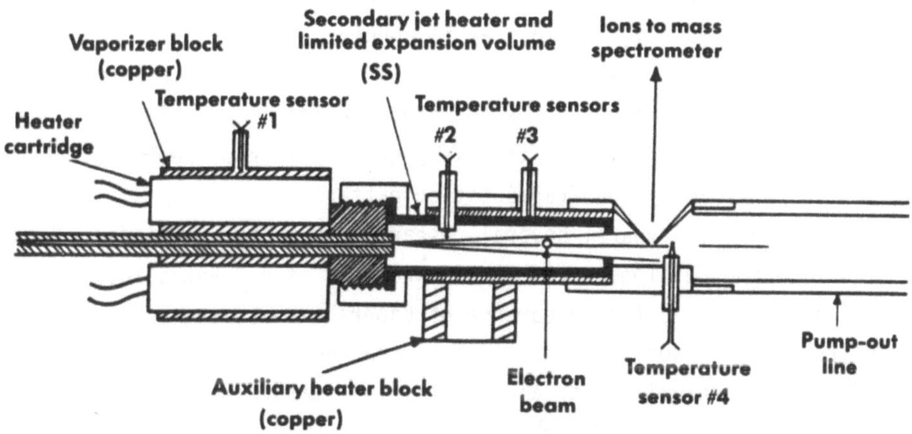

Fig. 6. Schematic representation of a thermospray LC/MS interface (courtesy of Vestec)

The direct liquid introduction (DLI) with an external sample splitter is an alternative approach (40,42) and is illustrated in Fig. 5. The entire effluent passes by a pin hole which is located close to the ionization chamber and through which approximately 1% of the effluent passes. Heat from the ion source vaporizes the solvent leaving free solute molecules in the gaseous state. Ionization is initiated by a high energy electron beam followed by ion-molecule reactions characteristic of chemical ionization. Because solute molecules are not evaporated from a surface, thermally labile/highly polar substances may be analyzed. A variation of the DLI method uses a micro LC column in place of a conventional, narrow bore column (43). Since the optimum flow rate of such a column is only 2-40 L/min, the entire effluent may be passed directly into the mass spectrometer. This method eliminates the need for an effluent splitter but does suffer from a reduction in chromatographic performance.

A unique and particularly promising LC/MS interface, called thermospray, has recently been described by Blakley and Vestal (44). In this method, the entire LC effluent passes through a heated block via a small bore (0.015 mm inside diameter) tube. The temperature of the heated block, or vaporizer, is adjusted so that the effluent is completely volatilized just prior to its depature from the vaporizer. The effluent is then a supersonic jet of gas in which droplets of solvent and solute are entrained. Ions ejected from the droplets pass through a sampling orifice and are mass analyzed. This method relies on the existence of charged solute molecules in the effluent and does not use an electron beam for ionization. While details of the method are not yet completely understood, thermospray is conceptually very simple and is particularly useful for analysis of very polar substances. For example, an intense $M+H^+$ ion was observed for a tetradecapeptide at m/z 1757. The major limitation of the method at this time is a general lack of reproducibility of ionization efficiency and fragmentation pattern.

During the last ten years, several research groups have expended substantial effort in the development of LC/MS. While it is not yet a fully developed routine technique, the appearance of papers emphasizing applications over instrumentation suggests that we have reached a point where we now know that LC/MS will eventually rank along with other mature mass spectrometric techniques such as GC/MS and high resolution accurate mass measurements.

Acknowledgment: Preparation of this manuscript was supported by National Institute of Health Grant GM 21584.

References

1. M. Barber, R.S. Bordoli, R.D. Sedgwick, and A.N. Tyler, J. Chem. Soc. Chem. Commun., p. 325 (1981).
2. D.J. Surman and J.C. Vickerman, ibid, p. 324
3. J.A. McCloskey in "Basic Principles in Nucleic Acid Chemistry" (Tso, P.O.P., Ed.) Vol. I, Academic Press, Chapter IX (1974).

4. L. Grotjahn, R. Frank, and H. Blocker, Nucl. Acids Res. 10, 4671 (1982).
5. W. Alberth, K.M. Straub, A.L. Burlingame, Anal. Chem. 54, 2029 (1982).
6. C.N. McEwen, Anal. Chem. 55, 967 (1983).
7. R.A. McDowell and H.R. Morris, Int. J. Mass Spec. Ion Phys. 46, 443 (1983).
8. M. Arita, M. Iwamori, T. Higuchi, and Y. Nagai, J. Biochem. (Tokyo) 93, 319 (1983).
9. M. Barber, R.S. Bordoli, G.J. Elliott, R.D. Sedgewick, and A.N. Tyler, Anal. Chem. 54, 645A (1982).
10. R.A. Johnston, I.A. Lewis, M.E. Rose, Tetrahedron 39, 1597 (1983).
11. M. Kaminsky in "Atomic and Ionic Impact Phenomena on Metal Surfaces", Academic Press (1965).
12. A.W. Czanderna in "Methods of Surface Analysis", Elsevier Scientific Publishing (1975).
13. M. Barber, R.S. Bordoli, R.D. Sedgwick, A.N. Tyler, G.V. Garner, D.B. Gordon, L.W. Tetler, and R.C. Hider, Biomed. Mass Spec. 9, 265 (1982).
14. D.H. Williams, C.V. Bradley, S. Santikarn, and G. Bojesen, Biochem. J. 201, 105 (1982).
15. H.R. Morris, M. Panico, M. Barber, R.S. Bordoli, R.D. Sedgwick, and A.N. Tyler, Biochem. Biophys. Res. Commun. 101, 623 (1981).
16. K.L. Rinehart, L.A. Gaudioso, M.L. Moore, R.C. Randey, J.C. Cook, M. Barber, R.D. Sedgwick, R.S. Bordoli, A.N. Tyler, and B.N. Green, J. Am. Chem. Soc. 103, 6517 (1981).
17. S.K. Sethi, D.L. Smith, and J.A. McCloskey, Biochem. Biophys. Res. Commun. 112, 126 (1983).
18. G.R. Fenwick, J. Eagles, and R. Self, Biomed. Mass Spectrom. 10, 382 (1983).
19. H.E. May and D.M. Desiderio, J. Chem. Soc. Chem. Commun. 2, 72 (1983).
20. R.W. Kondrat and R.G. Cooks, Anal. Chem. 50, 81A (1978).
21. R.A. Yost and C.J. Enke, Anal. Chem. 51, 1251A (1979).
22. F.W. McLafferty, Science 214, 280 (1981).
23. R.W. Kondrat, G.A. McClusky, and R.G. Cooks, Anal. Chem. 50, 2017 (1978).
24. R.J. Perchaslski, R.A. Yost, and B.J. Wilder, Anal. Chem. 54, 1466 (1982).
25. D.M. Desiderio, I. Katakuse, and M. Kai, Biomed. Mass Spectrom. 10, 426 (1983).

26. K.B. Tomer, F.W. Crow, and M.L. Gross, Int. J. Mass Spectrom. Ion Phys., 46, 375 (1983).
27. F.W. McLafferty, I.J. Amster, M.A. Baldwin, M.P. Barbalos, M. T. Cheng, S.L. Cohen, P.O. Danis, G.H. Kruppa, C.J. Proctor, and F. Turecek, Int. J. Mass Spectrom. Ion Phys. 45, 323 (1982).
28. J.H. Futrell, K.R. Ryan, and L.W. Sieck, J. Chem. Phys. 43, 1832 (1965).
29. R.G. Cooks, J.H. Beynon, R.M. Caprioli, and G.R. Lester, in "Metastable Ions", Elsevier Scientific Publishing, Amsterdam, 1973.
30. R.K. Boyd and J.H. Beynon, Org. Mass Spectrom. 12, 163 (1977).
31. F.W. McLafferty (Ed.), "Tandem Mass Spectrometry", John Wiley, New York, 1983.
32. A.P. Bruins, K.R. Jennings, and S. Evans, Int. J. Mass Spectrom. Ion Phys. 20, 395 (1978).
33. M.L. Gross, P.A. Lyon, F.W. Crow, and S. Evans, Paper presented at 28th Annual Conference on Mass Spectrometry and Allied Topics, New York 1980, p. 107.
34. W.R. Sherman, R.H. Bateman, B.N. Green, and I. Lewis, Paper presented at 31st Annual Conference on Mass Spectrometry and Allied Topics, Boston 1983, p. 356.
35. R.A. Yost, C.G. Enke, D.C. McGilvery, D. Smith, and J.D. Morrison, Int. J. Mass Spectrom. Ion Phys. 30, 127 (1979).
36. R.H. Bateman, B.N. Green, and D.C. Smith, Paper presented at 30th Annual Conference on Mass Spectrometry and Allied Topics, Honolulu 1982, p. 516.
37. A.E. Schoen, P. Dobberstein, J.W. Amy, J.D. Ciupek, and R.G. Cooks, Paper presented at 31st Annual Conference on Mass Spectrometry and Allied Topics, Boston 1983, p. 56.
38. W.H. McFadden, J. Chromatogr. Sci. 18, 97 (1980).
39. D.S. Millington, D.A. York, and P. Burns, Adv. Mass Spectrom. 8B, 1819 (1980).
40. C.N. Kenyon and A. Melera, J. Chromatogr. Sci. 18, 103 (1980).
41. J.D. Henion, B.A. Thomson, and P.H. Dawson, Anal. Chem. 54, 451 (1982).
42. N.J. Alcock, C. Eckers, D.E. Games, M.P.L. Games, M.S. Laut, M.A. McDowall, and M. Rossiter, J. Chromatogr. 251, 165 (1982).
43. J.D. Henion, J. Chromatogr. Sci. 19, 57 (1981).
44. C.R. Blakley and M.L. Vestal, Anal. Chem. 55, 750 (1983).

New Reagents and Methods for the Synthesis of Peptides, β-Lactams and Oligonucleotides

A Progress Report

Ivar Ugi, Helmut Aigner, Gudrun Glahsl, Rainer Herges, Gabriele Hering, Rudolf Herrmann, Gerd Hübener, Josef Götz, Peter Lemmen, Roland Obrecht, Dieter Rehn, Rainer Schwarz, Klaus Seeholzer, Franz Siglmüller, Jutta Stackebrandt, Ruth Stöcklein-Schneiderwind, Sekou Touré Amadou[1], and Hasso v. Zychlinski

Institut für Organische Chemie der Technischen Universität München, D-8046 Garching, FRG

[1]Department of Chemistry, University of Abidjan, Abidjan, Ivory Coast

Abstract: Several schemes of peptide and β-lactam synthesis by 4 CC are discussed, and the chemistry of some associated reagents is described: The resolution of racemic α- ferrocenyl isobutylamine by (-)-diacetone -2-keto-gulonic acid yields either enantiomer, depending on the solvent used. 1-cyclohexenyl isocyanide is obtained from 2-cyclohexenyl isocyanide by base catalyzed-isomerization. Some isocyanides are presented whose 4 CC products contain carbonamide groups that can be converted into activated carboxyl groups. The preparation of L-N -phthaloyl α,β-diaminoproprionic acid, a precursor of nocardicine and analogs, from L-asparagine is presented.

For the synthesis of oligonucleotides some new reagents and techniques are introduced: The TCBOC group, cleavable by $Co^{I}Pc^{\ominus}$, serves well to protect amino groups of the nucleotides. A variety of highly reactive five-membered cyclic phosphorylating reagents are described as well as some sequences of reactions in which the transition P^{III}- P^{V} is used to generate reactive phosphorylating reagents.

---

*Permanent affiliation: Dept. of Chemistry, University of Abidjan, Ivory Coast

Natural Product Chemistry
Ed. by Atta-ur-Rahman

Key words: Chiral α-ferrocenyl alkylamines, 1-cyclohexenyl isocyanide, activateable carboxy protection, L-N_α-phthaloyl α,ß-diaminopropionic acid, TCBOC-protection, dimethylamino TCB-chlorophospite, five-membered cyclic phosphorylating reagents.

Abbreviations: CAP = cyclic acid phosphate; CDI = carbonyl diimidazole; CEP = cyclic enediol phosphate; $Co^{I}Pc^{\ominus}$ = Co^{I}-phthalo-cyanine anion; 4 CC = from component condensation; DMAP = 4-dimethylamino pyridine; Fc = Ferrocenyl; IGOR = interactive generation of organic reactions; Im = imidazolyl; TCB = trichloro tert.-butyl = 2-trichloromethyl-2-propyl; TCBOC = TCB oxycarbonyl; TMS = trimethylsilyl; TFA = trifluoroacetyl.

1. Introduction

This paper covers a variety of reactions and reagents which were developed at our institute in recent years, or whose development is still in progress. Some of these will hopefully be useful for the synthesis of oligonucleotides, peptides, ß-lactam antibiotics and their derivatives.

The chemistry in this contribution seems heterogeneous, it has, however, grown from a common root which is isocyanide chemistry[1] and in particular the 4 CC of the isocyanides.[1-4] A brief genealogy of our diversified endeavours will clarify the relations between the various subtopics that are presented here.

The first post-doctoral research project of the present speaker was the chemistry of the pentazoles[5]. In order to compare the

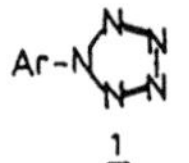

1

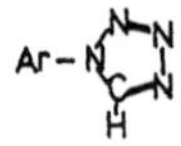

2

pentazoles 1 with some analogous tetrazoles 2, these were synthetized from the corresponding isocyanides[6]. When it became clear that no efficient methods for the preparation of isocyanides were available, the synthesis of isocyanides from the respective formamides was developed in 1958[7]. With the isocyanides now readily available, we began to explore their chemistry and discovered the 4 CC [1-4]. As soon as the 4 CC was known, it was clear to us that one of its main applications would be the synthesis of peptides and other α-amino acid derivatives [2,3], such as the β-lactam antibiotics [4]. This led directly to the investigation of asymmetric syntheses [8-10] and their chiral templates, in particular the ferrocene derivatives [11], and more recently the sulfoxides. Both classes of compunds have unique advantages in such asymmetric syntheses. It was also understood that stereoselective 4 CC peptide syntheses will require serious engagement in conceptual and experimental stereochemistry [12].

The optimum use of the 4 CC in the synthesis of peptides and related compounds requires an additional type of protective groups. These must be "orthogonal" to the known protective groups [13], i.e. protective groups which are stable towards the deprotective conditions of the other protective groups, and are removed under conditions which do not affect the other protective and auxiliary groups. The β-halogenated alkyl groups are such protective groups [13-15]. These are cleaved by Co^{I} supernucleophiles. The result was not only the "Stereochemical Analogy Model" [9] and the concept of "Chemical Chirality"[10] which serves the purposes of chemistry better than the purely geometrical chirality of Lord Kelvin [16] and Prelog [17] but also a permutational approach to stereochemistry [18] and the theory of the chemical identity group [12),19], a unified treatment of rigid and flexible molecules. The latter work included an analysis of the stereochemical behaviour of pentavalent phosphorus compounds [20),21] and the postulate that Berry Pseudorotation [22] is not the only conceivable isomerization mechanism of pentacoordinate molecules. This was done in collaboration with F. Ramirez and his research group. One of the results of the latter joint effort was the insight, that cyclic acyl phosphates and cyclic enediol phosphates are potentially useful in the synthesis of oligonucleotides [23]. While the 4 CC peptide syntheses were under development it was realized that the preparation of large peptides by a combination of 4 CC with the classical methods could be achieved in very many ways - for example there are 139 000 000 such syntheses conceivable for the A chain of insulin. Therefore the Leverkusen peptide synthesis program was developed by Kaufhold and Ugi in 1966 - 68 for computer aided design of optimized synthesis of peptides by 4 CC and classical methods [24]. Together with Corey's early version of LHASA [25] and Gelernter's SYNCHEM I [26] the Leverkusen program was the beginning of computer assisted synthesis [27]. In the meantime Dugundji and Ugi [28] have formulated an algebraic model of the logical structure of chemistry, a general theory of chemistry on which a great variety of computer programs for the deductive solution of chemical problems have been based [29]. The latest is IGOR [30], a program for the interactive generation of organic reactions, which has already invented various new reactions; some of the reactions proposed by IGOR have been veryfied in the laboratory [31].

2. Peptide Syntheses

The 4 CC can be used in various ways in peptide synthesis.

2.1 Asymmetric Peptide Segment Synthesis with Chiral α-Ferrocenyl Alkylamines

The synthesis of peptide segments by asymmetric 4 CC according to Scheme 1 is still one of the most interesting applications of the 4 CC [2,3].

Scheme 1

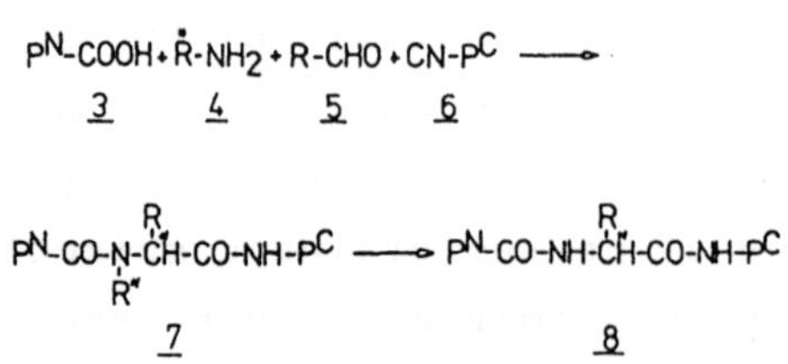

Although some peptide segments have thus been successfully prepared [3], various problems of this reaction scheme remain to be solved.

The synthesis of the required isocyanides is still under investigation. Our hopes that the dehydration of suitable formamides by protonated CDI [32] will generally afford the synthesis of the chiral α-isocyano esters have been disappointed; prolonged exposure of the isocyanide to the reagents lowers the yield and racemizes the product. Joullié and co-workers [33] have also pointed out the defiencies of this method; it seems not to be superior to our old "diphosgene" method [34].

Among the potential reagents for the synthesis of sensitive isocyanides N, N'- dibutyl CDI dication is one of the best candidates.

The α,β-unsaturated isocyanides 11, in particular 1-cyclohexenyl isocyanide, may be used to prepare α-amino acid derivatives with a C-terminal carbonamide group [35].

Scheme 2

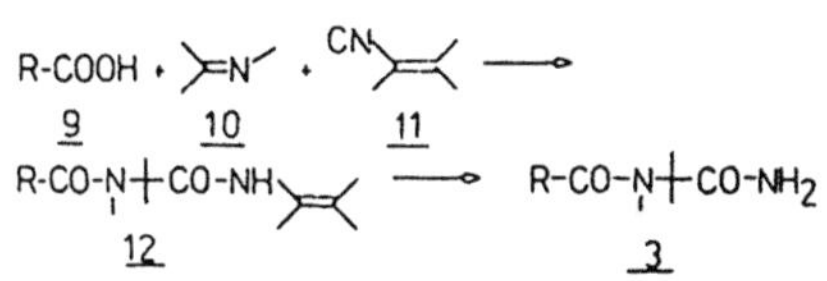

The first synthesis of 16 has been accomplished by treatment of 18 with $POCl_3$ in the presence of potassium tert. butanolate [35]. Recently we found an efficient access to 16 by tert.-butanolate catalyzed isomerization of 15 [36] which is easy to prepare from cyclohexene.

Scheme 3

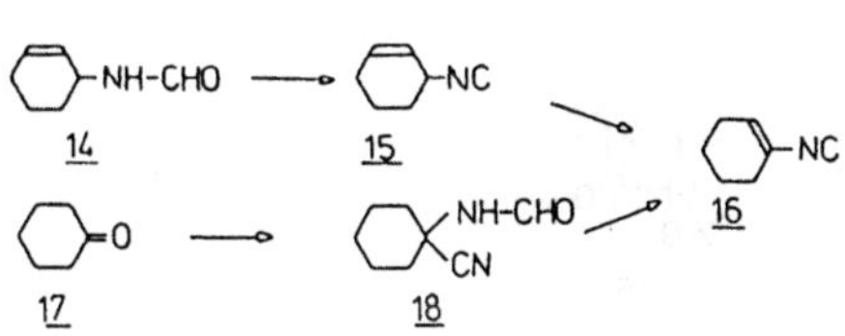

The best chiral templates for peptide syntheses by asymmetric 4 CC are α-ferrocenyl isobutylamine 19, R=Pr^i and other α-ferrocenyl alkylamines 19 whose group R is a secondary alkyl group [2,3].
Their use is illustrated by scheme 4.

Scheme 4

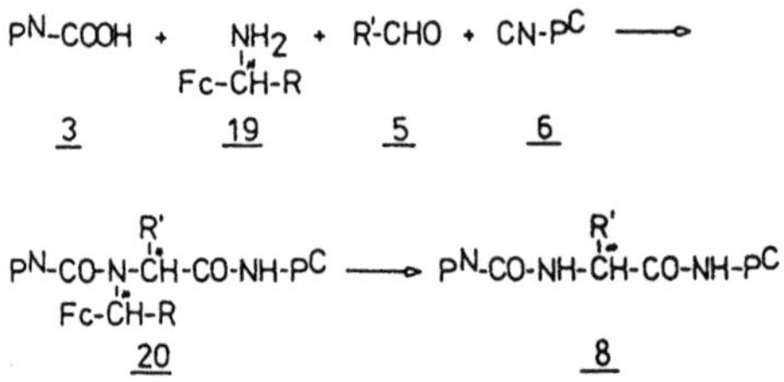

The preparation of such amines 19 was quite cumbersome until recently. Now α-ferrocenyl isobutylamine is readily available according to scheme 5 through the condensation of ferrocene with isobutylaldehyde in strong acid,and scavengeing the carbocation with ammonia [37]. The resolution of 19, R=Pr^i and

Scheme 5

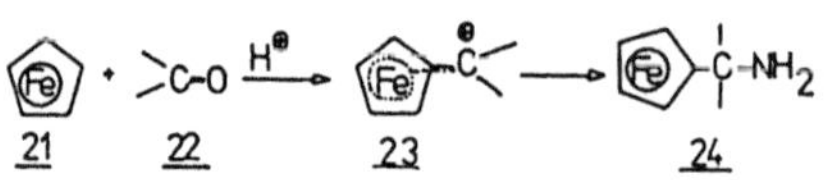

many analogs succeeds only with one of the many resolving reagents [38], namely (-)-diacetone-2-ketogulomic acid which was recommended to us by Gil-Av and Charles [39,40]. In fact in a two-step resolution involving a change of solvent 25 affords both anantiomers of 19, R=Pr^i, in analogy to the resolution of α-ferrocenyl N,N-dimethyl ethylamine by D-tartaric acid [41].

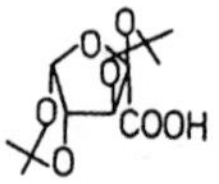

25

Our numerous attempts to synthetize a chiral amine 19 from inexpansive chiral natural products like α-pinene, menthol and carboxylic acid derived from pinane and menthane were most successful with menthol as the starting material. Our syntheses of ferrocenyl menthyl and neomenthyl carbinylamine, 43 and 44, is shown in scheme 6 [42]. The assumed structure of 43 is presently being confirmed by X-ray chrystallography.

Scheme 6

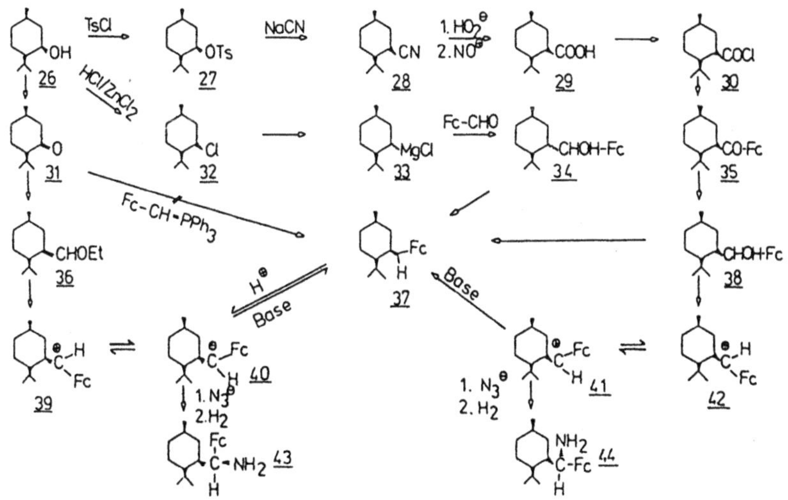

Although the peptide segment syntheses by stereoselective 4 CC have been studied most extensively, the results are often still conflicting. Partly this is due to the fact that for 4 CC yield, purity and diastereomer ratio of the product changes sometimes extremely with small changes in reaction conditions, and that the dependence of yield and stereoselectivity upon the reaction conditions seem to change drastically with the structure of the components [43]. Thus the data from our model experiment [44] can only provide some general guidance in the search for optimum reaction conditions of a 4 CC with given components. Therefore we are now developing methods for the optimization of 4 CC. We are trying to find methods which are quick and convenient enough to be applied to any individual 4 CC of interest. The 4 CC according to scheme 7 serves as a model reaction in the development of such

Scheme 7

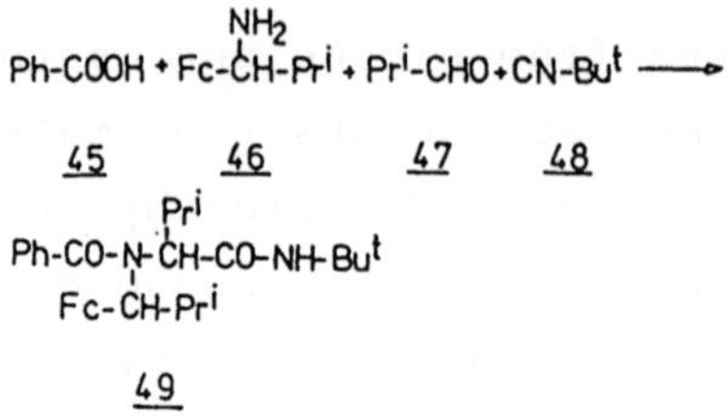

optimization methods. This model reaction is carried out under simultaneous variation of many parameters representing reaction conditions. The products are analysed by HPLC and the results are evaluated by the simplex method [45] on the response surface method [46]. The research group of S. Wold and R. Carlson [47], Umeå, Sweden, are kind enough to help us in the design and evaluation of these experiments.

2.2 Activated Peptide Segments

The selective conversion of a carbonamide group into carboxylic group plays an important role in the synthesis of ß-Lactam antibiotic by 4 CC[4]. One of the methods to achieve this is in essence described by scheme 8 [48].

Scheme 8

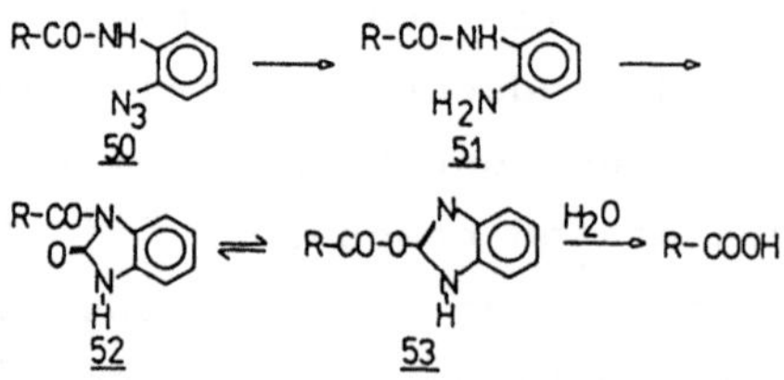

This stimulated a study whether or not such a sequence of reactions is usable in peptide syntheses according to scheme 9 which may be advantageous in some cases.

Scheme 9

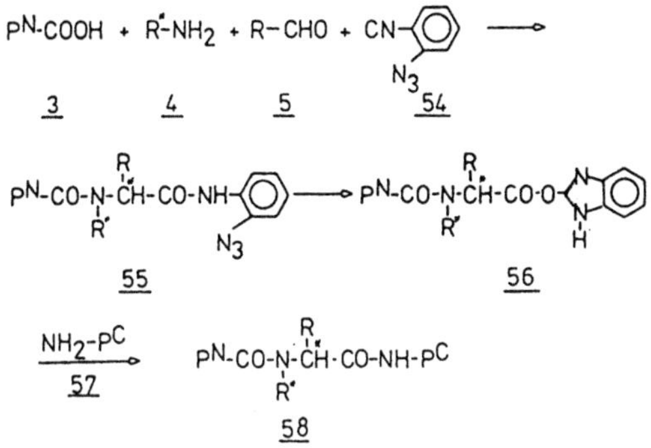

As model experiment for such peptide syntheses and the use of the o-azido anilide as protected carboxyl groups which are convertible into activated carboxyls, we are investigating the synthesis of Bz-CO·Ala·OMe from o-azidophenyl acetanilide, 50,R=Bz, and H-Ala-OMe as well as the synthesis of the Izumya peptide [49] 60 according to scheme 10, including a racemization study.

Scheme 10

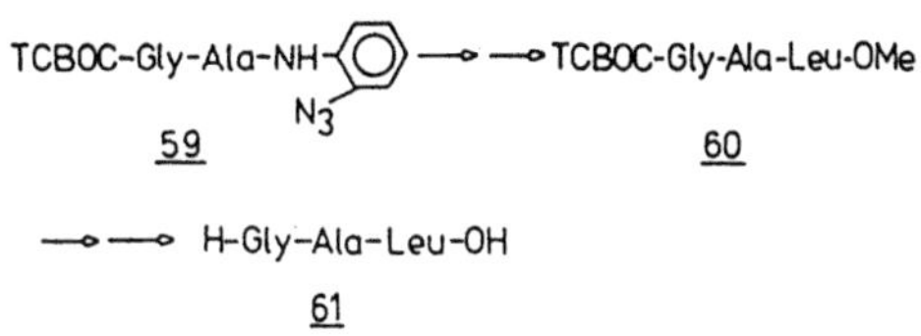

2.3 The Sulfoxido Group as a Chiral Reference System in 4 CC Peptide Syntheses

We are studying the p-tolylsulfoxido group, a chiral auxiliary group, in peptide syntheses by 4 CC according to scheme 11.

Scheme 11

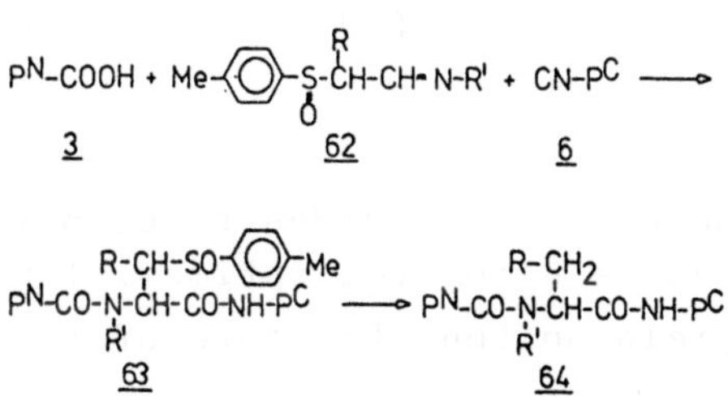

Such syntheses are particularly interesting, if a chiral amino acid unit is formed which is not accessible by the customary asymmetric 4 CC (see 2.1) [3], such as proline, phenyl alanine and the N-methyl α-amino acids.

The 4 CC according to scheme 12 is being tested as a model experiment

Scheme 12

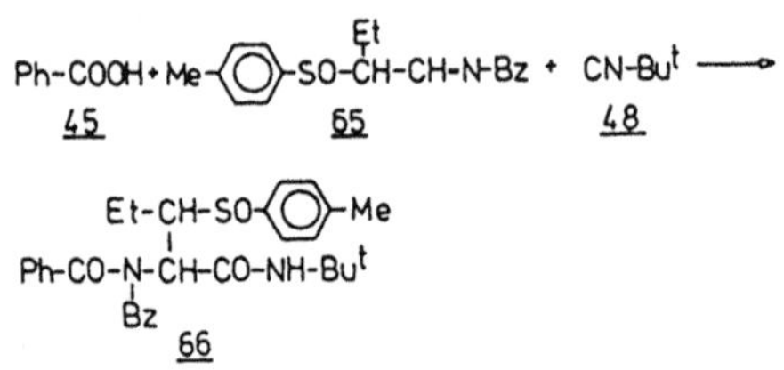

This 4 CC proceeds smoothly in concentrated methanol solution at 0°C and yields with high preference one of the diastereomers of 66 [50].

2.4 Activation of N-protected Amino Acids and Peptides by 4 CC

The 4 CC of enamines with N-protected amino acid derivatives and isocyanides yields the diacylimides as is illustrated by scheme 13.

Scheme 13

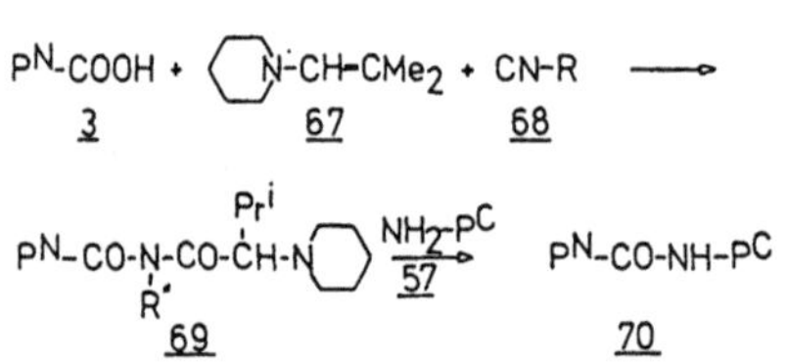

We have found that this synthesis of diacylimides can be used for peptide coupling.

As seen from Table 2 the aryl isocyanides react particularly well, with phenyl isocyanide being the optimum. Table 1 shows the dependence of the yield on the structure of the peptide derivatives [51].

2.5 Segment Coupling by 4 CC

We have tested numerous aldehydes as components of 4 CC peptide segment coupling according to scheme 14. Many of them were

Scheme 14

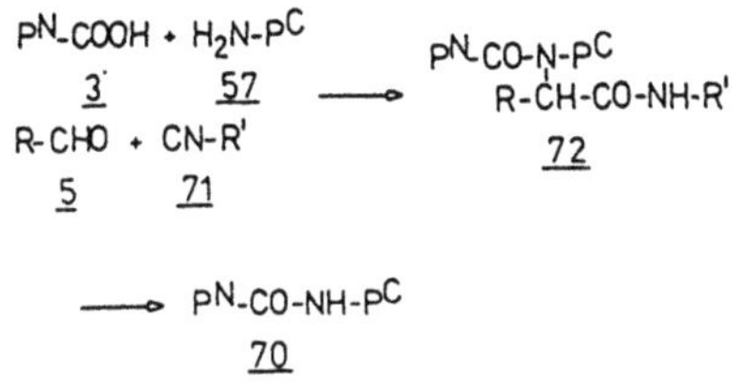

usable for such segment coupling, but none of them satisfies at both steps, the 4 CC and the cleavage step 72→70. The only aldehyde which is not yet eliminated from our screening program is 73.

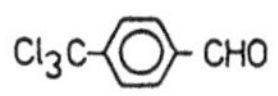

73

The critical question is whether or not the 4 CC products of 73 are well enough cleavable by $Co^{I}Pc^{\ominus}$ [13].

3. ß-Lactam Antibiotics

The 4 CC is one of the few methods by which the ß-amino acides and a variety of their derivatives are readily converted into the corresponding ß-lactams[52]; the 4 CC is now a widely used method for the synthesis of the ß-lactam antibiotics and their derivatives [4].

There are several syntheses of the nocardicines and related compounds whose key step is a 4 CC [53].

Since quite a few years we have done preliminary experiments for synthesis by 4 CC of ß-lactams of the nocardicine type. We planned to prepare an α-protected elimization of α,ß-diamino propionic acid from TCBOC asparagine by Hofmann degradation, but instead of the desired product 76 was formed.

Table 1. Peptide syntheses by coupling an amino ester with a diacylimide from 4CC of piperidinoisobutene and an N-protected amino acid derivative with phenyl isocyanide

Peptide	Yield (%)
BOC-Gly-Ala-Val-OMe	69
2-Val-Ala-OEt	52
Z-Val-Gly-OMe	83
Z-Val-Val-OMe	45

Table 2. Synthesis of Z-Val-Gly-OMe with piperidinoisobutene and various Isocyanides

Isocyanide	Z-Val-Gly-OMe Yield (%)
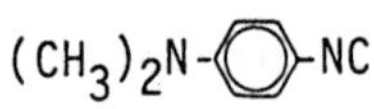	66
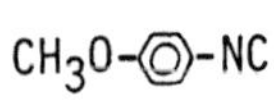	54
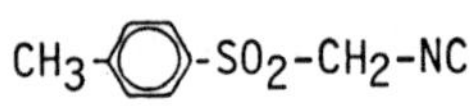	31
CH_3-NC	56
CN-CH_2-$COOCH_3$	64
CN-CH_2-$COOCH_3$	62[a]

a) Z-Val-Ala-OEt

Scheme 15

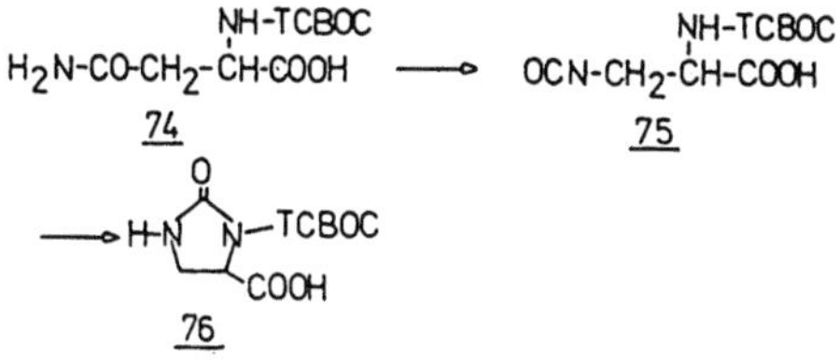

Such a cyclization could be avoided by using tosyl for N-protection. However, a cumbersome change of protective groups is necessary, before the ,ß-diamino propionic acid derivative can be used in ß-lactam syntheses.[54)] Recently Radhakrishna et al. [55)] have recommended 78 as a reagent for Hofmann degradation under non-alkaline conditions. This reagent converts N-phthaloyl asparagine smoothly into L-α-N-phthaloyl-α,ß-diamino propionic acid 79, an

Scheme 16

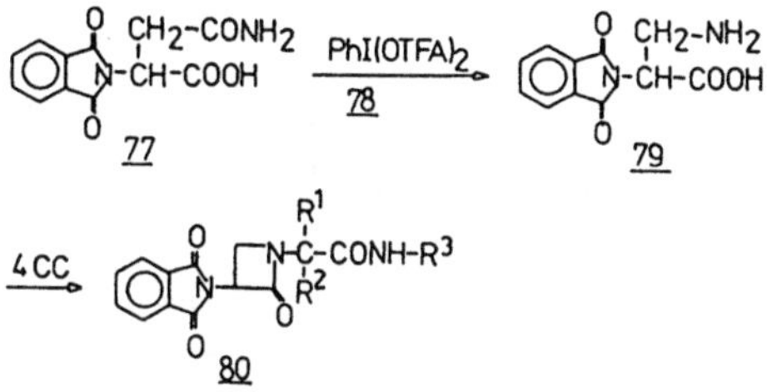

excellent starting material for 80 which is easy to convert into ß-lactams like nocardicine.

The 4 CC synthesis of a ß-lactam antibiotic or an analog requires the selective conversion of a carbonamido group into a carboxyl group in the presence of a labile ß-lactam system [4,48)]. One of the best ways to achieve this is illustrated by scheme 17 [56,57)].

4. Oligonucleotide Syntheses

Since about 10 years we are developping reagents and methods for the synthesis of oligonucleotides. At the early stages this work was done in close collaboration with Ramirez and his group [23)].

Scheme 17

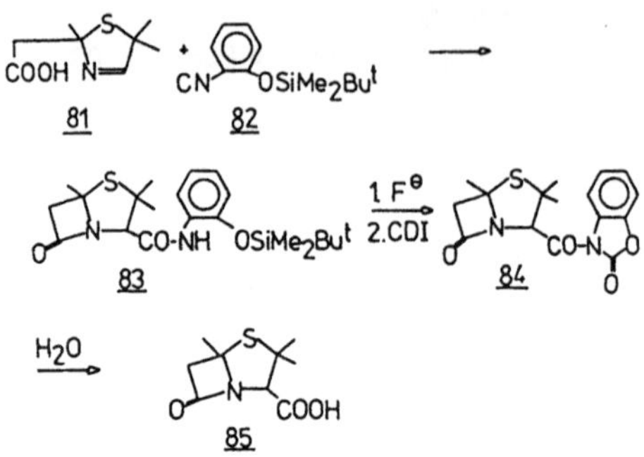

Our efforts in this field are characterized by the use of ß-halogenated alkyl groups as protective groups which are cleaved under extremely mild conditions by $Co^{I}Pc^{\ominus}$, [13-15] and a search for phosphorylating reagents with a five-membered unsaturated ring containing phosphorus as a ring member. The five membered cyclic phosphates are 10^{5} - 10^{7} times more reactive than their cyclic analogs [21].

4.1 The TCBOC Group as an N-protecting Group of Nucleosides

The TCBOC group was originally introduced as a N-terminal blocking group of peptide syntheses [14]. Its stability towards a variety of conditions encountered during syntheses, its "orthogonality" to other protecting groups, and its mild, selective removal by $Co^{I}Pc^{\ominus}$ at 0 - 20°C in practically neutral solution are its distinguishing advantages [13].

The reagent for introducing the TCBOC group is TCBOB-Cl 89 which is prepared from chloretone and phosgene or diphosgene.[14,58] It is meanwhile commercially available from Fluka A.G.

Scheme 18

$Cl_3C\text{-}CMe_2\text{-}OH + COCl_2$ (or $Cl\text{-}CO\text{-}O\text{-}CCl_3$) ⟶
86 87 88

$Cl_3C\text{-}CMe_2\text{-}O\text{-}CO\text{-}Cl$
89

The N-TCBOC nucleosides are readily obtained from the nucleosides by complete acylation with TCBOC-Cl in pyridine, followed by O-deacylation (scheme 19)[59].

Scheme 19

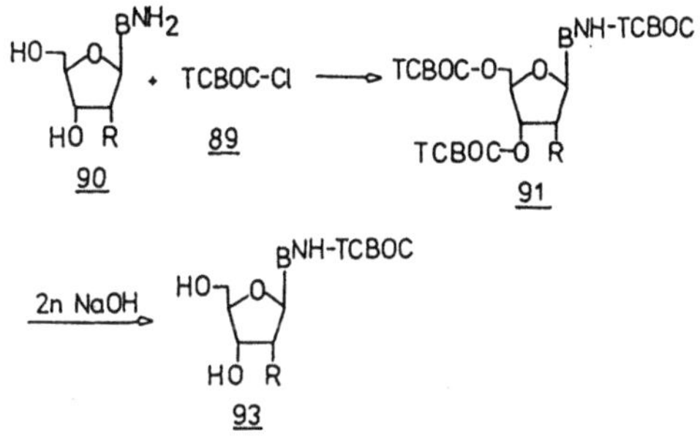

4.2 TCB Chlorophosphites and Chlorophosphates

Chloreton reacts with PCl_3 under a variety of conditions to form mixtures of 95 and 96 [80].

Scheme 20

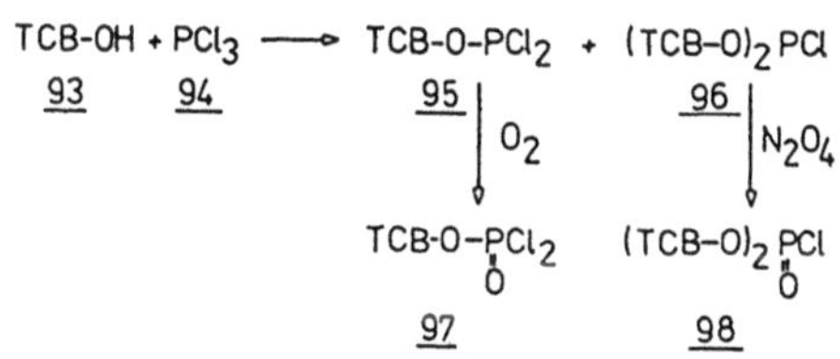

Recently we have found conditions under which practically pure 95 or 96 can be obtained from 93 and 94 [61]. Oxidation of 95 and 96 leads to the chlorophosphates 97 and 98.

Among the known phosphorylating reagents 6 excels by its ability to discriminate between the 5' and the 3'-OH group of nucleosides. In the presence ot DMAP 6 [62] phosphorylates exclusively the primary 5'-OH group, without attacking the secondary 3'-OH group [63]. This selectivity can be attributed to the two bulky TCB groups in 98 which would cause too much crowding in the pentacoordinated transition state or intermediate 3'-phosphorylation. No such 5'-selectivity is observed with the chlorophosphite 96.

Letsinger et al. [64] have developed a solid phase phosphite type oligonucleotide synthesis based on 95.

We have observed that it is advantageous to form the corresponding triazolides first from 95 [61]. The latter derivatives have less tendency to form the 3',3'-dinucleotides.

Scheme 21

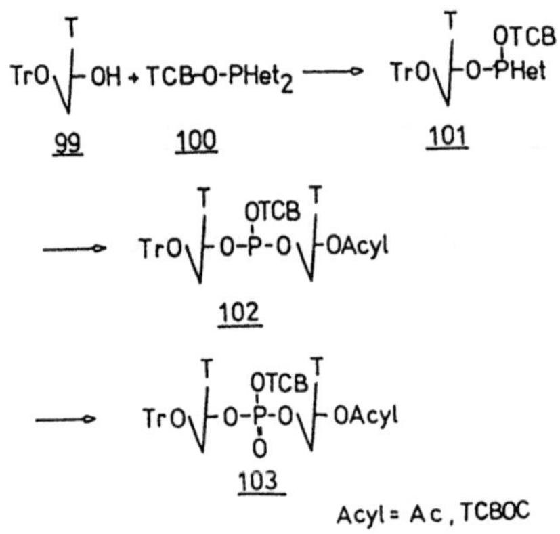

The phosphite dichloride 95 reacts with two equivalents of dimethylamine in dry ether almost quantitatively to form the amidochloride 105 [65](f.p. 36-38°).

Scheme 22

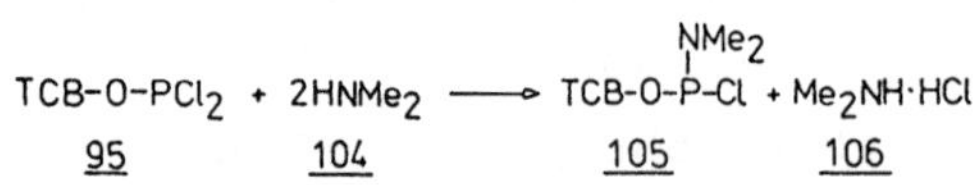

The amido chlorophosphite 105 can be used in analogy to Caruthers' reagent [66] (dimethylamino-methoxy-chlorophosphine). As a phosphite type reagent for solid phase oligonucleotide synthesis, 105, has the advantage that it leads to oligonucleotides with the easy to remove TCB phosphate blocking group instead of methyl protection which does not have the excellent deblocking conditions of TCB. The reagent 105 has been tested in a homogeneous phase dinucleotide synthesis in analogy to scheme 21.

4.3 Five-membered Cyclic Phosphorylating Reagents

Phosphorylating reagents with a five-membered ring containing phosphorus and at least one sp^2 carbon atom are extremely reactive [21]. Certain compounds of this class have the potential to become the key reagents of some new methods for the synthesis of oligonucleotides, methods in which the advantages of the phosphite methods and the phosphate methods are combined, namely the high acylation rates of the phosphites, and no oxidation step the advantage of the phosphate methods. The research group of Ramirez and ours spent prolonged efforts to prepare CAP and CEP derivatives which can transfer a CAP or a CEP group into a hydroxy group of a nucleoside[67] (=Nu).

Scheme 23

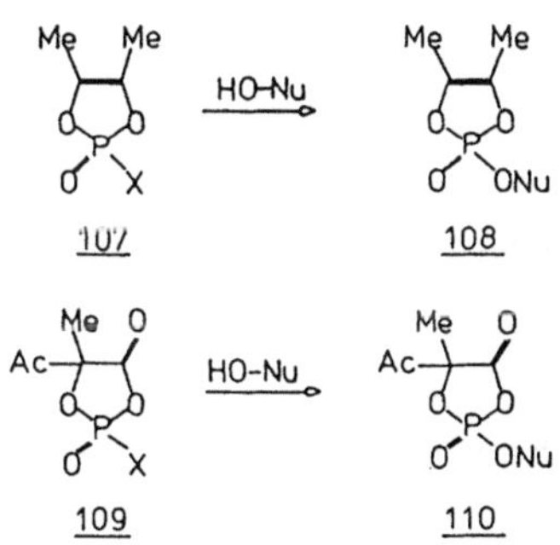

With the exception of the hard to prepare CEP imdazolide, a CEP transferring reagent[68], and the following model sequence of reactions [69], no reagents and methods were found for the preparation of the desired CAP and CEP derivations.

Scheme 24

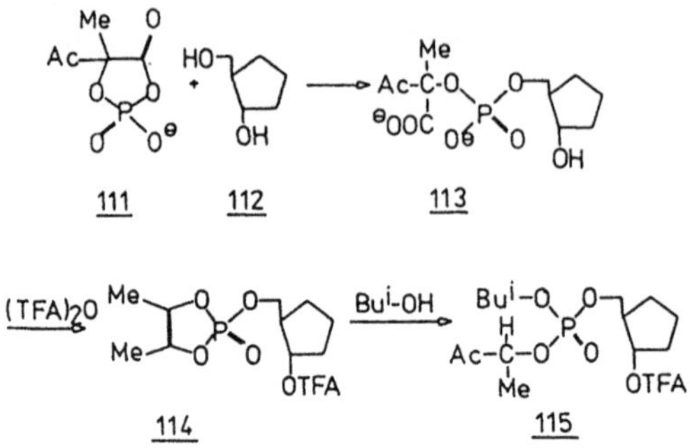

The 3-keto-2-butyl group is readily removed at pH = 8-9 from phosphotriesters like 115 containing this moiety. Thus the formation of phosphotriesters from CAP or CEP derivatives yields products with a well-cleavable blocking group.

After realizing that the desired five-membered cyclic phosphates with a good leaving group (see scheme 22) can neither be prepared via the phosphoranes, nor cyclizing phosphorylation of alkali enediolates and related compounds, we carried out some extensive exploratory experiments in which a variety of suitable O-TMS compounds were reacted with diphoro and dichloro alkylphosphates [70]. The results of this investigation indicated that alkyl phosphates with a reactive five-membered ring could indeed be prepared by this route. However, the corresponding cyclic phosphates with a good leaving group were not directly accessible by this route.

A study of Gozman et al. [71] on cyclic amido phosphites guided our following synthesis of CEP-Cl 107, X=Cl [72].

Scheme 25

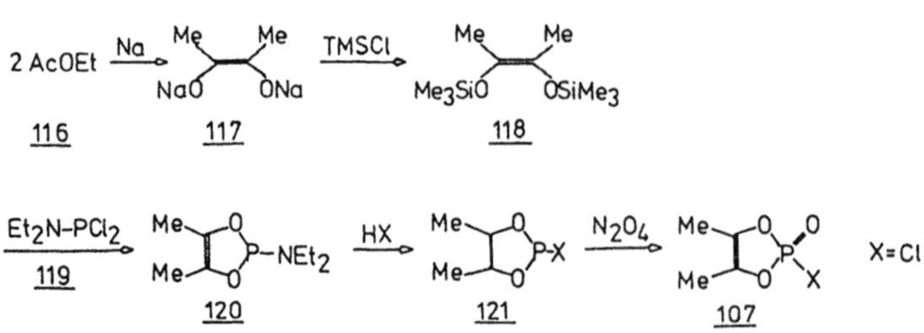

The cyclic chlorophosphite 121, X=Cl reacts with alcohols smoothly and fast to form the corresponding alkoxy conpounds 121, X=OAlk, which can be oxidized to the respective alkoxy-CEP, highly reactive phosphorylating reagents.

The research group of Chattopadhaya and ours are collaborating in the development of solid phase oligonucleotide syntheses based on the chemistry of 107 and 121.

The fluoro analog CEO-F 107, X=F of CEP-Cl 107, X=Cl has been prepared in low yield according to scheme 26 [73].

Scheme 26

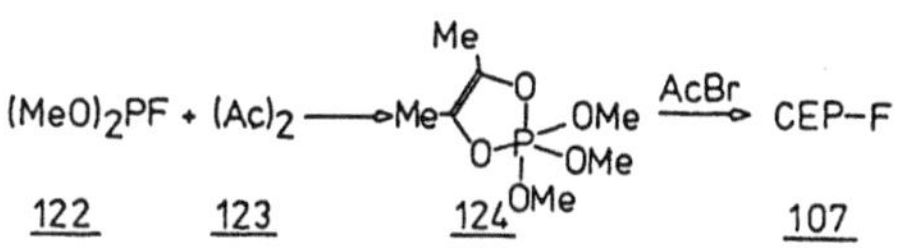

If CEP-F were needed in appreciable quantity, it would be better prepared from 120 by treatment with HF and subsequent oxidation.

The above cyclic phosphorylating reagents with a strong leaving group have all been designed for the synthesis of phosphodiesters; 127 is a CEP derivative which can be used for the preparation of phosphomonoesters as well as the diesters.

It is prepared according to scheme 27.

Scheme 27

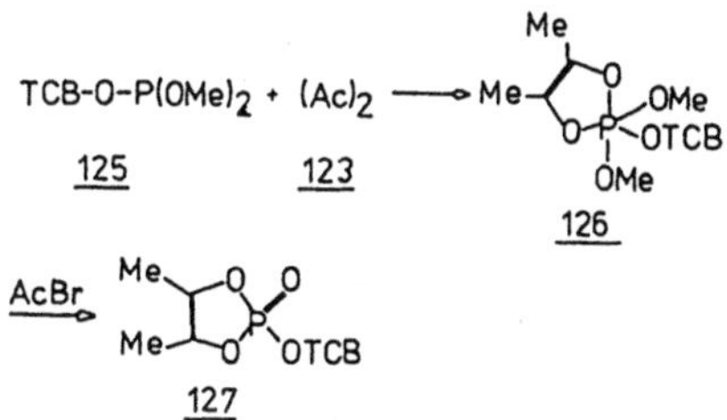

The reagent 127 discriminates quite well between primary and secondary alcohols. The system of reactions in scheme 28 represents a model for 5'-phosphorylation of nucleosides.

Scheme 28

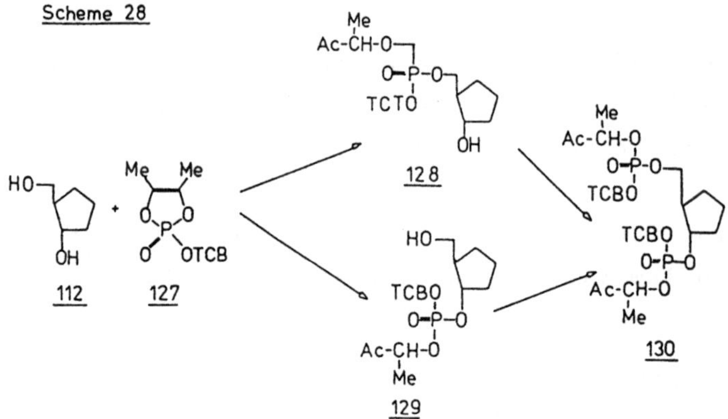

In this scheme productive selectivity is combined with destructive selectivity [10,74]. When 112 is reacted with less than one equivalent 127 a mixture of 128 and 129 is formed; in this mixture 128 dominates ($c_{128}: c_{129} \approx 10$). If, however, 112 is used in 15 - 20% excess a mixture of 128 and the non-isomeric easy-to-separate 130 results, because 129 with its primary hydroxyl group is phosphorylated ca. 10 times faster than 127. The competitive phosphorylation of 127 and 128 is a destructively selective process in which the ratio $c_{128}: c_{129}$ changes as an exponential function of the ratio $k_{128} : k_{129}$ of the rate constants and time [74].

The reaction of the TMS derivative of glycolic acid and 95 and subsequent oxidation leads to 131, which reacts smoothly with alcohols and nucleoside derivatives forming 132. The cyclic phosphite does not react in a uniform manner with alcohols [65].

Scheme 29

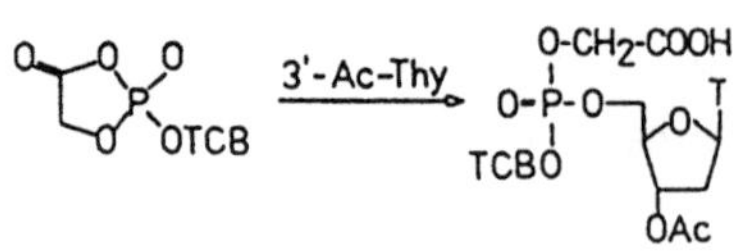

131 132

Within the framework of our search for new fast-reacting and 3'-5' selective phosphorylating reagents some model experiments with ethylene glycol chlorophosphite 134 have been carried out which have not yet led to usable preparative procedures or reagents. However, the chemistry involved seems to be interesting enough to be recorded here.

With diacetyl 123 undergoes a clean reaction for form 134, presumably via oxyphosphorane and phosphonium intermediates [75];

Scheme 30

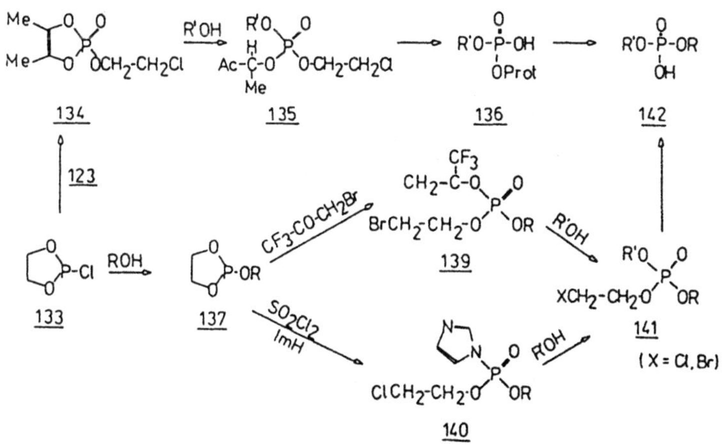

alcohols react with 134 to yield 135 which carries two mutually orthogonal protective groups. Thus 135 can be easily converted into the unsymmetrical phosphordiesters 136.

The phosphite 137 and the bromo ketone 138 undergo a Perkow type reaction [69]. Our hopes that the trifluorovinyl group would be a reactive leaving group have not been fulfilled; 139 is rather sluggish to react with alcohols [76]. Another way to convert 137 into a phosphorylating reagent is to treat it with sulfuryl chloride and imidazole. The resulting product 140 is indeed an active phosphorylating reagent [76]. The latter pathway seems to give us a lead to new ways to activate phosphite esters by a redox reaction.

Acknowledgement: We wish to express our gratitude to Deutsche Forschungsgemeinschaft (Normalförderung und SFB 145), Fonds der chemischen Industrie and Bund der Freunde der TUM (Wacker-Spende) for financially supporting our work.

References

1. I. Ugi ed.,"Isonitrile Chemistry", Academic Press, New York 1971.

2. I. Ugi in: "The Peptides" Eds.: M. Meienhofer, E. Gross Academic Press, New York 1979, Vol. 2,p. 365.

3. I. Ugi, D. Marquarding, R. Urban in: "Chemistry and Biochemistry of Amino Acids, Peptides and Proteins", Ed.: B. Weinstein, Marcel Dekker, New York, 1982, p. 245

4. I. Ugi, Angew. Chem. 94, 826, (1982); Angew. Chem. Int. Ed. 21, 810 (1982),

5. R. Huisgen, I. Ugi, Chem. Ber. 90, 1914 (1957), I. Ugi, H. Perlinger, L. Behringer, Chem. Ber., 91, 2324, (1958), 92, 1864 (1956); I. Ugi, Tetrahedron 1963, 1801 (1963), in: "Advances in Heterocyclic Chemistry" Vol. III, Ed.: A.R. Katritzky, Academic Press, New York 1964; p. 373; in "Comprehensive Heterocyclic Chemistry" Vol. 4, Pergamon, New York 1984 (in press).

6. E. Olivery-Mandalá,B. Alagna, 177 Gozz.chim. Ital. 40 II,(1919)

7. I. Ugi, R. Meyr, Angew.Chem. 70, 702 (1958)

8. J.O. Morrison, H.S. Mosher, "Asymmetric Organic Reactions", Prentice-Hall, Englewood Cliffs, N.J. 1971.

9. E. Ruch, I. Ugi, Theoret.Chim.Acta (Berl.), 4, 287 (1966), Topics in Stereochemistry" Vol. IV Eds.: N.L.Allinger, E.L. Eliel,Wiley, New York, 1969 , p. 99 ; see also:I. Ugi, Z.Naturforsch. 20b, 405 (1965)

10. J. Dugundji, R. Kopp, D. Marquarding, I. Ugi, Topics Curr. Chem. 75, 165· (1978),

11. I. Ugi, Rec. Chem.Progr., 30, 289 (1969)
D. Marquarding, P. Hoffmann, H. Heitzer, I. Ugi, J.Amer.Chem. Soc., 92 , 1969 (197)
G. Eberle, I. Lagerlund, I. Ugi, R. Urban, Tetrahedron, 34, 977 (1978)

12. J. Dugundji, R. Kopp, D. Marquarding, "Perspectives in Theoretical Stereochemistry", Springer, Heidelberg 1984 (in press).

13. H. Eckert, W. Breuer, J. Geller, I. Lagerlund, M. Listl, D. Marquarding, S. Stüber, I. Ugi, S. Zahr, H.v.Zychlinski IUPAC, Org. Synth. Symp. Proceedings, Jerusalem, Sept. 78 Pure Appl.Chem., 51 1219 (1979)

14. H. Eckert, M. Listl, I. Ugi,
Angew. Chem., 90, 388 (1978);
Angew. Chem. Int., Ed., 17, 361 (1978).

15. R.G.K. Schneiderwind und I. Ugi, Z.Naturorsch. 36b, 1173 (1981), T . Kaminura, T. Masegi und T. Hata, Chem.Lett. 1982,965; R.G.K.Schneiderwind, I. Ugi Tetrahedron 39, 2207 (1983)

16. Lord Kelvin, "Baltimore Lectures", Clay + Sons, London 1904.

17. V. Prelog, (Nobel Lectute), Science 193,, 17 (1976),

18. I. Ugi, D. Marquarding, H. Klusacek, G. Gokel, P. Gillespie, Angew.Chem. 82, 741 (1970); Angew. Chem.Int.Ed., 9, 703 (1970).

19. J. Dugundji, J. Showell, R. Kopp, D. Marquarding, I. Ugi, Isr.J.Chem. 20, (1980)

20. I. Ugi, D. Marquarding, H. Klusacek, F. Gillespie, F. Ramirez Accounts Chem. Res., 4, 288 (1971);
P. Gillespie, P. Hoffmann, H. Kusacek, D. Marquarding, S.Pfohl, F. Ramirez, E.A. Tsolis, I. Ugi,
Angew. Chem., 83, 691 (1971); Angew.Chem. Int. Ed., 10, 687 (1971),
J. Dugundji, P.D. Gillespie, D. Marquarding, I. Ugi, F.Ramirez in: "Chemical Applications of Graph Theory", Ed.: A.T.Balaban, Academic Press, London 1976, p. 107.

21. D. Marquarding, F. Ramirez, I. Ugi, P.D. Gillespie, Angew. Chem., 85, 99 (1973), Angew.Chem.Int.Ed., 12, 91 (1973).

22. R.S. Berry, J.Chem. Phys. 32, 933 (1960).

23. F. Ramirez, S. Glaser, P. Stern, P.D. Gillespie, I. Ugi Angew.Chem. 85, 39 (1973);
Angew.Chem. Int. Ed. 12, 66 (1973)
F. Ramirez, P. Stern, S.L. Glaser, I. Ugi, P. Lemmen Phosphorus 3, 165 (1973);
G.P. Smith, C.N. Caughlan, F. Ramirez, S. Glaser, P.Stern, J.Amer. Chem. Soc. 96, 2698 (1974);
F. Ramirez, I. Ugi, "Phosphorylating Reagents for Oligonucleotide Synthesis", Recent. Dev. Oligonucleotide Synth.Chem. Minor Bases tRNA , Ed. Paryzek, Zdzislaw, Univ. im Adama Mickiewicza, Poznan, Pol. (1974);
F. Ramirez, J.F. Marecek, I. Ugi, Synthesis, 1975, 99;
F. Ramirez, I. Ugi , Phosphorus Sulfur, 1, 231-245 (1976).

24. I.Ugi, Rec.Chem.Progr.30, 289 (1969).

25. E.J.Corey, W.T.Wipke, Science 166, 178 (1969)

26. H.Glernter, Topics, Curr.Chem. 41, 113 (1973).

27. "Computer-Assisted Organic Synthesis", ACS Symposium Ser. 61

28. J. Dugundij, I. Ugi
Top. Curr. Chem., 39, 19 (1973).

29. I.Ugi, J. Bauer, J. Brandt, J. Friedrich, J. Gasteiger, C. Jochum, W. Schubert, Angew. Chem., 91, 99 (1979); Angew. Chem. Int. Ed., 18, 111 (1979);
I. Ugi, J.Bauer, J. Brandt, J. Friedrich, J. Gasteiger, C. Jochum, W. Schubert, J. Dugundji, in: Ed.: J. Bargon Plenum Press, New York 1980, p. 275.

30. J. Bauer, I. Ugi
J.Chem. Res. 1982 (S), 298;
J.Chem. Res. 1982 (M) 3101, 3201.

31. R. Herges, Diss. Techn. Univ.München 1984; R. Herges, J.Bauer, J.Ugi, G. Horsmann, Tetrahedron (in preparation).

32. G: Giesemann, E.v.Hinrichs, I. Ugi, J.Chem.Res. (S), 1982, 79.

33. M. Joullié et al. (priv.comm.)

34. G. Skorna, I. Ugi,
Angew.Chem. 89, 267 (1977); Angew.Chem.Int.Ed. 16, 259 (1977): A. Efrati, I. Feinstein, L.Wackerle, 7.Org.Chem. 45, 4059 (1980)

35. I. Ugi, F.K. Rosendahl,
Liebigs Ann. Chem., 666, 65 (1963)

36. H.v.Zychlinski, I.Ugi (to be published soon).

37. R. Herrmann, I. Ugi
Angew.Chem. 91, 1023 (1979);
Angew. Chem. Int. Ed., 18, 956 (1979);
Tetrahedron 37, 1001 (1981).

38. J.Jacques, A. Collet, S.H. Wilen,
"Enantiomers, Racemates and Resolutions",
Wiley, New York 1981.

39. R. Charles, E. Gil-Av, (priv.Comm.)

40. G.Hübener R. Herrmann, K. Seeholzer, F. Siglmüller, Chem. Ber. (in preparation)

41. G. Gokel, I. Ugi, J.Chem. Educ. 49, 294 (1972).

42. R. Hermann, F. Siglmüller, I. Ugi (to be published soon).

43. D. Rehn, Diss. Techn. Univ. München 1975

44. I. Ugi, G. Kaufhold, Liebigs Ann. Chem. 709, 11- (1967),

45. S. Deming, paper given at the NATO-ASI Conference on chemomatrics at Cosenza, Italy, Sept. 1983. See also ref. 46,47.

46. G.E.P.Box, W.G.Hunter, J.S.Hunter, "Statistics for Experiments", Wiley, New York 1978.

47. R.Carlson, R. Phan-Tan-Lun,D. Mathieu, F.S.Ahouande, A. Babadjanian, J. Metzger, Acta Scand.Chem. B 32, 335 (1978); R. Carlson, T. Lundstedt, R. Phan-Tan-Lun, D. Mathieu, Nouveau J. de Chemie 7,315 (1983).

48. J.Geller, I. Ugi Chem. Scripta 22, 85 (1983)

49. N. Izumya, M. Muruoka, J. Amer.Chem. Soc. 91, 239 (1969).

50. K. Seeholzer, I. Ugi (unpabl. results).

51. H. Aigner, Diss. Techn.Univ.München 1983.

52. I. Ugi, C. Steinbrückner, Chem. Ber., 94, 2802 (1961).

53. H.P. Isenring, W.Hofheinz, Synthesis, 1981, 385, ESOC II, Stresa, June 1981, Abstract No.p. 31C.

54. D.H. Rich, R.D. Josensky, J.Med. Chem. 24, 567 (1981).

55. A.S. Radhakrishna, M.E. Parham, R.M. Riggs, G.M. Loudon, J.Org.-Chem. 44, 1746 (1979).

56. R. Obrecht, I. Ugi (unpubl.); see also: ref. 48 and R. Obrecht, S.T. Amadou, I. Ugi, Heterocycles (in press).

57. See also: I. Ugi und E. Wischhöfer, Chem. Ber. 95, 136 (1962)

58. H. Eckert, Diss. Techn. Univ.München 1976.

59. R.G.K. Schneiderwind, I. Ugi, Tetrahedron 39, 2207 (1983).

60. W. Gerrard, W.J. Green, R.J. Phillips, J. Chem. Soc. 1954, 1148, and proceeding communications, V.S. Abramov, V.K. Khairullin, J. Gen. Chem.. USSR 27, 501 (1957); H.A. Kellner, I. Ugi, Z.Naturforsch. 34B, 1159 (1979).

61. R.G.K. Schneiderwind-Stöcklein, Diss. Techn. Univ.München 1984, see also: R.G.K.Schneiderwind Stöcklein, I. Ugi, Z. Naturforsch (submitted).

62. G. Höfle, W. Steglich und H. Vorbrüggen, Angew.Chem. 90, 602 (1978); Angew.Chem. Int.Ed.Engl. 17 569 (1978).

63. H.A. Kellner, R.G.K. Schneiderwind, H. Eckert, I.K.Ugi Angew.Chem. 93, 581 (1981), Angew. Chem. Int. Ed. 20, 577 (1981).

64. R.L. Letsinger, E.P. Groody, T. Tanaka,J. Amer. Chem.Soc. 104, 6805 (1982)

65. G. Mering, I. Ugi (in preparation)

66. S.L.Beaucage, M.H.Caruthers, Tet. Lett. 22, 1859 (1981).

67. F. Ramirez, J.F. Marecek, I. Ugi, J. Amer.Chem.Soc. 97, 3809 (1975).

68. F. Ramirez, H. Okazaki, J.F. Marecek, Synthesis 1975, 637; J. Amer.Chem. Soc. 97, 7181 (1975).

69. P. Lemmen (unpublished).

70. R. Schwarz, I. Ugi Angew. Chem. 93, 836 (1981) Angew. Chem. Int. Ed. 20, 789 (1981).

71. V.A. Kukhtin, I.S. Gozman, Dokl. Akad. Nauk USSR 158, 157 (1964); Z. Obshch. Khim. 37, 881, 1644 (1967); II.Gozman ibid. 37, 2732 (1967); V.E. Beliski, N.N. Bezzubova, I.I. Gozman, ibid. 38, 1330 (1968); Izv. Akad. NaukUSSR 1968,2362. 2362.

72. J. Götz, I. Ugi, Heterocycles (in press).

73. R. Schwarz, Diss. Techn. Univ. München 1982.

74. J. Brandt, C. Jochum, I. Ugi, P. Jochum, Tetrahedron 33, 1353 (1977). ; see also: I. Ugi et. al, Proceedings of the Waters Symposium at Bad Soden, Feb. 1984, Waters, Königstein, W.-Germ.

75. G. Meggendorfer, R. Schwarz, I. Ugi, Tetrahedron Lett, 21 2493 (1980).

Hypothalamic Hormones, Aspects of Structure, Synthesis and Biological Activity

Wolfgang Voelter

Abteilung für Physikalische Biochemie am Physiologisch-chemischen Institut der Universität, Hoppe-Seyler-Str. 1, D-7400 Tübingen, FRG

1. Introduction

Since many decades there existed strong evidence from biochemical physiological, anatomical and clinical studies that the hypothalamus is a center of neurosecretion. However, the fundamental basis for the field of hypothalamic hormones was laid by the isolation of the thyrotropin-releasing hormone (TRH) by Schally[1] and Guillemin[2] and its structure elucidation by Folkers[3].

The hypothalamic hormones are biosynthesized in the hypothalamus of the brain and transferred in small blood vessels to the pituitary causing there the release of the pituitary hormones (Figure 1).

After the structure elucidation of TRH Schally[4] and Guillemin[5] succeeded to isolate another hypothalamic releasing hormone from porcine respectively bovine hypothalamic fragments: the luteinizing hormone-releasing hormone (LH-RH) or follicle-stimulating hormone-releasing hormone (FSH-RH).

In 1973 the isolations and structure elucidations of the growth hormone release-inhibiting hormone (somatostatin)[6,7] and in 1981 the structural characteristics of the corticotropin-releasing hormone[8] were reported.

Natural Product Chemistry
Ed. by Atta-ur-Rahman

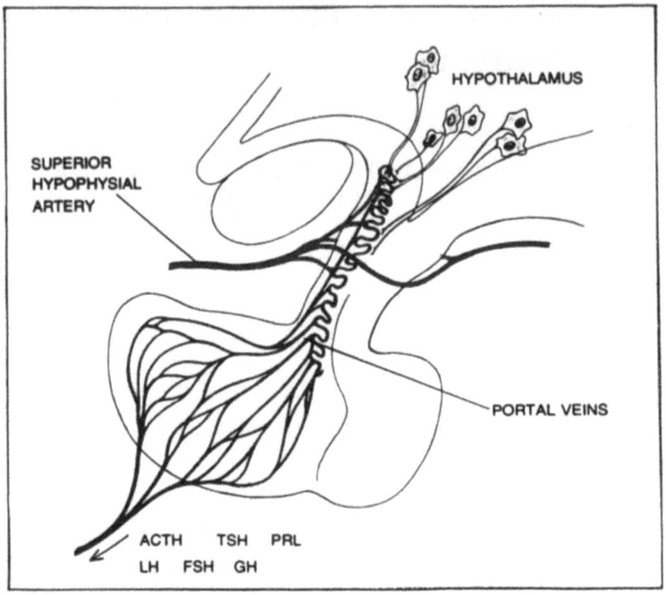

Figure 1. Schematic representation of the hypothalamus and pituitary region.

Table 1 lists the names , structures of the already known and specifications of the hypothalamus-releasing hormones. In Table 2 the names, their abbreviations, structures and regulatory effects of hypothalamus-inhibiting hormones are given.

The enormous amount of work necessary for the isolation of the hypothalamic hormones becomes obvious from Table 3 listing the scheme of purification of ovine TRH followed by Burgus et al.[9]: By the applications of ten purification procedures 1 mg of almost pure TRH could be isolated from 294 000 hypothalamic fragments.

2. Thyrotropin-releasing hormone (TRH)

The synthesis of TRH was achieved in numerous ways[10]. Also, an enormous number of TRH analogues was synthesiszed to elucidate the structure-activity relationship[11]. As an example the synthesis of Phe^{2}-TRH is given (Figure 2):

Table 1. Names (with abbreviations), structures and regulatory effects of hypothalamus-releasing hormones

Hormone	Structure (suggested)	Regulatory effect
Corticotropin-releasing hormone (CRH)	Ser-Gln-Glu-Pro-Pro-Ile-Ser-Leu-Asp-Leu-Thr-Phe-His-Leu-Leu-Arg-Glu-Val-Leu-Glu-Met-Thr-Lys-Ala-Asp-Gln-Leu-Ala-Gln-Gln-Ala-His-Ser-Asn-Arg-Lys-Leu-Leu-Asp-Ile-Ala-NH_2	ACTH* production
Follicle-stimulating hormone-releasing hormone (FSH-RH)	Pyr-His-Trp-Ser-Tyr-Gly-Leu-Arg-Pro-Gly-NH_2	FSH* production
Luteinizing hormone-releasing hormone (LH-RH)	Pyr-His-Trp-Ser-Tyr-Gly-Leu-Arg-Pro-Gly-NH_2	LH* production
Melanocyte-stimulating hormone-releasing hormone (MRH)	Cys-Tyr-Ile-Gln-Asn	MSH* production
Prolactin-releasing hormone (PRH)		PRL* production
Thyrotropin-releasing hormone, Thyroliberin (TRH)	Pyr-His-Pro-NH_2	TSH* production
Growth hormone-releasing hormone (TRH)	Val-His-Leu-Ser-Ala-Glu-Glu-Lys-Glu-Ala (not confirmed)	GH* production

* For abbreviations see table 2.

Table 2. Names (with abbreviations), structures and regulatory effects of hypothalamus-inhibiting hormones

Hormone	Structure (suggested)	Regulatory effect
Melanocyte-stimulating hormone release-inhibiting hormone (MRIH)	Pro-Leu-Gly-NH_2 Pro-His-Phe-Arg-Gly-NH_2 Cys-Tyr-Ile-Gln-Asn-Cys-OH (disulfide bridge Cys–Cys)	Supression of MSH* release
Prolactin release-inhibiting hormone (PIH)		Supression of PRL* release
Growth hormone release-inhibiting hormone (GIH)	Ala-Gly-Cys-Lys-Asn-Phe-Phe-Trp-Lys-Thr-Phe-Thr-Ser-Cys-OH (disulfide bridge Cys–Cys)	Supression of GH* release

* Abbreviations used in Tables 1 and 2.

ACTH: Adrenocorticotropic hormone; FSH: Follicle-stimulating hormone; GH: Growth hormone (somatotropin); LH: Luteinizing hormone; MSH: Melanocyte-stimulating hormone; PRL: Prolactin; TSH: Thyrotropin (thyroid-stimulating hormone).

Table 3. The scheme of purification of ovine TRH used by Burgus et al.[9]

	Stage	Number of hypothalamic fragments	Weight	TRF units/mg
1.	Lyophilized sheep hypothalami	294 000	25 kg	
2.	Alcohol-chloroform extract	294 000	294 g	1
3.	Ultrafiltration 'UM-3'	294 000	71 g	3
4.and 5.	Double gel filtration (Sephadex G-25, 0.5 M acetic acid)	286 000	16 g	16
6.and 7.	Double partition chromatography, 0.01 per cent acetic acid-n-butanol-pyridine (11:5:3)	280 000	246 mg	800
8.and 9.	Double adsorption chromatography, Norit/H_2O-ethanol-phenol	275 000	4.2 mg	30,500
10.	Partition chromatography, n-butanol-acetic acid-H_2O (4:1:5), (upper phase)	273 000	2.0 mg	58,500
11.	Repeat 10	270 000	1.0 mg	57,000

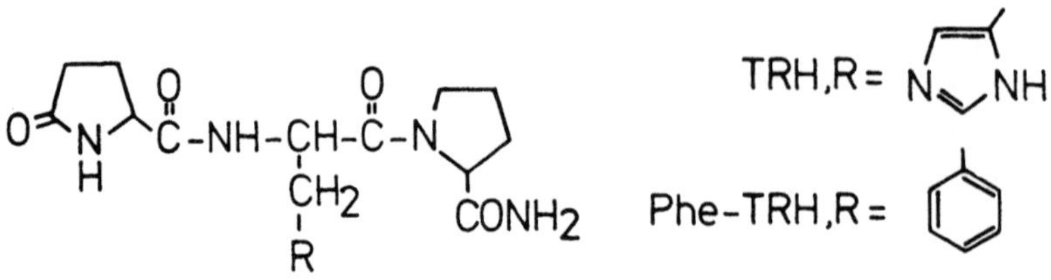

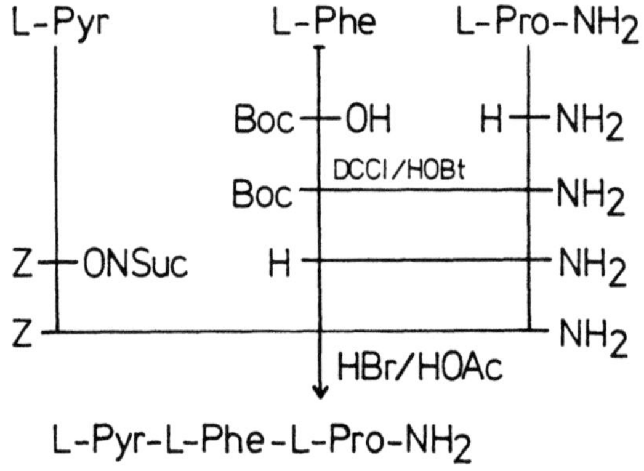

Figure 2. Synthesis of Phe^2 - TRH [12].

L-Polinamide is coupled by DCCI/HOBt to Boc-phenylalanine; the Boc group is cleaved by HCl/HOAc from the protected dipeptide amide and reacted with the benzyloxycarbonylpyroglutamyl succinimide ester.The benzyloxycarbonyl group is finally split off to yield the desired Phe^2-TRH[12].

From the synthesis of different TRH analogues and determination of their biological activity the following conclusions can be drawn:

1. Substitution of the ring of the pyroglutamyl residue by a pyrrolidine, tetrahydrofuran or cyclopentane moiety yields biologically inactive TRH analogues. Replacement of the pyroglutamyl by the open chain glutamine residue results also in loss of activity (Figure 3a).

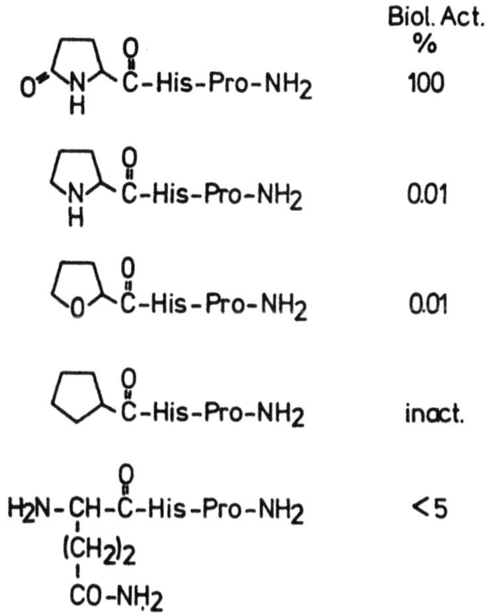

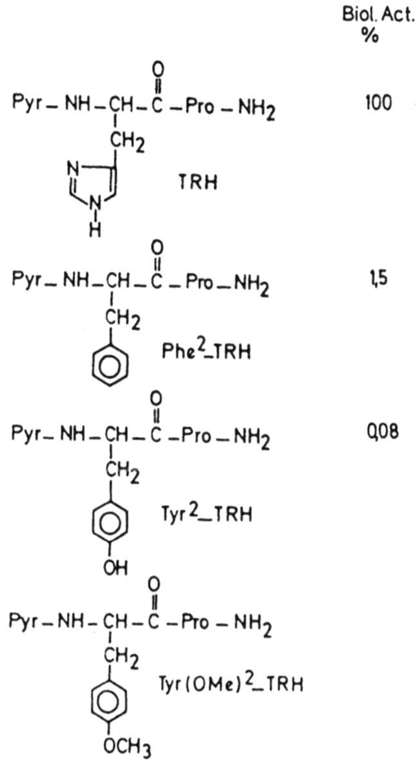

Figure 3a. Biological activity of X^1-TRH compounds with a modified ring of the pyroglutamyl residue (from upwards down: Pyr-(=TRH), pyrollidine carboxylic acid, tetrahydrofuran carboxylic acid, cyclopentane carboxylic acid, Gln-).

Figure 3b. Biological activity of X^2-TRH compounds with aromatic amino acids at position 2.

2. Replacement of the imidazole ring in TRH by aromatic residues like phenyl, hydroxyphenyl, or methoxyphenyl residues leads to TRH analogues with decreasing biological activity; (Figure 3b). Striking changes in the biological activities are observed if the imidazole ring of histidine is methylated at positions 3 and 1: (3-Me-L-His)^2-TRH is

	pKa Imid.	Biol. Act. %
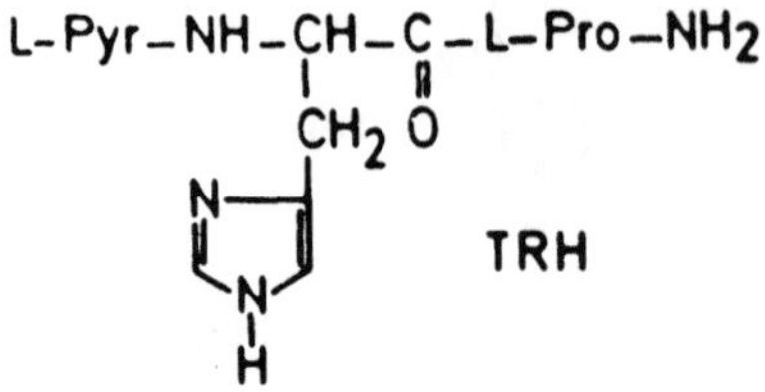 TRH	6,25	100
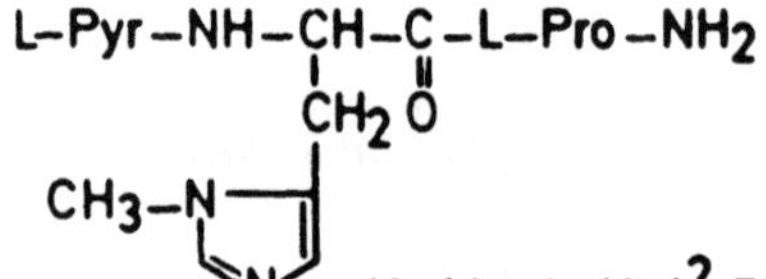 (3-Me-L-His)^2-TRH	5,95	800
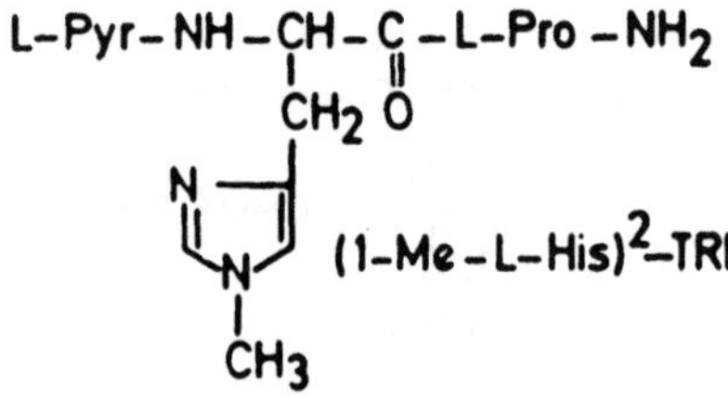 (1-Me-L-His)^2-TRH	6,60	0,04
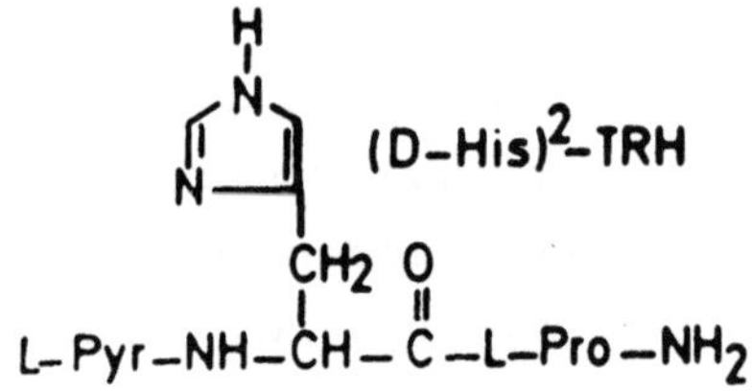 (D-His)^2-TRH		

Figure 3c. Biological activities of His^2-TRH compounds with a modified His residue.

eight times more potent than TRH while (1-Me-L-His)^2-TRH shows almost no biological activity. A change of the L-hystidyl by the D-histidyl residue also causes complete loss of biological activity (Figure 3c).

3. Replacement of the imidazole residue of histidine by an isopropyl, dimethyl thioether, hydroxyl group or by a hydrogen atom leads to TRH analogues of only low biological activity (Figure 3d).
4. Substitution of the histidine moiety by basic amino acids causes almost complete loss of the biological activity (Figure 3e).
5. Also the replacement of the prolinamide residue by those of amino acids with aliphatic or aromatic side chains leads to biologically inactive compounds (Figures 3f and 3g).
6. Changes of the proton of the amino group of the prolinamide residue shows a quite interesting relationships: Replacement of the amide hydrogen atom by a n-alkyl residue yields TRH analogues with increasing biological activity up to the n-pentyl derivative; TRH derivatives with an isopropyl, tert.-butyl or n-hexyl residue, however, have low biological activity only. While a change of the amide by the carboxyl group drastically reduces the biological activity. The methyl ester however, shows considerable bioactivity (Figure 3h).

It follows from Figure 3c that $(3\text{-Me-L-His})^2$-TRH is the so far found most biologically active TRH-analogue. The basis for an efficient synthesis of that particular TRH derivative is a convenient route to 3-methylhistidine. Conventional methylations of histidine result in low yields of a mixture of 3-methyl and 1-methylhistidine, which is difficult to separate[13]. Also total syntheses of D,L-3-methylhistidines are known[13,14,15], which so far could not be separated into its isomers. Therefore regioselective syntheses for 3-alkyl

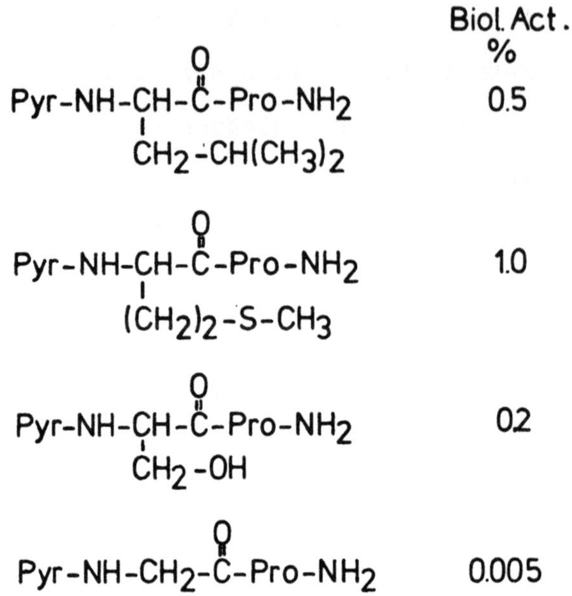

Figure 3d. Biological activities of (from upwards down) Leu^2-, Met^2-, Ser^2- and Gly^2-TRH.

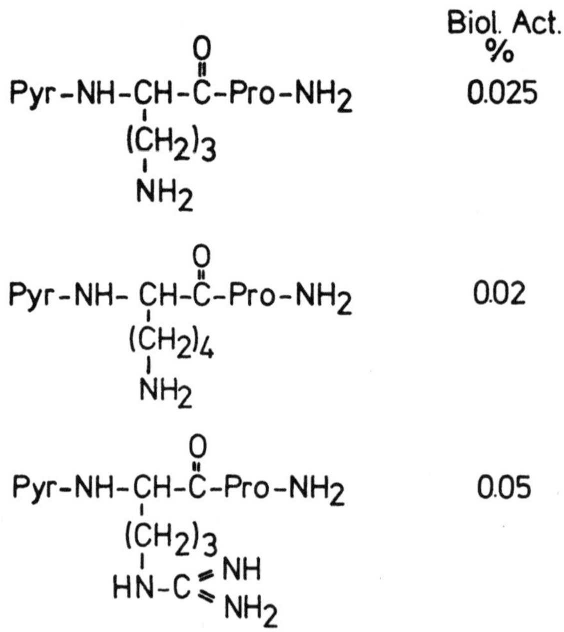

Figure 3e. Biological activity of X^2-TRH compounds (X^2= basic amino acids, from upwards down: Lys^2-TRH, Orn^2-TRH, Arg^2-TRH).

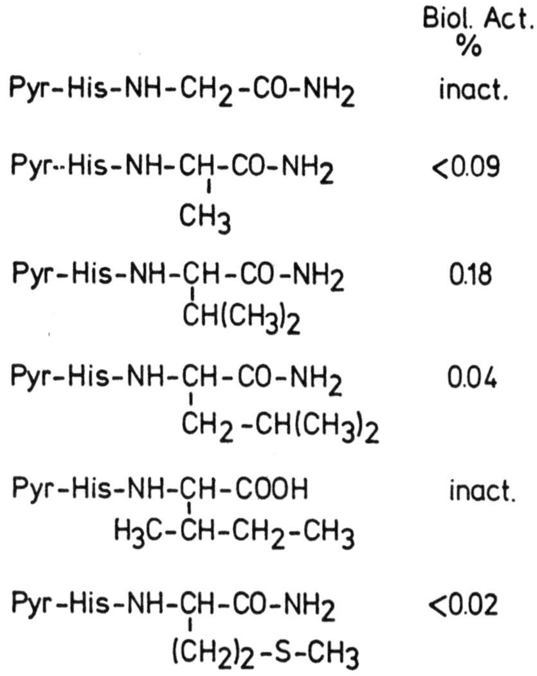

Figure 3f. Biological activities of X^3-TRH compounds (X^3= amino acid residues with aliphatic side chains; X^3= Gly, Ala, Val, Leu, Ile, Met).

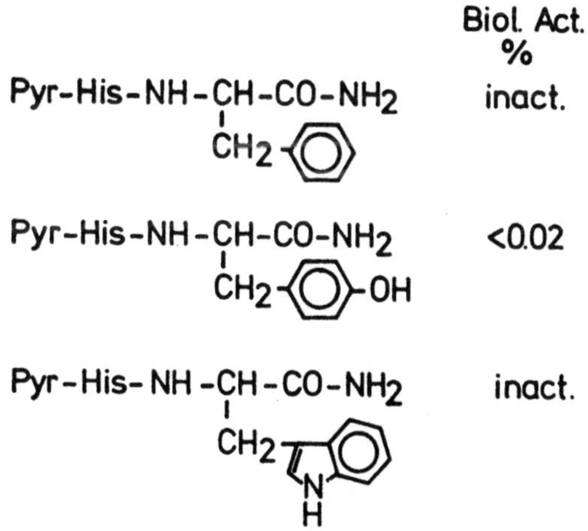

Figure 3g. Biological activities of X^3-TRH compounds (X^3= amino acid residues with aromatic side chains, from upwards down: X^3= Phe, Tyr, Trp).

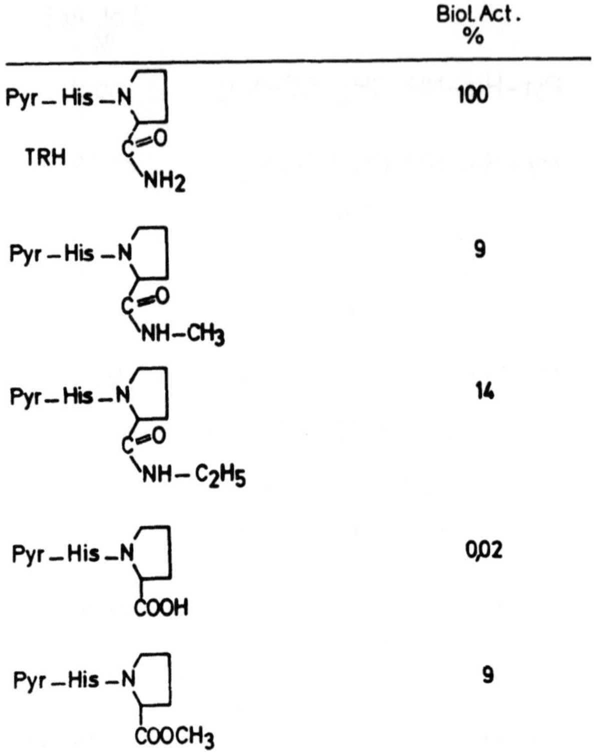

Figure 3h. Biological activity of X^3-TRH compounds with a modified amide function.

histidines were developed [16,17]. The synthesis of Beyerman et al.[16] starts from L-histidine methyl ester and N,N'-carbonyl diimidazole; our recently developed route [17] uses the much cheaper phosgene as one of the starting materials (Figure 4). Phosgene forms with histidine methylester a cyclic urea derivative ((7S)-2-alkyl-5,6,7,8-tetrahydro-7-(methoxycarbonyl)-5-oxoimidazo-[1,5-c]pyrimidine which can be alkylated by alkyl iodide. Hydrolysis of the iodide with 6N hydrochloric acid affords the 3-alkylhistidine.

Numerous spectroscopic investigations on TRH and its analogues have been undertaken to determine their conformations and con-

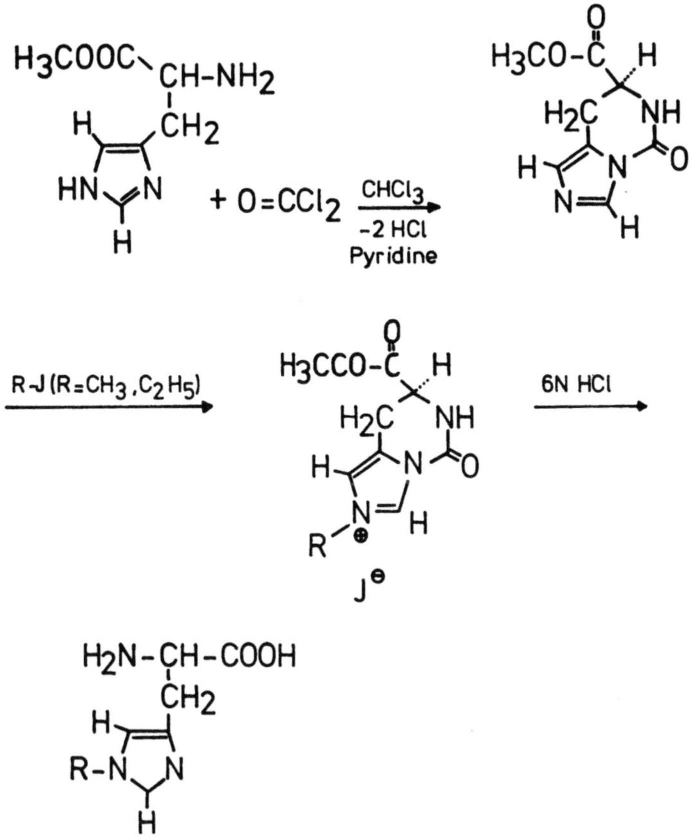

Figure 4. Convenient synthesis of alkylhistidines from histidine methylester via (7S)-5,6,7,8-tetrahydro-7-(methoxycarbonyl)-5-oxoimidazo [1,5-c] pyrimidine[17].

figurations. As we discovered first in 1971[18] ^{13}C-NMR spectroscopy is an excellent tool to study the cis and trans isomerism of proline peptides [19-26]. By this technique it can be unequivocally proven that TRH in water solution exists in a cis (≈ 15%) and trans (≈85%) isomer[27-30] (Figures 5 and 6). From proton NMR studies (Figure 7) conformational models were suggested [11]. The ^{1}H-NMR spectrum (Figure 7) was interpreted as follows[31]:

1. The trans and the cis protons of the $CONH_2$-group of proline resonate at 7.94 and 6.89 ppm respectively.
2. The imidazole C_2H and C_4H signals have shift values of 7.51 and 6.89 ppm respectively.
3. The imidazole C_4 proton resonance overlaps with that of the cis $CONH_2$ proton signal.
4. The C_4H signal of imidazole broadens upon the addition of a drop of water, while its CH_2 signal remains unchanged.
5. At 7.77 ppm appears the singlet of the pyroglutamyl residue.
6. The histidyl amide proton signal appears as a doublet at 8.07 ppm with a coupling constant of 7.5 to 8 Hz and this doublet coalesces upon irradiation at 4.64 ppm.

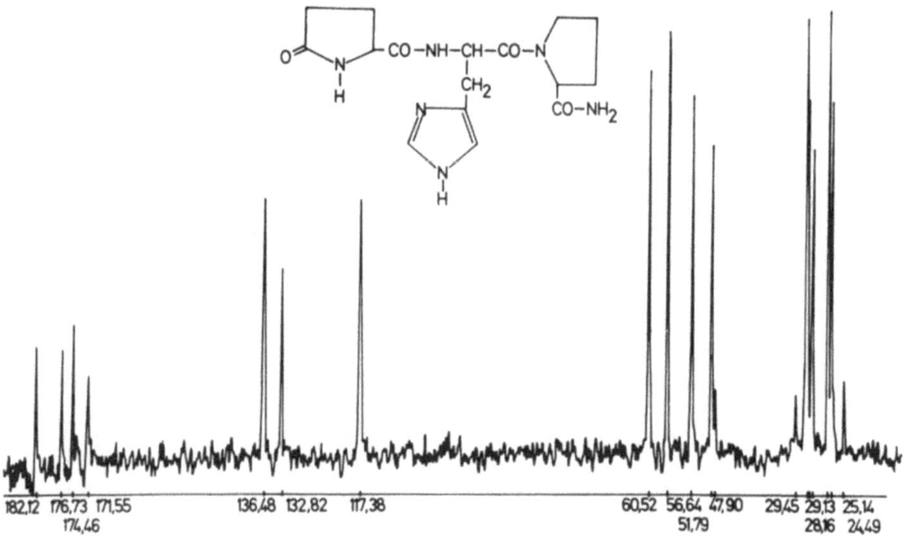

Figure 5. Pulse Fourier transform ^{13}C NMR spectrum (proton broad-band decoupled, referred to TMS=0) of TRH, 200 mg in 1.5 ml D_2O, accumulation of 16000 pulse interferograms [29].

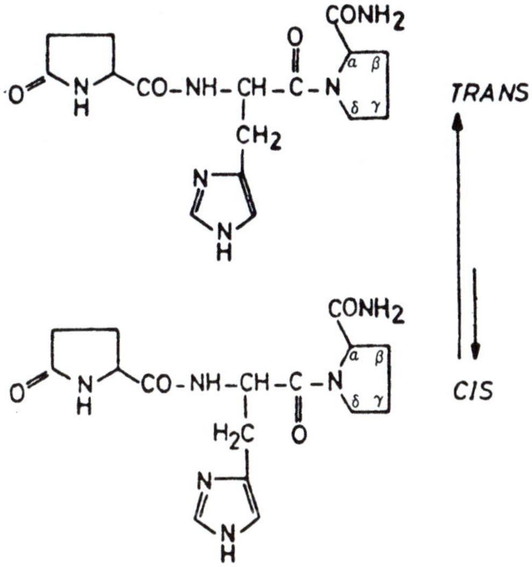

Figure 6. Cis-trans isomerism of TRH[10].

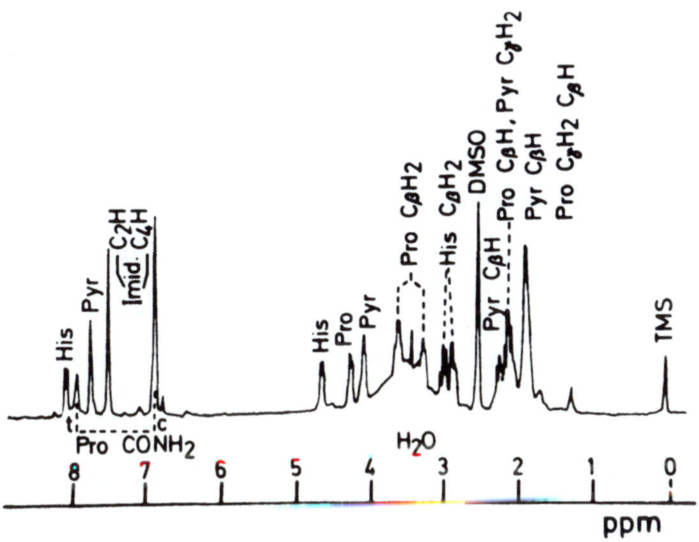

Figure 7. 150 MHz ^1H NMR spectrum of TRH, dissolved in DNSO-D_6[31].

7. Spectral comparison and decoupling experiments lead to the following signal assignments: Prolyl $C_\alpha H$ (4.22 ppm), prolyl $C_\beta H_2$ (2.10 and 1.82 ppm), prolyl $C_\gamma H_2$ (1.82 ppm), pyroglutamyl $C_\alpha H$(4.04 ppm) and pyroglutamyl $C_\beta H_2$(1.82 and 2.20 ppm).

From the C_αH-NH coupling constant JC_αH-NH(7.5-8 Hz) for histi-

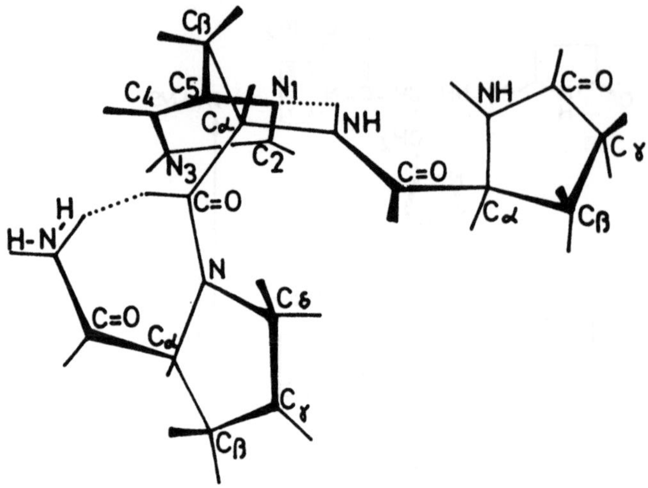

Figure 8. Suggested model of TRH, based on ^{1}H-NMR spectroscopy[31)].

dine, three possibilities for Ø can exist: Ø=-150°, Ø=-90° and Ø=+60°. It follows from Dreiding models that maximum stability exists for Ø=-150°, and therefore interaction between $N^{1\ im}$ and the histidyl amide aproton is possible." The anomalous pK value (6.25) of the imidazole ring, the low biological activity of L-Pyr-1-Me-L-His-L-Pro-NH_2 and the high TSH-releasing activity of L-Pyr-3-Me-L-His-L-Pro-NH_2 give strong evidence for the model of Figure 8."

"The strong downfield shift on the C-terminal trans amide proton gives strong evidence for a hydrogen bond between this proton and the histidyl carbonyl group."

Though ^{1}H NMR spectroscopy is a powerful tool for the determination of conformations of peptides there is no doubt that X-ray structural analysis gives a much more exact picture of the conformation of a peptide at least in its crystalline state. Owing to the difficulty in getting suitable crystals, however, for a relative long time no X-ray structural analysis of TRH could be performed. Our group was the first one reporting a X-ray analysis of a TRH analogue, namely that of Phe^2-TRH (Figure 9). In 1980 an X-ray crystal study of TRH tartrate was reported (Figure 10)[32]. From Figures 9 and 10 it becomes obvious that TRH and Phe^2-TRH have similar conformations in the crystalline state and that the overall conformation of TRH indicated by these studies is quite different from the model based on ^{1}H-NMR measurements.

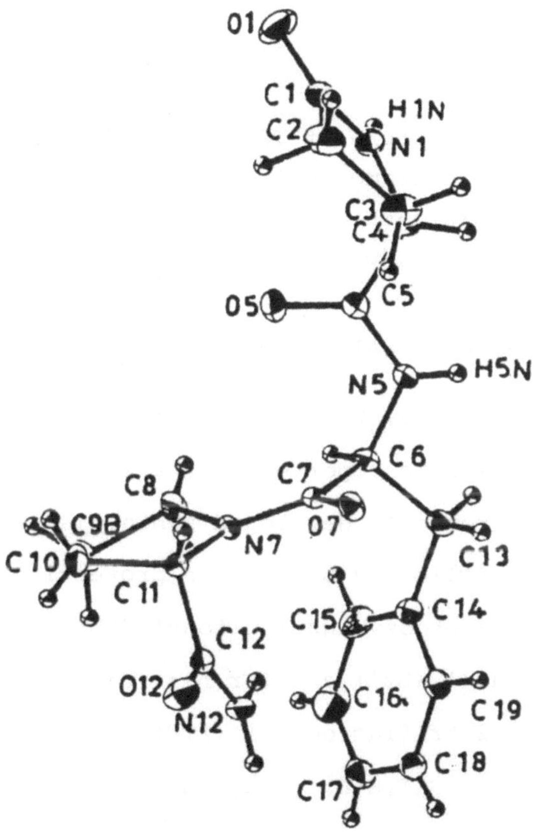

Figure 9. X-ray structure of Phe^2-TRH[12)].

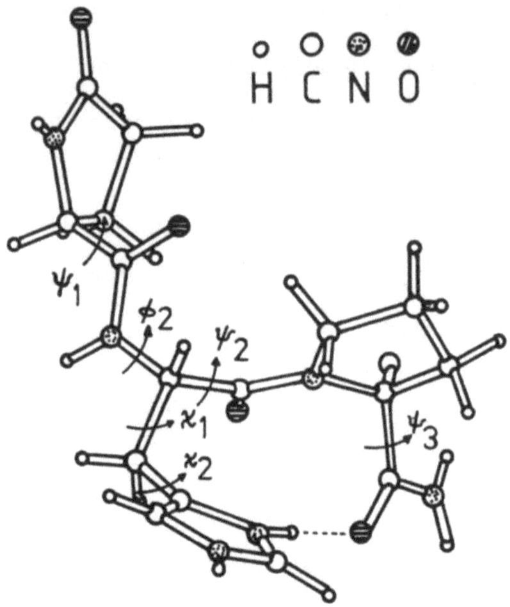

Figure 10. X-ray structure of TRH showing the torsional angles and the intramolecular hydrogen bond between the imidazole 1-NH of histidine and the carbamoyl oxygen of prolinamide [32)]

3. Luteinizing hormone-releasing hormone (LH/FSH-RH)

Also a large number of synthetic approaches to the LH/FSH-RH and its derivatives were reported in the literature[10,11)]. As an example our recently developed synthesis is shown in Figure 11[33)]. The following side chain protections are used: for arginine the p-methoxybenzolsulfonyl (Mbs), for tyrosine and serine the benzyl (Bzl) and for histidine the 1-benzyloxy-carbonylamino-2,2,2-trifluoroethyl (ZTE) group. Except for pyroglutamic acid DCCI/HOBT couplings are performed. The side chain groups are cleaved by HF/EDT/anisole.

Structural changes in the LH/FSH-RH molecule yield molecules with agonistic or antagonistic activities. A survey is given in Figure 12.

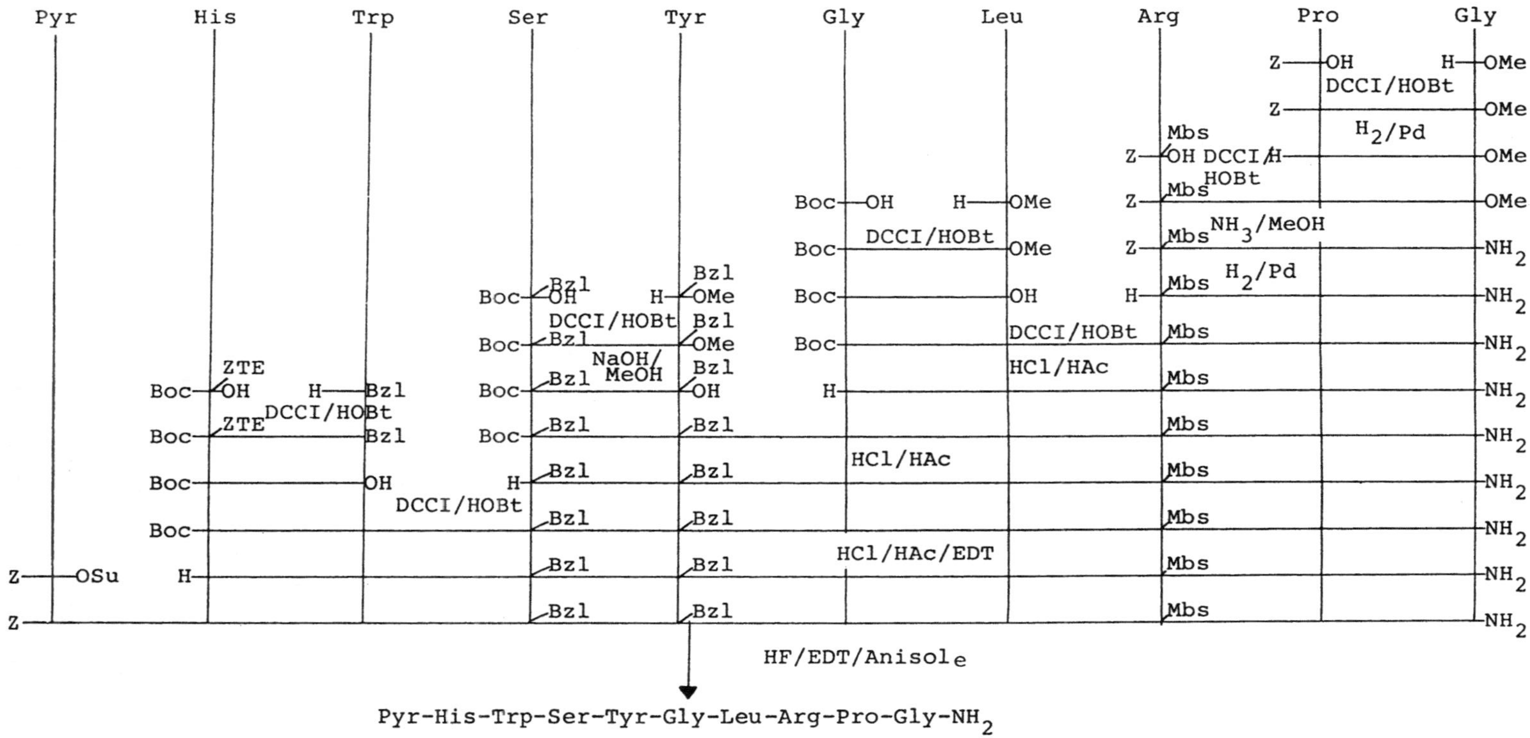

Figure 11. Synthesis of LH/FSH-RH[33].

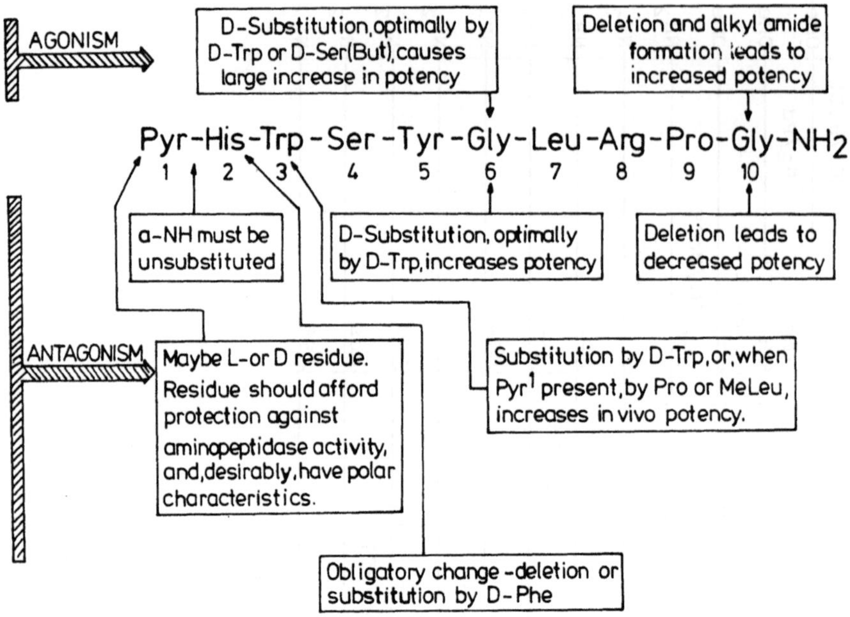

Figure 12. A survey of the structural changes of the LH/FSH-RH molecule leading to agonistic or antagonistic activities.

So far no LH/FSH-RH analogues with ß-amino acids have been reported in the literature. For the incorporation of homo - pyroglutamic acid instead of pyroglutamic acid and homoproline instead of proline these two ß-amino acids had to be synthesized.

For the synthesis of homo-pyroglutamic acid succinic anhydride is reacted with S-(-)-α-phenylethylamine. The resulting ω-carbinollactams are reacted with diethylphosphonoacetic acid methylester and sodium hyride to two diastereomeric esters and after hydrolysis the pure S-isomer of homopyroglutamic acid is received (Figure 13)[33].

Z-Proline is the starting material for the synthesis of Z-homoproline. With oxalyldichloride Z-proline yields the corresponding acid chloride which forms with diazomethane the

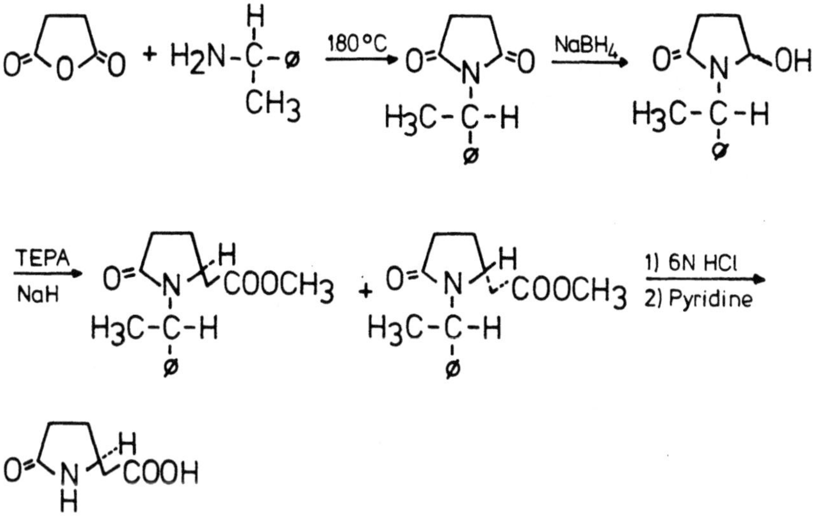

Figure 13. Synthesis of homo-pyroglutamic acid [33)].

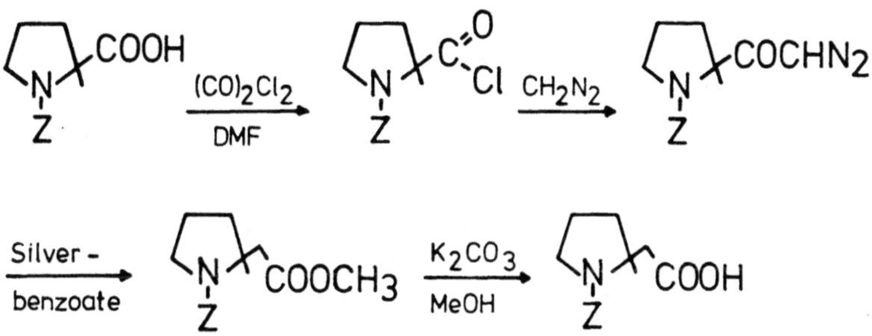

Figure 14. Synthesis of Z-homo-proline[33)].

diazoketone. Wolff rearrangement yields the desired Z-homo-proline methylester (Figure 14)[33].
Neither (h-Pyr^1)-LH/FSH-RH nor (h-Pro^9)-LH/FSH-RH show, however, much biological activity [33].

References

1) A.V.Schally, C.Y.Bowers, T.W.Redding and J.F.Barrett, Biochem. Biophys.Res.Commun. 25, 165 (1966).

2) R.Burgus and R.Guillemin, in J.Meites(ed.): Hypophysiotropic Hormones of the Hypothalamus: Assay and Chemistry, p.227, Williams & Wilkins, Baltimore 1970.

3) K.Folkers, Intra-Sci.Chem.Rep. 5, 263 (1971).

4) A.V.Schally, A.Arimura, Y.Baba, R.M.G.Nair, H.Matsuo, T.W.Redding, L.Debeljuk and W.F.White, Biochem.Biophys.Res. Commun.43, 393(1971).

5) M.Amoss, R.Burgus, R.Blackwell, W.Vale, R.Fellows and R.Guillemin, Biochem.Biophys.Res.Commun.44, 205(1971).

6) P.Brazeau, W.Vale, R.Burgus, N.Ling, M.Butcher, J.Rivier and R.Guillemin, Science 179, 77(1973).

7) R.Burgus, N.Ling, M.Butcher and R.Guillemin, Proc.Nat.Acad. Sci. 70, 684(1973).

8) W.Vale, J.Spiess, C.Rivier and J.Rivier, Science 213, 1394 (1981).

9) R.Burgus, T.F.Dunn, D.Desiderio, D.N.Ward, W.Vale and R.Guillemin, Nature 226, 321(1970).

10) W.Voelter, Fortschritte der Chemie organischer Naturstoffe 34, 439(1977).

11) D.Gupta and W.Voelter, Hypothalamic Hormones- Structure, Synthesis and Biological Activity, Verlag Chemie, Weinheim-New York, 1975.

12) J.J.Stezowski, C.Bürvenich and W.Voelter, Angew.Chem.91, 243 (1979).

13) H.H.Tallan, W.H.Stein and S.Moore, J.Biol.Chem 206, 825 (1954).

14) H.Rinderknecht, T.Rebane and V.Ma,J.Org.Chem. 293, 1968 (1964).

15) P.K.Martin, H.R.Matthews, H.Rapoport and G.Thyagarajan, J.Org. Chem. 33, 3758 (1968).

16) A.Noordam, L.Maat and H.C.Beyerman, Rec.,J.Roy.Netherlands Chem. Soc. 97, 293 (1978)

17) W. Voelter and H.Wollmann, in preparation.

18) W.Voelter, G.Jung, E.Breitmaier and E.Bayer, Z.Naturforsch.26b, 213 (1971).

19) W.A.Thomas and M.K.Williams, Chem.Commun.1972, 994.

20) R.Deslauriers, R.Walter and I.C.P.Smith, Biochem.Biophys.Res. Commun. 48, 854 (1972).

21) K.Wüthrich, A.Tun-Kyi and R.Schwyzer, FEBS Letters 25, 104 (1972).

22) W.Voelter and O.Oster, Chemiker-Ztg. 96, 586 (1972)

23) W.Voelter and O.Oster, Org.Magn. Resonance 5, 547 (1973).

24) W.Voelter, O.Oster and E.Breitmaier, Z. Naturforsch. 28b, 370 (1973).

25) W.Voelter, S.Fuchs, R.H.Seuffer and K.Zech, Monatsh.Chem.105 1110 (1974).

26) O.Oster, E.Breitmaier and W.Voelter, in T.Axenrod and G.A.Webb (eds.): Nuclear Magnetic Resonance Spectroscopy of Nuclei Other Than Protons, p.233, John Wiley & Sons, New York, London, Sidney, Toronto 1974.

27) R.Deslauriers, R.Walter and I.C.P.Smith, Biochem.Biophys.Res. Commun. 53, 244 (1973).

28) R.Deslauriers, C.Carrigou-Lagrange, A.-M.Bellocq and I.C.P.Smith, FEBS Letters 31, 59 (1973).

29) W.Voelter, O.Oster and K.Zech, Angew.Chem. 86, 46 (1974); Angew.Chem.Internat.Ed. 13, 131 (1974).

30) R.Deslauriers, W.H.McGregor, D.Sarantakis and I.C.P.Smith, Biochemistry 13, 3443 (1974).

31) S.Fermandjian, P.Pradelles and P.Fromageot, FEBS Letters 28, 156 (1972).

32) K.Kamiya, M.Takamoto, Y.Wada, M.Fujino and M.Nishikawa, J.Chem.Soc. 1980, 438.

33) W.Voelter and H.Echner, in preparation.

Structural Studies on Unusual Peptides

Dudley H. Williams
University Chemical Laboratory, Lensfield Road, Cambridge, CB2 1EW, Great Britain

Introduction

The structure elucidation of organic molecules can be regarded as falling into three areas. One area is that of X-ray crystallography; this approach requires single crystals, and therefore quantities of the substance of the order of mg or greater. It is also required that the crystals diffract X-rays satisfactorily, that they do not rapidly deteriorate in the X-ray beam, and further that the structure can be solved from the acquired data. In the absence of a heavy atom derivative, this last problem can be formidable if the asymmetric unit contains more than around 100 atoms.

A second area is that of solution methods to determine the structures of the standard biopolymers, particularly DNA, RNA, and proteins (or peptides). Astonishingly sensitive and rapid methods have been developed to determine the sequence of bases (in DNA or RNA) or of amino acids (in peptides and proteins). These methods determine only the primary structure (sequence), although it is true that in a limited number of cases, secondary and tertiary structure (i.e. the whole 3-dimensional picture) has been determined by the powerful method of X-ray crystallography.

Natural Product Chemistry
Ed. by Atta-ur-Rahman

The third area has to attempt to encompass all the remaining problems. It includes all compounds that are not standard biopolymers and, additionally, suitable crystals cannot be obtained. Thus, we must include not only those compounds which, no matter how much is available, refuse to crystallise, but also those cases where crystallisation is effectively impossible because of the small amount of substance available. This area will include an enormous number of physiologically important compounds such as hormones, vitamins, neuropeptides, toxins, and antibiotics (and their metabolites). Whereas, two decades ago, such problems had to be solved by tedious and demanding chemical degradations, recently a number of relatively rapid and powerful methods for their solution have evolved. Most notably, these methods include mass spectrometry and nuclear magnetic resonance. It is with this area of structure elucidation that the present article is concerned. The techniques will be illustrated by the solution of two structures.

The Structure Elucidation of Unusual Linear Peptides

In the case of a structure which is not only unknown but also novel, it is rare that mass spectrometry alone will provide extensive structural information. This is because the fragmentation patterns of complex molecules are enormously varied, and are more often rationalised with the value of hindsight rather than being used to determine structure. A notable exception is the sequence determination of linear peptides.[1] It is established that such molecules, when converted to their N-acetylpermethyl-derivatives (1), fragment at each amide bond in a relatively reproducible manner. This cleavage, occurring with charge retention by the N-terminal portion is shown below; it has arbitrarily been selected between residues 3 and 4, but occurs also at the other peptide bonds.

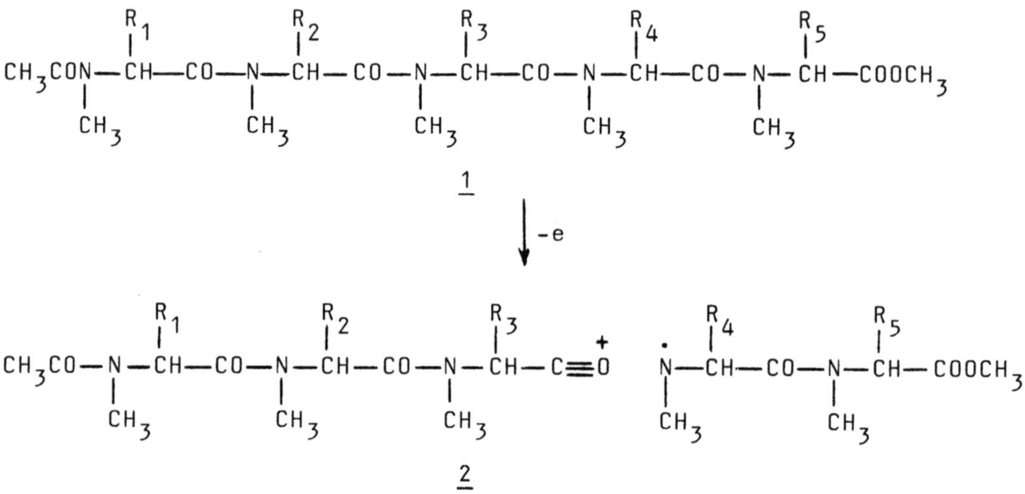

This fragmentation, shown above for the case of electron impact (EI) mass spectrometry, also occurs under chemical ionisation (CI) conditions. Given the masses of the twenty common amino acids, the ions corresponding to fragments containing the amino acids associated with R_1, R_1R_2, $R_1R_2R_3$, etc. are sufficient to sequence the peptide.[1] The method is limited (in EIMS) to peptides containing up to *ca* 10 amino acids, and requires ~25 nmol of peptide. In our own laboratory, and others, it has been used to sequence, or partially sequence, proteins. The proteins are broken into suitable fragments by enzymic digestion, and the fragments sequenced as above. Long sequences are then generated by repeating the process with enzymes of differing specificity, and overlapping (by means of portions of common sequence) all the fragments so obtained.[2] However, the same results can be achieved with comparable or greater efficiency by solution methods. The unique advantage of mass spectrometry in this area lies in studying unusual linear peptides. The unusual features include the identification of naturally occurring N-terminal blocking groups (which preclude the use of the Edman degradation). For example, an N-terminally blocked peptide was isolated <u>via</u> enzymic digestion of an alcohol dehydrogenase from <u>Drosophila melanogaster</u> N-11. It was acetylated with $(CD_3CO)_2O$ prior to permethylation; the mass spectrum of the product established the structural unit CH_3CO-Ser-Phe-.[3] Thus, the natural blocking group in the enzyme is determined to be acetyl; being already present in the enzyme, it precludes further reaction at the N-terminus with the deuterated analogue of acetic anhydride.

A more complex example is found in the recent determination of the structure of an unusual peptide isolated from larvae of the sawfly.[4] The poisoning of cattle by ingestion of larvae of an Australian species of sawfly (<u>Lophyrotoma interrupta</u>) is a serious problem in several of the grazing areas of Queensland.[5,6] Generally, the disorder occurs in late winter or spring when there is a heavy infestation of the silver-leaf ironbark tree (<u>Eucalyptus melanophloia</u>), the principal host of the sawfly larvae. Large accumulations of dead larvae may collect on the ground. Animals eat such material, and develop a considerable liking for it. It has been suggested that it may act as a protein source. However, the end result is poisoning, and critically affected animals usually die within two days.

The LD_{100} of the essentially pure toxin is 2 mg/kg. Amino acid analysis shows the presence of the following amino acids (molar ratios given in parenthesis): Asp(1.89), Glu(1.93), Ala(1.00), Val(1.16), Ile(0.82), Phe(0.92) (taking Ala = 1.00). Initial work, relying on the principle of electron-impact sequencing outlined above established the presence of benzoyl as an N-terminal blocking group, and led to a partial sequence PhCO-Ala-Phe-Val-Ile... . However, a typical problem of EI sequence determination, that of the higher mass ions being of low or negligible abundance (a particularly severe problem in this case), precluded extension of the sequence. Fortunately, at this period of the study, the technique of fast atom bombardment mass spectrometry (FAB MS) became available.[7] In this technique, a few µg (say 1-100 µg, depending on the amount of sample available, and the suitability of the compound for the technique) of sample are dissolved in a viscous, low-volatility, polar solvent such as glycerol or thioglycerol. The solution, on a mass spectrometer probe tip, is bombarded with xenon atoms of 4-10 keV translation energy. Although the technique is not a panacea for non-polar molecules, polar molecules in the molecular weight range (say) 300-5000 Daltons are projected (as ions) into the gas phase with astonishing efficiency. These ions, normally MH^+ in positive ion spectra, or $M\text{-}H^-$ in negative ion spectra (M being the molecule of interest), are then detected in the mass spectrometer in the usual way. In FAB spectra, MH^+ (or $M-H^-$) are normally of high abundance, with fragment ions being of lower abundance. The observed fragmentation processes correspond to low energy cleavages and, as for CI, give important sequence information in the case of linear peptides. For example, underivatised peptides give molecular weight information that is readily obtained up to 3000 Daltons; and, with the appropriate apparatus and technique, to 5000 Daltons. In favourable cases, underivatised peptides of up to ca 15 residues have been shown to give sequence information. Some of the most useful fragmentations are given in 3.

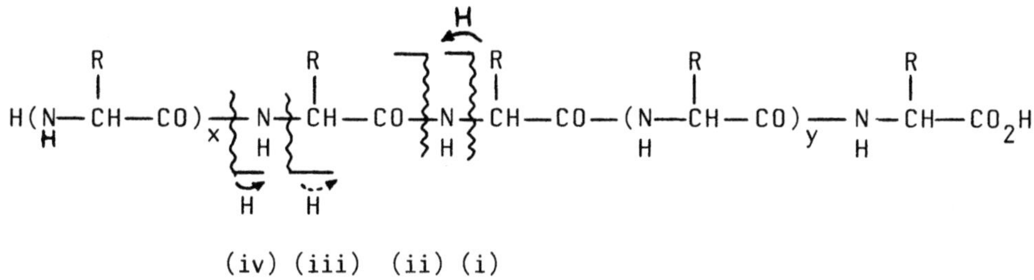

3

where the horizontal line leaving the "wiggley" line does so to the left, these fragmentations occur with charge retention by the N-terminal fragment; where this horizontal line is to the right of the "wiggley" line, charge retention is by the C-terminal fragment. The symbol (H with arrow) (or H with arrow) indicates that the cleavage is associated with hydrogen migration to the charged fragment. The symbol (H with dotted arrow) indicates that the cleavage is sometimes observed with associated hydrogen migration, and sometimes without. The mass differences associated with these cleavages have been tabulated.[8]

These points are illustrated by the positive and negative ion FAB spectra of the toxin produced in Figs. 1 and 2, respectively. Cleavage (ii) (see _3_) gives ions corresponding to the acylium ions _4_, occurring at _m/z_ 105, 176, 323, 422, 535, 650, 765, and 894 (and MH^+ at 1040, Fig. 1) and indicate the sequence PhCOAlaPheValIleAspAspGluGln; or the C-terminal Gln could equally be replaced by the isomeric amino acid carrying the amide on the backbone carbonyl and the carboxyl function in the sidechain (iso-Gln; see _5_ and _6_).

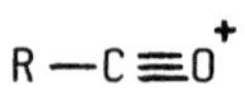

_4_

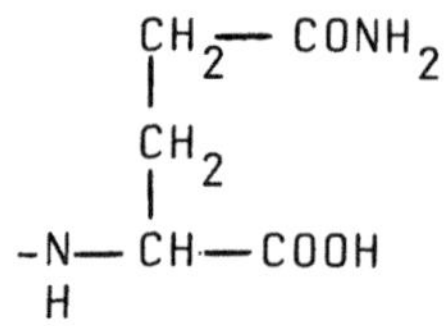

_5_, C-terminal Gln

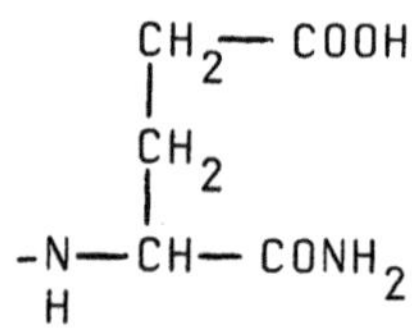

_6_, C-terminal _iso_-Gln

Fragments formed _via_ cleavage (i) (see _3_), and corresponding to fragment ions of the general formula _7_, at _m/z_ 667, 782 and 911 (in conjunction with MH^+, Fig. 2) confirm the nature of the three amino acids constituting the C-terminal portion.

$(RCONH_2)H^+$	$(H_2NR)H^+$	$RCONH^-$
_7_	_8_	_9_

Figure 1. The positive ion FAB mass spectrum (_m/z_ 320-1050) of the toxic peptide from sawfly larvae. The peaks marked G represent protonated oligomers of glycerol (the solvent used in the determination)

Figure 2. The negative ion FAB mass spectrum (_m/z_ 400-1060) of the toxic peptide from sawfly larvae

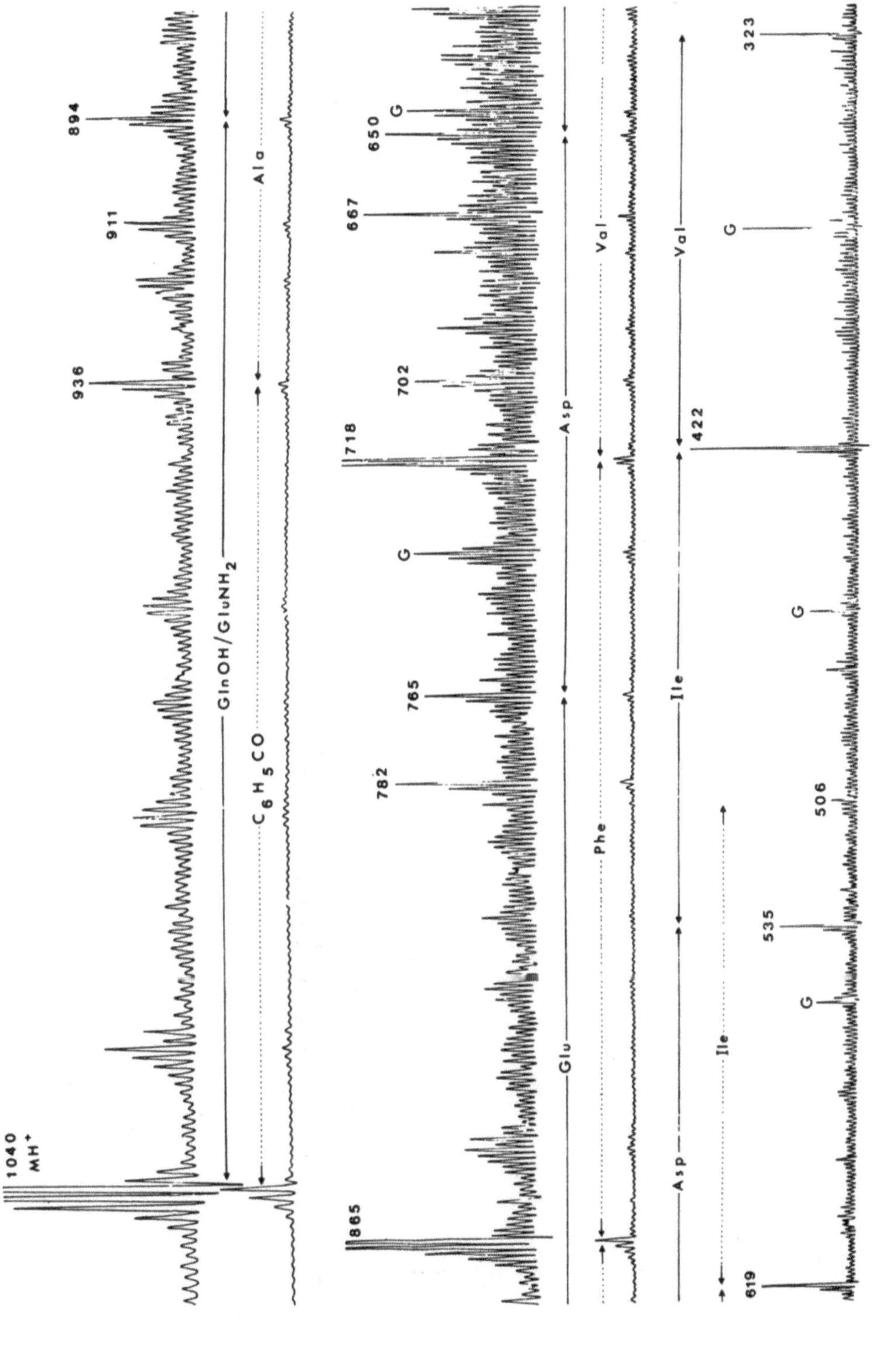
1040
MH+
936
911
894
GlnOH/GluNH2
C6H5CO
Ala
865
782
765
G
718
702
667
650
G
Glu
Asp
Phe
Val
Asp
Ile
Val
Ile
619
G
535
506
G
422
G
323

Figure 1

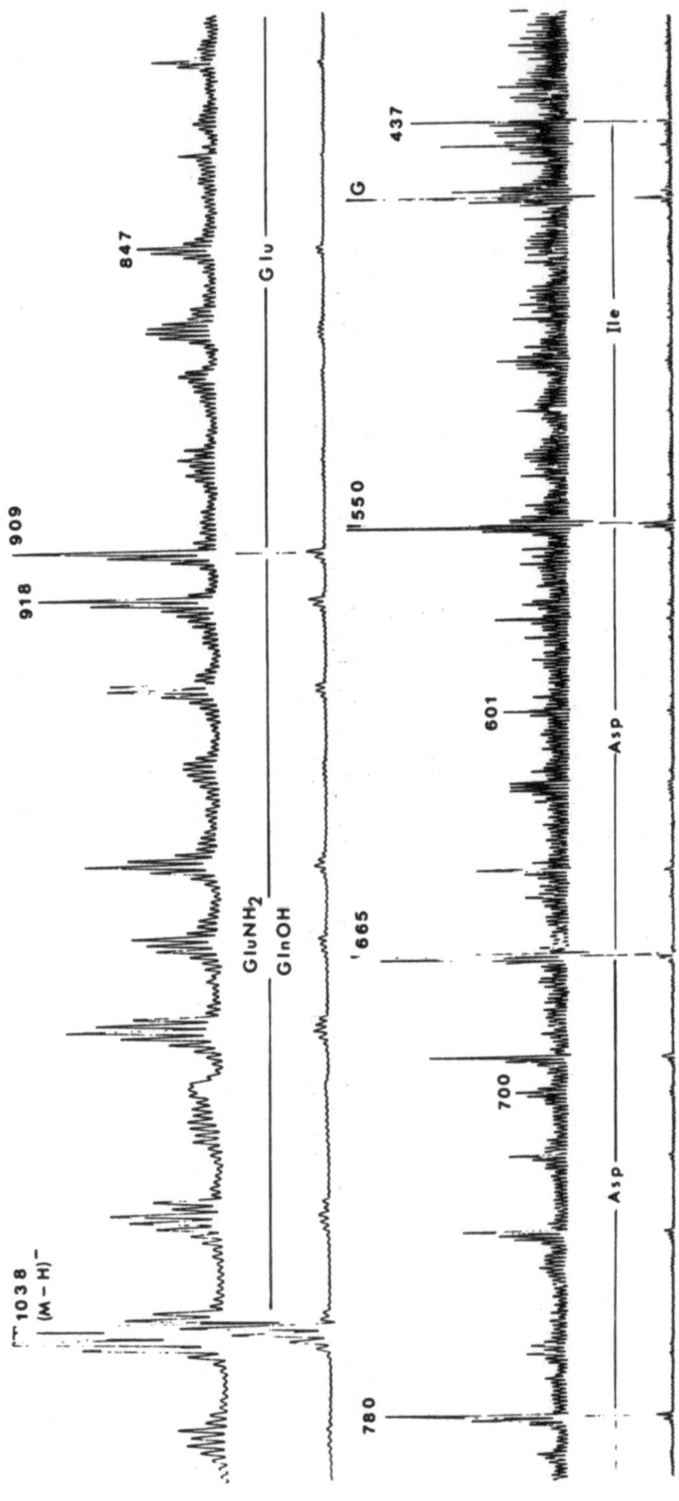
1038
(M - H)⁻
918
909
847
GluNH2
GlnOH
Glu
780
700
665
601
550
G
437
Asp
Asp
Ile

Figure 2

Those sequence ions which contain the C-terminal portion of the peptide are formed by cleavage (iv) (see 3). Cleavage (iv) gives rise to protonated amine fragments (8), which occur at m/z 865, 718, 619 and 506 in Fig. 1; they indicate the sequence XPheValIleY, where the masses of X and Y are in accord with the sequence proposed above. The negative ion FAB spectrum (Fig. 2) shows an extremely abundant $(M-H)^-$ ion at m/z 1038. A series of sequence ions corresponding to cleavage (i) (see 3), and most generally expressed in terms of the anion 9, occur at m/z 338, 437, 550, 665, 780 and 909 These,[5] together with the molecular weight information, indicate the sequence XValIleAspAspGluGln (or, as before, the C-terminal residue may be iso-Glu). These data attest to the power of FAB MS not only in determining the molecular weights of relatively large polar molecules, but also in sequence determination of peptides (which is especially valuable when they are N-terminally blocked).

The above established sequence might be regarded as giving a "structure" if the peptide were a product of ribosomal synthesis of a peptide. In such cases, it can be reasonably assumed that Asp, Glu and Gln are incorporated as such, and not as their iso-structures; and further, that all the amino acids are L. However, in our early experiments on the toxin (prior to the availability of FAB MS), an enzymic digestion with chymotrypsin had been attempted, in the expectation that cleavage at the C-terminal side of the Phe residue would occur, thus giving a C-terminal fragment which might conceivably have been sequenced by EI MS. The attempted enzymic digestion failed. One possible explanation for this failure is that one or more of the amino acids in the toxin might have D-stereochemistry.

Mass spectrometry cannot throw light on this question, but it can be probed using gas chromatography on a chiral column. The toxin was hydrolysed (6 M-HCl at 110°C for 72 h), and the resulting amino acids converted to their N-trifluoroacetyl isopropyl ester derivatives; these volatile derivatives were chromatographed on a chiral GC column ('Chirasil-Val'). Comparison of the retention times of the components with those of authentic samples established that Val, Ile, one Asp, and a Glu (or Gln) have the L-configuration; and that Ala, Phe, one Asp, and a Glu (or Gln) have the D-configuration. The ambiguity between Glu and Gln arises because Gln is converted to Glu by the total hydrolysis procedure.

The above results increased the fascination and importance of the problem. Prior to this study, no peptide isolated from an animal had contained more than a single D-amino acid. In "higher" organisms (the eukaryotes, including the animals), the synthesis of peptides and proteins occurs on ribosomes; in this process, only the 20 standard amino acids, in the L-configurations, are known to participate. However, in lower organisms (the procaryotes, including bacteria), peptides synthesis can, on occasions take place without ribosomes; this type of synthesis is effected on multi-functional enzymes, and can include the utilisation of unusual amino acids (e.g. phenylglycine), and amino acids in their D-configuration. Subsequently, we will return to this problem of the procaryotic or eucaryotic origin of the peptide toxin although, unfortunately, we will not answer it!

The above data give a further twist to the problem. The occurrence, in the hydrolytic products, of one L-Asp, one D-Asp, one L-Glu, and one D-Glu gives rise to no less than 4 combinations of stereochemistry at the C-terminus; viz.,

...D-Asp-L-Asp-D-Glu-L-Gln, ...L-Asp-D-Asp-D-Glu-L-Gln,
...D-Asp-L-Asp-L-Glu-D-Gln, ...L-Asp-D-Asp-L-Glu-D-Gln.

Additionally, since the peptide is "non-standard", we have to consider that Asp and Glu may be incorporated not only as they normally are in proteins (<u>10</u>; see also <u>5</u>), but as their iso-structures, in which the usual sidechain becomes part of the peptide backbone (<u>11</u>; see also <u>6</u>). Thus, determination of the amino acid sequence still leaves a total of 64 possible structures. The determination of a unique structure will now be considered.

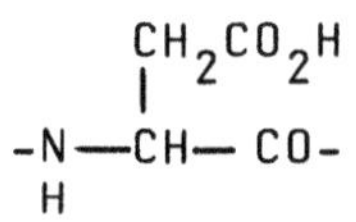

-<u>Asp</u>-

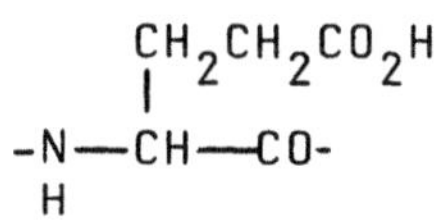

-<u>Glu</u>-

<u>10</u>

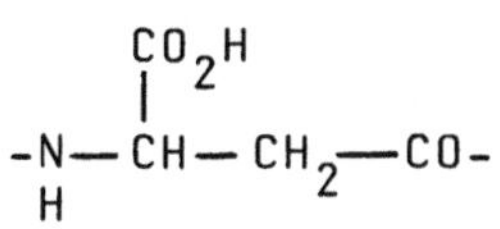

-<u>iso</u>-<u>Asp</u>-

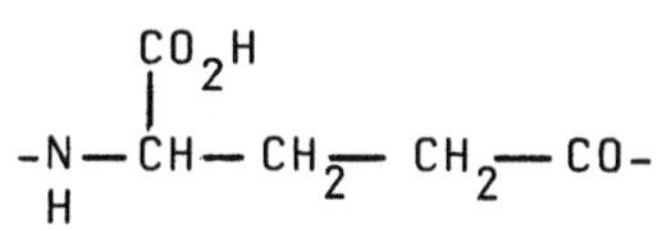

-<u>iso</u>-<u>Glu</u>-

<u>11</u>

The stereochemistries of the Asp residues were determined first. A possible strategy is to isolate a peptide fragment containing only one of the Asp residues, and then to determine the stereochemistry of that residue. Fortunately, it is known that acid hydrolysis of peptides occurs selectively at Asp residues, and therefore the toxin was subject to relatively mild acid hydrolysis (10% aqueous HCl; 100°C; 1.5h), and the resulting peptides fragments were isolated _via_ preparative HPLC. The molecular weights of the isolated fragments were then determined by FAB MS.[4] One fraction contained abundant ions only at _m/z_ 493 and 247 in the positive ion FAB spectrum. These ions correspond to MH^+ signals from the peptides PheValIleAsp and IleAsp. Since it can be argued reasonably that one number does not guarantee securely these sequences, a few µg of the fraction was subjected to N-acetylation (Ac_2O/H_2O), 1:4, room temperature 4h); and to esterifaction (1% HCl in MeOH, room temperature, 10h). The resulting derivatives, upon FAB MS analysis showed MH^+ peaks at _m/z_ 535 and 289; and _m/z_ 521 and 275, respectively. These peaks correspond to the anticipated mono-N-acetyl derivatives (+42 Daltons) and dimetnyl ester (+28 Dalton) derivatives, and therefore support the structure assignments. Note that these reactions occur in good yield, and the excess of reagents can be removed under vacuum, thus avoiding any tube-to-tube transfers or other manipulations in the derivatisations. Therefore, funcational group determination can be carried out on a µg scale by FAB MS; the importance and power of this technique is worthy of emphasis.

The above peptide fraction, containing only the aspartic acid residue at position 5 (Asp^{5}, numbering amino acids of the toxin from the N-terminus) was then hydrolysed. When volatile derivatives (prepared as given earlier) of the resulting amino acids were analysed on the chiral GC column, Asp^{5} was shown to have the D-configuration. Therefore, Asp^{6} has the L-configuration.

The stereochemical assignments of Glu^{6} and Gln^{7} can be determined if, prior to hydrolysis, Gln^{7} is modified such that the product of this modification does not produce a Glu residue upon hydrolysis. A suitable modification was achieved *via* a Hofmann degradation,[4] which brings about the conversion $RCONH_2 \rightarrow RNH_2$. Thus, Gln will afford 2,4-diaminobutyric acid (12) upon hydrolysis, whereas iso-Gln would transiently produce 4,4-diaminobutyric acid (13), and hence the ninhydrin-negative 4-oxobutyric acid (14).

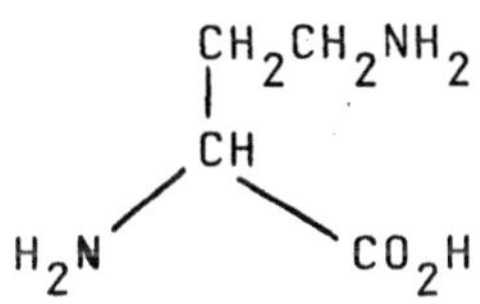

12

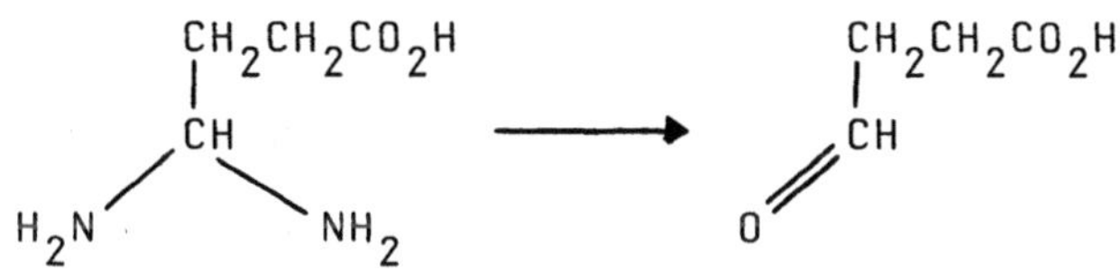

13 14

Following Hofmann degradation of the toxin, and acid hydrolysis of the product, amino acid analysis showed the production of 12. Moreover, analysis of a volatile derivative of 12 on a chiral GC column established it to possess the L-configuration. Therefore, the C-terminal residue is Gln (and not iso-Gln) and has the L absolute configuration. It follows that the penultimate residue, producting Glu upon hydrolysis, has the D configuration.

The ambiguity between normal and iso-residues at position 5, 6, and 7 in the toxin was resolved by successively exposing the toxin to (i) Hofmann degradation and (ii) Ac_2O/pyridine/D_2O. The latter treatment causes racemisation of, and deuterium incorporation into, the α-position of those amino acids which have free carboxyl groups attached to the α-carbon,[9] i.e. those in-chain amino acids which are incorporated as iso-structures (see 11), and normally constituted C-terminal amino acids (see, for example, 5, which has already been shown to be present in the toxin).

The chemical basis for the racemisation and deuterium incorporation is given in Fig. 3. When the free carboxyl group is attached to the α-CH, the mixed anhydride formed upon reaction with Ac_2O is subject to intramolecular nucleophilic attack by the amide function incorporating the carbonyl group of the adjacent amino acid. As a result, the acetate anion is displaced, and a 5-membered ring oxazolone is formed (Fig. 3). In this ring system, the hydrogen of the α-CH group is relatively acidic, being not only adjacent to a carbonyl group, but also to an olefinic double bond. In the presence of a base (pyridine) and D_2O, it will therefore be replaced by deuterium. Additionally, the intermediate enolate anion is sp^2 hydridised at the original α-carbon of the amino acid. Thus, base catalysed hydrolysis of the oxazolone (Fig. 3) produces an amino acid which is not only deuteriated at the α-position, but also racemic.

Toxin treated in the above manner yielded 2,4-diaminobutyric acid which incorporated deuterium and was racemised. Such deuterium incorporation and racemisation was not observed for the Asp residues, nor for the Glu residue. Therefore, these residues are not incorporated as iso-structures. The complete structure of the toxin is therefore that reproduced in Fig. 4.

In summary, the slowly-uncovered structural features of the toxin, which dictated those methods which could successfully tackle the problem are (i) the presence of the N-terminal blocking group, precluding the use of the Edman degradation, (ii) the presence of D-amino acids, precluding the use of enzymic digestion, (iii) the presence of a linear peptide, allowing the application of EI and, particularly in this case, FAB MS sequencing. Note that stereochemical problems are rarely amenable to analysis by mass spectrometry, and the successful use of a chiral column required standards of known absolute configurations. Nuclear magnetic resonance (NMR) was not called upon significantly in the structure elucidation because the structural sub-units were of established types. Additionally, "communication" between these sub-units, although in principle possible by means of the nuclear Overhauser effect (NOE, see later), could not be successfully exploited in the above example. In the linear molecule, the conformation was too floppy, and the molecular weight in "no-man's-land", between lower and higher values which could have been exploited.

The biosynthetic origins (prokaryotic or eukaryotic) of the toxin remain an enigma. It has recently been established that a fungus lives symbiotically with the sawfly larvae, in the gut.[10] It is possible that the fungus is responsible for the production of the D-amino acids, or even later stages of the biosynthetic pathway; but this is speculation, and a definitive conclusion will have to await further experimentation.

The Structure Elucidation of Unusual Monocyclic Peptides.

The previously discussed sequencing of (unusual) linear peptides by mass spectrometry is a useful general method. The strategy is not directly applicable to monocyclic peptides because fragmentation of an amide bond produces a linear peptide and no fragments. It might be thought that the internal energy of the linear product would be sufficient to produce fragments *via* further amide bond cleavages; but such is not generally the case in FAB spectra (low internal energy of ions). Further fragmentation often occurs in EI spectra, but the spectra are more difficult to interpret, and high resolution data are usually needed (see later).

An alternative general strategy has therefore to be sought. One approach is to cleave chemically an amide bond of the monocyclic peptide, and then use EI and/or FAB MS sequencing of the linear product. Such an approach was applied successfully to determine the structure of a peptide toxin from the blue-green algae, *Microcystis aeruginosa*.[11,12]

Since the end of the last century, heavy blooms of the blue-green algae *Microcystic aeruginosa* have been known to cause deaths among livestock consuming heavily contaminated water. The lethal doses (LD_{50}) of the purified toxins are *ca* 50 µg/kg for mice on interperitoneal injection.[13] The toxic peptides of *M. aeruginosa* are normally contained within the algal cells, and are released only when the cell is damaged. It has recently been concluded[13] that an increase in liver damage was indicated in a human population during the period of a bloom of M. aeruginosa in the water-supply reservoir.

Although enzymatic experiments and amino acid analysis (equi-molar ratios of L-leucine, D-glutamic acid, erythro-β-methyl-D-aspartic acid, L-alanine and D-alanine)[14] established the peptidic nature of the toxin, no sequence ions were seen in the FAB mass spectrum (which established a molecular weight of 909 Daltons). Additionally, it was shown that the peptide lacked a free basic N-terminus because the molecular weight was unchanged following treatment with Ac_2O/H_2O. The lack of fragmentation in the FAB spectrum, and the absence of a free N-terminus, together suggest the possibility of a cyclic peptide. However, it is clear that the identified amino acids fall far short of accounting for the observed molecular weight. Since the toxin was available in a quantity of ca 2 mg, it was possible to throw light on the non-standard amino acid components of the toxin by high-field ^1H nmr at 400 MHz.

Before discussing the results of the NMR study, it is necessary to outline the principle of an NMR phenomonon which has assumed great importance recently. It is the nuclear Overhauser effect (NOE). Although this effect has been known for many years, and was first exploited in organic chemistry in the 1960's, it is the advent of difference spectroscopy which has resulted in its widespread use during the last few years.[15]

If the intensity of the resonance of one proton (H^1) is normally I, then if $r_{1,2}$ (the internuclear distance between protons H^1 and H^2) is relatively small (as a useful guide for

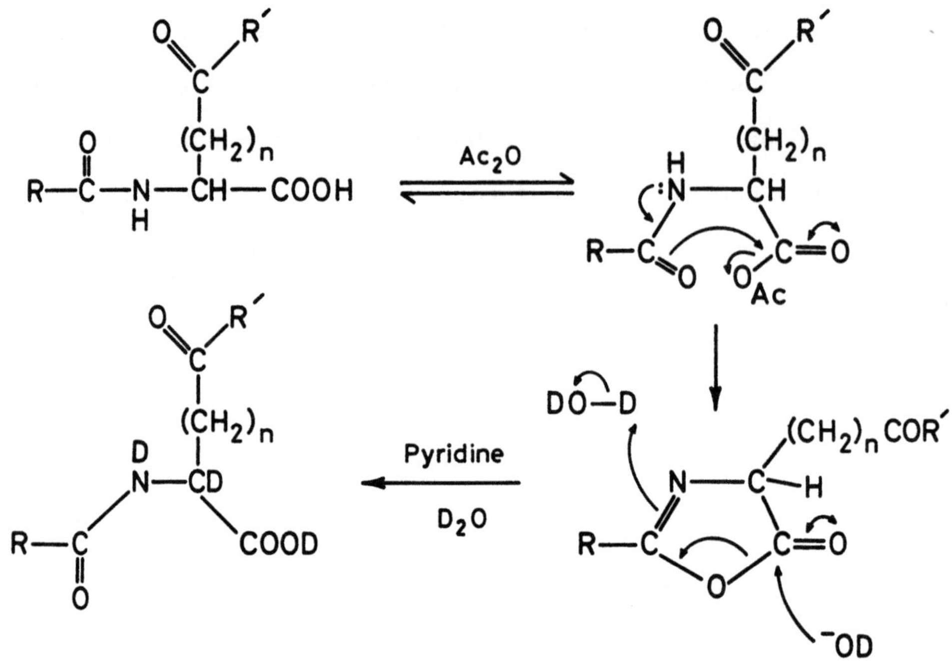

Figure 3. The chemical basis for racemisation of, and deuterium incorporation into, amino acids possessing a free carboxyl group attached to the α-CH

our present purposes say, <0.3 nm), the effect of irradiating a second proton (H^2) before recording the intensity of the resonance of H^1 is to change its intensity to I', i.e. $I \neq I'$. This change in intensity is called the NOE. The effect is indicated schematically in 15-18. Three hydrogen nuclei in a molecule are indicated in 15. The signals which arise in the ^1H NMR spectrum of 15 in the absence of pre-irradiating any of the protons are given in 16. Upon pre-irradiation of H^1 at its resonance frequency ν_1, the ^1H NMR spectrum is modified to 17. Note that in 17, the resonance of H^1 has disappeared, because its resonance is "saturated", i.e. the populations of upper and lower energy levels are equalised, by the pre-irradiation. The H^2 resonance intensity is reduced due to the NOE. The NOE difference spectrum (NOEDS, 18) is obtained by on-line computer subtraction of 17 from 16.

15

16

17

18

Note that since H^3 is distant from H^1, no observable NOE occurs to H^3. The value of the difference spectrum is evident for molecules containing many protons; the NOEDS then contains only signals due to the irradiated proton H^1, and less intense signals due to those protons which are proximate to H^1. Such experiments are extremely valuable in determining structures of molecules in solution.

Analysis of the ^1H 400 MHz spectrum of the toxin in D_2O established from chemical shift values and spin decoupling experiments the presence of the N-methyldehydroalanine (19) and of units 20 and 21. The connectivity between 20 and 21 was established to occur *via* the C-C bond indicated by a broken line (using NOES, determined in $[^2H]_6$-DMSO solution). Thus, pre-irradiation of the δ5.53 resonance led to a 12% reduction in the intensity of the δ1.66 resonance [normally expressed in the concise form {5.53} → 1.66(-12%)]. Similarly, {6.27} → 5.48(-15%) and {5.48} → 6.27(-15%).

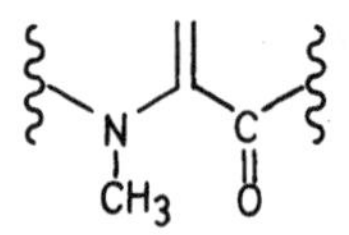

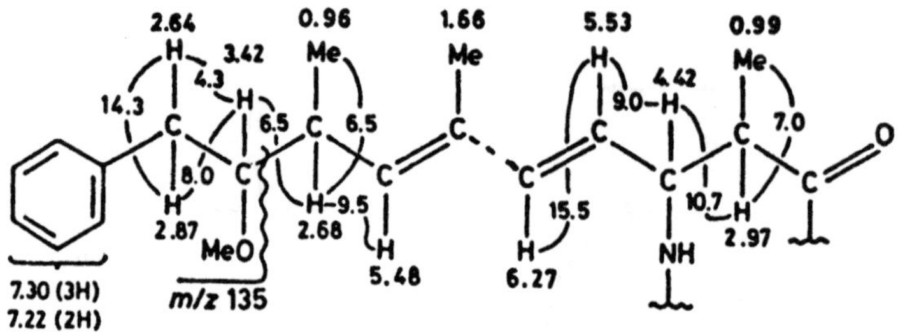

19 20 21

The above combination of the two units provides an excellent example of how connectivity can be established from NOES when evidence is not available from coupling constants (because four or more σ-bonds occur between adjacent protons).

Note that in the above case, the NMR experiment does not determine the nature of the electronegative substituents (OCH_3, NH, C=O) on the chain; but only that some electronegative substituent must be attached to the carbons bearing the protons at δ3.42, 4.42, and 2.97. The nature of these substituents was established from fragment ions occurring in the EI mass spectrum of the toxin and/or permethylated toxin. These spectra afforded ions at m/z 135 ($C_6H_5CH_2CH{=}\overset{+}{O}CH_3$, with or without permethylation), m/z 326 (22, following permethylation) and m/z 272 (23, following permethylation). The atomic compositions of the fragments 22 and 23 are supported by high resolution measurements. When the permethylation procedure utilises CD_3I instead of CH_3I, the m/z values of 22 and 23 shift to 329 and 275, respectively. Thus, the nitrogen atom is methylated in the derivatisation.

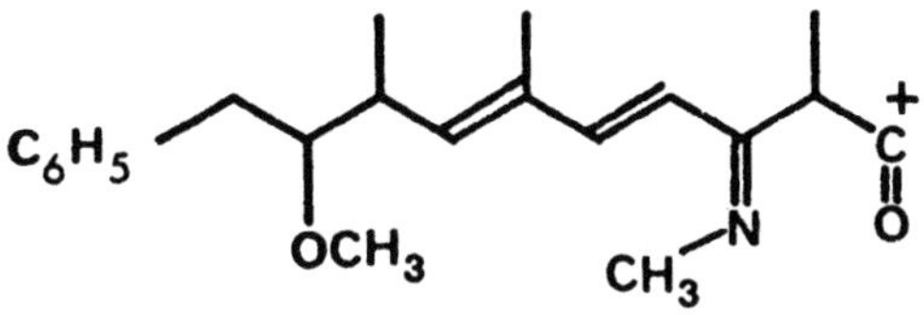

m/z 326.2112
(calc 326.2120)

22

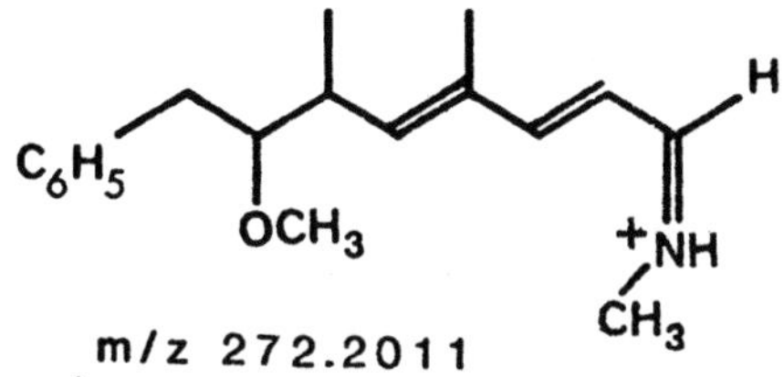

m/z 272.2011
(calc 272.2014)

23

In view of the presumed cyclic nature of the toxin, the hydrolysis of a single amide bond in the peptide might be attempted by limited acid hydrolysis.

Since the N-methyldehydroalanine unit (19) might be modified by even quite mild acid treatment (Michael addition or enamine hydrolysis), it was deemed wise to reduce this residue prior to the pulse acid hydrolysis treatment. The reduction of 19 was effected by conjugate addition of hydride ion, from $NaBH_4$, to give N-methylalanine. This procedure has the convenience that $NaBD_4$ can alternatively be used, and thus provide an isotopic label for the reduced residue. The reduced toxin was pulse hydrolysed (6 M HCl, 5 min, 100°C) and a product of molecular weight 812 Daltons (by FAB MS) isolated. Clearly, if the toxin has been ring opened, a portion of it has also been lost. Indeed, when the two-step procedure was repeated using $NaBD_4$, a product of molecular weight 812 Daltons was again obtained. The product therefore corresponds to reduction (+2), ring opening (+18), loss of N-methylalanine (-85) and methanol (-32 Daltons by elimination from 20). Micro-derivatisation experiments followed by FAB MS (see earlier) show the product to contain one amino and 3 carboxyl groups. Acetylation and permethylation of the product, followed by EI MS sequencing established the sequence Ala-Leu-"Glu"-Ala-βaa'-"Glu", where βaa' is the novel β-amino acid 20/21 minus methanol; and "Glu" represents Glu or β-Me-Asp (or their iso-structures), since these units are all isomeric. When the linear peptide was subjected to one cycle of the Edman degradation, N-terminal alanine was cleanly removed (as shown by FAB MS);[16] and chiral column GC-MS experiments (see earlier) on the amino acids derived from the product of Edman degradation established the absence of D-Ala and the presence of L-Ala. As expected, a second Edman degradation removed L-Leu, but the third Edman cycle failed. Evidently, the third amino acid, in the linear peptide, is not incorporated as a standard amino acid. However, in the product of the second Edman cycle, the amino group of the third amino acid is present as a free, basic group. This group can therefore be converted to the sulphonamide by reaction with dansyl chloride. Acid hydrolysis of the product produced the dansyl derivative of β-methylaspartic acid.[11] Since this amino acid is not incorporated as a standard amino acid in the linear peptide, it must be incorporated as β-methyl-iso-aspartic acid. Note that this does not necessarily imply that this acid is present as the iso-structure in the original toxin; since the Asp → iso-Asp conversion can be brought about by either mild acid or mild base, the isomerisation could have occurred in the course of the conversion of the toxin to the linear peptide. Since residue

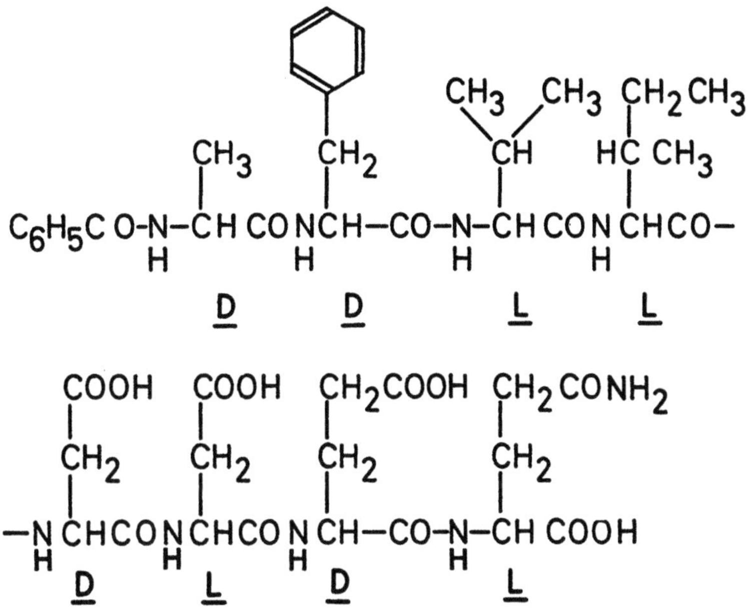

Figure 4. Structure of the peptide toxin from the sawfly *Lophyrotoma interrupta*

3 on the linear peptide is β-Me-*iso*-Asp, it follows that the C-terminal residue is Glu (or *iso*-Glu).

It remains to establish the points of attachment of the N-methyl-dehydroalanine residue which, following $NaBH_4$ reduction of the toxin, was lost during pulse hydrolysis. This was achieved[12] by exposing the reduced toxin to a milder pulse hydrolysis (moist trifluoroacetic acid at room temperature for 24h). The negative ion FAB mass spectrum of this hydrolysis product showed $[M-H]^-$ at *m/z* 928, indicating that the dihydro-toxin (molecular weight 911 Daltons) had undergone simple hydrolytic addition of water, without loss of an amino acid residue. Moreover, N-terminal amide sequence ions [see **3**, cleavage (i)] in this spectrum occurred at *m/z* 856 and

713, establishing that N-methyl-Ala is attached to a Glu (or iso-Glu) residue (24).[12]

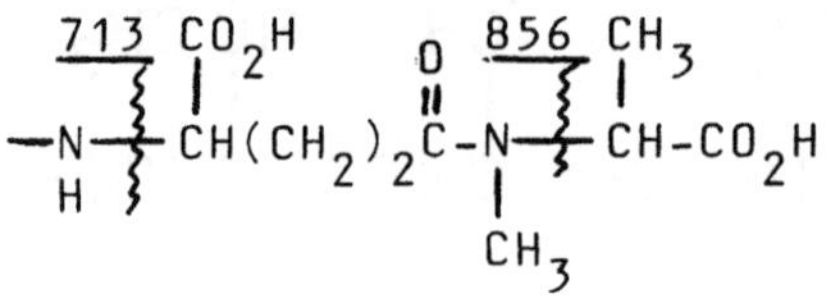

24

The presence of iso-Glu in the toxin was indicated not only by the change in chemical shift of the α-CH proton upon titration of the carboxyl group of the glutamic acid residue, but also by the chemical methods outlined in Figure 3.[12] Similar chemical experiments on the unreduced toxin demonstrated that β-methyl-Asp is incorporated as the iso-structure in the toxin.[12]

Thus, a monocyclic structure for the toxin can be completed by attachemnt of the carbonyl of N-Me-dehydro-Ala to the amino group of D-Ala (Figure 5). Note that in the published work,[11] only a partial structure for BE-4 had been assigned, and that the tentative attachment[11] of N-Me-dehydro-Ala as an exocyclic residue is now shown[12] to be incorrect. Experiments to determine the stereochemistry of the asymmetric centres in the β-amino acid have yet to be carried out.

As in the case of the sawfly liver toxin, note how complex details of stereochemistry and connectivity can be proved by use of GC, enzymic methods, and MS, so long as the sub-units have standard structures. Where a sub-unit of novel structure was involved, it was necessary to turn to high field ^{1}H NMR; and in particular, to spin decoupling and NOEs.

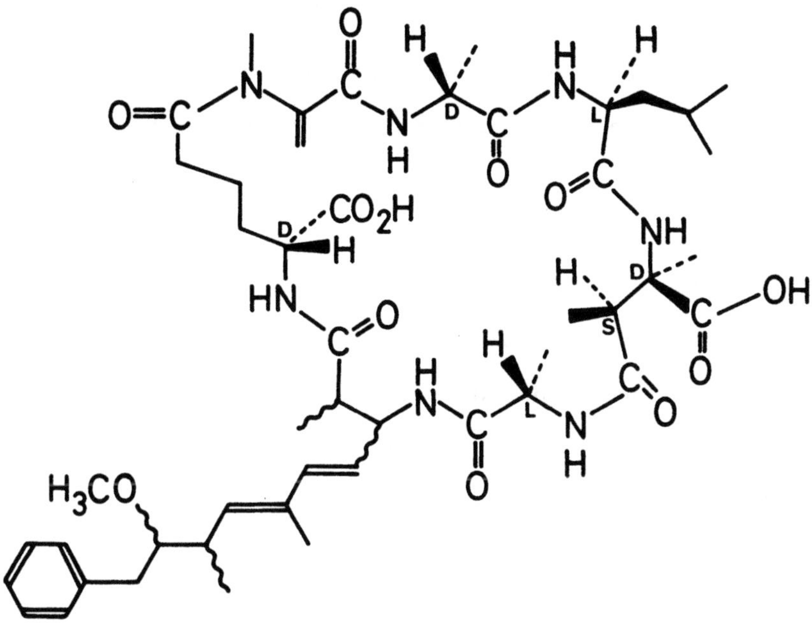

Figure 5. Structure of the peptide toxin BE-4 from M. aeruginosa

Conclusion

During the last 20 years, mass spectrometry and nuclear magnetic resonance have evolved to become extremely powerful methods for the elucidation of complex organic structures. In particular, the recent advent of FAB mass spectrometry and nuclear Overhauser effect difference spectroscopy is of importance; and it appears that two-dimensional NMR spectroscopy[15] is another important advance in extracting structural information from the spectra of complex molecules.

It is concluded that if the above techniques are combined with the powerful chromatographic methods of GC and HPLC, and with chemical derivatisations carried out on a micro-scale, the outlook for the structure elucidation of complex molecules in the future is a very promising one. There is no shortage of challenging and important problems.

Acknowlegements

I wish to thank my colleagues, named in the co-authored references, who have made the described work possible. Additionally, I thank the SERC and the Royal Society for financial support.

References

1. H.R. Morris, D.H. Williams, G. Midwinter, and B.S. Hartley, Biochem. J., 1974, 141, 701.
2. A. Auffret, T.J. Blake, and D.H. Williams, Eur. J. Biochem., 1981, 113, 333.
3. A.D. Auffret, D.H. Williams, and D.R. Thatcher, FEBS Letters, 1978, 90, 324.
4. D.H. Williams, S. Santikarn, F. DeAngelis, R.J. Smith, D.R. Reid, P.B. Oelrichs, and J.K. MacLeod, J. Chem. Soc., Perkin Trans. I, 1983, 1869.
5. F.H.S. Roberts, Queensl. Agric. J., 1932, 37, 41.
6. L.L. Callow, Queensl. Agrig. J., 1955, 81, 155.
7. M. Barber, R.S. Bordoli, R.D. Sedgwick, and A.N. Taylor, J. Chem. Soc, Chem. Commun., 1981, 325.
8. D.H. Williams, C.V. Bradley, S. Santikarn, and G. Bojesen, Biochem. J., 1982, 201, 105.

9. G.N. Holcomb, S.A. Janes, and D.N. Ward, Biochemistry, 1968, 7, 1291.

10. Professor H. Kleinkauf, Technical University of Berlin, private communication.

11. S. Santikarn, D.H. Williams, R.J. Smith, S.J. Hammond, D.P. Botes, A. Tuinman, P.L. Wessels, C.C. Viljoen, and H. Kruger, J. Chem. Soc. Chem. Commun., 1983, 652.

12. D.P. Botes, A. Tuinman, P.L. Wessels, C.C. Viljoen, H. Kruger, S. Santikarn, D.H. Williams, R.J. Smith and S.J. Hammond, in preparation.

13. I.R. Falconer, A.M. Beresford, and M.T. C. Runnegar, Med. J. Aust., 1983, 511.

14. D.P. Botes, C.C. Viljoen, H. Kruger, P.L. Wessels, and D.H. Williams, Toxicon, 1982, 20, 1037.

15. See, for example, J.K.M. Sanders and J.D. Mersh, in "Progress in Nuclear Magnetic Resonance Spectroscopy", Vol. 15, pp. 353-400, Pergamon Press, Oxford, 1983.

16. C.V. Bradley, D.H. Williams, and M.R. Hanley, Biochem. Biophys. Res. Commun., 1982, 104, 1223.

Primary Structure of Hemoglobin from Reptile *Uromastix hardwickii* and its Status in Molecular Evolution

Sabira Naqvi[1], H. Jornvall[2], and Z. H. Zaidi

[1]H. E. J. Research Institute of Chemistry, University of Karachi, Karachi-32, Pakistan
[2]Department of Chemistry 1, Karolinska Institutet, S-10401 Stockholm, Sweden

1. Introduction

Much work has been carried out on the hemoglobins of different classes of vertebrates and invertebrates. Reptiles occupy an important position in the animal kingdom. However there is little data on the hemoglobin from this class.

In this study we are reporting studies carried out on the tropical lizard Uromastix hardwickii. This hemoglobin is interesting as isolation of component is very difficult and so is the chain separation. Tryptic peptides were isolated and sequence study carried out on these peptides and corresponding peptides to α or β chain were put together

2. Experimental and Results

2.1 Isolation of Hemoglobin and Carboxymethylation

Uromastix hardwickii were caught alive from the District of Karachi. The blood was collected from the neck in 3.8% sodium citrate used as anticoagulant. Red cells were separated by centrifugation, washed three times with physiological saline and lysed with distilled water. The cell debris were removed by centrifugation and hemoglobin solution was lyophilized. The globin was prepared by the method of Anson and Mirsky [1930].

Natural Product Chemistry
Ed. by Atta-ur-Rahman

For carboxymethylation, the globin was reduced with dithiothreitol (185 nmoles/mg protein) for 2 hours at 37°C, and reacted with neutralized labelled iodoacetic acid (1.26 mole/mg protein). The reaction was carried out for 2 hrs. at 37°C in the dark. Reagents were removed by extensive dialysis against distilled water.

2.2 *Characterization of Hemoglobin on Electrophoresis*

(i) SDS-polyacrylamide gel electrophoresis

Electrophoresis of the crude hemoglobin was carried out on 12% polyacrylamide gel (4 x 80 mm) according to Maurer (1969). The electrode buffers used were Tris glycine pH 8.9 and Tris-HCl pH 8.0 for the cathode and anode chambers respectively. Both buffers contained 0.1% sodium dodecyl sulphate. The sample was prepared by treating with 2-Mercaptoethanol for 10 mins. at 100°C, and subjected to the gels. A constant current of 7mA/tube was applied for 2½ hrs. The gels were stained with 2% amido black in 7% acetic acid and destained with 7% acetic acid. The electrophoretic result shows that hemoglobin is separated in four bands (Fig.1a).

Electrophoresis on charge basis was performed with the above described conditions, except that sodium dodecyl sulphate was not used. The result shows six bands (Fig.1b)

(ii) Isoelectric focusing

Isoelectric focusing was carried out on polyacrylamide gels according to Wrigley (1969) using ampholyte in the pH range 3 to 10. 0.4% H_2SO_4 and 0.4% ethanol amine were used as the anode and cathod buffers respectively. An electric current of 2mA/tube was applied for 3 hrs. The gels were fixed in 12.5% trichloro acetic acid. Staining was carried out with 0.25% Coomassie brilliant blue R 250 according to St.Groth et-al. (1963), and destained with ethanol:acetic acid: water (3:1:6). (Steel et al. 1978). Hemoglobin was separated in fourteen bands of different concentrations, as shown in Fig. 1c

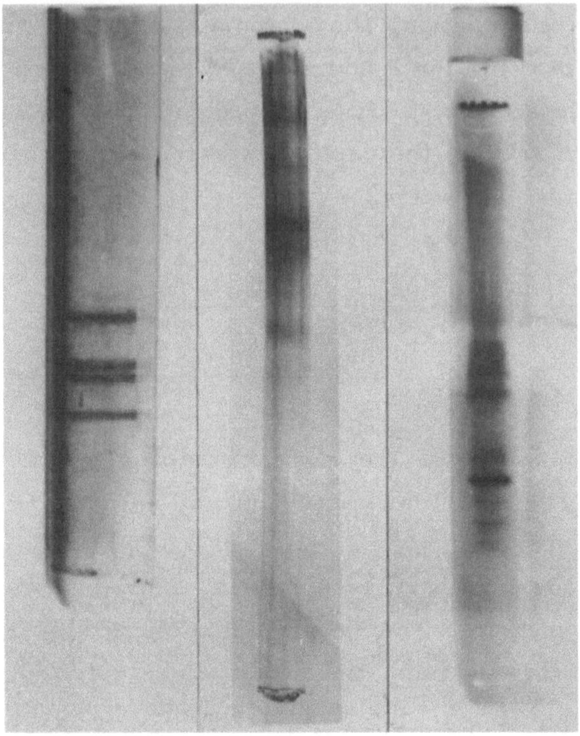

Fig. 1. Disc electrophoresis of Uromastix hardwickii hemoglobin on polyacrylamide gel (12%, 4 x 80 mm) current 7 mA/tube.

a) SDS-gel electrophoresis
b) On charge basis
c) Isoelectric focusing on 6% polyacrylamide gel (0.4 x 12 cm).

2.3 Separation of Uromastix Hemoglobin/Globin

(i) Separation on Sephadex G-75

Hemoglobin was chromatographed on Sephadex G-75 column (130 x 1.5 cm) for separation of components. Elution was carried out with 1.8 M acetic acid at a flow rate of 15 ml/hr and 3 ml fractions were collected. The ultraviolet absorption was recorded at 280 nm. The hemoglobin was separated in four peaks. The elution of the first peak started before the first void volume of the column (Fig.2)

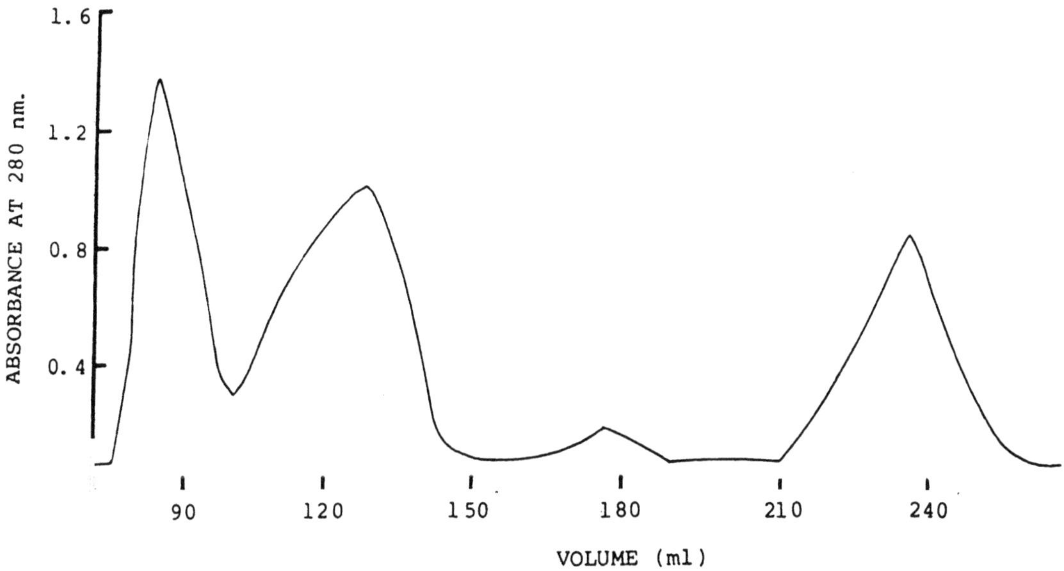

Fig. 2. Chromatography of crude hemoglobin on a Sephadex G-75 column (1.5 x 130 cm). Eluted with 1.8 M acetic acid at a flow rate of 12 ml/hr. Fraction volume 3 ml.

(ii) CM-Cellulose chromatography

Chain separations were carried out on CM-Cellulose CM 52 in 8 M urea in the pH range 4.6 to 7.6 (Clegg et al.1966). At pH 4.6 and 5.6 the starting buffer consisted of 8M urea, 0.02M sodium acetate, 0.05M 2-Mercaptoethanol. The globin was dialysed against this buffer and subjected to a column (15 x 1.5 cm). The elution was carried out with a linear NaCl gradient from (O-0.2M). The separation was also performed with 8 M urea, 0.05M 2-Mercaptoethanol, 0.05M Na_2HPO_4 at pH 6.6 and 7.6. Linear Na^+ ion gradient from 0.005 M to 0.03M Na_2HPO_4 was used for elution.

The fractions eluted were pooled, dialysed against 5% acetic acid and desalted on a Sephadex G-50 column (180 x 1.5 cm).

The result shows that the peaks were not completely resolved (Fig. 3) and were similar at all pH values and buffer conditions tried. N-terminal sequence analysis revealed the presence of α and β chains in all fractions.

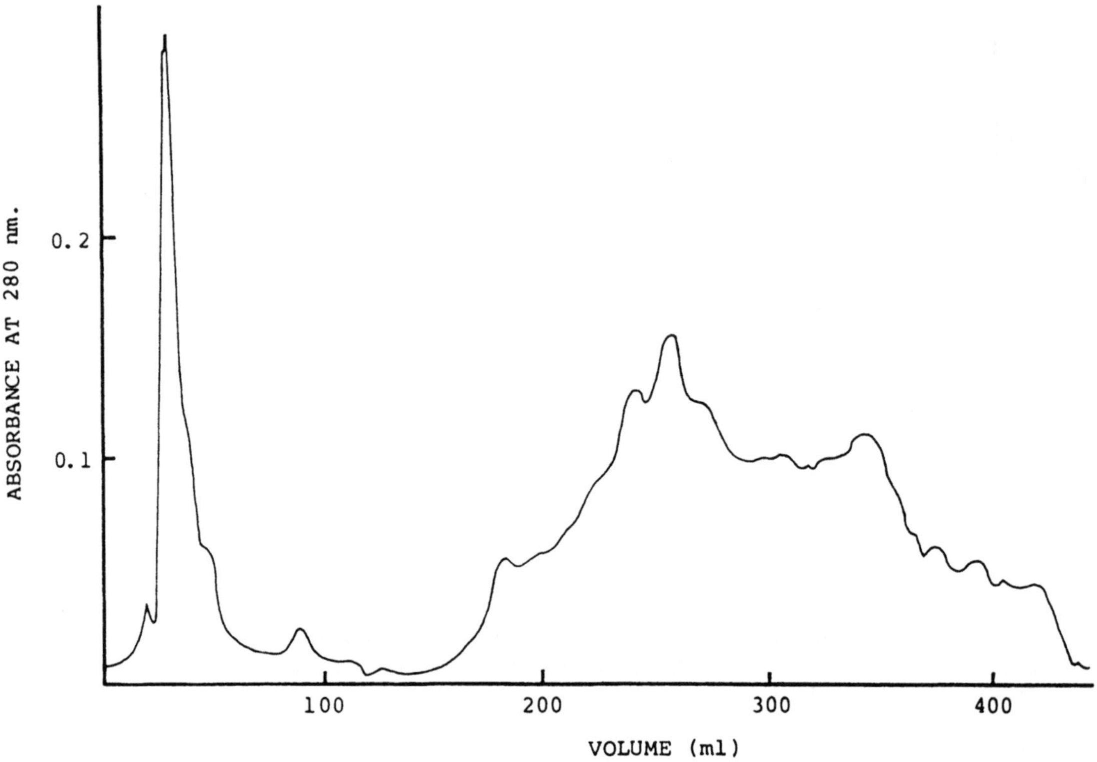

Fig. 3. Chromatography of globin on CM-Cellulose (1.5 x 15 cm) column at a flow rate of 20 ml/hr. Starting buffer: 8M Urea + 0.02M sodium acetate + 0.05M 2-Mercaptoethanol PH 5.6. Final buffer: Starting buffer + 0.2M NaCl.

2.4 Isolation of Tryptic Peptides

The carboxymethylated globin was dissolved in 0.1M ammonium bicarbonate and TPCK-trypsin was added in an enzyme to protein ratio of 1:50. pH was maintained at 8.5. The reaction was carried out for 4 hours at 37°C.

Peptides were fractionated on Sephadex G-50 column (180 x 1.5 cm) in 5.4 M acetic acid. The tryptic digest was separated into two fractions (A and B, mainly composed of large and small peptides respectively) Fig.4. The peptides from both fractions were purified by reverse phase-high performance liquid chromatography. The separation was carried out on a μ-Bondapak C_{18} column and peptides were eluted in 0.1% TFA with increasing concentration of acetonitrile Fig. 5.

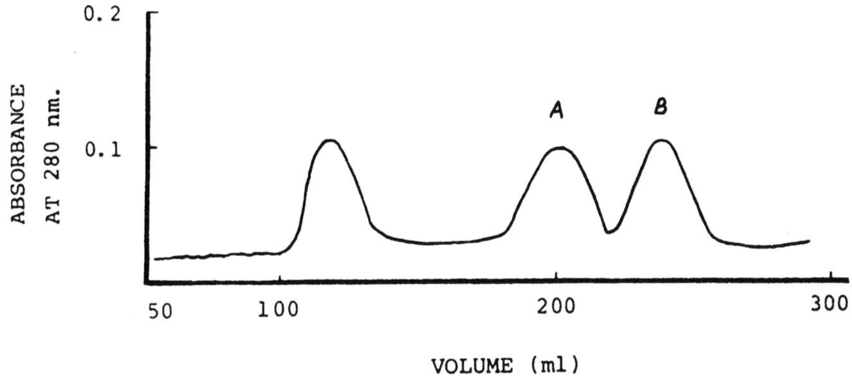

Fig. 4. Gel chromatography of tryptic peptides on a Sephadex G-50 column (1.5 x 180 cm). Eluted with 5.4M acetic acid at a flow rate of 12 ml/hr.

2.5 Amino Acid Analysis

Amino acid compositions were determined on a Beckman 121M and Biotronik LC 6001 analyzer after hydrolysis in evacuated tubes with 6M HCl for 24 hours at 110°C.

2.6 Amino Acid Sequence of Tryptic Peptides

The tryptic peptides purified on RP HPLC were sequenced by manual degradation according to Edman method (Edman and Begg 1967) and by DABITC method (Von Bahr Lindstrom et-al. 1982), and some peptides were analyzed with liquid phase sequencer. The peptides are numbered with the letter A or B which indicate the two fractions obtained by pre-separation on Sephadex G-50. Numbers are according to elution order from RP-HPLC. The sequence of the peptides are given as follows:-

Peptide A1
Thr Asp Asp Lys Asn His Val Arg.

Peptide A2
Thr Tyr Phe Pro His Phe Asp Ala Ser Ala Gly Ser Asn Glu Leu Lys.

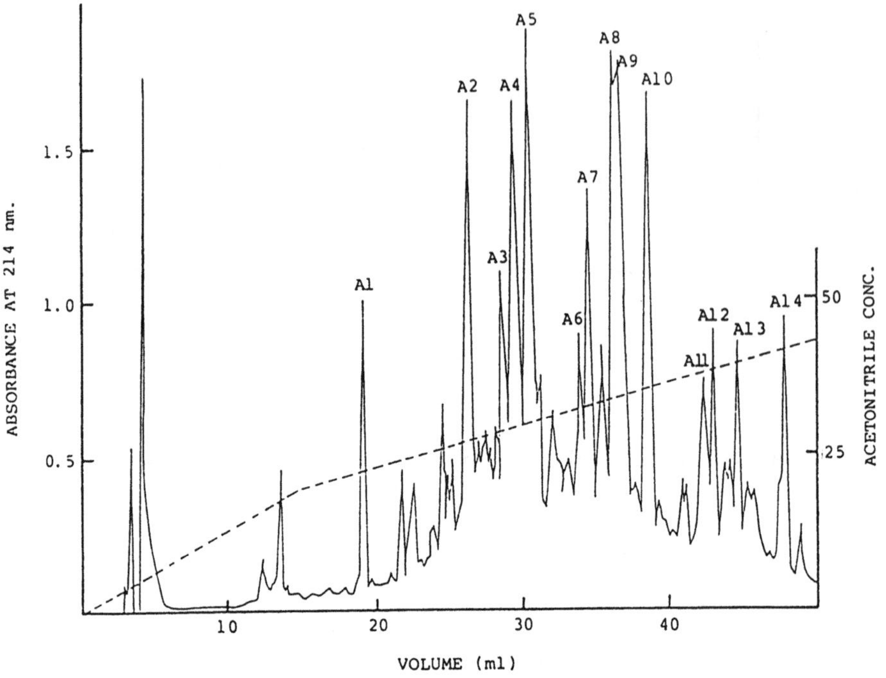

Fig. 5a. Purification of tryptic peptides by Reverse phase-HPLC on a μ-Boundapak C_{18}. Eluted with 0.1% TFA with a gradient of acetonitrile. The separation is from the pool A. Flow rate 60 ml/hr.

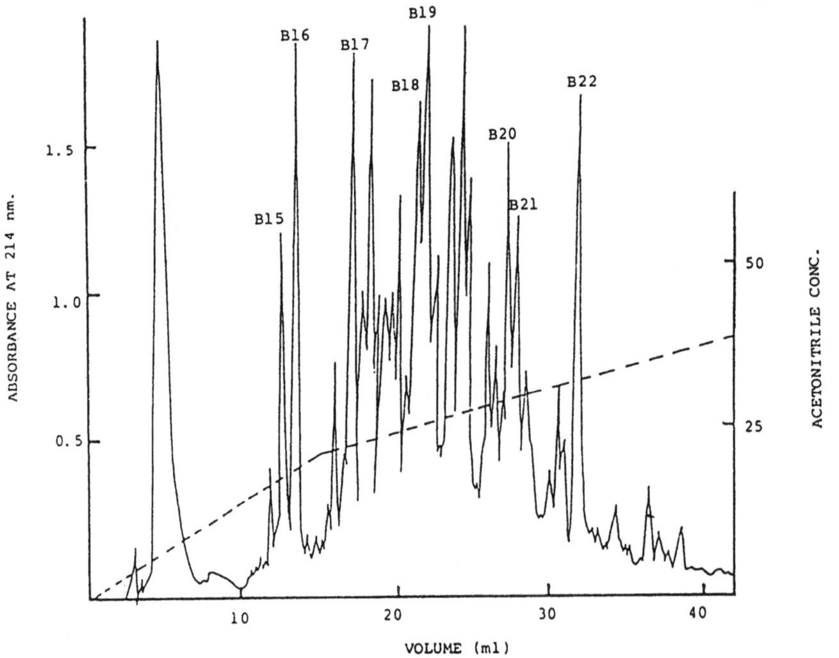

Fig. 5b. Purification of tryptic peptides by Reverse phase-HPLC on a μ-Boundapak C_{18}. Eluted with 0.1% TFA with a gradient of acetonitrile. The separation is from the pool B.

Peptide A3

Leu Val Gly Val Val Ala His Ala Leu Ala Arg.

Peptide A4

Val Gly Pro His Leu Asp Asp Tyr Gly Gly Glu Ala Leu His Arg.

Peptide A5

Thr Tyr Phe Ser His Phe Asp Leu His Glu Asn Ser Ala Gln Ile Arg.

Peptide A5*

Leu Ser Glu Tyr His Cys Asn Lys.

Peptide A7

Phe Glu Asp Phe Gly Asn Leu Ser Thr Pro Ser Ala Ile Leu Asn Asn Pro Lys.

Peptide A8

Ala Ile Trp Gly His Val Ser Asn Asn Pro Glu Ala Phe Gly Ala Glu Leu Tyr Arg.

Peptide A9

Phe Phe Gly Asp Phe Gly Asn Ile Ser Ser Ala Ala Ala Ile Thr Gly Asn Pro Lys.

Peptide A10

Phe Phe Glu Asp Phe Gly Asn Leu Ser Thr Pro Ser Ala Ile Leu Asn Asn Pro Lys.

Peptide A11

Leu Leu Gly Asp Val Leu Ile Thr Leu Ser Ala Ala Asn Phe Gly Lys.

Peptide A12

Lys Val Val Asp Ala Leu Thr Gln Ala Val Asn Asn Leu Asp Asp Leu Ser Gly Ala Ile Ser Lys.

Peptide A13

Val Val Asp Ala Leu Thr Gln Ala Val Asn Asn Leu Asp Asp Leu Ser Gly Ala Ile Ser Lys

Peptide A14

Val Asp Val Gly Ser Val Gly Gly Glu Thr Leu Ala Asn Leu Leu Val Val Tyr Pro Trp Thr Gln Arg.

Peptide B15

Arg Tyr His

Peptide B16

Asn Leu Asp Asn Val Lys.

Peptide B17

Leu Ser Asp Leu His Ala Glu Lys

Peptide B18

Leu Phe Thr Ala His Pro Ala Ser Lys.

Peptide B19

Val His Trp Thr Ala Glu Glu Lys

Peptide B20

Asn Phe Glu Val Tyr Pro Gln Thr Lys

Peptide B21

Val Ile Thr Ser Phe Gly Asp Ala Leu Lys

Peptide B21*

Leu His Val Asp Pro Val Asn Phe Arg.

Peptide B22

Ala Leu Ile Asn Ala Tyr Trp Gly Lys.

										10										20					
α_1	Val	Leu	Thr	Asp	Asp	Asp	Lys	Asn	His	Val	Arg	Ala	Ile	Trp	Gly	His	Val	Ser	Asn	Asn	Pro	Glu	Ala	Phe	Gly
						A1															A8				
α_2																	Val	Gly	Pro	His	Leu	Asp	(Asp)	Tyr	Gly
																					A4				
					30										40										50
α_1	Ala	Glu	Ala	Leu	Tyr	Arg	Leu	Phe	Thr	Ala	His	Pro	Ala	Ser	Lys	Thr	Tyr	Phe	Ser	His	Phe	Asp	Leu	His	Glu
											B18										A5				
α_2	Gly	Glu	Ala	Leu	His	Arg	Asn	Phe	Glu	Val	Tyr	Pro	Gln	Thr	Lys	Thr	Tyr	Phe	Pro	His	Phe	Asp	Ala	Ser	Ala
											B20										A2				
										60										70					
α_1	Asn	Ser	Ala	Gln	Ile	Arg					Lys	Val	Val	Asp	Ala	Leu	Thr	Gln	Ala	Val	Asn	Asn	Leu	Asp	Asp
																					A12				
																						A13			
α_2	Gly	Ser	(Asn	Gln	Leu	Lys)																			
					80										90										
α_1	Leu	Ser	Gly	Ala	Ile	Ser	Lys	Leu	Ser	Asp	Leu	His	Ala	Gly	Lys										
											B17														

Fig. 6. Amino acid sequence of the α-chain from the hemoglobin of Uromastix hardwickii.

The peptides are discussed briefly as:
No peptides corresponding to the C-terminal region of the α-chain (after position 90) were recovered (Fig.6). This is probably merely because not all peptides were recovered. However, in relation to possibly lower recovery of C-terminal segments, it may be of interest that these parts of some hemoglobin chains appear susceptible to proteolytic removal. Regions of both chains were now recovered in more than one peptide from the Uromastix preparation. For example peptides A2 and A5 (Fig.6) are homologous, both corresponding to positions 41-56 of the viper α-chain, and peptides A9 and A10 both appear to correspond to a region around position 50 of the β-chain. Relative amounts differ but, from the peak heights in Fig. 5 none of these peptides can be eliminated as an obvious contaminant. Consequently, the preparation is concluded to contain at least two types of both α- and β-chain, explaining the difficulties in attempts at chain separations.

3. Discussion

A clear separation of chains was not possible, although the purity of the hemologbin was of acceptable level. The separation on SDS gel electrophoresis shows the presence of four components and the electrophoresis gave six chains. However the isoelectric focusing depicted as many as fourteen bands. It has been noted that in IEF, there are more bands than the number of components present. It is also possible that six bands in electrophoresis may be deceptive and that the chains may have been separated during electrophoresis. The separation of the components and chains could not be achieved and this may be because of aggregation of molecules. This also explains the elution of a peak before the void volume as shown in fig. 2.

The peptide analyses reveal that more than one peptide matches single region in the hemoglobin chain suggesting the multiplicity of chains.

Some of this multiplicity is explained by alternative enzymatic cleavage (each of the peptide pairs A7/A9 and A12/A13, Fig.6 and 7).

```
                                       10                                                      20
β1  Val  His  Trp  Thr  Ala  Glu  Gln  Lys  Ala  Leu  Ile  Asn  Ala  Tyr  Trp  Gly  Lys  Val  Asp  Val  Gly  Ser  Val  Gly  Gly
    ──────────────────── B19 ──────────  ───────────────────── B22 ─────────  ──────────────────────────────────────
                        30                                       40                                            50
β1  Glu  Thr  Leu  Ala  Asn  Leu  Leu  Val  Val  Tyr  Pro  Trp  Thr  Gln  Arg  Phe  Phe  Glu  Asp  Phe  Gly  Asn  Leu  Ser  Thr
    ─────────────────── A14 ──────────────────────────────────────────────  ─────────────────────────── A10 ──────────

β2                                                                          Phe  Phe  Gly  Asp  Phe  Gly  Asn  Ile  Ser  Ser
                                                                            ─────────────────────────── A9 ───────────
                                                                            ─────────────────────────────── A7 ───────
                                            60                                               70
β1  Pro  Ser  Ala  Ile  Leu  Asn  Asn  Pro  Lys                                   Val  Ile  Thr  Ser  Phe  Gly  Asp  Ala  Leu
    ────────────────────────────────────────────                                  ─────────────────── B21 ──────────────

β2  Ala  Ala  Ala  Ile  Thr  Gly  Asn  Pro  Lys
    ────────────────────────  ──────────────────
    ────────────────────────────────────────────
                        80
β1  Lys  Asn  Leu  Asp  Asn  Val  Lys                      Leu  Ser  Glu  Tyr  His  Cys  Asn  Lys  (Leu) His  Val  Asp  (Pro
    ───  ────────── B16 ──────────                          ──────────────────────── A5* ──────   ──────────── B21* ───
                                       110                                                  120
β1  Val) Asn  Phe  Arg  Leu  Leu  Gly  Asp  Val  Leu  Ile  Thr  Leu  Ser  Ala  Ala  Asn  Phe  Gly  Lys
    ──────────────────  ───────────────────────────────────────── A11 ─────────────────────────────
                        130                                      140
β1                                     Leu  Val  Gly  Val  Val  Ala  His  Ala  Leu  Ala  Arg  Arg  Tyr  His
                                       ──────────────────────── A3 ─────────────────────  ──── B1 ────
```

Fig. 7. Amino acid sequence of the β-chain from the hemoglobin of Uromastix hardwickii.

									10										20					
Human	Val	Leu	Ser	Pro	Ala	Asp	Lys	Thr	Asn	Val	Lys	Ala	Ala	Trp	Gly	Lys	Val	Gly	Ala	His	Ala	Gly	Glu	Tyr	Gly
Chic-ken	Val	Leu	Ser	Asn	Ala	Asp	Lys	Asn	Asn	Val	Lys	Gly	Ile	Phe	Thr	Lys	Ile	Ala	Gly	His	Ala	Glu	Glu	Tyr	Gly
Viper	Val	Leu	Ser	Glu	Asp	Asp	Lys	Asn	Arg	Val	Arg	Thr	Ser	Val	Gly	Lys	Asn	Pro	Glu	Leu	Pro	Gly	Glu	Tyr	Gly
Caiman	Val	Leu	Ser	Gln	Glu	Asp	Lys	Ser	His	Val	Lys	Ala	Ile	Trp	Gly	Lys	Val	Ala	Gly	His	Leu	Glu	Glu	Tyr	Gly
Uromas-tix	Val	Leu	Thr	Asp	Asp	Asp	Lys	Asn	His	Val	Arg	Ala	Ile	Trp	Gly	His	Val	Ser	Asn	Asn	Pro	Glu	Ala	Phe	Gly
"																	Val	Gly	Pro	His	Leu	Asp	Asp	Tyr	Gly

					30										40										50
Human	Ala	Glu	Ala	Leu	Glu	Arg	Met	Phe	Leu	Ser	Phe	Pro	Thr	Thr	Lys	Thr	Tyr	Phe	Pro	His	Phe	Asp	Leu	Ser	His
Chic-ken	Ala	Glu	Thr	Leu	Glu	Arg	Met	Phe	Ile	Gly	Phe	Pro	Thr	Thr	Lys	Thr	Tyr	Phe	Pro	His	Phe	Asp	Leu	Ser	His
Viper	Ser	Glu	Thr	Leu	Thr	Arg	Met	Phe	Ala	Ala	His	Pro	Thr	Thr	Lys	Thr	Tyr	Phe	Pro	His	Phe	Asp	Leu	Ser	Ser
Caiman	Ala	Glu	Ser	Leu	Glu	Arg	Met	Phe	Cys	Ala	Tyr	Pro	Gln	Thr	Lys	Ile	Tyr	Phe	Pro	His	Phe	Asp	Met	Ser	His
Uromas-tix	Ala	Glu	Ala	Leu	Tyr	Arg	Leu	Phe	Thr	Ala	His	Pro	Ala	Ser	Lys	Thr	Tyr	Phe	Ser	His	Phe	Asp	Leu	His	Glu
"	Gly	Gl-	Ala	Leu	His	Arg	Asn	Phe	Glu	Val	Tyr	Pro	Gln	Thr	Lys	Thr	Tyr	Phe	Pro	His	Phe	Asp	Ala	Ser	Ala

										60										70					
Human	Gly	Ser	Ala	Gln	Val	Lys	Gly	His	Gly	Lys	Lys	Val	Ala	Asp	Ala	Leu	Thr	Asn	Ala	Val	Ala	His	Ala	Asp	Asp
Chic-ken	Gly	Ser	Ala	Gln	Ile	Lys	Gly	His	Gly	Lys	Lys	Val	Ala	Leu	Ala	Ile	Thr	Asn	Ala	Ile	Glu	His	Ala	Asp	Asp
Viper	Gly	Ser	Pro	Asn	Leu	Lys	Ala	His	Gly	Lys	Lys	Val	Ile	Asp	Ala	Leu	Asp	Asn	Ala	Val	Glu	Gly	Leu	Asp	Asp
Caiman	Asn	Ser	Ala	Gln	Ile	Arg	Gly	His	Gly	Lys	Lys	Val	Phe	Ala	Ala	Leu	His	Asp	Ala	Val	Asn	His	Ile	Asp	Asp
Uromas-tix	Asn	Ser	Ala	Gln	Ile	Arg					Lys	Val	Val	Asp	Ala	Leu	Thr	Gln	Ala	Val	Asn	Asn	Leu	Asp	Asp
"	Gly	Ser	Asn	Gln	Leu	Lys																			

							80						90
Human	Met	Pro	Asn	Ala	Leu	Ser	Ala	Leu	Ser	Asp	Leu	His	Ala	His	Lys
Chic-ken	Ile	Ser	Glu	Ala	Leu	Ser	Lys	Leu	Ser	Asp	Leu	His	Ala	His	Lys
Viper	Ala	Val	Ala	Thr	Leu	Ser	Lys	Leu	Ser	Asp	Leu	His	Ala	Gln	Lys
Caiman	Leu	Ala	Gly	Ala	Leu	Cys	Arg	Leu	Ser	Asp	Leu	His	Ala	His	Asn
Uromas-tix	Leu	Ser	Gly	Ala	Ile	Ser	Lys	Leu	Ser	Asp	Leu	His	Ala	Glu	Lys

Fig. 8. Comparison of the α-chain from Uromastix with Human, Chicken, Viper and Caiman α-chain.

									10										20					
Human	Val	His	Leu	Thr	Pro	Glu	Glu	Lys	Ser	Ala	Val	Thr	Ala	Leu	Trp	Gly	Lys	Val	Asn	Val	Asp	Glu	Val	Gly	Gly
Chicken	Val	His	Trp	Thr	Ala	Glu	Glu	Lys	Gln	Leu	Ile	Thr	Gly	Leu	Trp	Gly	Lys	Val	Asn	Val	Ala	Glu	Cys	Gly	Ala
Caiman	Ser	Pro	Phe	Ser	Ala	His	Glu	Glu	Ser	Leu	Ile	Val	Asp	Leu	Trp	Ala	Lys	Val	Asp	Val	Ala	Ser	Cys	Gly	Gly
Uromastix	Val	His	Trp	Thr	Ala	Glu	Glu	Lys	Ala	Leu	Ile	Asn	Ala	Tyr	Trp	Gly	Lys	Val	Asp	Val	Gly	Ser	Val	Gly	Gly

					30										40										
Human	Glu	Ala	Leu	Gly	Arg	Leu	Leu	Val	Val	Tyr	Pro	Trp	Thr	Gln	Arg	Phe	Phe	Glu	Ser	Phe	Gly	Asp	Leu	Ser	Thr
Chicken	Gly	Ala	Leu	Ala	Arg	Leu	Leu	Ile	Val	Tyr	Pro	Trp	Thr	Gln	Arg	Phe	Phe	Ala	Ser	Phe	Gly	Asn	Leu	Ser	Ser
Caiman	Asp	Ala	Leu	Ser	Arg	Met	Leu	Ile	Ile	Tyr	Pro	Trp	Lys	Arg	Arg	Tyr	Phe	Glu	His	Phe	Gly	Lys	Leu	Ser	Thr
Uromastix	Glu	Thr	Leu	Ala	Asn	Leu	Leu	Val	Val	Tyr	Pro	Trp	Thr	Gln	Arg	Phe	Phe	Glu	Asp	Phe	Gly	Asn	Leu	Ser	Thr
"																Phe	Phe	Gly	Asp	Phe	Gly	Asn	Ile	Ser	Ser

									60										70					
Human	Pro	Asp	Ala	Val	Met	Gly	Asn	Pro	Lys	Val	Lys	Ala	His	Gly	Lys	Lys	Val	Leu	Gly	Ala	Phe	Ser	Asp	Gly	Leu
Chicken	Pro	Thr	Ala	Phe	Leu	Gly	Asn	Pro	Met	Val	Arg	Ala	His	Gly	Lys	Lys	Val	Leu	Thr	Ser	Phe	Gly	Asp	Ala	Val
Caiman	Asp	Gln	Asp	Val	Leu	His	Asn	Glu	Lys	Ile	Arg	Glu	His	Gly	Lys	Lys	Val	Leu	Ala	Ser	Phe	Gly	Glu	Ala	Val
Uromastix	Pro	Ser	Ala	Ile	Leu	Asn	Asn	Pro	Lys								Val	Ile	Thr	Ser	Phe	Gly	Asp	Ala	Leu
"	Ala	Ala	Ala	Ile	Thr	Gly	Asn	Pro	Lys																

					80										90										
Human	Ala	His	Leu	Asp	Asn	Leu	Lys	Gly	Thr	Phe	Ala	Thr	Leu	Ser	Glu	Leu	His	Cys	Asp	Lys	Leu	His	Val	Asp	Pro
Chicken	Lys	Asn	Leu	Asp	Asn	Ile	Lys	Asn	Thr	Phe	Ser	Gln	Leu	Ser	Glu	Leu	His	Cys	Asp	Lys	Leu	His	Val	Asp	Pro
Caiman	Lys	His	Leu	Asp	Asn	Ile	Lys	Gly	His	Phe	Ala	His	Leu	Ser	Lys	Leu	His	Phe	Glu	Lys	Phe	His	Val	Asp	Cys
Uromastix	Lys	Asn	Leu	Asp	Asn	Val	Lys						Leu	Ser	Glu	Tyr	His	Cys	Asn	Lys	Leu	His	Val	Asp	Pro

									110										120					
Human	Glu	Asn	Phe	Arg	Leu	Leu	Gly	Asn	Val	Leu	Val	Lys	Val	Leu	Ala	His	His	Phe	Gly	Lys	Glu	Phe	Thr	Pro	Pro
Chicken	Glu	Asn	Phe	Arg	Leu	Leu	Gly	Asp	Ile	Leu	Ile	Ile	Val	Leu	Ala	Ala	His	Phe	Ser	Lys	Asp	Phe	Thr	Pro	Glu
Caiman	Glu	Asn	Phe	Lys	Leu	Leu	Gly	Asp	Ile	Ile	Ile	Val	Val	Leu	Gly	Met	His	His	Pro	Lys	Asp	Phe	Thr	Leu	Gln
Uromastix	Val	Asn	Phe	Arg	Leu	Leu	Gly	Asp	Val	Leu	Ile	Thr	Leu	Ser	Ala	Ala	Asn	Phe	Gly	Lys					

					130										140						
Human	Val	Gln	Ala	Ala	Tyr	Gln	Lys	Val	Val	Ala	Gly	Val	Ala	Asn	Ala	Leu	Ala	His	Lys	Tyr	His
Chicken	Cys	Gln	Ala	Ala	Trp	Gln	Lys	Leu	Val	Arg	Val	Val	Ala	His	Ala	Leu	Ala	Arg	Lys	Tyr	His
Caiman	Thr	His	Ala	Ala	Phe	Gln	Lys	Leu	Val	Arg	His	Val	Ala	Ala	Ala	Leu	Ser	Ala	Gln	Tyr	His
Uromastix								Leu	Val	Gly	Val	Val	Ala	His	Ala	Leu	Ala	Arg	Arg	Tyr	His

Fig. 9. Comparison of the β-chain from Uromastix with Human, Chicken and Caiman β-chain.

However, remaining pairs have unique peptides that are homologous to hemoglobin chains. The globin homology of both alternatives, as well as the peak heights in Fig. 5 suggest that both alternatives represents true hemoglobin chains. Relative amounts are difficult to estimate from the separation patterns, but top alternative in Fig. 6 and 7 appear to be major constituents (Fig. 5 where peak areas are: A7+A10 > A9 for the peptides corresponding to position 50 in the β-chain). The differences in amino acid sequences are surprisingly large with each chain type around 50%.

Thus the multiplicity of each type is due to corresponding Uromastix hardwickii gene multiplicity, the ancestral gene duplication must be distant. The globin studied was a pool from 200 animals, the variations detected might include some species or strain multiplicities. In any case, the results reveal structural heterogeneity in Uromastix globin apart from establishing the extent of a general relationship between Uromastix and other hemoglobins.

The hemoglobin chains studied are aligned with other species (human chicken, crocodile and viper) and results show large structural variations, (about 50% homology) Fig. 8 and 9.The relationship hardly appear closer to viper than to birds or mammals indicating large evolutionary changes between these two reptilian species. (Naqvi et-al. 1983).

References

Anson, M.L and Mirsky, A.F. (1930) **J.Gen.Physiol.** 13, 469.

Clegg, J.B. Naughton, M.A. and Weatherall, D.J. (1966), **J.Mol.Biol.**, 19, 91.

Edman, P. and Begg, G. (1967) **Eur. J.Biochem.** 1, 80.

Maurer, H.R. (1969) Disc Electrophoresis and related techniques of poly-acrylamide electrophoresis. **Pub.Walter-de-Gruyter** Berlin, New York, 1071

Naqvi, S., Zaidi, Z.H., Von Bahr-Lindstrom, H., Carlquist M. and Jornvall, H. (1983), **FEBS Lett.**, 162, 290.

Steel, Jr. JCH, Nielson thB (1978) **Anal.Biochem.**, 84, 218

St. Groth, FS., De Webster RG and Datyner, A. (1963) **Biochem.Biophys. Acts.**, 71, 377.

Von Bahr-Lindstrom, H., Hempel, J. and Jornvall, H. (1982) **J.Prot. Chem.**, 1, 257.

Wrigley, C.W. (1969) **Science Tools** 15, 17.

Subject Index

MIX
Papier aus verantwortungsvollen Quellen
Paper from responsible sources
FSC® C105338

If you have any concerns about our products,
you can contact us on
ProductSafety@springernature.com

In case Publisher is established outside the EU,
the EU authorized representative is:
Springer Nature Customer Service Center GmbH
Europaplatz 3, 69115 Heidelberg, Germany

Printed by Libri Plureos GmbH
in Hamburg, Germany